THE EFFECTS
OF NOISE
ON MAN

ENVIRONMENTAL SCIENCES

An Interdisciplinary Monograph Series

EDITORS

DOUGLAS H. K. LEE
National Institute of
Environmental Health Sciences
Research Triangle Park
North Carolina

E. WENDELL HEWSON
Department of
Atmospheric Science
Oregon State University
Corvallis, Oregon

DANIEL OKUN
University of North Carolina
Department of Environmental
Sciences and Engineering
Chapel Hill, North Carolina

ARTHUR C. STERN, editor, AIR POLLUTION, Second Edition. In three volumes, 1968

L. FISHBEIN, W. G. FLAMM, and H. L. FALK, CHEMICAL MUTAGENS: Environmental Effects on Biological Systems, 1970

DOUGLAS H. K. LEE and DAVID MINARD, editors, PHYSIOLOGY, ENVIRONMENT, AND MAN, 1970

KARL D. KRYTER, THE EFFECTS OF NOISE ON MAN, 1970

R. E. MUNN, BIOMETEOROLOGICAL METHODS, 1970

M. M. KEY, L. E. KERR, and M. BUNDY, PULMONARY REACTIONS TO COAL DUST: "A Review of U. S. Experience," 1971

THE EFFECTS
OF NOISE
ON MAN

KARL D. KRYTER
Stanford Research Institute
Menlo Park, California

1970

ACADEMIC PRESS   New York and London

ACADEMIC PRESS, INC.
111 Fifth Avenue, New York, New York 10003

United Kingdom Edition published by
ACADEMIC PRESS, INC. (LONDON) LTD.
Berkeley Square House, London W1X 6BA

Library of Congress Catalog Card Number: 74-117112

Second Printing, 1971

Printed in the United States of America

For my wife Grace

and

daughters Dianne, Victoria and Kathryn

TABLE OF CONTENTS

PREFACE

Several years ago, when I was at Bolt Beranek and Newman Inc. in Cambridge, Massachusetts and was under contract to the Office of the Surgeon General of the Army, I began to update a monograph, published in 1950, titled "The Effects of Noise on Man." This effort was continued and eventually completed at Stanford Research Institute, under contract to the National Aeronautics and Space Administration. Some of the sections concerning concepts and data related to noise-induced deafness are from papers prepared under a research grant from the National Institutes of Health.

An attempt has been made to provide a critical and historical (dating from 1950) analysis of the relevant literature in the field and, as warranted, to derive new or modify existing techniques for the evaluation of environmental noise in terms of its effects on man. In Parts I and II of this book, fundamental definitions of sound, its measurement, and concepts of the basic functioning and attributes of the auditory system are provided. These chapters also present, along with their experimental basis, procedures for estimating from physical measures of noise its effects on man's auditory system and speech communications. Part III is devoted to man's nonauditory system responses and includes information about the effects of noise on such things as work performance, sleep, feelings of pain, vision, and blood circulation. It is clear that some of the more complex, and perhaps more important from a health viewpoint, effects of noise have to do with these somewhat second-order reactions. Tolerable limits of noise with respect to its effects on man's auditory and nonauditory systems are suggested at various places.

The bibliography consists of those items referred to in the text, plus some additional items that are particularly pertinent to given points or that are important general sources of relevant information. In the preparation of this book some 4000 articles were, except for many of the non-English publications, read or reviewed. A limited number of copies of the original draft bibliography, which is organized around 24 subtopics, can be obtained by writing to me at the Stanford Research Institute. I regret that space does not permit the inclusion of a larger bibliography.

ACKNOWLEDGMENTS*

Grateful acknowledgment is given to the numerous authors and editors of journals and books for permission to reproduce their figures. I owe many thanks to Miss Dorian Doebler, my secretary, for her skills and great patience during innumerable typings of the manuscript and to Mrs. Margaret Troy and Mrs. Natalie McDonald for editing the major portion of the bibliography. Michael Hecker, Manfred Heckle, Eric Rathe and Henning von Gierke graciously translated or furnished me translations of portions of some of those articles reviewed that were published in German.

I am also deeply indebted to Drs. Frank R. Clarke and James R. Young, Stanford Research Institute, Professor James P. Egan, University of Washington, and Professor Walter A. Rosenblith, Massachusetts Institute of Technology, for reviewing various portions of the manuscript. Their detection of errors and constructive suggestions were invaluable; some of their suggestions could not be utilized however because of limited time and ability on my part. Finally, I wish to express gratitude for the encouragement and support I received in this endeavor from Colonel William Hausman and Dr. Glen R. Hawkes of the Office of the Surgeon General of the Army, Dr. Gilbert Tolhurst of the Office of Naval Research, and Mr. Harvey Hubbard and Mr. Phillip Edge of the National Aeronautics and Space Administration. I can only hope that this book partially justifies the support they have given me.

*We have reproduced, with their permission, some figures and tables from the documents of the American National Standard Institute. Copies of those documents may be purchased from that organization at 1430 Broadway St., New York, New York.

THE EFFECTS
OF NOISE
ON MAN

PART I

AUDITORY SYSTEM RESPONSES TO NOISE

Introduction

In the fields of electronics, neurophysiology, and communication theory, noise means signals that bear no information and whose intensities usually vary randomly in time. The word noise is used in this sense in acoustics, but more often it is used to mean sound that is unwanted by the listener, presumably because it is unpleasant or bothersome, it interferes with the perception of wanted sound, or it is physiologically harmful. Noise, as unwanted sound, does not necessarily have any particular physical characteristic (such as randomness) to distinguish it from wanted sound. For example, an information-bearing signal such as speech may be so intense that it is subjectively unwanted and may even be harmful to the ear of the listener, whereas a sound such as so-called "white" noise that is random, or nearly so, in the physical sense may be subjectively quite acceptable, particularly if it serves to mask other sounds that, if audible, would be bothersome. As far as man's auditory system is concerned, there is no distinction to be made between sound and so-called noise, and in the text to follow the word "noise" is often used in place of "sound" merely to draw attention to the theme of the book.

There are certain unwanted effects of sounds that appear to be related rather precisely to physical characteristics of the sound in ways that are more or less universal and invariant for all people. The effects we refer to are (a) the masking of wanted sounds, particularly speech, (b) auditory fatigue and damage to hearing, (c) excessive loudness, (d) some general quality of bothersomeness or noisiness, and (e) startle.

These unwanted effects of sound upon man's peripheral and subjective auditory response system are mainly what this book is about. Because the effects are (a) similar for all people, (b) neither primarily dependent on learning, nor, except to some extent for startle, able to be unlearned, and (c) quantitatively related to the physical nature of sounds, they deserve the attention and

1

understanding of persons interested or involved in the design of devices that generate sound, in the control of the sound during its transmission, or in the protection of the health and well-being of people exposed to the sound. Indeed, nearly all measurements made of sound by acoustical engineers are made for the immediate or ultimate purpose of evaluating or controlling the effects of the sound on man.

Chapters 1 through 3 of this book are concerned with basic and somewhat academic (except to the research worker) information concerning noise and functioning of the ear. Some readers may wish to turn immediately to Chapter 4, the start of the material on noise damage to the ear, or even to Part II of the book, *Subjective Responses to Noise*. However, all parts are interrelated and a fuller understanding of the state-of-the-art and problems in this field is to be had from reading the whole book.

Chapter 1

Analysis of Sound by the Ear

Definitions of Sound

For the human listener, sound in the frequency domain is defined as acoustic energy between 2 Hz and 20,000 Hz, the typical frequency limits of the ear. The lowest frequency of sound that has a pitch-like quality is about 20 Hz and the upper frequency audible to the average adult is about 10,000 Hz. Hertz (Hz) is the name, by international agreement, for the number of repetitions of similar pressure variations per second of time; this unit of frequency was previously called "cycles per second" (cps or c/s). The decibel (dB) is the common unit of measurement of sound pressure. It is $20 \log_{10}$ of the ratio between the root-mean-square (rms) pressure of a given sound and usually, and for this document, the reference rms pressure of 0.0002 microbar (μbar). While the unit μbar, and another unit, dynes per square centimeter (dyne/cm^2), are in common use, the unit newtons per square meter (N/m^2) is becoming the international standard unit of sound pressure. These units are related to each other as follows: 0.1 N/m^2 = 1 dyne/cm^2 = 1 μbar.

In the temporal domain, the rise time of a sound is the time required for a sound to go from ambient air pressure to the first occurrence of its peak pressure. The duration of a sound is the time in seconds from the start of the rise in pressure to the time the pressure envelope starts again to stay at ambient.

In the intensity-time dimensions, sounds are labeled as being either "impulsive" or "nonimpulsive." Impulse sound, for this document, is defined as a change in rms air pressure greater than 40 dB per 0.5 sec; all other 0.5-sec intervals of sound are nonimpulsive. Nonimpulsive intervals may be described as changing in level or steady-state. Sound is here said to be steady-state when the rms pressure remains relatively constant (within ±5 dB) for successive periods of 0.5 sec. A sound, unless shorter in total duration than 0.5 sec, can go from impulsive to nonimpulsive and vice versa during its existence.

The amplitude-phase relations between frequency components, and the

3

number of components (bandwidth) of a sound will determine the moment-to-moment fluctuations to be expected in rms pressure taken over all frequencies. Therefore, the specification of a ±5 dB tolerance for steady-state level is suggested as a practical range. Theoretical calculations and actual measurements of random noise show that variations of ±5 dB or less are to be expected with 95% certainty for frequency bands wider than about 10 Hz, and ±2 dB or less for bandwidths wider than about 100 Hz (see Fig. 1 from Galloway [272]). Because of this unavoidable fluctuating of relatively narrow bands of random noise, it is the practice to use a sound level meter which has a pressure averaging time constant of 0.5 sec for making band spectral analysis of noise in order to achieve reasonably reliable measurements.

Pressure-Temporal-Spectral Response Characteristics of the Ear to Sound

Figure 2 is a schematic drawing showing the path and means by which sound enters the human ear, where it is transduced into motion in the fluid (perilymph) of the cochlea. This fluid action causes nerve fibers on the basilar membrane to send impulses to higher nerve centers where the impulses are perceived or interpreted as sound.

The analysis of sound, in classical auditory theory, takes place in the ear with respect to the physical dimensions of frequency and intensity. Except for special circumstances, some of which will be discussed later, phase information appears to be of little significance to the subjective response to sounds. The primary psychological aspects of frequency and intensity are pitch and loudness. Psychological dimensions other than pitch or loudness — for example, density, volume (size), and perceived noisiness — have also been related to the frequency-intensity characteristics of sounds. These will be discussed later.

The attribute of pitch has been ascribed in the past to possible cues of place of stimulation on the basilar membrane of the cochlea and rate of firing of peripheral neural units; the attribute of loudness has been ascribed to the rate at which neural impulses are generated in the cochlea. It is, however, the perception of the patterns of complex pitch and loudness in the flow of time that makes audition such a useful and pervasive part of man's consciousness.

Critical Bandwidth of the Ear

The abilities of the ear to perceive pitch as a function of frequency and loudness as a function of intensity, and to detect small changes in these attributes, were well mapped out prior to 1950. However, the ability of the ear

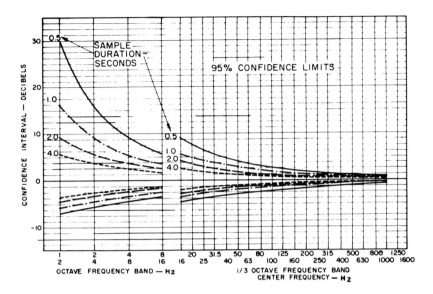

FIGURE 1.   Confidence intervals, in decibels, for spectrum analysis of short-duration random samples of noise. From Galloway (272).

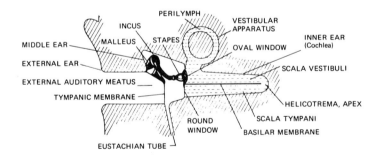

FIGURE 2.   Schematic drawing of the human ear. Sound waves enter the external meatus, and move the tympanic membrane (eardrum) which sets the three ossicles (malleus, incus, and stapes) in motion. When the stapes footplate moves inward, the perilymph inside the cochlea flows in the direction of the helicotrema and makes the round-window membrane bulge outward. The cochlea is actually coiled in the human ear and not straight as shown in this diagram. The aural muscles, not shown, are located in the middle ear. After Békésy (49).

to behave as a bandpass filter was not extensively delineated until after 1950.

The concept, as conceived by Fletcher (232), of filters having what are called critical bandwidths within the ear has proved to be most significant — it has furnished a basis for explaining some auditory behavior with respect to speech perception, auditory fatigue, loudness, pitch perception, and masking. Crudely put, the cochlea and its associated nerve nets often seem to behave as a very large set of overlapping, bandpass filters connected in parallel. These filters, like most filters, have skirts that are not sharp (see Shafer *et al.* [724]). Their bandwidths change as a function of frequency, and they appear to become somewhat broader when the signal intensity is increased, in particular, it seems that at high intensity levels the upper skirt of the filter becomes much less steep than the lower skirt. The critical band concept holds that increasing the bandwidth of a masking noise beyond a certain width will not increase the degree to which a pure tone located at the center of the band is masked; only the energy at frequencies nearer the center frequency contribute to the analysis and masking of the tone at the center.

Fletcher (232), Hawkins and Stevens (355), Egan and Hake (206), and Bilger and Hirsh (66) all conducted experiments in which the intensity level of a pure tone, presented with a very broad band random noise, was adjusted until the tone was just audible. This process was repeated with pure tones of different frequencies. Swets *et al.* (789) conducted a similar experiment in which they masked a narrow band of noise with a wider band. The results (see, for example, the bottom two curves on Fig. 3) show that the width of the critical band varies as a function of frequency. For the two lowest curves in Fig. 3, critical band is defined as the ratio between the spectrum level of white noise and that of the pure tone at masked threshold, i.e., the critical band was defined as being the band of noise around a pure tone at center frequency whose acoustic power equaled that of the pure tone when at masked threshold. DeBoer (190) wonders if the upturn in lower curves of Fig. 3 below about 200 Hz may be due to the statistical fluctuation in level of narrow bands of random noise which would possibly adversely affect detection of the pure tone.

Other investigators (Schafer *et al.* [724], Gassler [292], Zwicker *et al.* [907], Hamilton [334], Greenwood [324], Scharf [725], and deBoer [190]) made a more direct attack on the measurement of the critical band. In these studies various procedures were used: the bandwidth of masking noise was varied, a band of noise was masked by two tones bracketing a tone, and the loudness of narrow bands of noise and the loudness of tones separated by different distances along the frequency scale were measured. Greenwood (324) notes that Mayer in 1894 found that two tones, separated sufficiently to avoid all roughness or beats, were separated by about what has since been measured as the critical band for masking and loudness. Plomp and Levelt (629) propose and demonstrate that musical consonance appears to be based on patterns of tones with harmonics

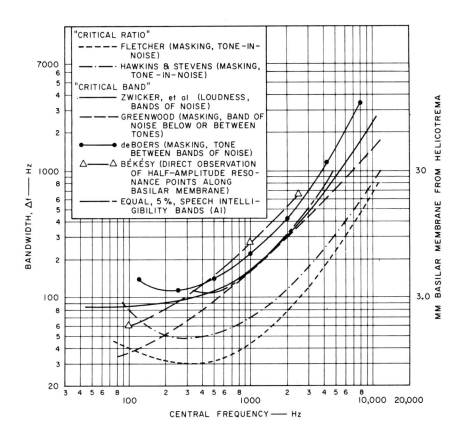

FIGURE 3. Bandwidth and position on basilar membrane of sounds as function of band center frequency for various parameters.

separated by critical bandwidths. Green (319) finds that the detectability of multiple pure tone signals increases linearly when the components are separated by critical bandwidths, although Marill (531), in an earlier study similar to Green's, did not find this.

It was found by the above investigators that when the critical band was measured directly, its bandwidth varied more or less as a function of frequency in the same fashion as did the critical band measured indirectly by the masking of a pure tone by a white noise, but that its width was about 2½ times as wide (see Fig. 3). Zwicker *et al.* (907) suggest that the width calculated from the indirect measurements be called the "critical ratio," and the term "critical band" be applied to the width measured directly (see Table 1). This suggestion seems logical, since the original assumption that the tone and noise should be of equal power at masked threshold is arbitrary.

Shown on Fig. 4, and also Fig. 3 at the 6 dB downpoints, is the bandwidth of the resonance functions for the basilar membrane of the cochlea as measured by Békésy (49a, 50). Greenwood (325) concludes, from an analysis of perceptual data on the critical band and from Békésy's direct measurements of basilar membrane resonance, that one critical bandwidth (not ratio) extends about 1 mm along the basilar membrane in the frequency region from about 400 to 6000 Hz (see right-hand vertical ordinate on Fig. 3).

It should be recognized that the auditory system is capable of perceiving pitch changes, "trills," "beats," etc. from frequency changes in an acoustic stimulus that are much less in width than the critical band, critical ratio, or the gross hydromechanical patterns resulting in the cochlea from stimulation by sounds of different frequencies. It is probable (see Licklider [503]) that neural mechanisms present in the brain stem are responsible for some of this further sharpening action.

When sound pressure is continued at a steady level for longer than about 200 msec, the pressure or turbulence at any place on the basilar membrane is perhaps proportional only to the rms pressure of the sound presented to the ear. This latter point is indirectly demonstrated by Fig. 5 which shows that the loudness of a sound does not change appreciably, provided its duration exceeds 200 msec. Perhaps, in this regard, the ear can be thought of as a leaky condenser, in electrical analog, where the energy put into the system is removed at a discharge rate such that, after about 200 msec, the initial energy is dissipated and the system reaches an equilibrium, provided that energy continues to be applied at a more or less steady rate. In effect, the ear responds to the band spectral sound pressure level averaged over intervals of time.

The hydromechanical behavior of the inner ear is such that the low frequencies cause turbulence (the presumed stimulator in some manner of the receptor cells on the basilar membrane) toward the part of the cochlea called the apex, farthest from the place where sound vibrations enter the inner ear. Also,

TABLE 1

Subdivision of Audible Frequency Range into Critical Bands
From Zwicker (902).

Number	Center frequencies Hz	Cut-Off frequencies Hz	Bandwidth Hz
1	50	20-100	80
2	150	100-200	100
3	250	200-300	100
4	350	300-400	100
5	450	400-510	110
6	570	510-630	120
7	700	630-770	140
8	840	770-920	150
9	1000	920-1080	160
10	1170	1080-1270	190
11	1370	1270-1480	210
12	1600	1480-1720	240
13	1850	1720-2000	280
14	2150	2000-2320	320
15	2500	2320-2700	380
16	2900	2700-3150	450
17	3400	3150-3700	550
18	4000	3700-4400	700
19	4800	4400-5300	900
20	5800	5300-6400	1100
21	7000	6400-7700	1300
22	8500	7700-9500	1800
23	10,500	9500-12,000	2500
24	13,500	12,000-15,500	3500

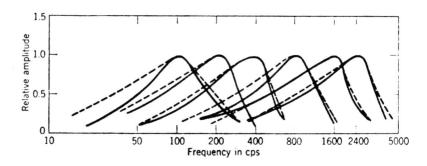

FIGURE 4. Resonance curves for six points on the cochlear partition. The solid curves are measured values (Békésy); the dashed curves are theoretical values calculated by Zwislocki (1948). From Békésy and Rosenblith (50).

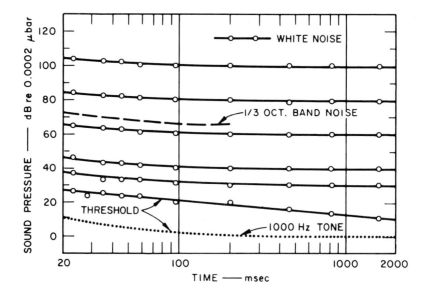

FIGURE 5. Equal-loudness contours for white noise as a function of stimulus duration. The sound pressure of a short burst of noise that sounded equal in loudness to a standard 1.0-sec burst is plotted as a function of the duration of the short burst. After Miller (544) and Port (652).

this turbulence is asymmetrical, being spread out toward the base of the cochlea and truncated toward the apex; see Fig. 6. These two facts contribute to a characteristic effect of sound that will be apparent in the text to follow; namely, that low frequency sounds tend to stimulate many more receptor fibers on the basilar membrane than do high frequency sounds.

Model of Inner Ear

The model of the inner ear we are subscribing to for the purposes of the text to follow is that (a) the time-pressure envelope of a sound is displayed, by hydromechanical means related to the construction of the inner cochlea and the fluids it contains, as a time-varying pressure-frequency pattern along the basilar membrane within the cochlea, (b) the neural receptors on the basilar membrane can respond to the changes in pressure and turbulence of the cochlear fluid, (c) the auditory nervous system is capable of interpreting neural firings from the basilar membrane with respect to the number of neural firings, the place on the basilar membrane initiating the firings, and the periodicity of the firings, and (d) when the rms pressure of the sound stays steady for a period longer, within limits, than about 200 msec, the rate and/or periodicity of neural responses from the basilar membrane becomes stabilized. For reasons to be discussed later, it will be proposed that this temporal interval be taken as 500 msec rather than 200 msec.

Outer and Middle Ear

The outer and middle ear appear to have the function not only of transmitting to the inner ear the pressure waveform of the sound, but also of protecting the inner ear from having to operate on sounds outside its capacity. In regard to the latter, it is noted that (a) the middle ear can prevent the transmission to the inner ear of pressure waves having rise times longer than 200 msec by means of the action of the eustachian tube (or even by a rupturing of the eardrum), (b) when presented with high intensity pressure waves, small muscles in the middle ear can contract, stiffening the ossicular chain and thereby attenuating the transmission of sound (also, with very intense sound, the ossicular chain appears to rotate from its normal axis in a way that limits or even reduces the pressure level reaching the inner ear), and (c) the mass and stiffness of the ossicular chain are such as to prevent transmission of a pressure wave with a rise time of less than 50 μsec. These time durations, 200 msec to 50 μsec, correspond, of course, to the period of the frequencies of 2 Hz to 20,000 Hz, matching the frequency band limits of the inner ear which are set, according to

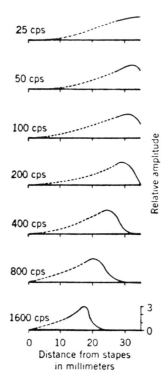

FIGURE 6. Displacement amplitudes along the cochlear partition for different frequencies. The stapes was driven at a constant amplitude, and the amplitude of vibration of the cochlear partition was measured. The maximum displacement amplitude moves toward the apex as the frequency is increased. From Békésy (49a).

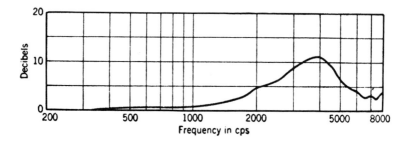

FIGURE 7. Effects of "resonance" in the external meatus. The ordinate shows the ratio in decibels between the sound pressure at the eardrum and the sound pressure at the entrance to the auditory canal. From Wiener (874).

Békésy, largely by the dimensions of the inner ear and nature of the basilar membrane.

The acoustic resonance of the ear canal and of the outer ear contributes to the frequency response characteristics of the ear. It is significant, as seen in Fig. 7, that, relative to frequencies below 1000 Hz, the ear canal effectively amplifies higher sound frequencies, particularly around 4000 Hz. Although this may at times contribute to overstimulation of the inner ear at the higher frequencies, this "resonance" may be a useful compensation for the nonlinear attenuation of these higher frequencies during the transmission of sound through air and most other media.

Methods for Measuring Sounds in Order to Predict Their Effects on Hearing

It is customary and useful to describe the spectra of sound according to frequency bands that are (a) one Hz wide (called spectrum level), (b) one-third octave wide, and (c) one octave wide. In addition, sound pressure measurements are often made of the spectra over all frequencies, using a sound level meter with either a uniform weighting for all frequencies or with certain differential weighting given to different frequencies. Weightings, as a function of one-third and full octave bands, are specified in Table 2 and, as a function of frequency, in Fig. 8.

As will be discussed later, sound pressure measurements made with a sound level meter with a frequency weighting network do not measure sound pressure level per se but attempt to measure the physical correlate of a quantity or attribute of sound such as loudness. Sound level meters give readings in decibels relative to 0.0002 microbar, integrating (with a nominal integrating time constant of 0.2 sec [200 msec] when set on "fast" meter action, and 0.5 sec [500 msec] on "slow" meter action) the sound pressure over all frequencies from, for most meters, about 50 to 10,000 Hz. In recent years, precision sound level meters have been developed that extend that range. These units of measurement, when used as predictors of some human response quantities, will be designated in this document as dB(A), dB(B), dB(C), and dB(D), depending upon the particular frequency weighting employed. In the general literature, and in this report, when sound pressure levels are reported as unqualified dB values, it is to be understood that the weighting network of the meter was set on flat, with equal weighting at all frequencies.

On occasion it is appropriate to convert band levels, measured in terms of one bandwidth, to the levels that would have been present had the measurements been made with respect to other bandwidths. Means for making these conversions are shown in Fig. 9. Such band spectral conversions are justified

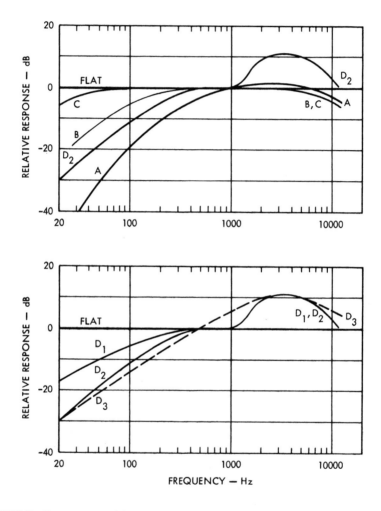

FIGURE 8. Frequency weightings for sound level meters. Upper Graph: Standard weightings A, B, and C (25), and newly-proposed (462), and herein recommended, D-weighting. Lower Graph: Recently-proposed D_1 (458), D_2 (same as D of upper graph), and D_3 (898) weightings. D_3 adjusted upward by 6 dB to better show relation to D_2. In the text to follow, unless otherwise specified, D-weighting will refer to D_2.

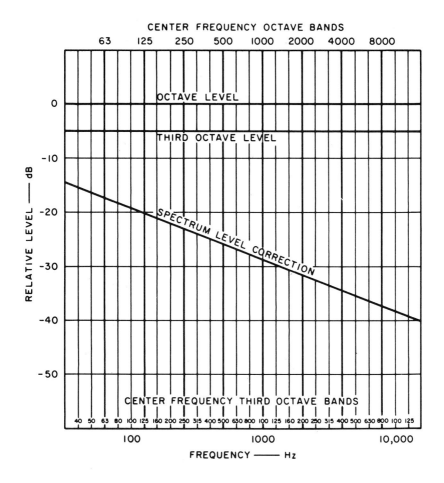

FIGURE 9. Relative differences between spectrum level (bandwidth of 1 Hz), third and full octave band levels for broadband sounds of continuous spectrum.

TABLE 2

Cut-off Frequencies and Center Frequencies of Preferred 1/3 Octave Band Filters, and A, B, C, and Proposed D Frequency Weightings for Sound Level Meters

Cut-Off Frequencies	Center Frequencies	dB(A)	dB(B)	dB(C)	dB(D_1)	dB(D_2)	dB(D_3)
(45-56) Hz	50 Hz	-30.2	-11.7	- 1.3	-12	-19	-26
(56-71)	63	-26.1	- 9.4	- 0.8	-11	-17	-24
(71-90)	80	-22.3	- 7.4	- 0.5	- 9	-14	-22
(90-112)	100	-19.1	- 5.7	- 0.3	- 7	-11	-20
(112-140)	125	-16.2	- 4.3	- 0.2	- 6	- 9	-18
(140-180)	160	-13.2	- 3.0	- 0.1	- 5	- 7	-16
(180-224)	200	-10.8	- 2.1	0.0	- 3	- 5	-14
(224-280)	250	- 8.0	- 1.4	0.0	- 2	- 3	-12
(280-355)	315	- 6.5	- 0.9	0.0	- 1	- 2	-10
(355-450)	400	- 4.8	- 0.6	0.0	0	0	- 8
(450-560)	500	- 3.3	- 0.3	0.0	0	0	- 6
(560-710)	630	- 1.9	- 0.2	0.0	0	0	- 4
(710-900)	800	- 0.8	- 0.1	0.0	0	0	- 2
(900-1120)	1000	0.0	0.0	0.0	0	0	0
(1120-1400)	1250	+ 0.5	- 0.1	- 0.1	+ 2	+ 2	+ 2
(1400-1800)	1600	+ 1.0	- 0.1	- 0.1	+ 6	+ 6	+ 3
(1800-2240)	2000	+ 1.2	- 0.2	- 0.2	+ 8	+ 8	+ 4
(2240-2800)	2500	+ 1.2	- 0.3	- 0.3	+10	+10	+ 4.5
(2800-3550)	3150	+ 1.2	- 0.5	- 0.5	+11	+11	+ 5
(3550-4500)	4000	+ 1.0	- 0.8	- 0.8	+11	+11	+ 5
(4500-5600)	5000	+ 0.5	- 1.3	- 1.3	+10	+10	+ 4.5
(5600-7100)	6300	- 0.2	- 2.0	- 2.0	+ 9	+ 9	+ 4
(7100-9000)	8000	- 1.1	- 3.0	- 3.0	+ 6	+ 6	+ 3
(9000-11,020)	10,000	- 2.5	- 4.3	- 4.3	+ 3	+ 3	0

only when it can be assumed or shown that the energy in a sound is more or less uniformly distributed over an appropriate range of frequencies. It is customary to express the sound pressure level of relatively intense impulses, (e.g., sonic booms) in terms of pounds per square foot (psf) or pounds per square inch (psi). Figure 10 can be used to convert sound pressures to a common unit (dB, psf, or psi) of intensity.

The determination of the spectrum of nonimpulsive sounds is readily accomplished by commercially available bandpass filters and meters having relatively rapid response characteristics. The spectral analysis of impulsive type sounds is more difficult, except by computer-aided techniques; however,

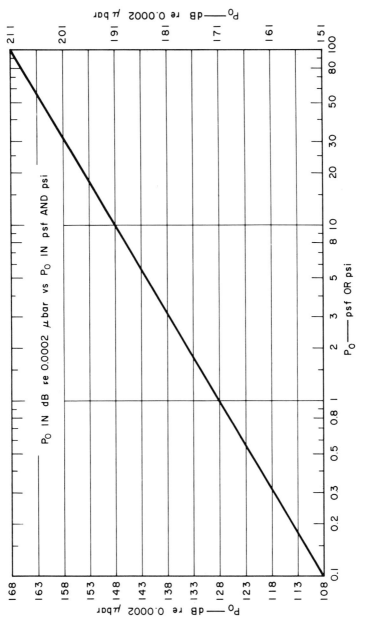

FIGURE 10.   Graph for converting pressures in pounds per square foot (psf), left-hand ordinate, or pounds per square inch (psi), right-hand ordinate, to dB re 0.0002 μbar.

approximations to impulse spectra can be achieved through the use of rather simple graphs. J.R. Young (personal communication) has described graphically some of the basic relations between the physical parameters of certain impulses and band spectral levels. Figures 11 and 12 for "one period" impulses, and Fig. 13 for impulses having the form of exponentially damped sinusoids, in conjunction with information about overall sound pressure level, duration, rise time, and period, can be used for determining the approximate band levels of general types of impulsive sounds. Examples of the use of Figs. 11-13 will be given in Chapters 5 and 9.

It might be noted that the spectrum of nonimpulsive sounds is the average spectral distribution of energy per unit of time (called the power spectrum). Since the general response time of the ear is long compared with the duration of most acoustic impulses, it is more appropriate to report the energy than the power spectrum of impulsive sounds.

It is proposed for basic and detailed psychoacoustic purposes that all sounds, regardless of their temporal or spectral nature, be reduced to octave, or preferably, to one-third octave band spectra in each 0.5-sec intervals of time from the start, the end of a sound. It is also suggested that certain groupings of one-third octave and full octave band spectral measures below 355 Hz be added together on a power basis, particularly, as will be discussed later, for purposes of estimating subjective loudness and perceived noisiness in order to make all the bandwidths used in the measurements proportional to critical bands. This is done on the assumption that the critical bands are more valid for purposes of describing the audition of broadband sounds than are octave bands or fractions thereof when the latter differ from critical bandwidths, as they do below about 355 Hz. These relations and proposed adjustment procedures are shown in Fig. 14.

In order to add, on a power basis, the sound pressure levels of different frequency bands, it is first necessary to divide the decibel levels in each band by 10 to find the antilogs of each scaled decibel level; sum these antilogs, convert back to a logarithm to the base 10 and multiply by 10. We will refer to this summation in the text as addition on a "10 $\log_{10}$ antilog basis." A convenient, but approximate, method for accomplishing the power addition of sound pressures when given in decibels is shown in the nomograph of Fig. 15.

As noted earlier, when the sound pressure of a given sound is increased, its level in dB increases on a 20 $\log_{10}$ antilog basis and not the 10 $\log_{10}$ antilog basis that occurs when two different sounds (sounds that contain different frequency components or components that are randomly out of phase) are added together. The addition of the pressures of in-phase sounds is referred to in this text as 20 $\log_{10}$ antilog summation. Figure 16 provides a nomograph useful for summing measures that are to be added together on a 20 $\log_{10}$ antilog basis.

At various places in the text to follow, procedures will be described that call

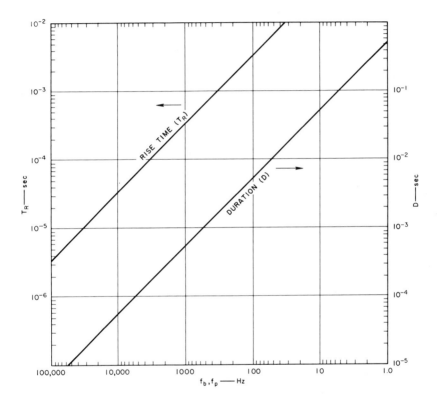

FIGURE 11. Shows the lower cut-off frequencies as function of duration (D) and the upper cut-off frequency as function of rise time (T_R) of an impulse. f_b is the frequency, determined by T_R, of the point at which the slope of the spectrum "breaks" from -6 to -12 dB/oct. f_p is the frequency determined by D, at which the spectrum reaches its peak intensity. After J.R. Young (personal communication).

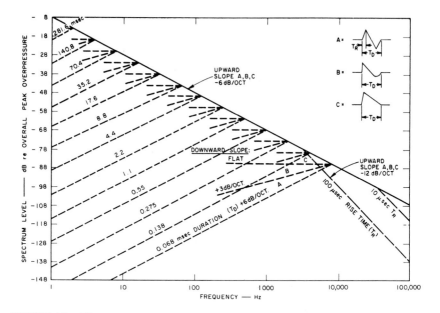

FIGURE 12.   The general spectrum level envelope of impulses having various waveforms.
            After J.R. Young (personal communication).

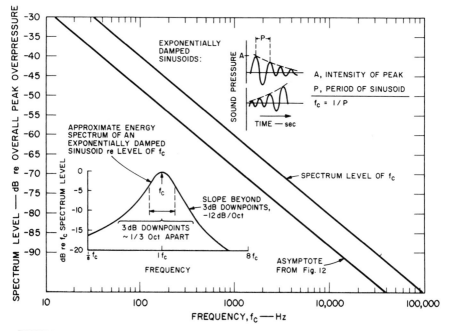

FIGURE 13.   The approximate spectrum level of impulses having waveform of an expo-
            nentially damped sinusoid. After J.R. Young (personal communication).

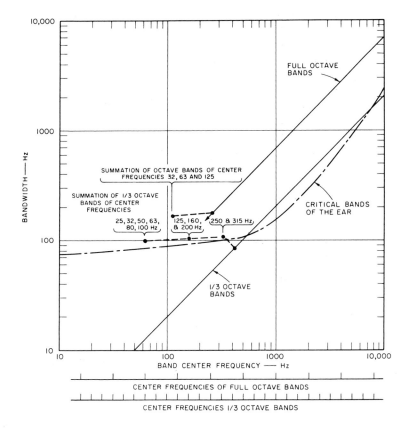

FIGURE 14. Showing relation between critical bands of the ear and widths of full octave and 1/3 octave bands. Also shown are suggested groupings of octave and 1/3 octave bands below 355 Hz to make these bands proportional to critical bands of the ear. From Kryter (462).

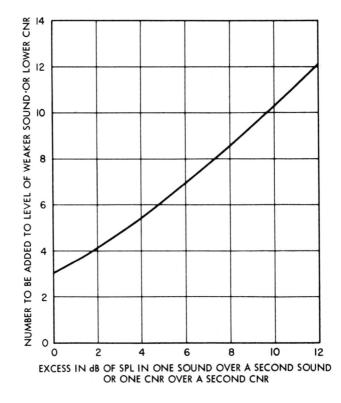

FIGURE 15.   Graph for summing SPLs in dB of two sounds or for summing two CNRs. (The CNR is to be defined in Chapter 9.) This summation is said in the text to be on a $10 \log_{10}$ antilog basis. Example: Assume a SPL of 32 dB for one sound and a SPL of 30 dB for a second sound of different spectrum or different phase relations; their sum will be 34 dB. A CNR of 32 plus a CNR of 30 will equal a CNR of 34.

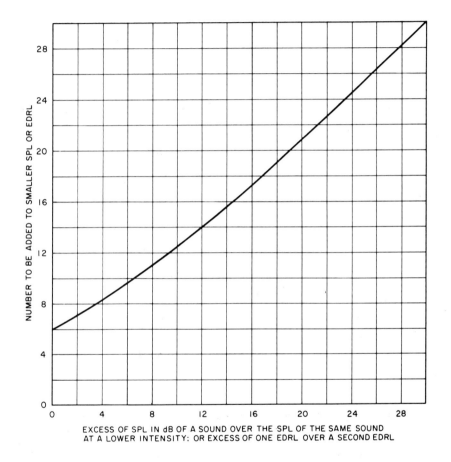

FIGURE 16.   Graph for summing SPLs in dB of the same sound, or for summing EDRLs. (The EDRL unit is to be defined in Chapter 6.) This summation is said in the text to be on a 20 $\log_{10}$ antilog basis. Example: Assume a SPL of 32 dB for one sound and a SPL of 30 dB for a second sound of the same spectrum and phase as sound one; their sum will be 37 dB. An EDRL of 32 plus an EDRL of 30 will give a CDR of 37.

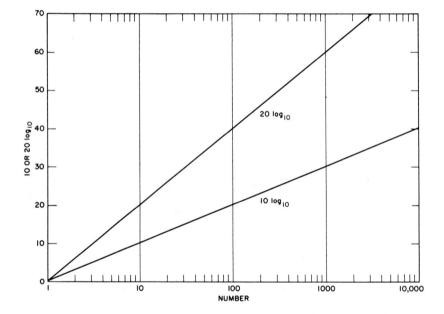

FIGURE 17.   Graph for summing on a 10 $\log_{10}$ or 20 $\log_{10}$ antilog basis. Example: the
SPL of 100 sounds from different sources, measured at a point where the
SPL of each sound is the same but their phase relations as a function of
frequency are random, will be 20 dB (10 $\log_{10}$) greater than the SPL of a
single source. Increasing by a hundred-fold the pressure of a given sound,
with its frequency components in phase, will cause a 40 dB (20 $\log_{10}$)
increase in its SPL.

for the summation of repeated occurrences of sound pressure levels or related measures that are equal to each other in magnitude. Figure 17 can be used for achieving this summation either on a $10 \log_{10}$ antilog or $20 \log_{10}$ antilog basis.

For certain psychoacoustic purposes it is appropriate to sum sound over relatively long (greater than 0.5 sec) intervals of time. This summation, or integration, can be achieved by the procedures outlined above for power, i.e., $10 \log_{10}$ antilog basis. Dividing this sum by some reference duration of time will provide what will be called the effective level for the chosen reference duration.

As noted earlier, because of phase relations between frequency components, as well as possibly other conditions, sounds of nominally the same intensity from one 0.5-sec interval to the next will fluctuate somewhat in their measured sound pressure level. As a practical and appropriate procedure for predicting some psychoacoustic effects of sound, it is proposed that a nominal sound pressure level be obtained for nonimpulsive sounds. The nominal sound pressure level of a nonimpulsive sound will be, for practical purposes, taken as the arithmetic average of the sound pressure level in dB found in each 10 successive 0.5-sec intervals. When this average level changes by 1 dB it is said that a new sound is present, provided, however, that the level in any 0.5-sec interval does not differ from this average by more than ±5 dB or does not fall below some functional threshold of hearing that may be specified. When the sound in a 0.5-sec interval is more than 10 dB different than the preceding 0.5-sec interval, a new sound, by definition, is said to start. When the sound is less than a specified threshold, there is said to be effective silence with respect to the particular auditory function represented by the specified threshold and the sound is said to stop.

The choice of 0.5 sec rather than 0.2 sec (the approximate average loudness time constant of the ear) is predicted on practical considerations related to the aforementioned band spectrum measurement of nonimpulsive sounds and upon psychological data that show that auditory discriminations are made most effectively when sounds have durations as long or longer than about 0.5 sec. The appropriateness of the choice of a time constant of 0.5 sec for the auditory system is shown in Figs. 18 and 19 where it is seen that optimum perceptual discrimination appears to occur only after the ear has about 0.5 sec of time to process sounds.

On the other hand, note should also be made of the fact that the spectra of sounds are evaluated, for auditory purposes and in accordance with the definition of the word sound, only after, or as though, the acoustic signals first passed through a 2 to 20,000 Hz band pass filter. When this is accomplished, the spectra that are obtained are the most meaningful as far as the ear is concerned.

The threshold of various auditory functions are important to the description of many effects of noise on man. In most of the descriptions to be given below, particularly the ability of the human ear to hear signals in the presence of noise

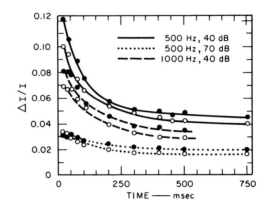

FIGURE 18.   The differential threshold for intensity as a function of the duration of the added increment. $\Delta I/I$ is the ratio of the sound-pressure increment to the standard sound pressure. Increments that last 0.5 sec are as readily detectable as longer ones, but shorter durations require greater intensity. The open and the closed circles distinguish two listeners. From Miller and Garner (545a).

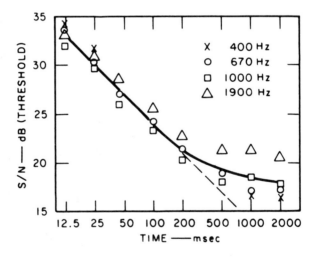

FIGURE 19.   The relation between the masked threshold of a pure tone and its duration. Noise energy for the S/N is in terms of energy per one-cycle bandwidths. Each plotted point is the average of 160 observations. The heavy line is a visual fit of the data, and the dashed line is the extension of the curve which represents linear integration. From Garner and Miller (291).

or after noise insult, the thresholds have been specified on the basis of the rule that 50% correct detection of a signal, when a signal is always present, represents a person's threshold of hearing for the particular hearing task at hand. When the signal is in one of two intervals of time, and the task of the observers is to detect which, 75% correct detection is said to represent the person's threshold.

From physical information meeting the specifications described above, it is usually possible, as will be discussed later, to calculate with good accuracy the response of man's auditory system to sound, and to calculate with an accuracy of practical significance important psychological and sociological behavior in response to noise.

Chapter 2

Masking and Speech Communication in Noise

Introduction

A major function of the auditory system is the analysis of acoustical signals so that wanted information bearing components in a sound wave can be discriminated or separated from the unwanted or noisy parts. In a sense, noise is always present during the hearing process — in the limiting case of quiet it is the internal noise floor of the auditory system, but usually noise is present in the acoustical signal along with wanted, acoustically-coded information. The interference or masking of wanted signals by noise is merely the converse of the analysis process.

The masking of speech is the most important masking effect of noise on man. But before discussing this effect we will briefly review some studies conducted, primarily since 1950, on the masking of pure tones and bands of noise. These data are of considerable interest and will provide some basis for understanding the effects of noise on speech. The effects of noise on masking and other aspects of the process of speech communication will be discussed later.

Masking of Pure Tones and Bands of Random Noise

The general method used for measuring masking of tones or bands of noise is as follows: using a pure tone or narrow band noise generator, the threshold of audibility is determined at a number of frequencies for the listener in the quiet. Then, while a masking pure tone or band of noise is presented, the listener redetermines his threshold of audibility by means of other (called "probe") tones or bands of noise. The increase in level required for the probe tone or band of noise to be audible at each frequency represents the amount of masking caused by the masking tone or band of noise.

Direct and Frequency-Spread Masking

Direct masking is that masking that occurs when the receptors in the cochlea that normally process a signal of a given frequency are not functionally available because they are being stimulated by another signal of the same frequency or because they are being affected by the upward and downward spread of masking along the basilar membrane from another signal. Ehmer (210), Small (750), and Carter and Kryter (127) confirmed and extended the masking patterns for pure tones that had been found earlier by Wegel and Lane (865). Egan and Hake (206), Ehmer (211), Zwicker (900), Saito and Watanabe (711), and Carter and Kryter (127) measured the masking pattern of narrow bands of noise. Typical examples of the results obtained are shown in Fig. 20. It is seen in Fig. 20 that the masking pattern from narrow bands of noise is much smoother, particularly at the vicinity of the center frequency of the masker, than those found with pure tones; the latter masking functions are disturbed by audible beats that occur between the probe tone and the masking tone and its harmonics. These harmonics are introduced by nonlinear distortion in the ear.

The curves in Fig. 20 reveal several interesting characteristics of direct masking:

1. The band of noise causes more masking around its center than does the pure tone. Increased masking near the center frequency of the masker band of noise would, of course, be expected from the integrative action of the critical bandwidth of cochlear functioning. In addition, as Egan and Hake and Ehmer suggest, the pure-tone masking may be only apparently lessened at the locus of the masking tone because beats between the probe and masking tone cause a false measurement of the threshold of the tone.

2. There is an asymmetrical upward spread of masking that becomes more severe at higher intensity levels. Békésy's observations of asymmetrical resonance patterns along the basilar membrane offer an apparent mechanism to explain the asymmetrical upward spread of masking.

Of particular interest are the masking patterns obtained by Finck (227) with intense low frequency tones ranging from 10 to 50 Hz (see Fig. 21). It is seen that the masking pattern for a tone as low as 25 Hz and an intensity level of 130 dB appears to extend almost flat to as high as 4000 Hz. Carter and Kryter (127) also tested the masking effects of tones as low as 50 Hz on tones and upon speech; their results, which will be discussed later, for a 50 Hz masking tone agree reasonably well with those of Finck.

It is of some interest to consider what effect, if any, masking noise has upon loudness and the ability of the ear to discriminate or detect changes in signal level — called the difference limen for intensity (see Harris [346] for a review of research on the difference limen for sound intensity). It is a generality, as is also

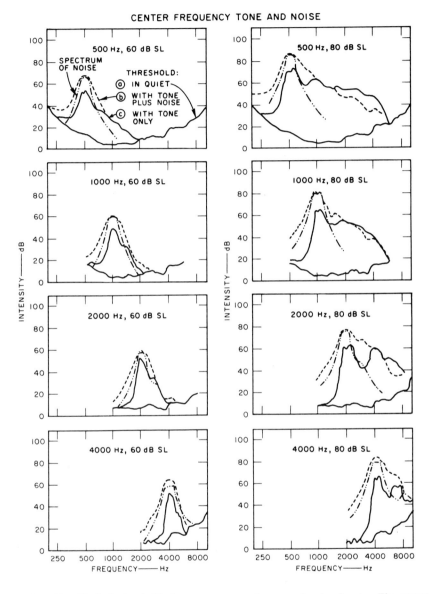

FIGURE 20.   Masked thresholds for pure tones and narrow bands of noise. The center frequency and sensation level (SL) of the tone and noise are parameters. From Ehmer (211).

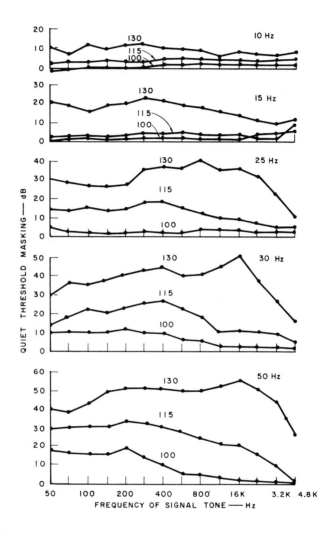

FIGURE 21. Masking with 5 low frequency pure tones (10, 15, 25, 30, and 50 Hz) and 3
intensity levels (100, 115, and 130 dB SPL). The ordinate shows masking in
dB relative to the quiet threshold. The abscissa shows the frequency of the
signal tone in Hz. The parameter is the SPL of the masking tone in dB, re:
0.0002 dyne/cm^2. Each point is the average masking experienced by 5
listeners. From Finck (227).

shown in experiments on the intelligibility of speech in noise, that it is the signal-to-noise ratio and not the absolute level of the masking noise, up to 110 dB or so, that determines the detectability of changes in signal intensity. Figure 22 from Small *et al.* (752) shows that the difference limen for intensity (ΔI) for an actave band of noise is essentially constant in sensation levels above 20 dB.

The general constancy of the difference limen is no doubt related to the well-established fact that the ear recruits loudness in the presence of noise so that a sound but a few decibels above a masking noise appears about as loud as it would were the masking noise not there. Figure 23 from Hellman and Zwislocki (365) demonstrates that the loudness of a sound grows much more rapidly above its threshold in noise than in the quiet. A matter of both theoretical and practical importance is that, in the quiet, loudness grows in the ear with a sensori-neural hearing loss as does loudness in the normal ear in the presence of a noise sufficient to cause a comparable threshold shift (see Fig. 24).

Pitch Changes with Direct Masking

A number of investigators (161, 207, 864) have reported that the pitch of a tone may change when heard in the presence of a band of noise. If the band of noise is of a higher frequency, the pitch decreases slightly; if the noise is of a lower frequency than the tone, the pitch increases. Both of these effects occur only when the loudness of the tone and the noise are not too different from each other.

Egan and Meyer (207) offer a convincing explanation of why these pitch changes may occur. The argument, put forth also by deBoer (190), is that the locus or central tendency of the area on the basilar membrane that has the highest signal-to-noise ratio determines what pitch is perceived. This concept, which is a general model for direct masking in the cochlea, is illustrated in Fig. 25. It is seen in this figure that the point on the frequency scale enjoying the maximum signal-to-noise ratio is not the center frequency of the pure tone, but is lower in frequency for the tone below the band of noise and higher for the tone above the band of noise.

Remote Masking

Remote masking was discovered and named by Bilger and Hirsh (66). Remote masking refers to the fact that a high frequency band of noise, provided it is sufficiently intense, will elevate the audibility threshold for pure tones of low frequency. This is shown in Fig. 26 from a study conducted by Spieth (760). It is usually presumed that this masking is direct masking caused by the presence of

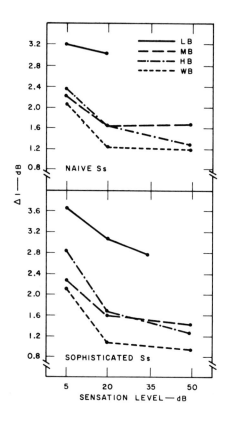

FIGURE 22. Mean ΔI (increase in intensity) for naive and sophisticated subjects to detect change in intensity. Each data point represents 44 threshold determinations in the upper panel and 24 in the lower panel. From Small *et al.* (752).

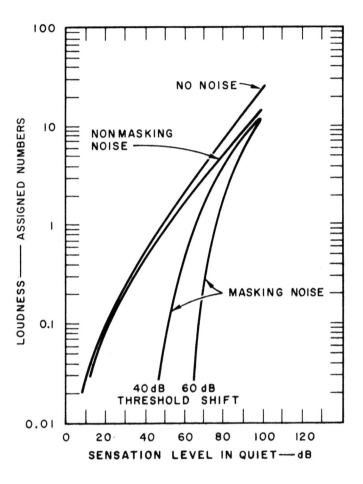

FIGURE 23.   Masked monaural-loudness curves, obtained by the method of numerical magnitude balance, compared to the curves of numerical magnitude balance without masking and in the presence of a nonmasking noise. From Hellman and Zwislocki (365).

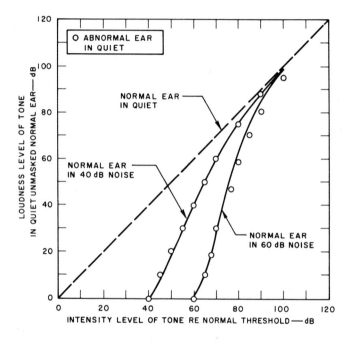

FIGURE 24.   Loudness-level curves of a partially masked tone, obtained by the method of adjustment, compared to loudness-balance data, collected and contributed by Miskolczy-Fodor (560), in ears with sensori-neural hearing loss exhibiting loudness recruitment. After Hellman and Zwislocki (365).

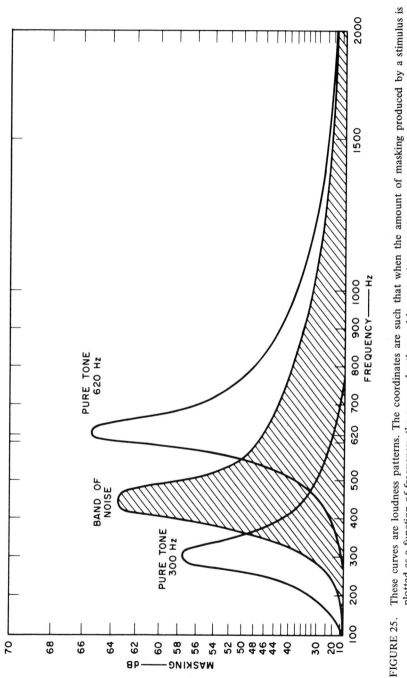

FIGURE 25.   These curves are loudness patterns. The coordinates are such that when the amount of masking produced by a stimulus is plotted as a function of frequency, the area under the resulting curve is proportional to the loudness of that masking stimulus. From Egan and Meyer (207).

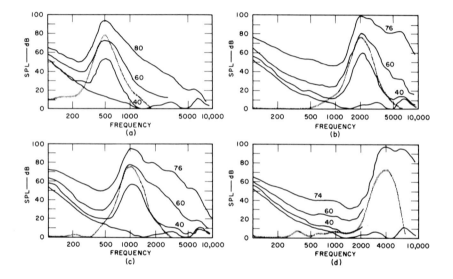

FIGURE 26.   Solid curves represent the average binaural pulsed pure-tone thresholds in
quiet (bottom curves) and in the presence of several levels of noise centered
at 500, 1000, 2000, and 4000 Hz. The spectrum SPL (sound pressure level
per Hz) of the noise producing each threshold curve is shown as the
parameter in each figure. The hatched curve in each figure shows the
spectrum of the noise at the maximum level used. All data were obtained
from the same five individuals. From Spieth (760).

low-frequency distortion products that result from the amplitude distortion that occurs when the signal strength is sufficiently intense to overload the ear.

Bilger (64) has recently demonstrated remote masking with subjects whose intra-aural muscles had been cut. This result would appear to rule out the possible apparent masking as the result of attenuation of low frequency sounds due to reaction of the aural muscles to intense sound (see Chapter 3).

Central Masking

Central masking is said to occur when sound presented to one ear raises the threshold of sound presented to the opposite ear in a way that cannot be attributed to contralateral direct masking, action of the aural reflex, or binaural phase interactions. Contralateral direct masking is that due to sound presented to one ear reaching the other ear; this is usually small because the sound presented to one ear is usually attenuated due to transcranial conduction by about 50 dB upon reaching the opposite ear. The aural reflex, to be discussed in Chapter 3, which acts bilaterally, may cause a threshold shift at the lower frequency in the same and opposite ear — this phenomenon is sometimes called contralateral remote masking. Binaural phase interactions will also be discussed later. In general, central masking is a phenomenon that is relatively negligible and unexplored.

Ward (839), in a recent review of masking and in an earlier paper (826), summarized the several masking effects by means of Fig. 27. The curves are labeled in accordance with the type of masking that was affecting the threshold change at 500 Hz measured in the right ear (RE) of listeners. While direct ipsilateral masking is 30 to 100 dB more effective than the other types, these other types cannot be ignored; we shall see later, for example, that ipsilateral remote masking contributes to the masking of speech by high-frequency noise. However, the amount of central masking is apparently rather small and cannot be separated from direct contralateral masking or aural reflex effects.

Temporal Masking

In recent years, following the pioneer work of Samojlova (713), Pickett (620), and Chistovich and Ivanova (134), considerable attention has been given to the temporal pattern of masking. In these investigations, a probe tone of very brief duration is presented both before and after a masking tone or noise. Figure 28 from Elliot (214) gives typical results; the small amount of masking for dichotic listening (probe tone in one ear, masking tone in opposite ear) indicates that temporal masking is primarily of a direct or at least ipsilateral sort.

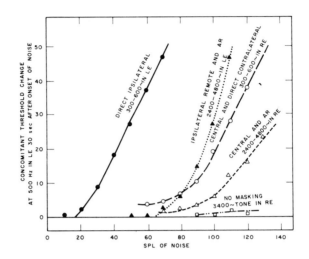

FIGURE 27.   Growth of direct and remote masking. The threshold at 500 Hz in the left ear was measured in the presence of low-frequency (300-600 Hz) or high-frequency (2400-4800 Hz) noise, or a 3400 Hz pure tone, in the left ear (LE) or right ear (RE). After Ward (826).

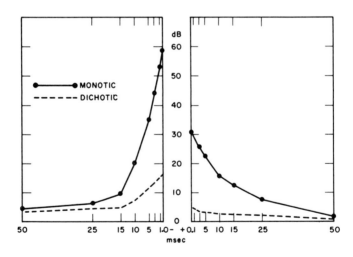

FIGURE 28.   Backward and forward masking under conditions of 90 dB masking and 5 msec probe duration. The abscissa represents the masking interval (maskers not present) with the positive values for forward masking on the right and the negative time values of backward masking on the left. The ordinate represents amount of masking in dB, i.e., the difference between masked threshold and unmasked threshold of the probe. Data on the monotic listening condition are shown by the solid line while the dotted line represents dichotic listening. From Elliot (214).

Forward masking in time is not surprising — it could be a manifestation of temporary auditory fatigue or some sort of refractory period due to the previous stimulation. But how can a masker elevate the threshold of a sound preceding it in time? Wright (890) and Zwislocki (911) suggest that the effect is due to a restriction in the time available for the auditory system to summate the energy and loudness of the tone or a click preceding the masking noise. Apparently a given length of time is required because of a stimulus-intensity-neural-response time factor, wherein it is hypothesized that the neural impulses from the much more intense masking noise reach the brain sooner than the impulses resulting from the test tone or click at threshold. Presumably the growth of the perception of a signal is the integral of the distribution of impulses in the various neural pathways from the cochlea to the higher centers, and since the weaker, preceding sound activates the slower pathways, its growth of loudness occurs at a slower rate than that of the more intense and later occurring masking sound. Miller (544) earlier hypothesized similar factors in an attempt to explain the longer duration required for bursts of noise to be heard when near their threshold.

Temporal masking is obviously a factor in the detection of temporal order of two stimuli. Hirsh (377) found that a 10-20 msec separation in time is required between two sounds for the human observer to correctly detect which of the two sounds came first. With only a 2-3 msec delay, a separation in time between two sounds was heard, but the order in which the two sounds came could not be identified.

Binaural Effects

We have already mentioned binaural masking attributable to the aural reflex, contralateral transmission of sound, and central masking. At this time those effects that are due to variations in phase relations between the ears for either the signal or the masking noise will be considered.

Jeffress and his colleagues (409), Pollack (633), and others following the earlier work of Licklider (501) and Hirsh (374) have extended the knowledge of how the two ears work together in terms of signal detection in noise. For some unknown reason, if both of two signals are presented simultaneously to both ears, they mask each other by the minimal amount if one signal is in phase with respect to itself at the two ears and the other is out of phase with respect to itself at the two ears; but if the phases are the same at the two ears for both signals (i.e., both in or both out of phase) mutual masking is increased from 0 to 16 dB or so, depending upon the frequency spectra involved. Intermediate degrees of phase correlations cause intermediate effects (Jeffress *et al.* [409]). Thus certain obvious advantages may be gained in communication systems operated in noise fields when control of the signal and noise phase relations at the two ears is possible; more will be said about this in the section on speech communications.

The conclusions described above are for the binaural presentation of the signal and the noise. Also of interest are the conditions in which the noise is presented sometimes to one ear and sometimes to both ears and the signal to but one ear. A startling finding here, as shown in Fig. 29, is that adding noise in phase to the ear opposite the ear receiving both the tone signal and the noise reduces the masking of the tone when the tone is at a sensation of more than about 10 dB above its threshold in the quiet; that is, the addition of noise at the opposite ear improves the detection of the signal. This phenomenon has been labeled Masking Level Difference (MLD), the difference in level of the tone at detection threshold when being masked by noise in the same ear and in both ears.

These binaural phase effects are clearly due to an analysis process going on in the central nervous system − a process that is also consciously recognized as a locus of the signal and the noise in so-called phenomenal space. By phenomenal space is meant the locus, on introspection, of the source of a sound; for example, when a recording of a musical instrument is played via earphones with the signal in phase at the two ears, one gets the impression that the sound is centered in the middle of one's head; changing the phase relations between the two ears tends to externalize the source placing it to the side of the head where the lower frequency components in the signal lead in phase.

Finally, under binaural listening, it should be mentioned that in addition to phase differences for a signal or signals at the two ears, intensity differences also make important contributions to the detection and localization of sound in space. As Gardner (276) has shown, phase differences between tonal signals in the presence of noise improve detection for frequencies only below 2000 Hz, and pressure level differences between the two ears increase detection for all frequencies above about 500 Hz, with increased effectiveness at the higher frequencies.

Localization, perhaps a better term is lateralization, of the source of impulsive sound with respect to the listener is, apparently, based on at least two cues: (a) there is the well-known precedence effect (first investigated by Wallach et al. [823] but known as the "Hass effect" in architectural acoustics, see recent review by Gardner [279]) where the position of the source of sound is ascribed to the side of person, or ear, first receiving the sound; and (b) phase and intensity differences between the sound at the two ears. The precedence and intensity cues were investigated by Freedman and Pfaff (266) who found that a 25 to 45 microsecond (depending on the experimental method used) temporal difference for a click at the two ears was equivalent to about a 1 dB difference in dichotic intensity with respect to lateralization of the source of the clicks.

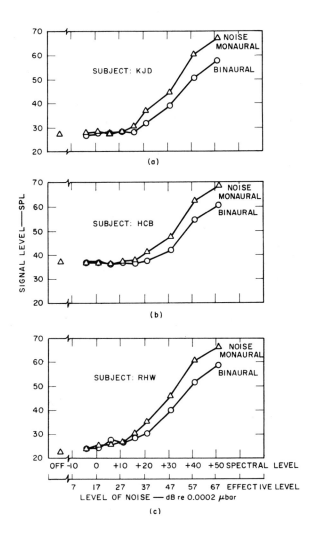

FIGURE 29. Masked thresholds (required level of signal to be audible in presence of noise) for one ear vs. noise levels in it, or in both ears. From Blodgett *et al.* (74).

Effects of Noise on Speech Communications

Before presenting information on the effects of noise on speech communications, including the masking of speech, it is in order to first describe the general physical characteristics of speech and methods used to measure speech masking.

Some Physical Characteristics of the Speech Signal

Considerable insight into the effects of noise on the reception of speech in noise is gained from an examination of the speech spectra shown in Fig. 30. We see in that figure that the rms level, measured every 1/8 sec of the acoustic speech wave, when uttered at a constant level of effort encompasses a range of nearly 30 dB. In Fig. 31 are shown the idealized long-time rms average spectrum and octave band levels one meter from male and female talkers using a normal level of effort for telephone communications or for face-to-face talking in a typical office or room with some noise present. Speech levels are usually measured and expressed in terms of the long-time (60 sec or so) rms pressure. The long-time rms level of speech can be approximated by averaging in a sentence the peak deflections for each word on a sound level meter and changing this average as follows, with the meter set on fast: +3 dB(A), -5 dB(B), -3 dB(C), +1 dB(D); and, if set on slow: +8 dB(A), 0 dB(B), +2 dB(C), and +6 dB(D). (See also Gardner [278].)

Increasing the level of background noise tends, of course, to cause a talker to increase his vocal effort. The increase in effort is usually not sufficient to completley override the increases in noise level. For example, Korn (435) found, with conversational vocal efforts, about a 3.5 dB increase in voice level for each 10 dB increase in room noise, whereas Webster and Klumpp (861) found as much as a 7 dB increase per 10 dB increase of noise when the talkers were using strong vocal effort; Kryter (440) and Pickett (619) found about a 3 dB increase in shouted vocal effort with a 10 dB increase in masking noise. The results of these studies and a more recent study by Gardner (278) are approximated by the curve shown in Fig. 32 (see also Table 3). It should be understood that the function shown in Fig. 32 presumes that the talkers are equally motivated to be intelligible in all noise conditions and to adjust their vocal effort on the basis of how much effort they believe is required for their speech to be understood.

Lane et al. (484), by varying the side-tone presented to the talker of his own voice, found about a 5 dB increase for a 10 dB decrease in side-tone. It seems fair to conclude that since voice level does not keep pace with increases in noise level or decreases in side-tone, a talker apparently hears his voice more through air and tissue paths within his head than through the external ear canal.

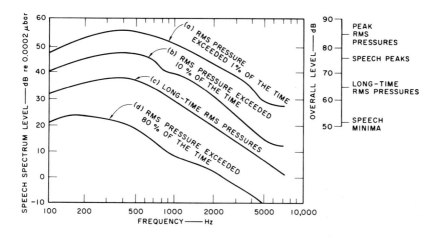

FIGURE 30.   Spectrum level of male speech measured over 1/8-sec intervals. From Dunn and White (198).

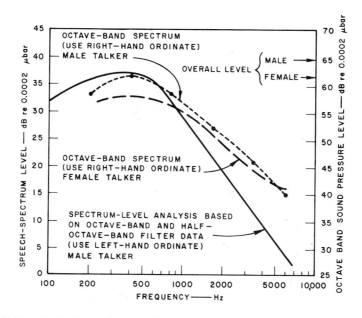

FIGURE 31.   Idealized speech spectra for male and female talkers one meter from talker. Normal level effort for typical, everyday talking conditions. After French and Steinberg (267) and Benson and Hirsh (53).

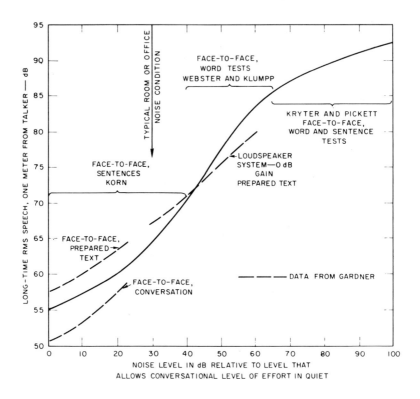

FIGURE 32.　Shows effects of room noise on speech level. The solid line represents and connects data points obtained by Korn (435), Webster and Klumpp (861), Kryter (440), and Pickett (619). The dashed line is based on data from Gardner (277,278).

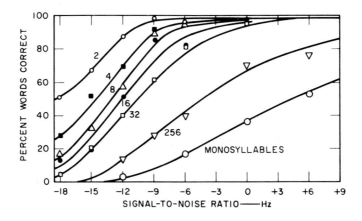

FIGURE 33.　The word intelligibility scores obtained at various speech-to-noise ratios for test vocabularies containing different numbers of alternative English monosyllables. The bottom curve was obtained with a vocabulary of approximately 1000 monosyllables. From Miller *et al.* (549).

TABLE 3

Talking Levels Employed under Various Test Conditions During Face-to-Face
Communication. From Gardner (278).

Environment	Separation of participants	Talking level as measured at 1 meter (dB-B)
(a) During face-to-face conversation		
Free-space room	39 in.	49.5
	12 ft	53.5
Quiet office (NC-23)	39 in.	58.0
	12 ft	62.5
(b) During face-to-face exchange of prepared text		
Free-space room	39 in.	57.0
	12 ft	58.5
Quiet office (NC-23)	39 in.	64.0
	12 ft	66.5

A number of studies concerned with a wide variety of acoustical and other factors upon speaking rate have been conducted. The general effect of noise on reading or talking rate is not pronounced until very high levels are reached when speech becomes slower due to increased effort by the talker. A review of these effects upon speech communication was issued by the Office of Naval Research (588).

Message Set

In the discussion of the masking of speech by noise, masking effectiveness, unless otherwise specified, will be in terms of the degradation in test scores of the understandability of speech in the presence of noise. These tests are variously called intelligibility or articulation tests; the distinction is usually made on the basis of how they are scored. If the sense or meaning of the word, phrase, or sentence is of interest, it is called an intelligibility test, whereas if communication performance is measured in terms of the individual phonemes or speech sounds in each word, it is called an articulation test.

In a study published in 1950, Miller *et al.* (549) demonstrated that the intelligibility or understandability of speech in noise is a strong function of the probability of occurrence of a given speech sound, word, or phrase. The larger the message set being used in a given communication system, the lower the probability of occurrence (the more information present), and the more susceptible is the communication process subject to interference from noise. As illustrated in Fig. 33, the understandability of the words is as much influenced by the message, or information, set size as the masking noise.

Test Materials

Speech materials and test procedures for measuring the effectiveness of speech communications under the stress of noise and other types of distortions imposed on the speech signal had been well developed and standardized by 1950 at the Bell Telephone Laboratories and the Harvard Psycho-Acoustics Laboratory (see Egan [203]). Several new contributions to speech testing procedures have been made since then.

Miller and Nicely (548) found that by scoring tests in terms of the actual confusions or substitutions made by the listener on individual speech sounds, they could gain important insights into what aspects or features of speech are affected by masking. For example, Fig. 34 shows which consonant sounds are confused with which other consonant sounds as a function of signal-to-noise (random, broadband noise) ratio; in Fig. 34, the sounds attached to lines that are close together are confused with each other at all signal-to-noise ratios lower than the level where they first come together. At a signal-to-noise ratio of -18 dB, none of the consonants can be distinguished from each other.

Fairbanks (222), and later House *et al.* (391), developed word tests that consisted of a number of subensembles (5-6 words) of a small total number (250 or less) words; K.N. Stevens (770) developed nonsense syllable tests of phonemes that likewise consisted of subensembles of a small number of phonemes. In the House *et al.* test, called the Modified Rhyme Test (MRT), and in Stevens' Nonsense Syllable Tests, the listener is presented with answer sheets on which are printed the words or phonemes that might possibly be heard in each test item. As a result, much of the learning and variability normally present in speech intelligibility testing is reduced with these new tests. Figure 35 shows how scores on the Rhyme Test compare with some other speech test scores.

Masking of Speech by Noise

As seen in Fig. 36, the masking effectiveness of different frequency bands of noise is different as a function of signal-to-noise ratio. Clearly, mere

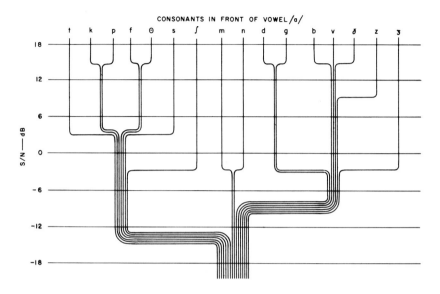

FIGURE 34. Masking of consonants by random, white noise. From Miller and Nicely (548).

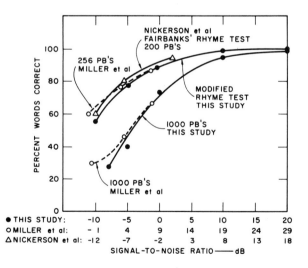

FIGURE 35. The relations between percentage of words correct and signal-to-noise ratio. Each solid point is the average of fifteen 50-word tests with one talker and eight listeners. The test results represent scores obtained after the listeners were thoroughly trained on the various word intelligibility tests involved. From Kryter and Whitman (469).

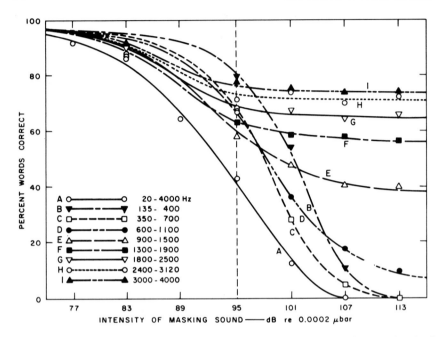

FIGURE 36.   Percentage of words correct as a function of the intensity of narrow bands of masking noise. This speech was not filtered and its level was held constant at 95 dB. From Miller (542).

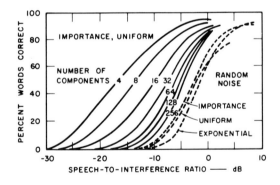

FIGURE 37.   Masking of unfiltered speech by line-spectrum interference (solid curves) and by continuous-spectrum interference (dashed curves) in regular articulation tests: Percentage word articulation vs. speech-to-interference ratio for various numbers of components. The density of spacing of the line components was governed by the importance function, and the lines were uniform in amplitude. The continuous spectra were shaped by filters. From Licklider and Guttman (504).

measurement of the ratio of the long-time speech-to-noise sound pressure level is not an adequate indicator of the masking of speech by noise. We shall see later that by taking into joint account the nature of the speech spectrum, the critical bandwidth for speech, and direct (including upward and remote) masking, it is possible to make general statements about the direct masking of speech by noise and to predict fairly well the kinds of results shown in Fig. 36.

The importance of considering the signal-to-noise ratio at a number of points along the frequency scale follows not only from the fact that different frequency regions of speech are somewhat more important than others to speech intelligibility, but also because the long-term speech spectrum is curved, falling off at the rate of about 9 dB per octave above about 500 Hz (see Fig. 31). The spectrum of the "instantaneous" peaks of speech is flatter than the rms pressure spectrum; however, it appears that the rms pressure spectrum is the effective spectrum with respect to the understanding of speech. For this reason, it is the lower speech frequencies that will be the last to be masked by noises whose spectra fall off less steeply than the speech spectrum as the signal-to-noise is decreased. French and Steinberg (267) demonstrated this fact by progressively reducing the level of filtered speech until it was made inaudible (was masked) by the threshold of hearing, and Webster and Klumpp (862) found a similar pattern by measuring speech intelligibility under a variety of noise spectra and levels.

Licklider and Guttman (504) did a study of the masking of speech by line (pure tone) and continuous spectra noise interference, in order to show how best, for a given amount of noise power, one could mask or reduce the intelligibility of speech. They varied the number and relative amplitude of the masking components. The density of spacing between components was varied in accordance with the critical bandwidth function of the ear over the range from about 200 to 6100 Hz, called the "importance function," (the critical bandwidth and the width of bands equally important to intelligbility are proportional to each other; see Fig. 31). Also these investigators masked the speech with random noise with the amplitude, as a function of frequency, being either uniform, negative exponential, or proportional to the critical bandwidth of the ear. It appears from the results shown in Fig. 37 that:

1. 256 components that are separated according to the relative importance function are 2 or 3 dB less efficient, in terms of noise power, than a continuous spectrum random noise in the band 200-6100 Hz that has a spectrum that declines as a function of frequency at the rate of 3 dB per octave (so called pink noise or negative exponential).

2. The masking effectiveness of but a few pure tones indicates that the upward spread of masking and remote masking as found with pure tones contributes significantly to speech masking.

A similar finding – the importance of the upward spread of masking – is demonstrated by the speech interference effects of pure tones of 50, 100, or 2000 Hz, see Fig. 38. It is also seen in Fig. 38 that pure tones of equal sensation level cause more equal amounts of speech masking than do tones of equal sound pressure level.

Effects of Vocal Effort on Intelligibility

The effect of noise level upon vocal effort and speech level was discussed above. It is of interest to know what the effect of vocal effort in talking might have upon the intelligibility of the speech, keeping signal-to-noise ratio constant. Pickett (618) conducted research on this problem and found the results shown in Fig. 39; clearly, speech uttered with very weak and very high levels of effort is not as intelligible as speech in the range from about 50 to 80 dB (measured at one meter from the talker) even though the speech-to-noise ratio is kept constant.

Effects of Speech Intensity on Intelligibility

In many noisy environments, the speech signal is often speech that has been spoken into a microphone at a normal or near normal level of effort and then amplified by electronic means to make it audible above the noise. Here the question, of course, is how intense can the speech be before it loses intelligibility, presumably because it is distorted due to overloading of the ear. Pollack and Pickett (647), see Fig. 40, found that in the quiet (signal-to-noise ratio of 55 dB) there was no loss in intelligibility even at speech levels of 130 dB; when there was some noise present, however, speech above 85 dB or so declined in intelligibility with signal-to-noise ratio kept constant.

These data demonstrate the interesting fact that the intelligibility tests, when near 100% correct (the case of speech of 130 dB in the quiet), are somewhat insensitive indicators of the true fidelity or undistorted nature of a given speech signal. The potentially degrading effects upon the intelligibility of overloading the ear with very intense speech are, of course, actually present when the speech is heard in the quiet, but can only be measured when the test scores have been lowered and made more sensitive by some additional stressful condition such as noise.

It was stated earlier that masking is usually not particularly affected, in the normal ear, by temporary auditory fatigue provided the signal (pure tones)-to-noise ratio remained constant. However, Pollack (641) found that the effective masking of speech did increase significantly during a 13-minute

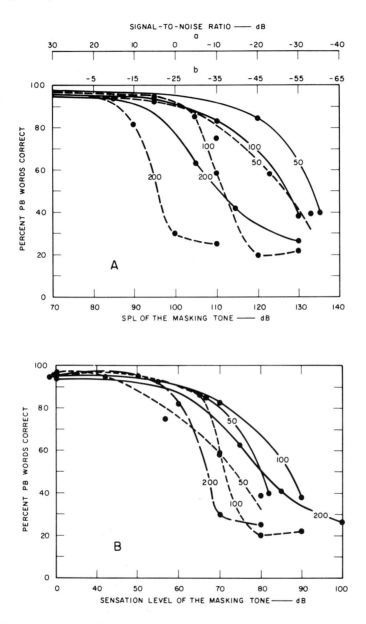

FIGURE 38. Percentage PB words correct as a function of the sound pressure level (upper graph) and the sensation level of the masking tone (lower graph). Solid curves are for speech at an SPL of 100 dB (upper abscissa [A]). Dashed curves are for speech at an SPL of 75 dB (upper abscissa [B]). Parameter is frequency in Hz of masking tone. From Carter and Kryter (127).

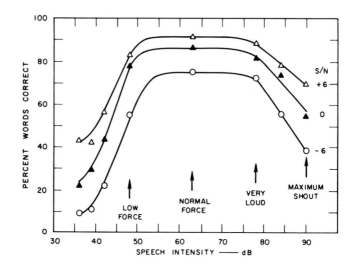

FIGURE 39.   Relations between speech intelligibility in noise and vocal force. Vocal force measured as speech intensity one meter from lips in a free field. Parameter, overall signal-to-noise ratio, dB. Noise, 70 dB, flat spectrum. From Pickett (618).

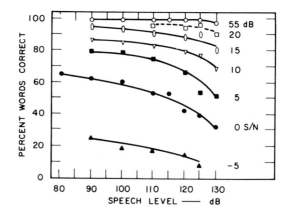

FIGURE 40.   The deterioration of intelligibility at high levels. The ordinate is the percentage of monosyllabic words correctly received by the listeners. The abscissa is the overall speech level (before addition of the noise). The parameter is signal-to-noise ratio (S/N). From Pollack and Pickett (647).

exposure to random broadband noise (flat 100 to 5000 Hz) at levels above 115 dB or so. This effect (see Fig. 41) is presumably due to an inability of the fatigued ear to discriminate among the speech sounds as well as the normal, unfatigued ear.

Interrupted Noise

When the speech signal is masked, either partially or completely, by a burst of noise, its intelligibility changes in a rather complex manner, as shown in Fig. 42. These functions are explained by Miller and Licklider (547) as follows: at interruption rates of less than about 2 per sec (which, for a noise-time fraction of 0.5 would make the duration of each burst of noise as long or longer than 0.25 sec) whole words or syllables within a word tend to be masked; at interruption rates between about 2 and 30 per sec, the noise duration is so brief that the listener is able to hear a portion of each syllable or phoneme of the speech signal, thereby tending to reduce the amount of masking; when the interruption rate is more frequent than 30 per sec, the spread of masking in time around the moment of occurrence of a burst of noise results in increased masking until by 100 interruptions per sec there is effectively continuous masking. This is in good agreement with the temporal masking results shown in Fig. 28 where it is seen that appreciable masking occurs only for 5-10 milliseconds before and after an intense sound.

That the increased masking of speech is due to a spread in time, presumably both forward and backward, is demonstrated in Fig. 43 where the intelligibility of the speech that is interrupted by turning it off and on in the quiet can be compared with that of speech that is turned off during noise bursts and is on between noise bursts. The signal-to-noise ratio refers to the signal in the quiet versus the noise alone. We see here that the temporal masking does not degrade the speech until the interruption rate exceeds 20 or 30 per sec. (The intelligibility of the speech represented in Fig. 43 is degraded somewhat by the switching transiengs and hence is lower for given signal-to-noise ratios than continuous speech interrupted by noise, as in Fig. 42.)

Miller and Licklider found that the above effects were the same for random or regularly spaced interruptions and that varying the speech-time fraction did not appreciably change the nature of the relation of interruption rate to intelligibility. Pollack (639) found that, over rather wide limits, varying the signal-to-noise ratio at rather slow rates provided intelligibility comparable to a steady-state signal-to-noise ratio equal to the average, and that varying the absolute levels but keeping the signal-to-noise ratio constant had no appreciable effect on measured speech intelligibility.

One of the most common noises that masks speech is speech itself — the

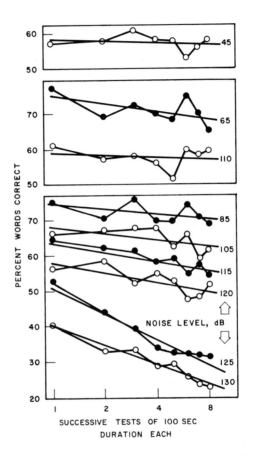

FIGURE 41.   Speech intelligibility in noise as a function of continuous noise (and speech)
exposure. The abscissa represents successive 25-item test lists, each of 100
sec duration. The ordinate represents the average percentage of monosyllabic
words correctly reproduced. The parameter on the curves represents the
overall noise level. The vertical axis has been broken to avoid overlap among
conditions. S/N ratio is 0 dB. Each point represents the average of 500
determinations – one 25-item test list read by each of 4 talkers to a testing
crew of 5 listeners. From Pollack (641).

babble of other voices. Figure 44 shows how speech intelligibility is affected as a function of the number of competing voices. By the time eight voices are present, the "noise" spectrum is apparently practically continuously present.

Binaural Factors in Speech Perception

As discussed previously, the two-eared listener is able to some extent to separate, by some central nervous system mechanism, a signal from noise on the basis of relative phase and temporal relations of the signal and the noise reaching the two ears. If there are some temporal or frequency differences at the two ears that are different for the signal than for the noise, it appears that the listener may direct his attention to the sound he wishes without conscious regard to localization or phenomenal space. This is particularly noticeable when the noise consists of other speech signals — what has become to be called the "cocktail party" effect — and when the competing signals differ somewhat in spectra (91, 501, 649, 870b).

A somewhat extreme situation for direct person-to-person communication, but one which demonstrates clearly the advantages of binaural listening vs. monaural in the presence of masking sounds, was studied by Pollack and Pickett (649). They presented, via earphones, a speech signal in phase at the two ears against one background of speech presented to one ear and another background of speech to the other ear; some of the results are shown in Fig. 45. The control condition in Fig. 45 was achieved by merely disconnecting one of the listener's earphones; it is obvious that some of the direct masking of the speech that takes place with monaural listening is appreciably overcome on the basis of cues available with binaural stereophonic listening.

Experiments conducted by Licklider (501) led to an understanding of binaural listening to speech in noise. Table 4 shows the effect on intelligibility of all combinations of monaural and binaural listening to speech and noise over earphones. It is seen in Table 4 that speech intelligibility is the greatest for binaural noise and speech when they are of opposite phase and is at a minimum with monaural noise and speech in the same ear. Hirsh (374) demonstrated that localization cues available in freefield listening, because of phase and intensity differences between the two ears, were responsible for increased intelligibility that occurred when noise and speech source were separated in space.

Combating the Interference with Speech by Noise

The importance of redundancy of information, or of reducing alternatives, to the intelligibility of speech was mentioned in the section on speech intelligibility

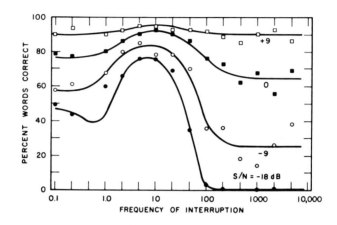

FIGURE 42.   The masking of continuous speech by interrupted noise. Word articulation is
plotted against the frequency of interruption of the noise, with the
speech-to-noise ratio in decibels as the parameter. Noise-time fraction is 0.5.
From Miller and Licklider (547).

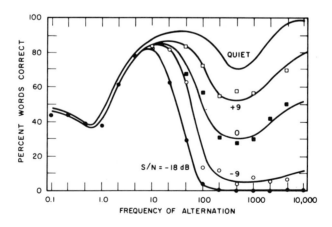

FIGURE 43.   Word articulation as a function of the frequency of alternation between
speech and noise, with signal-to-noise ratio in decibels as the parameter.
From Miller and Licklider (547).

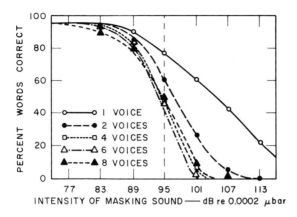

FIGURE 44.   Word intelligibility as a function of the intensity of different number of masking voices. The level of the desired speech was held constant at 94 dB. From Miller (542).

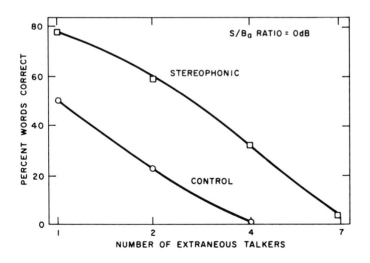

FIGURE 45.   Comparison between the average intelligibility scores of an adjusted speech-to-background noise (S/B_a ratio of 0 dB) for the binaural-stereophonic listening condition and for the monaural-control listening condition. From Pollack and Pickett (649).

TABLE 4

Monaural-Binaural Presentation and Interaural Phase Relations as Factors Influencing the Masking of Speech by White Noise *

Percent PB words correct. From Licklider (501).

		Binaural Noise			Monaural Noise	
		+	0	−	R	L
Binaural	+	18.0%	27.4%	35.4%	98.0%	99.0%
Speech	−	43.0	27.3	15.8	98.1	98.8
Monaural	R	30.3	13.2	20.1	16.6	98.7
Speech	L	18.1	8.3	15.2	98.4	15.4

*Key: + in phase; − out of phase; 0 random phase; R right ear;
        L left ear.

testing. This factor can be used to advantage in various ways in combating the masking of speech by noise.

A simple, and apparently the most beneficial, way of effecting this reduction of alternatives is to restrict the talkers to a limited number of specific words, phrases, or sentences that they can use when communicating by speech in a given situation. Such constraints have been found to be effective for many military operations. Moser (563) and his colleagues have contributed to the standardization of voice message procedures for the U.S. Air Force and international commercial aviation. Some benefit is, of course, also gained by prescribing the exact procedures — order of talking and how to talk. Standardizing the procedures, and the messages to be used, reduces the amount of information with which the listener must cope and thereby improves speech communications in noise (Pollack [640] Frick and Sumby [268]).

A second method of increasing redundancy of information is to have the talker repeat his words or messages. Thwing (801), for example, found that the intelligibility of single words increased by about 5-10 percentage points (equivalent to a reduction in the noise level of about 3 dB) when each word was repeated once. Further repetition caused little further improvement.

Increasing Signal Level

Increasing the level of the signal relative to that of the noise is the most effective way to avoid masking of the speech. This may not be possible for a variety of reasons:

1. With direct person-to-person talking, the noise may be so intense (or the listener so far away) that the talker cannot effectively override it.

2. When a communication system, such as a telephone or radio telephone is involved, the power available for amplification of the speech signal may be limited.

3. The masking noise may be mixed with the speech at the talker's microphone so that amplification of the signal likewise increases the noise, leaving the signal-to-noise ratio and intelligibility relatively constant.

4. The masking noise may be so intense at the listener's ears that increasing the speech level by means of an electronic amplifier system is not practical because making the speech more intense would overload the ear, causing distortion and possible pain to the listener.

Methods of alleviating the masking effects of noise 'for each of the above-listed conditions have been investigated and will now be presented.

Person-to-Person Talking

Pickett and Pollack (623) report that a small megaphone improved speech intelligibility relative to the unaided voice by an amount equivalent to a reduction of the noise level by 6.5 to 11.5 dB, depending on the noise spectrum — the least gain was found with a "flat," white noise, the greatest with a noise having a -12 dB slope above 100 Hz.

Peak Clipping

Research on the effects of amplitude-distortion, in particular peak-limiting, upon speech intelligibility and its application to a speech system, with limited power for use in noisy environments, was accomplished by Licklider (500), prior to 1950. Because this procedure is so uniquely applicable to speech communications in the presence of noise, a brief description of its effects is justified.

It is obvious from an examination of the amplitude waveform of a speech signal, as for example in Fig. 46, that certain portions of the speech wave, those

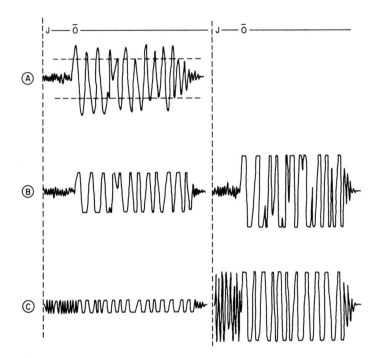

FIGURE 46.   Schematic representations of word "Joe." A is undistorted; B is after 6 dB clipping; and C is after 20 dB clipping. Clipped signals in B and C are shown reamplified until their peak-to-peak amplitudes equal the peak-to-peak amplitude of A. After Licklider (500).

parts usually associated with the consonant, are much less intense than the parts present when vowels are uttered. Adding noise masks the consonant sounds at a lower level than that required to mask the vowels.

The trick, in order to have the consonants override the noise and yet not increase the peak-power requirements of a transmission system, is to increase the level of the consonants relative to that of the vowel sounds. This can be done simply by passing the speech through a so-called "peak-clipper" and then reamplifying the result to whatever peak-power level is available — this process is illustrated in Fig. 46 where we see that the peak-to-peak amplitude of the consonant "J" is made equal to the unclipped and clipped vowel "o" by peak-clipping and amplifying the speech by 20 dB. Speech thusly clipped has a greater average speech power for a given peak power, and is more intelligible in noise than unclipped speech.

However, certain precautions must be kept in mind when peak-clipping is to be used; first, speech peak-clipped by more than about 6 dB sounds distorted and noisy due to the clipping of the vowel waveform when listened to in the quiet (when heard in noise, peak-clipped speech sounds relatively undistorted because the distortion products from the speech signal tend to be masked by the noise); and second, when there is noise mixed with the speech prior to peak-clipping, the amount of clipping that is beneficial is limited. This latter fact is revealed through a comparison of the top and bottom graphs on Fig. 47.

Noise Exclusion at the Microphone

One way to keep noise out of a microphone is to attach the microphone directly to tissues of the throat and head so that it will not pick up airborne noise but will pick up the speech signal through the body tissues. Such microphones are reasonably effective in excluding noise when attached to the throat, ear, teeth, and forehead, but tend to somewhat distort the speech signal (see Moser *et al.* [564]).

Placing an air-activated microphone in a shield that is held close to the talker's mouth will typically achieve noise exclusion as shown in Fig. 48. A third method is to use a close-talking-pressure-gradient microphone. Here both surfaces of the active element of the microphone are exposed to the air. Random incidence sound waves (the noise) will thus impinge on both sides of the element more or less simultaneously, depending on frequency wavelength; thus they tend to cancel each other, i.e., the microphone element does not move. The speech signal, on the other hand, is highly directional when the microphone is held close to the lips and correctly oriented, and therefore activates the moving element of the microphone. The amount of noise cancellation achieved is shown in Fig. 49 as a function of frequency.

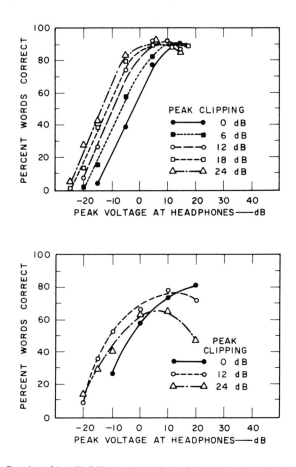

FIGURE 47.   Results of intelligibility tests conducted with the talkers in the quiet and the
listeners in ambient airplane noise (upper graph), and intelligibility tests
conducted with both the talkers and the listeners in the presence of
simulated airplane noise (lower graph). In the lower graph, note that when
the microphone picks up noise, clipping is not so beneficial; it is even
detrimental to speech intelligibility. A dynamic microphone (nonnoise-
canceling) was used. From Kryter et al. (475).

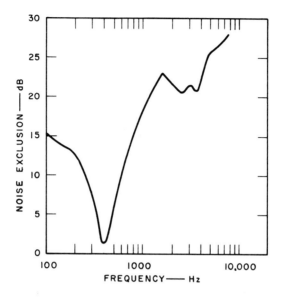

FIGURE 48. The noise exclusion of a noise shield. From Hawley and Kryter (357). (Used with permission of McGraw-Hill Book Company, 1957.)

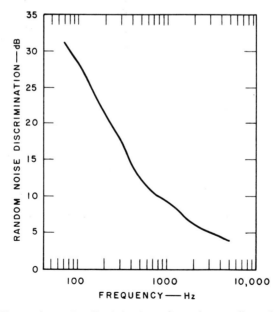

FIGURE 49. The random-noise discrimination of a noise-canceling microphone. From Hawley and Kryter (357). (Used with permission of McGraw-Hill Book Company, 1957.)

Noise Exclusion at the Ear

Earplugs and muffs for over the ears have received considerable attention as a means of protecting the ear against auditory fatigue from exposure to intense noise (see Fig. 50). However, these devices interact in various fortunate and unfortunate ways with the reception of speech by the users.

In the first place, earplugs or muffs would attenuate equally, at any one frequency, the speech signal and ambient noise passing through them; since the signal-to-noise ratio at any one frequency would remain constant at the listener's eardrum, we would expect speech intelligibility to be the same whether or not earplugs or muffs were worn. However, what happens is that (a) in high-level noise, speech intelligibility is improved when earplugs or muffs are worn because the speech and noise is reduced to a level where the ear is not overloaded and therefore discriminates the speech from the noise somewhat better; and (b) in low-level noise, on the other hand, speech intelligibility is decreased when earplugs or muffs are worn because the speech is reduced along with the noise to a level below the listener's threshold of hearing (see Fig. 51). Thus, persons who are suffering some hearing loss will not benefit, in terms of speech communication, from wearing earplugs or muffs in a low noise level as much as will the person with normal hearing. The effect of earplugs and muffs upon speech communication (or any other signal detection) in noise can only be predicted from a knowledge of the hearing of the listener, the spectrum of the noise at the listener's ears, and the sound attenuation characteristics of the earplug or muff. Some of these interactions are illustrated in Fig. 52.

Effect on Voice Level

It should also be noted that when one is in noise, plugging the ears results in a drop of one to two decibels in voice level. Apparently the earplugs or ear covering attenuate the ambient noise without lowering to as great an extent the speaker's own speech, which he hears both by tissue and bone conduction through his head and by airborne sound. When the speaker wears earplugs in the quiet, he raises his voice level by three to four decibels since his own voice now sounds weaker to him because of attenuation of the airborne components of the speech wave (see Fig. 53).

Nonlinear Earplugs

Earplugs or muffs appear to be counter-indicated, with regard to speech communication, in the situation where they are probably needed most —

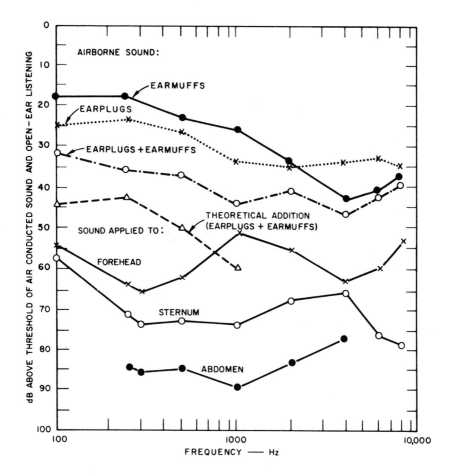

FIGURE 50. Measured attenuation curves of (a) earplugs, (b) earmuffs, (c) earplugs and earmuffs together, (d) theoretical addition of attenuations in decibels provided by earplugs and by earmuffs. After Zwislocki (910) (top 4 curves) and von Gierke (817) (bottom 3 curves). (Used with permission of McGraw-Hill Book Company, 1957.) The bottom three curves show the audibility of sound when applied to different parts of the body and conducted to the ear through tissues of the body.

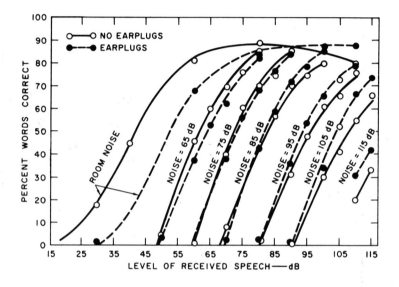

FIGURE 51.   The relation between intelligibility and speech level with noise level as the
parameter for listeners with normal hearing. In loud noise, the earplugs
improve intelligibility. From Kryter (440).

namely, in the presence of intermittent, impulsive noise, such as gunfire. Here, the wearer of earplugs or muffs cannot hear weak speech during the silent intervals between impulses. The ideal solution would be a nonlinear device that would let weak sounds through at full strength but would attenuate intense sounds.

Rüedi and Furrer (710), Zwislocki (908), and Collins (158) have described the theoretical basis for such nonlinear devices and have built and tested models of them. The devices, which are essentially acoustic filters, operate on a frequency selective basis, affording significant attenuation for frequencies above 1500 Hz but offering little or no attenuation to frequencies below 1500 Hz, the region containing the strongest speech components. Also, the device can be made to become significantly nonlinear only after the overall sound pressure level exceeds a certain level.

Noise Cancellation

Olson and May (594) developed a device for actively canceling ambient noise. In one version of this system a microphone, mounted close to the listener's ear, picks up environmental noise. The signal from this microphone is then amplified and fed into a loudspeaker (or to the listener's earphones) so that it is 180° out of phase with the noise signal and therefore acoustically cancels the noise at the listener's ears. This procedure is effective only for frequencies below 200 Hz or so, and for this reason will generally not help speech communication in noise.

This procedure was tried as a means of canceling the hum of powerful industrial transformers that were annoyingly audible in residential areas near the transformers. Since the noise was primarily 60 Hz hum, considerable — about 20 dB — cancellation of the noise could be achieved. However, wind could cause "drifting" of the transformer noise and make cancellation of the noise unreliable.

"Electrical" Stimulation of Hearing

It has been observed a number of times in the past that applying an electrical signal that has been modulated by a speech or other audible waveform to the skin near the ear can be heard by the subject as though the stimulus had been applied acoustically to the ear. It has been suggested that the electrical stimulus is conducted by means of fluid in the tissue around the ear to the auditory nerve or to the auditory nerve fibers within the cochlea, and that the nerve fibers are thereby stimulated. Although Simmons (742) has demonstrated that insertion of electrodes into the nerve endings of the auditory nerves within the cochlea can elicit hearing of a rather crude sort, it is highly probable that the hearing that occurs as the result of application of an electrical signal to the external skin near the ear is the result of the electrical stimulation being transduced into an acoustic signal by some mechanical or electro-mechanical process external to the nerve endings. This acoustic signal is then transmitted to the cochlea either via the ossicular chain or by conduction through the bone surrounding the cochlea. The fact that the electrical signal can be a modulated RF carrier lends credence to the notion that a mechanical detection system is involved in this type of "electrical" stimulation of the ear.

The practical application, assuming an efficient transducer instrument can be developed, regardless of the precise mechanism involved in "electrical" stimulation of the ear, is that it provides a possible means of avoiding the ambient noise that may be present in the listener's environment. In this case, the noise can be perhaps eliminated by the use of ear plugs or ear covering devices with the acoustic signal being applied electrically to the skin near the listener's ears.

Speech Systems for Use in High-Intensity Noise

The major interest in speech communication in noise has come from the military services where high noise levels are often present in operational situations. Many of the principles pertaining to speech communication in noise that could be applied to the design of optimum communication systems for use in noise have been outlined above. The general state-of-the-art is summarized in Fig. 54. An additional procedure that might be included in Fig. 54 is a noise-operated automatic gain control that would further protect the listener against extraordinarily intense speech signals presented in a fluctuating noise level. Pollack (639) has shown that some advantage in terms of comfort and protection of hearing is achieved, without necessarily reducing intelligibility, by means of a noise-activated gain control in the speech system, i.e., the gain is increased if the noise level increases, and is decreased below a specified level if the noise decreases sufficiently.

Estimating Speech Intelligibility on the Basis of Physical Measurement of the Speech and Noise

Articulation Index (AI)

On the basis of data related to the intelligibility of filtered speech and certain assumptions regarding the equivalence of bandwidth and signal level (when both

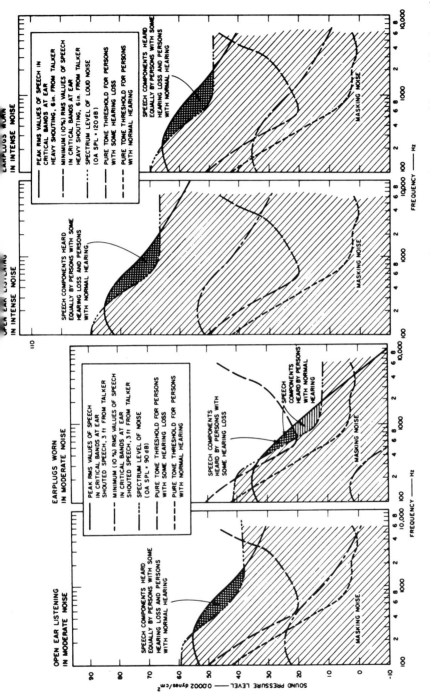

FIGURE 52. Left graphs: Show the amount of speech signal that is audible in presence of moderate noise by persons with normal hearing and persons with some degree of hearing loss. It is seen that with open ear listening, both listeners hear equal amounts of speech but that when wearing earplugs, persons with hearing loss hear less speech than persons with normal hearing. Right graphs: Show amount of speech signal that is audible in presence of an intense noise. It is seen that as with open ear listening, listeners hear equal amounts of speech when listening with ears open or with ears plugged.

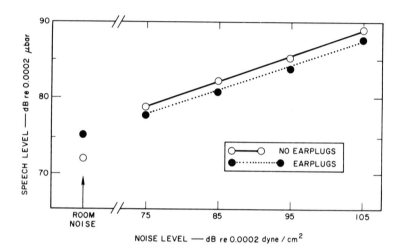

FIGURE 53. The effect of noise level on the average speech intensity used by the eight
speakers with and without earplugs. The speech and noise levels are
measured at the listener's position (7 ft from the speaker). When the residual
room noise was present, wearing earplugs caused the speakers to increase
their voice level by 3 dB whereas wearing earplugs in noise resulted in a slight
lowering of the voice level – one to two dB. From Kryter (440).

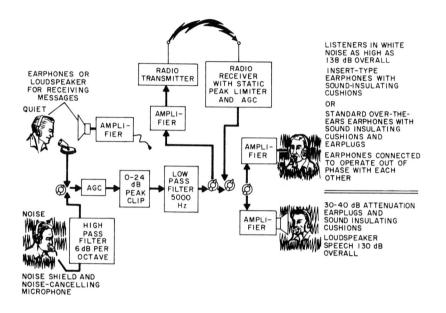

FIGURE 54.   Speech communication systems for use in intense noise. From Kryter (443).

are measured in equivalent power), French and Steinberg (267) outlined a procedure whereby one could calculate, from purely physical measurements, an index to the intelligibility of speech. They called this the Articulation Index or AI.

The AI concept holds that speech intelligibility is proportional to the average difference in dB between the masking level of noise and the long-term rms plus 12 dB level of the speech signal taken at the center frequency of twenty relatively narrow frequency bands. The masking spectrum of a noise may be different than the noise spectrum because of the spread of masking and remote masking. This proportionality holds, provided the difference falls between 0 and 30 dB. It is again noted that these twenty bands, which were chosen because they were found to contribute equally to the understanding of speech, are proportional to the critical bandwidth of the ear as determined from studies of loudness and masking (see Fig. 3).

The only significant modifications that have been made to the calculation procedures for AI as proposed by French and Steinberg have had to do with the specification of exact procedures to be followed by converting noise spectra to noise-masking spectra and methods for calculating AI from octave and one-third octave band speech and noise spectra (the original 20 band method requires spectrum level values). The steps to be followed in the calculation of AI have been further developed (448, 449) and are published as a Standard (S3.5, 1969) (31). Figure 55 gives the work sheet used for calculating AI from one-third octave band speech and noise spectra, and an example of the calculation of an AI. Figure 56 shows the general relation between AI and various speech intelligibility test scores.

Two points made in the Standard S3.5 that bear repetition here are (a) that the AI can be applied properly only to communication systems and noise environments as specified in the subject document, and (b) that there are types of communication systems and noise masking situations that can only be evaluated by direct speech intelligibility or other performance tests. In particular, speech communication systems that process speech signals in various ways in order to achieve speech bandwidth compression cannot be validly evaluated by the AI procedure.

Suggestions are given in S3.5 for refinements to AI to take into account such things as the vocal effort used by the talker, interruption in the noise, face-to-face talking, and reverberation present in the listening situation. In this regard it should be noted that other procedures for making allowances for reverberation besides that used in S3.5 have been proposed: Bolt and MacDonald (77) suggest that reverberation effects could be properly accounted for by adding to the measured noise level an amount that depends upon the reverberation time; more recently Janssen (407) recommended that the measured level of the speech signal be reduced to an effective level by an amount

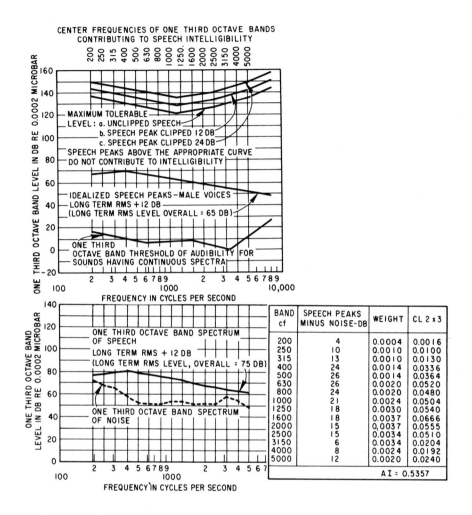

FIGURE 55. Upper Graph: Work sheet for AI, one-third octave band method. Lower graphs: Examples of the calculation of an AI by the one-third octave band method. From (31).

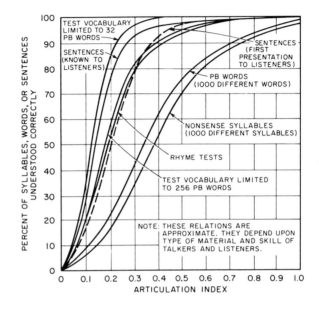

FIGURE 56.   Relation between AI and various measures of speech intelligibility. From
(31).

that depends upon the reverberation time. Also Levitt and Rabiner (499) have proposed that the effects of binaural phase relations of speech and noise upon speech intelligibility, as discussed earlier, can be predicted by AI when certain adjustments are made to measured speech-to-noise ratios present in different frequency bands.

Other AI Procedures

Procedures similar to those used for finding AI have been proposed in several countries over the past 15 to 20 years (307, 408, 673, 707). The variations are principally in terms of the width of the frequency bands in which the signal-to-noise ratios are to be determined.

Cavanaugh *et al.* (129) have suggested the use of a graphical procedure for the estimation of AI. In their method one plots the spectrum of the noise on the same graph paper as the peak instantaneous levels reached by speech signals. The area between the noise spectrum and the speech peaks, adjusted for the relative importance assigned to different speech frequencies, is proportional to the AI for that speech and noise condition (see also the section on "Speech Privacy," Chapter 9).

Validity of the AI Procedure

Data collected by French and Steinberg, Miller (542), Egan and Wiener (208), and others, with respect to masking of speech by noise of various bandwidths and spectra shapes, provide a basis for demonstrating the ability of AI to predict the relative proficiency of given communication systems or conditions. Figure 57 shows some of these findings. Other figures and tables to be given later will also show that the AI is a reasonably accurate method for predicting intelligibility of speech in the presence of noise.

It should be emphasized that the value of the scores obtained on speech intelligibility tests are influenced by the proficiency and training of the talker and listening crew involved in any given test, as well as the difficulty of the speech material being used. Therefore, one cannot expect that a given communication system will provide identical test scores when tested in different laboratories and particularly with different groups of listeners and talkers, even though the AI of the system remains constant. In fact the inherent variability in speech intelligibility testing (see House *et al.* [391]), while often not very large when similar test materials are used, is a recommendation for the use of AI whenever appropriate.

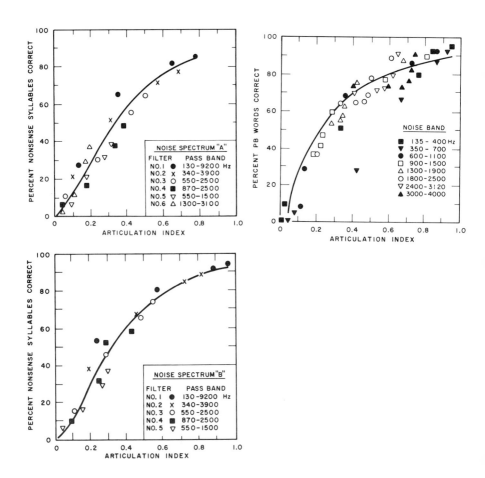

FIGURE 57.   Left graphs: Comparison of obtained and predicted test scores for speech
passed through a bandpass filter and heard in the presence of a broadband,
negatively sloped spectrum noise set at various intensity levels. From Kryter
(449) after Egan and Wiener (208). Right graph: Comparison of obtained
and predicted test scores for broadband speech in the presence of narrow
bands of noise set at various intensity levels. From Kryter (449) after Miller
(542).

Criteria of Acceptable Noise Levels for Speech Communications

Intelligibility tests and related calculation procedures are of paramount value in the evaluation, selection, and design of the components of speech communication systems, and in the control of environmental noise conditions for the operation of such systems. However, the assessment of the masking effects of noise on speech, either by intelligibility testing, or by hand or automated calculation procedures do not, of course, directly indicate how bothersome this masking will be in a given communications situation. A communications system, including the noise environment in which it is operated, that gives satisfactory performance when used with a special vocabulary and trained operators (for example, air traffic control by radio [268]) would probably be judged as completely unacceptable if used by untrained operators or in situations where flexible, nonstandardized speech communication is permitted.

Additional variables are the importance of the messages and the standards of the users. No study, to my knowledge, relating the scores of speech intelligibility tests to performance ratings of given communications systems for various classes of uses and communications requirements, has been reported, although some laboratory tests of ratings of effort required to use a telephone system in free conversation and general satisfaction ratings of telephone communications have been obtained (Richards and Swaffield [674]).

Nevertheless, some standards of expected satisfaction with speech communication systems have evolved. Beranek (54), for example, suggests that a communication system that has an AI of less than 0.3 will usually be found unsatisfactory or only marginally satisfactory, an AI of 0.3 to 0.5 will generally be acceptable, an AI of 0.5 to 0.7 will be good, and a system with an AI higher than 0.7 will usually be considered very good to excellent.

Relations Between AI and SIL

Beranek (55) proposed a simplified version of AI to be used in predicting the effectiveness of person-to-person speech communication in the presence of noise. Beranek estimated what the average speech level would be in the octave bands 600-1200, 1200-2400, and 2400-4800 Hz at various distances from a talker using various vocal efforts; next, he estimated what noise levels would be required in these same octave bands to give an AI of about 0.5 assuming the noise spectrum was a relatively continuous broadband noise. The average of the decibel levels in the three octave bands from 600-4800 Hz were tabulated for this condition, as shown in Table 5, and called SILs (Speech Interference Levels). The SILs in Table 5, presumably equivalent to AI of 0.5, should allow sentence intelligibility scores of about 95% correct and PB word scores of about 75% correct (see Fig. 56). As we shall see later, the noise present in various other combinations of octave bands than those between 600 and 4800 Hz have been proposed as a means of calculating the SIL.

TABLE 5

Speech Interference Levels That Permit Barely Reliable Conversation, or the Correct Hearing of Approximately 75% of PB Words, At Various Distances and Voice Levels. From Beranek (55)

Distance between talker and listener (ft)	Voice Level			
	Normal	Raised	Very Loud	Shouting
	Speech Interference Level (dB)			
0.5	71	77	83	89
1	65	71	77	83
2	59	65	71	77
3	55	61	67	73
4	53	59	65	71
5	51	57	63	69
6	49	55	61	67
12	43	49	55	61

It has been found that various other ways of measuring sound will provide reasonable estimates or indices to AI, provided that the energy in a masking noise is spread predominately only over the frequency region covered by the normal speech spectrum. The methods of principal interest, in addition to SIL, are: (a) overall sound level meter readings with either socalled dB(A), and dB(D) frequency weighting (see Table 2 and Fig. 8), (b) loudness level in Phons (see Chapter 7) and perceived noise level in PNdB (see Chapter 8), and (c) the noise rating contour (NR, NC, or NCA) procedure to be described in Chapter 9.

While these various methods, particularly SIL, have been found to be a reasonably accurate method for evaluating speech communication in many noises, as will be shown below, they should not be applied to noise spectra that have intense low- or high-frequency components. Other limitations of SIL and these other procedures, compared to AI, are that certain broad assumptions must be made in their use regarding the interactions between room acoustics, the noise present, the vocal effort used by the talker, and the level of the speech

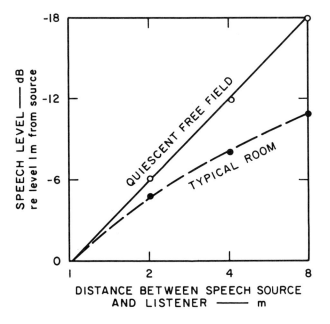

FIGURE 58. The approximate effect of acoustic environment and distance (in meters) between speech source and listener upon received speech level. After Beranek (56).

received by the listeners. In this regard, Fig. 58 is provided as a means of roughly estimating the effects of distance between the talkers (or source of the speech) and listeners and acoustic conditions upon speech level at the listeners' ears.

As a practical matter it is perhaps important to be able to measure and report a noise environment in units other than AI; not only might the instrumentation and methods be more convenient, but there is also the possibility of establishing criteria of acceptability for noise with respect to more than one effect of noise (masking of speech, annoyance, damage to hearing, etc.) on the basis of a common single measurement unit. The next two sections present experimental test data that evaluate the relative effectiveness with which other noise measurement techniques estimate the effect of noise on the intelligibility of speech.

Recent Studies of Masking of Real-Life Noises

Three rather extensive experiments have been conducted in recent years that provide a significant amount of intelligibility test scores obtained in the presence of a wide variety of noises (432, 470, 878). These test scores reveal the masking effects of the particular noises involved and also afford significant opportunities for evaluating the relative abilities of AI and other methods of noise measurement to predict the masking of speech by noise.

Klumpp and Webster (432), and Webster and Klumpp (862) report the band spectra of 16 noises they found to provide equal interference with speech (Rhyme word test scores of about 50% correct), and overall values calculated or measured from these spectra for the following frequency and/or bandwidth weighting procedures: AI, SIL (3 and 4 octave band methods covering frequencies from 175 to 4800 Hz), dB(A), dB(B), dB(C), DIN 3 (a frequency weighting available on Sound Level Meters made in Germany), NC, NCA (and two variations thereof), Phon (Stevens), and PNdB. Figure 59 illustrates their findings. In Fig. 59 we have omitted a few of the variations of a given procedure used by Klumpp and Webster when these variations provided no significant improvement in the predictions of test scores over the procedure on which they were based. Since the noise spectra reported are those present when the speech test scores were about equal (50% correct), the value obtained by a given measurement procedure should be the same for all 16 of the noises if that measure is to be considered a good index, for 50% rhyme word intelligibility, to the speech interference effects of the noises. The greater the deviations, given in terms of equivalent dB units, of the values of each noise from the mean of all 16 noises, the less well is the test score 50% predicted.

Kryter and Williams (470) determined the relation between peak levels reached between noise and average intelligibility scores and also the relation between the peak levels and the average number of words (see Table 47, Chapter

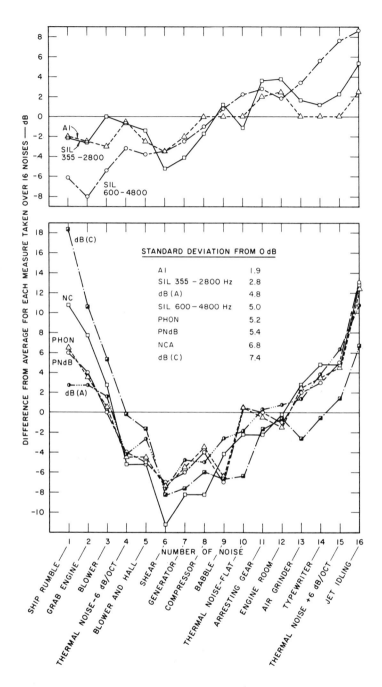

FIGURE 59. Shows difference in dB units between average for each measurement procedure taken over the 16 noises and the value for each of the noises when speech intelligibility equaled 50% words correct. After Klumpp and Webster (432) and Webster and Klumpp (862).

9) that would be masked during an aircraft flyover, given a certain peak level reached by the noise, and a given conversational level and rate of speaking for the speech. Williams *et al.* (878) later determined the understandability of speech and the level of an aircraft fly-over noise present when a test word was also present. The results of these two investigations are shown in Fig. 60.

As can be seen from a visual comparison of Figs. 59 and 60, the results of these three studies are in reasonable, but not complete, agreement with respect to the rank order of merit with which the various physical measures predict the rank order of the speech test scores. It appears that AI, as one might expect, is consistently superior to the other methods of predicting the interference effects of noise on undistorted speech.

In the Klumpp and Webster study (432) SIL 355-2800 Hz, the arithmetic average of the octave bands centered at 500 Hz, 1000 Hz, and 2000 Hz, or SIL 300-2400 Hz (octave band center frequencies of 425, 850, and 1700 Hz) gave reasonably good predictions, and SIL 600-4800 Hz (octave band center frequencies of 750, 1700, and 3400 Hz) gave a generally poorer prediction of the speech interference effects of the noises. Most of the noises were recorded aboard Naval vessels and 10 of the 16 noises contained most of their energy in the frequency region below 1000 Hz. Webster and Klumpp (862) recommend SIL 355-2800 Hz over SIL 300-2400 Hz, even though the latter performed slightly better in their study, because the octave bands involved correspond to those now specified as preferred by various standardization groups (23). On the other hand, in the Kryter and Williams (470) and Williams *et al.* (878) studies with aircraft noises, some of which had most of their energy at or above 1000 Hz, the reverse was true – SIL 600-4800 Hz appeared to be better correlated with the speech test scores than other SILs (Williams *et al.* did not, however, report valuse for SIL 300-2400 Hz). Presumably this difference is partly attributable to the greater predominance of lower frequency noise in the Klumpp and Webster study than in the other two studies.

In addition it should be noted that a possibly serious restriction to the generality of the Klumpp and Webster deductions lies in the use of 50% Rhyme Test scores as their measure of speech communication performance. As seen in Fig. 56, this represents a generally low level of communication proficiency (equivalent to an AI of about 0.2) and one which could be achieved only at noise levels where, particularly for their noises, the portions of the speech signals above 2000 Hz or so were completely obliterated by the noise. At perhaps somewhat more realistic speech-to-noise ratios, the SILs based on higher frequency bands than those recommended by Webster and Klumpp would probably be more accurate predictors of speech interference. In any event, altogether the results of speech masking experiments clearly demonstrate that in order to best predict the speech interference effects of a given noise (rather than the average of some type of noise spectrum) a unit of measurement such as AI rather than SIL or dB(A), PNdB, or Phon, is required. It should be noted that

NC (or NCA), which has been extensively used for noise evaluation, does not indicate very well the effects of noise on speech communication.

Regardless of the lack of generality of certain of the simple (dB[A], dB[D], dB[C]) or complex, from a noise measurement point of view, (SIL, PNdB, Phons, or NC) procedures for estimating the speech interference effects of noise, it is likely that some of these procedures will often be used for this purpose. Conversion of these noise measures into an estimated equivalent AI value can be accomplished by means of Table 6. Table 6 shows the levels a noise, shaped like the long-term speech spectrum (see Fig. 55), must be, if measured in terms of the various units designated, in order to obtain a given AI value with a given intensity of speech at the listener's ears.

An Evaluation of AI and Other Units with Representative Noises

When a noise to be evaluated, has a speech-like spectrum, and many so-called low frequency room and office noises do (see Chapter 9), Table 6 can be used with considerable accuracy. In an attempt to provide a perhaps more general and realistic estimate of the accuracy with which AI will be estimated by the various measures of noise, we have chosen from the research literature seven noise spectra, five representing five different but relatively common sources and two tailored from a random noise generator, as shown in Table 7. Figure 61 shows how these seven noises are related to AI. The average deviations of the levels of the seven noises, as measured by each of the units cited when the noises were set to the same AI value, are given in the lower right-hand column of numbers on Fig. 61. These columns of numbers show, for example, that the dB(A) level of the seven noises in question would, on the average, deviate 2.6 dB, dB(D), 2.2 dB, and SIL 600-4800 0.7 dB from their average level when the AI value of the seven noises were all the same.

Also indicated on Fig. 61 are qualitative statements concerning the acceptability of speech communication systems having certain AIs, and the AIs to be achieved, on the average, for these seven noises when the speech at the listener's ears is of a specified level. The horizontal lines are speech levels one meter from a talker, and can be converted to those present at other distances by means of Fig. 58. The use of these seven noise spectra, as shown in Fig. 61, must be interpreted with some caution. For one thing, any small set of noise spectra may be unrepresentative in kind or numbers of the noise present in the real world. It is also true that noise control engineering is usually applied to given specific noise sources or environments one at a time, and, unless one is dealing with a homogeneous set of noises, it is somewhat risky, from an accuracy point of view, to make noise measurements that do not permit calculation of AI.

Webster (859) has plotted some of the general relations shown in Fig. 61 using distance between talker and listener as a variable and vocal effort as the

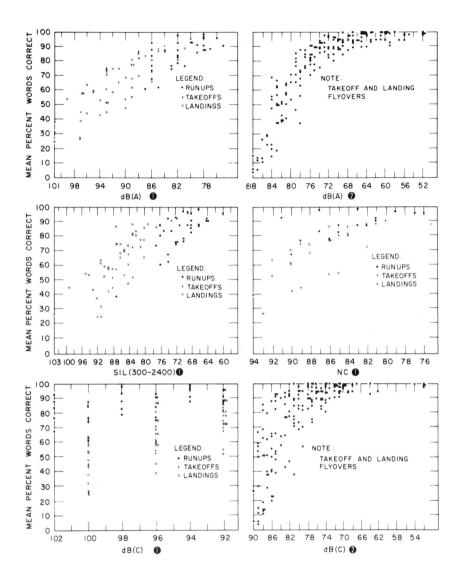

FIGURE 60.   The mean percent words correct plotted against peak levels of aircraft
flyover noises as measured by various means. (1) From Kryter and Williams
(470) and (2) from Williams *et al.* (878).

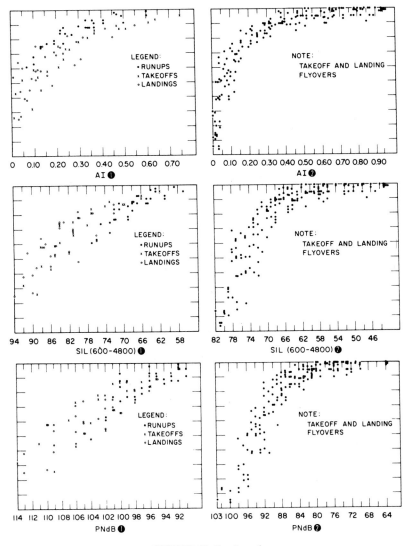

FIGURE 60. Continued.

TABLE 6

Relation of Various Speech and Noise Spectrum Levels to AI and Speech Test Scores

Noise having speech-shaped spectrum measured by units given on the right side of table will give the AI and word intelligibility scores given on top left side of table when the speech level is that at lower left side of table. For example, a noise with a dB(A) of 64 and speech level of 75 dB will provide an AI of 0.66, PB words of 88%, and MRT words of 88%, and sentences of 98% correct.

Long-time RMS level of speech (presented via loudspeakers, earphones, or as specified, direct person-to-person) required at listener's ears to obtain:				Noise measured at listener's ears. Spectrum of noise shaped like long-time spectrum of speech							
				SIL							
AI	0.0	0.33	0.66	1.00				Phon or PNdB	dB(A)	NC	dB(C)	dB(D)
PB words	0 %	50 %	88 %	99 %	350-2800 Hz	300-2400 Hz	600-4800 Hz					
MRT words & sentences	0 %	80 %	98 %	100 %								

Long-time RMS level of speech (presented via loudspeakers, earphones, direct person-to-person, 1 meter apart) at listener's ears

80 dB	90 dB	100 dB	110 dB	350-2800 Hz	300-2400 Hz	600-4800 Hz	Phon or PNdB	dB(A)	NC	dB(C)	dB(D)
80 dB	90 dB	100 dB	110 dB	83 dB	84 dB	78 dB	98 dB	89 dB	87 dB	92 dB	91 dB
75	85	95	105	78	79	73	93	84	82	87	86
65	75	85	95	68	69	63	83	74	72	77	76
55	65	75	85	58	59	53	73	64	62	67	66
45	55	65	75	48	49	43	63	54	52	57	56
35	45	55	65	38	39	33	53	44	42	47	46
25	35	45	55	28	29	23	43	34	32	37	36

Note 1: Person-to-person speech levels limited to those between 55 dB (conversation level in quiet) and 80 dB (shouting level).

Note 2: Percent of word or sentence intelligibility scores for a given noise measurement (SIL, Phons, PNdB, etc.) and speech level approximately correct only when: (a) noise has spectrum shaped like long-time RMS speech spectrum; (b) speech is relatively undistorted.

TABLE 7

Representative Octave Band Spectra to Be Used for the Evaluation of Various
Procedures for Estimating Some Effects of Noise on Man

Noise #1 - Thermal Noise (-6 dB/Oct. above 106 Hz)

Oct. Band cf-Hz	53	106	212	425	850	1700	3400	6800	OA
Flat		89	93	87	81	75	69	63	57	95

dB(C)	dB(A)	dB(D_1)	dB(D_2)	dB(D_3)
95	82	89	87	80

Noise #2 - Thermal Noise ("Flat") Klumpp & Webster[432]

Oct. Band cf-Hz	63	125	250	500	1000	2000	4000	8000	OA
Flat		60	64	68	69	71	78	75	72	82

dB(C)	dB(A)	dB(D_1)	dB(D_2)	dB(D_3)
81	82	90	90	85

Noise #3 - Motor Generator Klumpp & Webster[432]

Oct. Band cf-Hz	63	125	250	500	1000	2000	4000	8000	OA
Flat		71	70	71	72	65	71	68	60	79

dB(C)	dB(A)	dB(D_1)	dB(D_2)	dB(D_3)
79	76	83	83	78

Noise #4 - Commercial Jet A/C Landing, 610' Alt. Kryter & Williams[471]

Oct. Band cf-Hz	53	106	212	425	850	1700	3400	6800	OA
Flat		81	88	89	91	94	95	92	93	101

dB(C)	dB(A)	dB(D_1)	dB(D_2)	dB(D_3)
100	100	107	107	103

Noise #5 - Planer Karplus & Bonvallet[429]

Oct. Band cf-Hz	53	106	212	425	850	1700	3400	6800	OA
Flat		82	84	85	87	88	88	87	85	95

dB(C)	dB(A)	dB(D_1)	dB(D_2)	dB(D_3)
95	94	101	101	96

Noise #6 - Trolley Buses Bonvallet[78]

Oct. Band cf-Hz	53	106	212	425	850	1700	3400	6800	OA
Flat		68	72	74	73	69	64	58	52	79

dB(C)	dB(A)	dB(D_1)	dB(D_2)	dB(D_3)
79	73	78	77	72

Noise #7 - Automobiles Bonvallet[78]

Oct. Band cf-Hz	53	106	212	425	850	1700	3400	6800	OA
Flat		70	73	72	67	62	58	54	50	77

dB(C)	dB(A)	dB(D_1)	dB(D_2)	dB(D_3)
77	68	74	73	67

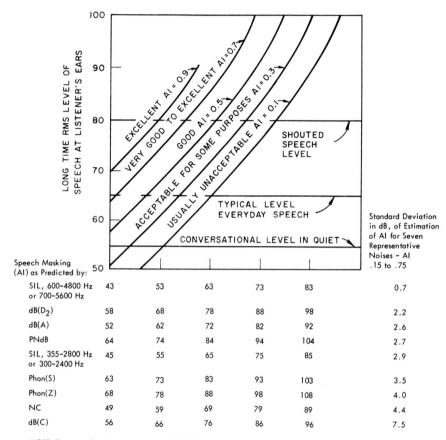

Speech Masking (AI) as Predicted by:						Standard Deviation in dB, of Estimation of AI for Seven Representative Noises – AI .15 to .75
SIL, 600–4800 Hz or 700–5600 Hz	43	53	63	73	83	0.7
dB(D₂)	58	68	78	88	98	2.2
dB(A)	52	62	72	82	92	2.6
PNdB	64	74	84	94	104	2.7
SIL, 355–2800 Hz or 300–2400 Hz	45	55	65	75	85	2.9
Phon(S)	63	73	83	93	103	3.5
Phon(Z)	68	78	88	98	108	4.0
NC	49	59	69	79	89	4.4
dB(C)	56	66	76	86	96	7.5

NOTE: Percent of word or sentence intelligibility scores for a given noise measurement
(SIL, Phons, PNdB, etc.) and speech level approximately correct only when:
(a) Noise has spectrum shaped like the average of the seven spectra given in Table 7;
(b) Speech is relatively undistorted

FIGURE 61.   Average speech interference of the seven noises of Table 7 as measured by
various units as predictors of AI, the parameter, as a function of the
intensity of speech reaching the listeners' ears. The right-hand column
indicates the standard deviation with which the various units estimate the AI
of seven representative noises. The overall SPL of each noise was adjusted
until the AI for each noise had the same value. The average value was then
found for each of the other measures of speech masking (SIL, PNdB, etc.);
these averages and the standard deviations of the seven noises from the
averages are tabulated on the abscissa. The deviations are approximately
correct for AI values from about 0.15 to 0.75. (See also Table 76.)

parameter, as shown in Fig. 62. Webster also shows some relations between AI and other physical units of noise measurement listed in descending order of statistical accuracy for evaluating the masking of speech. Again, the statistics shown in Fig. 62 are based on speech at a very low level of intelligibility in the presence of 16 shipboard noises, and the relative proficiency, except for AI, of the various units of noise measurement are not necessarily representations of their proficiency with other noises and at other levels of noise masking.

Devices for Evaluating Speech Systems

The AI procedures were proposed as a means of evaluating the performance of a communication system without requiring the administration of time-consuming and costly speech intelligibility or articulation test procedures. But it is often as impractical to make the physical measurement necessary to find the noise and speech spectrum at the listener's ears for the calculation of AI as it is to apply speech intelligibility test procedures.

Tkachenko (802) proposed a very simple, yet excellent, method for obtaining the AI of a speech communication system. He developed an artificial speech signal that consisted of 20 pure tones spaced at the center frequency of the 20 bands found by French and Steinberg to be equally important to the understanding of speech. Each tone is audited separately by the listener who adjusts an attenuator controlling the level of the tone until it is just audible. The level of the tone is read from the attenuator which is calibrated in terms of equivalent, normal speech level, and 12 dB is added to take into account the speech peak factor. This process is repeated for each of the 20 pure tones and the average of the levels required for the tones to be just audible is proportional to AI.

Although this method of Tkachenko's is not as automatic as some other methods, which will be described below, it has much to recommend it. For one thing it uses the human listener to determine the exact masking effect of the system noise, something which, short of conducting speech intelligibility tests, is only estimated in AI and related calculation procedures.

Automatic Devices

In 1959 Licklider et al. (507) developed an electronic device which, when applied to a speech communication system, will automatically provide a number that is, usually, proportional to AI. This machine is actually based on a somewhat different concept than that of AI, although it uses the frequency-weighting importance function and the signal-to-noise ratio weighting function developed for AI.

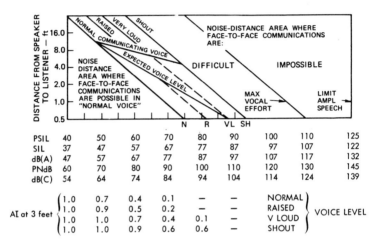

FIGURE 62.   Voice level and distance between talker and listener for satisfactory
face-to-face speech communication as limited by ambient noise level. Along
the abscissa are rank-ordered, from best to worst, various objective measures
of noise level; the top one, and the one that varies least for predicting speech
interference of Navy shipboard noises at PSIL of greater than 70, is the
average octave band level in the octaves centered at 500, 1000, and 2000 Hz;
next is SIL averaged over the three octaves from 600-4800 Hz; next the
A-weighted sound level meter reading; next the perceived noise level (PNL)
in PNdB; and finally, the C-weighted sound level meter reading. Below all
these noise measures is the articulation index (AI) calculation for various
assumed voice levels at 3 ft. After Webster (859).

Licklider *et al.* device repeatedly plays, over the communication system to be evaluated, a recorded brief (several seconds) sample of speech. At any one moment in time, the output of the speech system being evaluated is simultaneously compared, in a narrow frequency band, with the recorded speech input. This process is repeated over and over, each time the locus of the frequency band in which the input and output speech is to be compared is changed. Suitable integrating circuits average the measured correlations between the input and output signal.

As well as revealing the interference or disruptive effects of any noise present in a system, the average correlation calculated for a given system appropriately reflects the presence of distortions, such as frequency shifts and rapid fluctuations in noise and speech levels. These later distortions are usually ignored in the calculation of AI. On the other hand, this system sees only actual noise spectra present in a communication system and does not make allowances for upward and remote masking. Robertson and Stuckey (679) found this device predicted speech intelligibility test scores quite well for most, but not all, noise interference conditions they tested.

Goldberg (305) described a machine called a Voice Interference Analysis Set that attempts to measure or compute the AI for a communication system. This machine applied an amplitude-modulated tone (1000 Hz) to the system under test. The level of the received signal is compared to the noise level found at the receiver of the system under test in ten bands that are proportional in width to the 20 AI bands. Certain corrections are electronically determined when the noise that is present would cause a significant upward spread-of-masking. Performance data on this device have not, to our knowledge, been as yet published.

Kryter *et al.* (472) designed a Speech Communication Index Meter (SCIM) that is similar to the Goldberg machine in that it attempts to more or less directly calculate AI from measured signal-to-noise ratios. SCIM transmits a broadband signal that simulates normal speech with respect to long-term spectrum shape and, to some extent, amplitude variations. The simulated signal-to-actual noise ratios are found in 9 frequency bands and appropriately averaged to arrive at an approximate AI value. This device takes into account the effects on the speech intelligibility of direct masking (including the upward spread), frequency shifting, frequency distortion, and amplitude limiting of the speech signal. Preliminary comparison of SCIM-measured-AI with speech intelligibility scores indicates reasonable accuracy in the prediction of the test scores. SCIM has recently been modified to reflect the effectiveness of speech communication systems operating with a varying, in time, signal or noise level (Hecker, *et al.* [362]).

Simplified Measuring Set

It appears that the monitoring of the background noise levels in commercial telephone circuits can be adequately accomplished with a device that, like a sound level meter, measures the noise energy over all frequencies. The measuring instrument, called Bell 3A, weights the frequencies in accordance with judgments made by listeners of the speech interference, over a telephone, of 14 different pure tones. It also has an integration time of 0.2 sec in order to simulate the growth of loudness as judged by listeners. It should be noted that this instrument is based on judgments, not speech intelligibility tests, of how tones would interfere with speech on a standard telephone circuit.

Aikens and Lewinski (8) found the interesting fact that telephone users will accept a 3 to 4 dB lower signal-to-noise ratio (as measured on the Bell 3A message circuit noise set) when the speech level is at a modest level than when the speech is at a level only 10 dB higher. It is possible that the 3A device evaluates more the annoyance or loudness of circuit noise than its masking effect of speech, since a 3-4 dB increase in signal-to-noise ratio at the important speech frequencies should cause a noticeable reduction in speech intelligibility.

Chapter 3

The Aural Reflex

Introduction

Two small muscles in the middle ear, the tympani and stapedious, which are attached to the small ossicular bones that connect the eardrum with the cochlea, mediate the so-called aural reflex and thereby play a significant role in audition, particularly when noise is present. The aural reflex can influence the effects of noise with regard to masking, loudness, and auditory fatigue. The tympanic and stapedial muscles contract when the ear is exposed to a sound that is about 80 dB above threshold level. In man, the reflex action is inferred from (*a*) various perceptual auditory tests, (*b*) physical measurements of changes in the volume of the external ear canal, and (*c*) changes in the acoustic impedance of the eardrum.

A sound, if sufficiently intense, in one ear will activate the reflex in both ears, although, as Møller (562) has found, the ipsilateral contraction is somewhat stronger than the contralateral. The more intense the sound, the greater, up to a point, is the degree of contraction. It appears that the reflex is more responsive to random noise than to pure tones and to higher frequencies than to lower. These differences are due to the relative loudnesses of these stimuli — stimuli of equal loudness are apparently equally effective in eliciting the reflex.

It has been found (Ward [826]) that the reflex appears to adapt or relax in the presence of continued stimulation after about 15 minutes of exposure to an intense steady-state noise. That the muscles are not fatigued can be shown by the fact that the reflex can be reactivated by changing the acoustic stimulus (Wersall [870]). It is conceivable that the reflex gradually relaxes during continued stimulation in order to compensate for, or because of, a gradual decrease in the loudness of sounds with long duration stimulation, so-called perstimulatory fatigue. Perstimulatory fatigue will be discussed later in Chapter 7.

The aural reflex seems to be most readily activated and maintained by intermittent, intense impulses of noise. Latency of the reflex is about 35 to 150 msec, depending on the intensity of the stimulus, and relaxation time following an impulse of noise is reported to be as long as 2-3 sec for complete relaxation with most of the recovery probably occurring within about 0.5 sec. The effect of this reflex upon auditory fatigue from gunfire will be illustrated later.

The reflex is involuntary and, except to the specially-trained subject, its occurrence is not detected by the average person. It can apparently be conditioned to light and other stimuli and some people can cause it to contract voluntarily. Whether people can by volition cause the reflex to relax when active is another question, although it has been a suggestion to explain some phenomena related to threshold measurements. What mechanical changes take place in the ossicular chain of bones to which the muscles are attached as the result of their contraction is a matter of some conjecture. For one thing, the acoustic impedance of the eardrum is changed but the relative position of the eardrum is probably unaltered with total reflex activity because the two muscles appear to be antagonistic to each other.

It would seem reasonable to think that the aural reflex merely served to stiffen the eardrum and the bones of the middle ear so that they would not transmit sound as effectively as normal. Something like this must occur but it affects the transmission only of sounds below 2000 Hz or so and apparently does not attenuate the transmission of higher frequency sounds. Since the ear, as will be discussed later, is more susceptible to auditory fatigue to these higher than to the lower frequencies, the exact protection function appears somewhat mysterious.

An argument can be made that the aural reflex is designed to protect the ear from drastic changes in the velocity of movement of the eardrum-chochlear fluid system. It is conceivable that, when the velocity of movement of the cochlear fluid exceeds a certain critical value, the disturbance differs from the usual hydromechanical turbulence in the cochlea, and spreads both downwards and upwards along the cochlear partition. It would appear that the best way to protect the ear against this type of trauma is to reduce the transmission of frequencies below 2000 Hz or so because (a) the middle ear is most compliant in the frequency region from 300 to 1500 Hz; and (b) acoustic stimuli that occur at levels in excess of about 150 dB are invariably impulsive and have their major energy in the frequency region below 2000 Hz. It is beyond the scope of this document to discuss in more detail the physiology and theory of the aural reflex. Recent reviews are those of Fletcher (244), Jepsen (410), Møller (562), and Wersäll (870).

Masking, Loudness, and Auditory Fatigue

Studies concerned primarily with the perceptual-psychological effects of the aural reflex, as distinct from their physiological correlates, fall into three general categories (a) masking, (b) loudness, and (c) auditory fatigue.

Masking

The effects of the aural reflex on masking are investigated in the following way: A sound at an intensity sufficient to elicit the aural reflex is presented to one ear and at the same time the listener is tracking his threshold for pure tones in the opposite ear. When activated, the reflex, being bilateral, will cause a rise in the threshold at some frequencies in the "quiet" ear. It is, of course, necessary that the frequency content of the sound used to elicit the reflex be sufficiently different in frequency from the tones being auditorially tracked so as not to have present direct masking due to transcranial conduction.

This change in the contralateral threshold has been often, unfortunately, called Contralateral Remote Masking or CRM (826); the action, to be sure, is contralateral, but it is not necessarily "remote" (the contralateral threshold shift can be caused by low frequency sounds as well as high) and it is not masking, in the sense of direct or even remote masking, as those terms were originally used. In this text we will call the change in the threshold of hearing in the ear which is contralateral to the ear used for eliciting the reflex, contralateral threshold shift (CTS), and not CRM.

Figures 63 to 65 from Ward (826), Fig. 66 from Loeb and Riopelle (524), Fig. 67 from Reger (664), and Fig. 68 from Fletcher and Loeb (250) reveal many of the salient facts about CTS. First, it is to be noted in Fig. 63 that the greatest shift seems to occur for a 500 Hz tone regardless of the frequency content of the stimulus used to arouse the reflex; second, broadband noise and clicks cause a stronger reflex response than tones or narrow bands of noise; third, Fig. 64 shows that, up to a point, the CTS in dB increases linearly with stimulus intensity; and fourth, Fig. 65 shows that with continued exposure the reflex relaxes or adapts.

Loudness

Finding the threshold of hearing for a tone is, in a sense, finding its minimum loudness; Loeb and Riopelle (524) had subjects find not only the CTS but also the loudness (relative to prereflex loudness) of the 500 Hz test tone presented at suprathreshold levels (see Fig. 66). They found that when the contralateral reflex activating signal (a tone of 2200 Hz) was present, the loudness of a 500 Hz test tone at a sensation level of 105 dB decreased relative to its loudness as remembered from a preceding moment when the reflex activating signal was not present. The decrease in loudness for the test tone at 105 dB was equivalent to a decrease in 10-15 dB in the intensity of the test tone; at a sensation level of but 20 dB, the decrease in loudness of the 500 Hz test tone with the reflex present was only 3-5 dB.

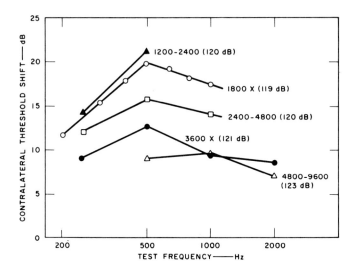

FIGURE 63.   Contralateral threshold shift 30 sec after noise onset as a function of
frequency. Adapting noise spectrum is the parameter; noises designated with
"X" have very narrow bandwidths (less than 1/3 octave) and center
frequency designated. From Ward (826).

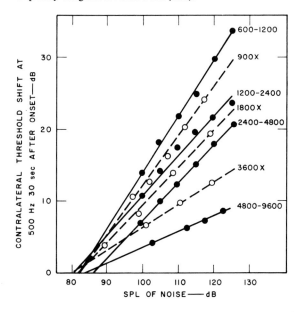

FIGURE 64.   Growth of contralateral threshold shift at 500 Hz with level of arousal noise.
Noises marked "X" have very narrow bandwidths (less than 1/3 octave) and
center frequencies designated. Filter setting is the parameter. From Ward
(826).

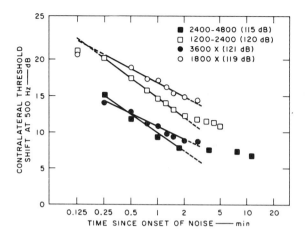

FIGURE 65.   Adaptation of contralateral threshold shift. From Ward (826).

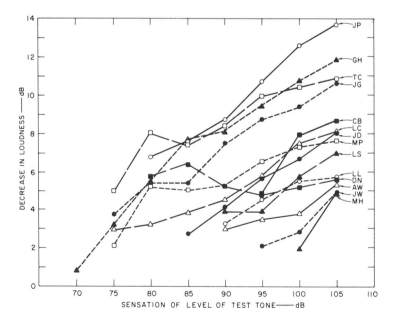

FIGURE 66.   Decrease in perceived loudness as a function of the SL of test tone. From
              Loeb and Riopelle (524).

This result led to the conjecture that the reflex possibly had a nonlinear "snubber" action for intense signals but would not attenuate weak signals. This conclusion must remain in doubt in view of the fact that CTSs of 10-15 dB, rather than only 3-5 dB, have been obtained by others (Ward [830]) for roughly comparable stimulus conditions.

Egan (202) found that putting the noise into one ear and the signal into the opposite ear increased the loudness of the speech by as much as 5 dB. On the other hand, Shapley (733) found noise in the contralateral ear reduced the loudness of a low-frequency tone in the opposite ear; and Prather (653) found some loudness decrement and some increment depending on the frequency of a tonal signal.

Voluntary Control of Reflex

It appears that some people are able to voluntarily activate their aural reflex (Reger et al. [665]). These people are apparently aware not only of a reduction in the loudness level of some sounds, but they also hear the sound made by the contraction and relaxation of the intraaural muscles. Reger et al. estimate that 1% to 2% of people have this ability and that others can be trained. Figure 67 shows the threshold shift observed at various frequencies during maximum voluntary contraction of the aural reflex.

Auditory Fatigue

It is seen in Fig. 68 that within a few seconds after the stimulus eliciting the aural reflex is removed, the threshold of audibility returns to normal in the contralateral ear—in short, as one would expect, the contralateral threshold shift does not signify the presence of any auditory fatigue. Actually of course, the activated reflex should serve to prevent the onset of auditory fatigue.

Simmons (742) measured, by means of electrodes placed on the cochlea, the hearing acuity in cats, some of whom had had the tendons from intraaural muscles, or the muscles themselves, severed. The animals were then exposed to intense noise for a period of time. It was found that the animals who had normal intraaural muscles and were not anesthetized when in the noise (the reflex is more or less absent in the anesthetized animal) suffered the least amount of hearing loss.

Fletcher and Riopelle (255) and Fletcher (241) showed that the aural reflex also protects man's ear from auditory fatigue as the result of exposure to gunfire. Fletcher and Riopelle elicited the reflex with a brief 1000 Hz tone at 98 dB in one experiment, and, in later experiments with a click and band of noise, 200

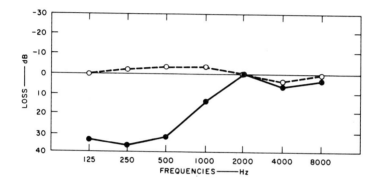

FIGURE 67. The dotted line shows the average hearing level of eight selected ears. The solid line shows the average hearing level of the same ears during maximum voluntary contraction of middle ear muscles. From Reger *et al.* (665).

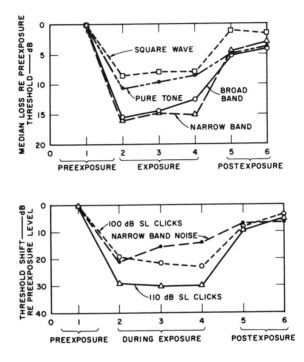

FIGURE 68.   Amount of contralateral threshold shift produced at 500 Hz by several acoustical stimuli. From Fletcher and Loeb (250).

msec before exposing the ear to the impulse from the firing of a gun. They measured the shifts in threshold, which were temporary, after 200 rounds of firing. A second experimental condition was the same as that just described except the tone, click, or band of noise was withheld prior to each round of firing. Presumably in the latter case the aural reflex was not active when the gun noise reached the ear. Figure 69 shows that the aural reflex afforded as much as 15 dB protection from temporary threshold shift (TTS).

Fleer (231) found that subjects who could voluntarily control their aural reflex, and who did so when being exposed to impulse noise, suffered 0 to 20 dB less TTS than subjects who could not voluntarily elicit their aural reflex. This is also shown in Fig. 69.

It has been known from physiological studies that the reflex relaxes following cessation of a stimulus that causes full contraction. Thus, whenever impulsive sounds are separated by more than a certain period, the reflex action presumably present as the result of the presence of each impulse provides no attenuation to the succeeding impulse.

Germane to this is an experiment of Ward's (827) in which he exposed subjects to acoustic impulses separated by 1-, 3-, 9-, and 30-sec intervals. As seen in Fig. 70, the threshold shifts are roughly the same for intervals up to 9 seconds. Ward concludes that the reflex was therefore inactive within 1 second after the impulse, else the shift with the 1-sec intervals should have been less than with 3- or 9-sec intervals. While this may be true, it does not necessarily follow from the data; it is possible that the shift immediately after each of the 1-sec interval pulses was less than after the 3- or 9-sec intervals, due to some residual action of the aural reflex. However, the greater recovery possible with the 3- and 9-sec pauses allowed the threshold shifts to subside in those cases to levels equal to the 1-sec interval condition before the insult from a succeeding impulse occurred. Other experimental results from exposure to impulse noise to be presented later suggest that there might be some residual reflex contraction up to 3 seconds in some people.

Relation between CTS and Reduced TTS

Loeb and Fletcher (518) conducted studies to determine whether there was a correlation in individuals, as one might expect, between CTS and reduced TTS from broadband noise as indicators of aural reflex action to clicks presented in the contralateral ear. Both CTS and TTS has test retest reliability coefficients in the region of 0.5, which would be considered as rather low reliability. Even so, the repeatability of the group averages for CTS and TTS is rather impressive. Notice in Fig. 71 the small range between the results obtained on the three separate test days.

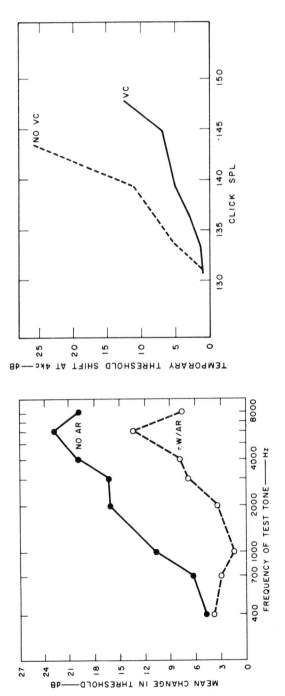

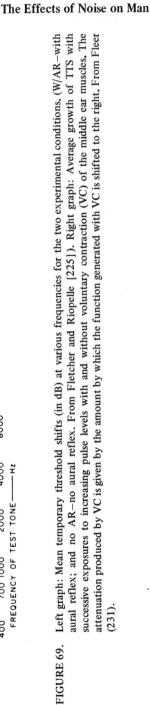

FIGURE 69. Left graph: Mean temporary threshold shifts (in dB) at various frequencies for the two experimental conditions. (W/AR—with aural reflex; and no AR—no aural reflex. From Fletcher and Riopelle [225]). Right graph: Average growth of TTS with successive exposures to increasing pulse levels with and without voluntary contraction (VC) of the middle ear muscles. The attenuation produced by VC is given by the amount by which the function generated with VC is shifted to the right. From Fleer (231).

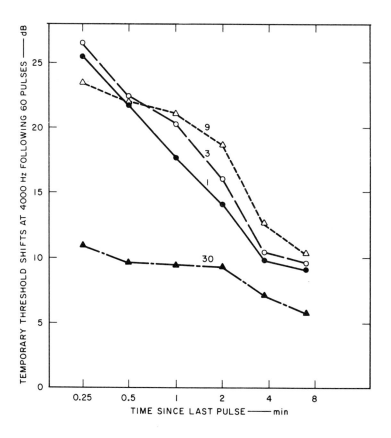

FIGURE 70. Recovery in time of the TTS at 4000 Hz produced by 60 high-intensity pulses. Interpulse interval, in seconds, is the parameter. From Ward (827).

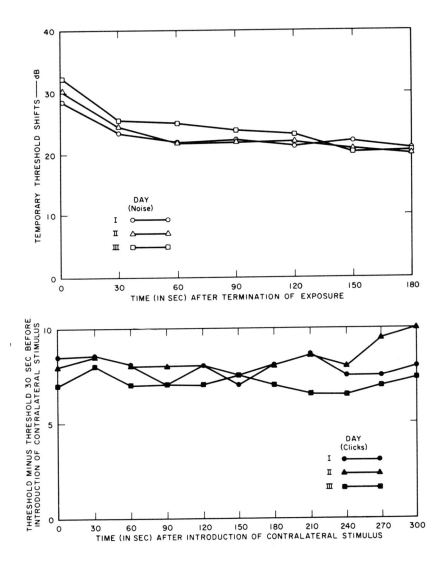

FIGURE 71.   Upper graph: TTS produced by the noise stimulus. Lower graph: CTS, vertical ordinate, produced by contralateral stimulus (clicks). Clicks continued for period of 300 seconds. From Loeb and Fletcher (518).

Aural Reflex in Persons with Hearing Loss

The aural reflex is used as a means of diagnosing certain types of hearing disorders (see Jepsen [410] and Zwislocki [913]). In particular, its absence is taken to indicate conductive difficulties in the middle ear.

Terkildsen (796) found that the persons exposed to traumatic noise in industry and having significant hearing losses had somewhat weaker aural reflexes than persons with normal hearing. Hecker and Kryter (359), on the other hand, found that soldiers with large permanent hearing losses (presumably due to exposure to gunfire) showed greater aural reflexes than soldiers with normal hearing. These investigators also found that the men with greater reflex activity showed less TTS when exposed to gunfire than was exhibited by men with lesser aural reflexes; however, this difference may have been due to the decreased sensitivity (larger permanent hearing losses) of the former group relative to the latter and not because of some increased protection afforded by the aural reflex, although this latter is a possibility. It might also be noted in this regard that an elevation in a person's threshold of hearing due to noise-induced sensori-neural deafness would not necessarily change the normal loudness of the sound of the gunfire (presumably the factor that would elicit the reflex) because of loudness recruitment found in this type of hearing disorder (see Fig. 24).

Chapter 4

Audiometry

Introduction

It has been known for many years that the ear is susceptible to a temporary and eventually a permanent loss in hearing acuity from exposure to intense sound. Important aspects of the problem have to do with (*a*) the measurement by quantitative and valid means of normal and impaired hearing, and (*b*) the relation between hearing impairment and the noise exposure conditions — spectrum level, duration, and years of exposure. An ancillary question is that of estimating what the relation is between hearing impairment and the handicap to everyday living caused by this impairment.

Although this chapter is centered on the effects of noise on hearing, these effects are understandable only in relation to normal hearing. Normal hearing is defined as the average auditory ability of persons who do not have pathological ears due to disease, injury from a blow, or exposure to intense sounds.

Since hearing is mainly used for the reception of speech, the major purpose of most hearing tests is to directly or indirectly evaluate that ability, although hearing is obviously much used for the reception of nonspeech signals, such as music, and for sensing the spatial locus of sound sources. However, the speech signal, in addition to its prevalence, is probably the most complex acoustical signal with respect to both the spectral and intensity factors with which man must cope. Testing hearing acuity directly with speech signals involves problems that have not as yet been completely solved or identified. Hearing ability is most generally measured by finding, in the quiet, the threshold of audibility for pure tones and then estimating whether or not a person is able to hear speech adequately.

Audiometric Zero

Figure 72 shows average thresholds of audibility of pure tones for normal young adults as measured under laboratory conditions. Also shown in Fig. 72 are upper limits of hearing as measured by sensations of tickle and pain elicited in the ear by intense pure tones. Figure 72 indicates that the pressure required for threshold is about 6 to 10 dB greater when the sound is presented by an earphone (MAP) than when the sound is presented in the field by a loudspeaker or other source away from the head (MAF). The reason for this difference is not completely understood but is probably primarily due to the impedance of the ear as a function of the small cavity under the earphone compared to the open field or open ear listening. This difference in sensitivity is not confined to hearing at threshold but is found at all intensities, and appropriate adjustments must be made when comparisons are attempted between open ear and earphone listening tests. For example, it appears that the ear is less susceptible to temporary threshold shift from a sound at a given sound pressure presented via an earphone than via a loudspeaker (Kryter *et al.* [476], see also the following discussion on impairment for the reception of speech).

The average and median thresholds of audibility for large groups of people with undamaged auditory systems have been determined in a number of studies as shown in Fig. 73. Indicated in Fig. 73 are the sound pressure levels, as measured on a coupler or artificial ear, for different frequencies that have been accepted by the International Organization for Standardization (ISO) as representing the median threshold for young, normal adults. It is seen that the ISO recommended (28) standard pressures are similar to those found by Sivian and White (745), as shown in Fig. 72. The range of the thresholds for the normal young adult ear (see Table 8 and Fig. 74) appears to be about 30 dB. Saunders (723) suggests that this variability be attributed to variations in the areal ratio (the ratio of the area of the eardrum to that of the oval window of the cochlea). He found in human cadavers the areal ratio differed over a range equivalent to a range of pressure gains in the middle ear of about 30 dB.

The results of several of the survey studies given on Fig. 73 are probably not representative of the average or median threshold of audibility of the normal, nonnoise, or diseased damaged young adult ear and deserve some discussion. One survey of hearing, conducted by the U.S. Public Health Service (USPHS) in 1935-36 (808) (Report No. 4, 1938) found average thresholds for young, normal adults that were about 10 dB higher than previous and subsequent medians, and averages found both in the laboratory and in other field type surveys. Why this difference was found has never been fully understood, but is usually ascribed to lack of precise physical and procedural control (see Davis and Usher [179]), and to the possibility that the listeners were given minimum instruction when the audiograms were taken. Further credence is given to this latter hypothesis by a

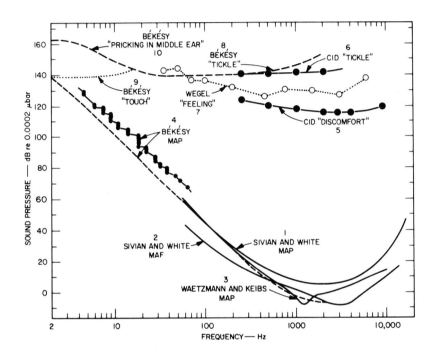

FIGURE 72. Determinations of the threshold of audibility and the threshold of feeling. Curves 1-4 represent attempts to determine the absolute threshold of hearing at various frequencies. MAP = minimum audible pressure at the eardrum; MAF = minimum audible pressure in a free sound field, measured at the place where the listener's head had been. Curves 5-10 represent attempts to determine the upper boundary of the auditory realm, beyond which sounds are too intense for comfort and give rise to nonauditory sensations of tickle and pain, etc. From Licklider (502).

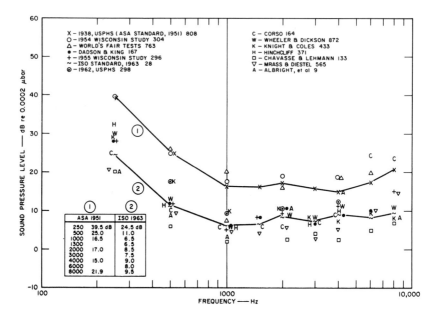

FIGURE 73.   Sound pressure levels for median threshold of audibility of pure tones found
for young adults free of otological disease or damage.

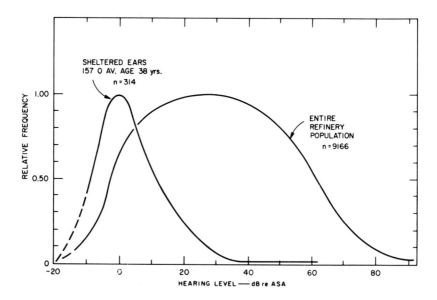

FIGURE 74.   Observed distribution of hearing levels at 4000 Hz for a reference group and
an entire refinery population. From Hermann (368).

TABLE 8

Variability of Audiometric Threshold Measurements

Standard deviations (estimated from Q_1 and Q_3) of thresholds in dB. Average of both ears for male and female in the total sample tested at Wisconsin State Fair under "clinical" conditions. From Glorig et al. (304).

Test Frequency--Hz

Age	500	1000	1500	2000	3000	4000	6000	No. of Persons
10–19	6.4	6.0	6.4	7.1	7.5	7.5	9.4	391
20–29	7.1	7.0	7.5	8.8	11.4	14.1	13.3	629
30–39	8.7	8.5	9.9	11.6	17.7	19.8	20.5	660
40–49	11.5	13.2	15.3	16.9	21.5	20.5	21.8	663
50–59	14.4	15.6	18.2	20.4	22.6	21.1	23.5	633
60–69	18.9	21.8	25.0	24.7	20.7	21.8	24.3	350

Test Frequency--Hz

Age	250	500	1000	1500	2000	3000	4000	6000	8000	No. of Persons
18–24	4.5	5.4	4.4	5.0	5.4	6.5	7.0	9.7	11.2	39
26–32	6.6	8.4	6.5	8.0	8.7	11.3	12.9	14.7	14.5	104
34–40	6.5	6.4	4.8	7.5	8.7	10.9	13.2	14.9	14.9	106
43–49	15.1	14.7	13.8	12.1	12.9	15.2	13.4	16.0	17.8	66
51–57	9.2	9.1	9.9	11.1	11.5	13.9	16.7	17.3	18.6	152
59–65	10.1	10.6	11.4	12.9	15.0	17.3	17.8	15.2	21.6	151

study conducted by Glorig (296) and Glorig et al. (304) at the Wisconsin State Fair. These investigators found that when the listeners were given the instruction and attention usually used at otological clinics found in industries or elsewhere (1955 Wisconsin State Fair Study), the median thresholds were similar to those specified in the ISO Standard, but when the audiograms were given hastily and with minimum instruction (1954 Wisconsin State Fair Study), the average thresholds were about 10 dB higher (see Fig. 73).

Steinberg et al. (763) obtained results from hearing tests conducted at the World's Fair in New York City that are in close agreement with the U.S. Public Health Service data of 1938 and the 1954 Wisconsin State Fair Study, as shown in Fig. 73. However, in the Steinberg et al. study, the people taking the test were unselected (volunteers walking past the test booth) and the subjects held the earphone to their ear by hand rather than by means of a headband as in normal practice of audiometry. These two procedures of holding the earphone to the ear give significantly different results as shown in Fig. 75 from Wheeler. Also, it is conceivable that some of people at the fairs had been exposed to crowd and other noises for several hours and were suffering some small amount of TTS when they took the threshold tests. In fact, the rather close agreement between the surveys conducted by the USPHS, 1938, at the Wisconsin State Fair in 1954, and at the World's Fair is evidence that the USPHS data of 1938 do not reflect average normal hearing.

The importance of not including persons exposed to rather intense noise in a sample of people presumed to have normal (nondiseased or damaged) ears is illustrated in Fig. 76 from Glorig et al. (304). The probable cause of the differences in thresholds of hearing of three groups of men shown in Fig. 76 is ascribed to difference in the noise environment in which they worked.

ASA 1951 Standard

The American Standards Association (ASA) (19) specified sound pressure levels for audiometers appropriate for normal, average hearing thresholds that are about those reported in one U.S. Public Health Survey (Report No. 4) (808). (The American Standards Association has undergone two changes of title since publication of its standard on Audiometers in 1951. First, to the United States of America Standards Institute, Inc., and then to its present title, American National Standards Institute, Inc.) The specifications for the maximum noise levels allowable for audiometric test booths were also set to permit the measurement of thresholds no more than 10 dB below this socalled audiometric zero dB hearing level (HL).

These actions were probably unfortunate for several reasons:

1. Surveys of the thresholds of hearing of persons exposed to noise would underestimate by about 10 dB the impairment to hearing due to noise or other factors.

2. The error is somewhat self-prepetuating for the following reasons.

(a) Some commercial fixed-frequency audiometers are designed so that the minimum level they will generate is no more than 10 dB less than the ASA Standard (about the average level of the ISO Standard). A person's threshold is usually, but not always, taken as the lowest sound pressure level below which he failed to consistently hear the tone. Since it is not possible to explore a person's hearing at levels below the −10 dB on audiometers meeting ASA specifications, a bias is introduced into the data that would tend to force the average, and to some extent, the medians, of large groups of subjects towards ASA normal.

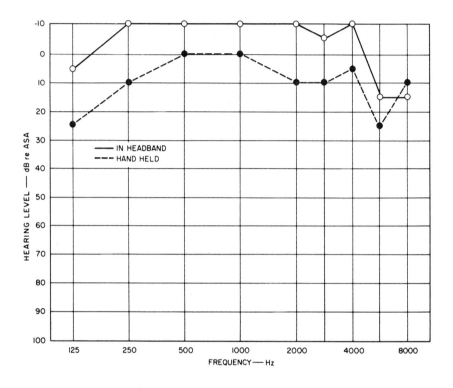

FIGURE 75. Contrast between audiograms obtained with earphone held in hand and in headband. From Wheeler (871).

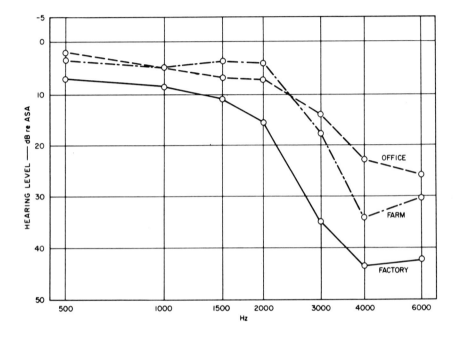

FIGURE 76.   Median hearing level, re ASA 1951, of men 30-39 years of age working on farms (n = 26), in offices (n = 87), and in factories (n = 85), for the left ear only. From Glorig *et al.* (304).

(b) In addition, since the ASA specification for noise in audiometric booths is geared to the ASA audiometer standards, the environmental noise in the test booths could mask the audiometer signals, even if they could be presented at lower levels, and thereby somewhat control the distributions of hearing levels measured in these environments.

The bias due to the use of audiometers designed to ASA specifications that was mentioned above is perhaps one reason that the data reported by Riley *et al.* (678) on the mean hearing threshold of nonnoise-exposed persons in one U.S. industry, and the results of U.S. Public Health Surveys (Glorig and Roberts [298]) do not fall on the ISO values for normal hearing. As seen in the top curves on Fig. 77, the mean thresholds in a young industrial population studied by Riley *et al.* (678) were several dB higher (less sensitivity) than would be required to meet the ISO Standard, but were about 6 dB lower than the ASA Standard; if the hypothesis about a bias being present when audiometers are used that meet ASA Standards is correct, the means found by Riley *et al.* could readily be 3 dB or so above their true value.

It has been proposed (677) that the problem would be eased if not solved by making the specifications for and the reading of audiometers in terms of actual sound pressure levels rather than in terms of a reference value for normal, so-called audiometric zero. While this suggestion may have merit for other reasons, it avoids the issue at stake — what sound pressure levels as a function of frequency are representative of normal hearing?

Presbycusis and Sociocusis

Figure 77 shows results of typical studies in which the threshold of hearing was examined in persons of different ages. It is invariably found that the threshold of audibility declines with age, more rapidly at high frequencies than at low, and more severely for men than women. This decline is called presbycusis; whether or not it is due to physiological aging or to wear and tear on the auditory system by the intense noises and sounds of everyday living is an open question. Glorig *et al.* (304) found evidence that the home environment (urban vs. rural, rural being noisier) does influence the amount of presbycusis experienced — they suggest that so-called presbycusis is probably a mixture of aging and "sociocusis." Sociocusis is a name suggested to describe loss in hearing due to the typical noises all, or nearly all, members of our society are exposed to in normal nonwork activities (Glorig and Nixon 297a).

Except for young adult persons, who are naturally lacking in many year's of exposure to the noise of their working environment, the thresholds measured in the health and public-fair surveys are probably somewhat influenced, particularly for males, by both noise-induced deafness and aging processes. Corso

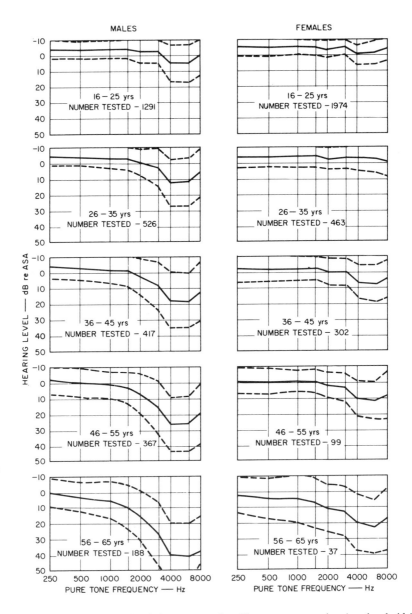

FIGURE 77.   Hearing thresholds by age grouping. The mean average hearing threshold in
dB of left and right ears have been plotted (solid line) against pure tone
frequency. The broken lines show plots of plus one and minus one standard
deviation. The subjects were from an American industry screened for known
otological disease. From Riley *et al.* (678).

(164) working with subjects from a nonindustrial, relatively noise-free environment found that there was a seemingly pure presbycusis of aging deterioration in the hearing of men and women. The deterioration was considerably less than found in general survey studies for men, as might be expected if the data from these general surveys included some amount of noise-induced deafness (see Fig. 78). Corso's data are probably the best available determination of pure presbycusis for the U.S. population because of the careful otological examination and screening applied to his subjects and because of the quiet living environment from which they were taken.

Rosen *et al.* (699) reported that in people living in a noise-free area in Sudan, Africa, there was no evidence of presbycusis, and concluded that so-called presbycusis in more modern societies is really noise-induced sociocusis. This conclusion, on the basis of these data, is not to be taken without serious qualification because the subjects (a small group of people who may or may not have been truly representative of their age group) appeared to be aging in all respects at a slower rate than what would be considered normal in the United States (also see Bergman [62]). For example, blood pressure was nearly the same in the young and old age groups of adults. Rosen *et al.* do not conclude that lack of noise kept blood pressure down, but that this was due to diet which of course may have also reduced presbycusis. Hinchcliffe (372) has noted the converse, that in an essentially noise-free society the auditory thresholds were somewhat depressed in persons with histories of infectious diseases that presumably did not directly involve the auditory system.

Presbycusis or sociocusis, whichever it is, is clearly a factor to be taken into account when attempts are made to evaluate the effects of exposure to industrial noise on hearing.

Relation between Pure-Tone Thresholds and Speech Perception

A second major issue in the evaluation of the effect of noise on hearing, besides that of determining what is normal hearing for pure tones, is that of the relation between pure-tone audiograms and the ability of persons to perceive speech. The committee on the Conservation of Hearing of the American Academy of Opthomology and Otolaryngology (AAOO) of the American Medical Association specified that this ability should be in terms of: "the ability to hear everyday speech under everyday conditions. The ability to hear sentences and repeat them correctly in a quiet environment is taken as satisfactory evidence of correct hearing for everyday speech" (510, p. 236). This would appear to be a highly questionable definition of everyday speech and everyday conditions. Everyday speech includes single word or phrase messages (which are generally less easily correctly understood than sentences), distortions due to

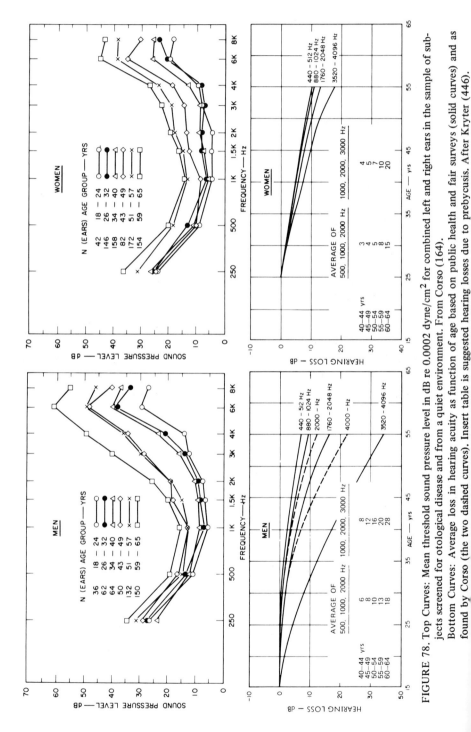

FIGURE 78. Top Curves: Mean threshold sound pressure level in dB re 0.0002 dyne/cm² for combined left and right ears in the sample of subjects screened for otological disease and from a quiet environment. From Corso (164).

Bottom Curves: Average loss in hearing acuity as function of age based on public health and fair surveys (solid curves) and as found by Corso (the two dashed curves). Insert table is suggested hearing losses due to prebycusis. After Kryter (446).

such things as the talker having a "cold," poor pronunciation, speaking from a distance, using a telephone, etc., and often, some ambient noise or competing music or speech from other persons. All of these conditions much more severely degrade the understanding of speech in the partially deafened person than in the person with normal hearing. In short, AAOO defined a type of speech material and listening condition that would be the least likely to show any impairment in the deafened person.

In addition, most of the speech tests and test conditions that have been devised and used for determining in the clinic a person's ability to understand speech are not suitable as tests of everyday speech or as a means of establishing the relation between pure-tone thresholds and the ability to understand every-day speech. The criticisms against these tests are:

1. The most widely used tests have been the so-called W1 and W2 tests consisting of 36 Spondee words, or the W22 PB (Phonetically–Balanced) lists consisting of 200 monosyllabic words (375). These tests consist of but a relatively few words, many of which can be correctly identified on the basis of the pattern of their vowel sounds. These word lists do not fully test for the relatively more difficult and important consonantal sounds as they appear in a broad sample of everyday speech.

2. The hearing ability for speech of the listeners is graded only in terms of their ability to correctly perceive 50% of the test words; this is called the threshold of speech reception or intelligibility. It is known that persons with noise-induced hearing loss may differ most from normal hearing persons in that they cannot achieve word test scores much above 50%. Further, the tests are administered at whatever levels of intensity are required for each person to reach his threshold for speech intelligibility even though the levels may be far in excess of the intensity level of everyday speech.

3. The tests are administered under quiet conditions with no distortions in the speech transmission system, whereas in everyday conditions there is usually some masking noise and other types of acoustic distortions imposed upon the speech signal.

Quiggle *et al.* (656), using the 36 spondee word lists, and Harris *et al.* (350), using the 200 word PB lists, compared the pure tone audiograms of persons with mild to very severe degrees of hearing loss with their speech-loss rating as measured by the respective word tests. They found, by statistical correlation techniques, that the hearing-loss-for-speech score could be adequately predicted from the pure-tone threshold measurements at 500, 1000, and 2000 Hz. Hearing losses for frequencies above 2000 Hz contributed little or nothing to the prediction.

The results of these two studies require careful evaluation for several reasons.

1. They possibly were given weight by the AAOO Committee on the Conservation of Hearing (510) in their recommendation that thresholds at only these three frequencies be used in evaluating hearing loss for speech.

2. They are inconsistent with nearly all other studies on the relative importance of frequencies above 2000 Hz to the perception of speech.

3. Typically the threshold of hearing at 500 Hz is not appreciably different among people, except for those with very profound hearing losses. The fact, therefore, that the threshold at 500 Hz contributed significantly to the speech test score in these two studies suggests that either the subjects were preponderantly persons with extreme hearing losses (as was particularly the case of Harris *et al.*'s subjects who, on the average, had an HL, re ASA, of 15 at 500 Hz) or the speech tests did not represent a fair test of all types of speech sounds (as was particularly the case of Quiggle *et al.* who used the 36 Spondee word test), or both.

On the other hand, Young and Gibbons (895) (see Table 9) demonstrated in hard-of-hearing persons that the speech reception threshold correlated more highly with their thresholds at 500, 1000, and 2000 Hz, than at 4000 Hz,

TABLE 9

Intercorrelations among Seven Variables Based on Measures Obtained from 100
Listeners with Speech Discrimination Scores of 94% or Poorer
From Young and Gibbons (895).

	SRT	Hz 500	Hz 1000	Hz 2000	Hz 4000	Age
PB* Word	0.52†	0.36	0.48	0.65	0.45	0.13
SRT (Speech Reception Threshold)		0.90	0.92	0.66	0.38	0.19
500 Hz			0.82	0.48	0.32	0.23
1000 Hz				0.64	0.28	0.12
2000 Hz					0.50	0.08
4000 Hz						0.20

* The correlations with this variable are positive because per cent incorrect was used in the computations.

† For these data, a correlation of 0.25 is needed for significance at the 1% level.

whereas speech intelligibility scores on PB word tests correlated more highly with thresholds at 1000, 2000, and 4000 Hz than at 500 Hz. Also Mullins and Bangs (566), Kryter *et al.* (477) (see Table 10), Elliot (216, 217) (considering only her test groups in which the coefficients of correlations differed significantly from zero), Harris (348), and Harris *et al.* (351) have found that in the quiet, and particularly with mild amounts of noise or frequency distortions, the scores obtained on speech tests by persons with sensori-neular hearing losses were better predicted from pure-tone thresholds taken at 1000, 2000, and 3000 Hz than at 500, 1000, and 2000 Hz.

TABLE 10

Intercorrelations between Hearing Losses and Average of Speech Tests
From Kryter *et al.* (477).

	Test Frequency, Hz					
	500	1000	2000	3000	4000	6000
Av of all speech tests	0.31	0.48	0.76	0.75	0.62	0.39
500	···	0.81	0.40	0.27	0.23	0.23
1000		···	0.59	0.43	0.32	0.32
2000			···	0.76	0.53	0.32
3000				···	0.78	0.45
4000					···	0.60
6000						···

All subjects except normals, N = 162

Webster (857) suggests that Kryter *et al.* (477) and Kryter (451) did not include a sufficient number of persons with profound deafness in the frequency region at 500 Hz to obtain the proper correlations (which according to Webster would include only threshold measurements at 500, 1000, and 2000 Hz) between pure-tone thresholds and speech test scores. However:

1. Approximately 10% of the nonnormal hearing subjects in the Kryter *et al.* tests had hearing levels of +30 dB or more at 500 Hz (see Fig. 79).

2. Most importantly, for humanitarian as well as scientific purposes, a good methods of measuring hearing loss must properly evaluate the loss for individuals from mild to severe, and not merely a group average.

3. There is almost always present, in cases of noise-induced deafness, progressively more loss at the higher frequencies than at the lower frequencies. Accordingly, one cannot expect to predict hearing losses of as much as 40 dB or so above 2000 Hz by measuring thresholds at 1000 and, particularly, 500 Hz; but one can predict reasonably well losses at 500 and 1000 Hz from thresholds taken at 2000 Hz and above. Therefore, if speech intelligibility is at all related to

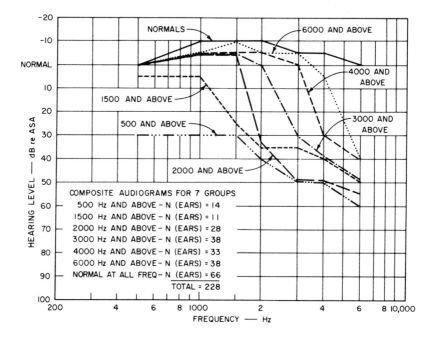

FIGURE 79. Average pure-tone audiograms, re ASA 1951, for subjects grouped according to the frequency region where hearing losses, if any, were measured. From Kryter *et al.* (477).

hearing at frequencies above 2000 Hz (as has always been found to be the case except for tests with Spondee words or other nonrepresentative speech conditions, or with a population of subjects who are predominately profoundly deaf) it is necessary to obtain audiograms at those higher frequencies. Indeed, as Kryter *et al.* (477) and Harris (348) have found, the least amount of information obtained from any one test frequency with respect to the ability of most people, except the profoundly deaf, to hear speech is that obtained from threshold tests at 500 Hz. Piese *et al.* (624) also recommend that hearing be tested at 3000 Hz for purposes of industrial hearing conservation.

The Committee on Hearing, Bioacoustics, and Biomechanics (CHABA) of the National Academy of Science and National Research Council is undertaking further study of speech test materials and test conditions that might serve as a realistic method for finding and validating the relations between pure-tone audiometery and the ability to understand speech by persons who are hard of hearing. The Modified Rhyme Test (MRT), recorded in quiet and low background noise, has been proposed for this purpose (Kreul, *et al.*, [439]).

Impairment of Hearing Speech

The procedures proposed by AAOO (510) for estimating from pure-tone audiograms the impairment for understanding everyday speech under everyday conditions are:

1. The hearing levels of each ear at 500, 1000, and 2000 Hz are averaged for each ear.

2. For each decibel that the average hearing level at 500, 1000, and 2000 Hz exceeds 15 dB re ASA (25 dB re ISO), allow 1.5% up to the maximum of 100%. (Although the differences between the ASA and ISO SPLs specified for normal hearing are not 10 dB different at all test frequencies, an average of 10 dB will be used in the discussions and calculations to be made on this matter [see Fig. 73] in this book.)

3. The smaller percentage (the ear with the better hearing) is multiplied by five, the result added to the percentage found for the other ear, and the total divided by six. The final percentage represents, according to the AAOO Guide, the evaluation of binaural hearing implairment. A rating of 100% impairment is achieved by the AAOO method when a person has an average HL of 82 dB or greater re ASA (92 re ISO) at 500, 1000, and 2000 Hz in each ear.

Handicap vs. Impairment

Certain laws and practices provide some monetary compensation to persons who have suffered a handicap because of hearing loss, as will be discussed in a later section. Whether "handicap" refers to a handicap to the performance of a person's normal occupation, to the performance of any normal work, or to the enjoyment of a normal social life is a matter of some debate.

Although the AAOO Guide supposedly defines an impairment threshold, it would appear that in reality it attempts to define a handicap threshold for persons who have worked in noisy occupations. The basis for the AAOO Guide discussions of this question with industrial otologists, was that these otologists are of the opinion that, in general, workers with hearing no worse than 0% impairment, according to the AAOO Guide, have no trouble with their hearing. This opinion is based at least in part by talking with the workers in the quiet of the otologist's office. These otologists also noted that at a near 100% impairment, according to the AAOO Guide, the person can understand a few words if maximum shouting level is used by the talker at a distance of a few inches from the listener's ear (at that distance and effort, speech levels reach the order of 120 dB). However, both of these speech conditions would appear to violate the stated AAOO criterion of impairment for "everyday speech under everyday conditions."

One of the possibly surprising things about occupational noise-induced deafness, and one which perhaps tends to reduce concern about it, is that persons with such deafness suffer no handicap from their hearing impairment when they are in the noise; the reason being that when the person is in the noise the masking due to the noise interferes more with the reception of speech or other signals than does the loss in hearing. This phenomenon follows from the fact that the amount of permanent shift eventually possible from exposure to a given noise will not exceed, usually will be somewhat less than, the masking that the noise can create in the normal ear. It is only when the person with the noise-induced hearing loss is placed in a less noisy environment than that which caused the loss, or is presented with distorted or weak speech, or attempts to understand speech in a crowd, or attempts to localize the place a sound is coming from, will the impairment to his hearing possibly become a handicap.

In this regard it might be noted that some military officers (suffering from some deafness probably due to gun noise) have testified to CHABA, that they are unable to satisfactorily perform their duties in conferences and meetings because of the inability to hear and understand normal speech even though their hearing is rated as no handicap by standards of the Veterans Administration (809) which, for all intents and purposes, are identical with those of AAOO. Nett et al. (576) found in a study of critical incidents of hearing handicap in a population of 378 hard-of-hearing persons that half of the persons had hearing

TABLE 11

Sound Pressure Levels for Pure Tone and Continuous Spectra Sounds According to AAOO (510) and ASA 1951 Standard (19)

Freq.-Hz	Threshold for pure tones re/ASA 1951		Threshold for sounds having continuous spectra re/ASA 1951		HL re/ASA 1951 for pure tone for person at threshold of hearing impairment for everyday speech according to AAOO‡	SPL of sounds, presented via loudspeakers or person-to-person, having continuous spectra at threshold of hearing impairment for speech according to AAOO
	SPL earphone*	SPL-F†	Width of critical band in dB**	(Col 3 -4) SPL-F†		(Col 5 -6) SPL-F†
1	2	3	4	5	6	7
125	54.5	48.5	-20	28.5	(0)	28.5
250	39.5	33.5	-20	13.5	(0)	13.5
500	25.0	19.0	-20	- 1.0	10	9.0
1000	16.5	10.5	-22	-11.5	15	3.5
1500	16.5	10.5	-23	-12.5	(17)	5.0
2000	17.0	11.0	-24	-13.0	20	7.0
3000	16.0	10.0	-26	-16.0	(25)	9.0
4000	15.0	9.0	-27	-18.0	(30)	12.0
6000	17.5	11.5	-29	-17.5	(40)	22.5
8000	21.0	15.0	-32	-17.0	(50)	33.0

* Sound Pressure Level (SPL) in 6 cc coupler or 6 cc earphone cushion.

† SPL in field at entrance to ear canal when speech is presented by loudspeakers or person-to-person. The difference, 6 dB, between Columns 2 and 3 is due to greater sensitivity of the ear to sound presented to the open ear compared to sound presented in a closed earphone-cushion combination.

** Spectrum level of continuous spectra sound at threshold re spectrum level of pure tone at threshold.

‡ Hearing level (HL) values in ( ) are considered to be typical when HL's are 10 dB at 500 Hz, 15 dB at 1000 Hz, and 20 dB at 2000 Hz re/ASA, 1951.

TABLE 12

Articulation Index (AI) Calculated from the Amounts, in dB, the Speech Signal Uttered at Various Intensity Levels Exceeds Various Thresholds of Audibility for Sounds of Continuous Spectra in the Field, i.e., by Loudspeakers or Person-to-Person

These thresholds estimated by means of pure-tone audiograms corrected for field vs. earphone listening (-6 dB), and the critical bandwidth of the ear (see Table 11).

Avg. HL at 500, 1000 and 2000 Hz		Avg. HL at 1000, 2000 and 3000 HZ		Weak Conversational Level in Quiet (Long-Term RMS = 50 dB)			Normal Conversational Level in Quiet (Long-Term RMS = 55 dB)			Everyday Speech Level (Long-Term RMS = 65 dB)			Shouting Level (Long-Term RMS = 80 dB)		
ASA	ISO	ASA	ISO	AI	% Sent.	% 1000 PB Words	AI	% Sent.	% 1000 PB Words	AI	% Sent.	% 1000 PB Words	AI	% Sent.	% 1000 PB Words
-5	5	5	15	0.81	99	94	0.84	100	95	0.98	100	98	1.0	100	100
5	15	15	25	0.56	97	81	0.72	98	92	0.84	100	95	0.98	100	98
15	25	25	35	0.34	87	52	0.47	95	73	0.72	98	92	0.84	100	95
25	35	35	45	0.17	36	17	0.26	68	35	0.47	95	73	0.72	98	92
35	45	45	55	0.03	5	2	0.09	15	8†	0.26	68	35	0.47	95	73
45	55	55	65	0	0	0	0	0	0	0.09	15	8	0.26	68	35
55	65	65	75	0	0	0	0	0	0	0	0	0	0.09	15	8
65	75	75	85	0	0	0	0	0	0	0	0	0	0	0	0

* See Sec.

† Sentences on first presentation to listeners.

§ The so-called threshold of speech intelligibility or discrimination (50% words correct), as measured by Spondee Word Tests (CID W-2), is equivalent to a score of about 20% correct when 1000 PB word tests are used (Hirsh[375]).

loss of 34% or less as estimated by the AAOO procedures, but 60% of the group considered the hearing loss their major health problem, and about 50% of the group estimated their loss as being more than a 60% loss in hearing ability. Many of the incidents of handicap occurred when the talker was 10-12 feet or so from the deaf person. Twenty-two percent of the group had conductive losses, 32% sensori-neural, and 46% mixed losses. Also, tests of speech intelligibility in the presence of mild amounts of noise or speech filtering as present in some telephone systems reveal a measurable loss in understanding when the hearing levels start to exceed 0 dB re ISO, as shown in Fig. 80.

Impairment of Speech Reception and Relation to Pure-Tone Hearing Levels

It should be instructive to determine the portions of typical speech signals that are audible to persons with depressed auditory tresholds and to calculate from this information an Articulation Index (AI) for estimating the reception of speech by these persons (see Chapter 2). As briefly discussed in Chapter 2, the reception and understanding of speech is very similarly affected by the elevation of the auditory threshold particularly for noise-induced, sensori-neural deafness as by the addition of comparable amounts of masking noise to speech perceived by the normal ear.

However, hearing levels are usually measured by pure tones or speech tests presented via earphones whereas everyday listening is under field conditions, i.e., typical person-to-person or loudspeaker. For this reason it is necessary to determine, as is done in Table 11, what the threshold sound pressure levels would be for speech heard in the field by persons having given pure-tone HLs determined by earphones. Also given in Table 11 are the field listening threshold sound pressure levels for sound of continuous spectra, such as speech, for specified pure-tone HLs as measured by an earphone-type audiometer. Figure 81 shows some of the relations between the sound pressure level and per cycle spectrum of tones and speech when at various thresholds.

Different AI values are, of course, obtained as the speech level and/or the hearing level of the listeners is varied. Table 12 gives the AIs and related percent word and sentence intelligibility scores for several speech levels and hearing levels when measured by an audiometer calibrated to ASA 1951 and to ISO Standards.

Figure 82, which is partly based on Table 12, shows some relations between HLs for pure-tones and the understanding of speech as heard under various everyday conditions. Although speech is often spoken in sentences and often at reasonably high intensities (the extreme right-hand curve on Fig. 82), the reception of individual words uttered at a rather weak level of effort, or when the

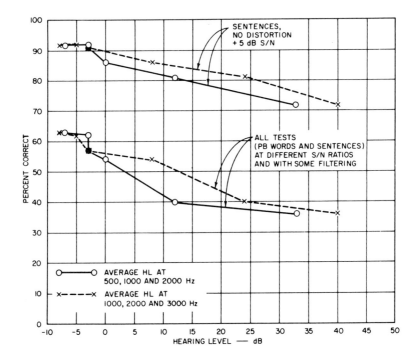

FIGURE 80.   Showing intelligibility test scores as a function of the average hearing level, re ASA, at 500, 1000, and 2000 Hz and at 1000, 2000, and 3000 Hz. From Kryter (451). (By permission of the Archives of Otolaryngology.)

listeners are eight to ten feet from a talker using an everyday intensity (the extreme left-hand curve in Fig. 82) is also on occasion important to people. An argument could be made that the impairment of the reception of difficult, weak-intensity speech starts at HLs that average 0 dB for 500, 1000, and 2000 Hz re ISO. However, the AAOO Guide recommends that impairment not be considered as starting until the average HL at 500, 1000, and 2000 Hz is 25 dB, re ISO. As shown in Fig. 82, this point provides the start of impairment of the reception of sentences uttered at an everyday level of intensity with the listener no more than one meter or so from the talker.

The most striking deficiency of the AAOO description of hearing impairment for speech is that it proposes that an average HL at 500, 1000, and 2000 Hz of 92 dB (re ISO) must be reached before 100% impairment for everyday speech is present. Figure 82 shows that 100% impairment for reception of sentences uttered at an everyday level of 65 dB occurs at an average HL of 500, 1000, and 2000 Hz of but 65 dB, and at 50 dB for weak monosyllabic words. Therefore it seems incorrect to say that impairment for the reception of everyday speech, or even handicap from that impairment, is not complete until average HLs of 92 dB are reached at 500, 1000, and 2000 Hz; 100% impairment for hearing everyday speech occurs at a much lower (less hearing loss) level.

Another shortcoming of the AAOO procedure is apparent when the procedure is applied to persons with sensori-neural hearing loss. The AAOO calculation procedure takes no recognition of the fact that the ear with sensori-neural deafness has lost some ability to understand speech even when sufficiently intense to be audible above its elevated threshold of hearing. This loss in ability to discriminate among speech sounds is illustrated in Fig. 83. It is seen in Fig. 83 that a person with a conductive type hearing loss (a condition wherein sound is abnormally attenuated during transmission through the middle or outer ear) is better able to understand intense speech than is a person with a sensori-neural loss, even though both have the same degree of loss measured at their threshold. Also note on Fig. 83 that this difference is ability is not well assessed, as mentioned previously, by the standard clinical speech tests that determine the threshold shift for the point at which 50% of words are correctly preceived.

Proposed New Method for Estimating Percent Hearing Impairment for Speech

Figure 82 offers perhaps a quantitative basis for setting criteria for the impairment of the reception of speech. To aid in the practical use of this figure we have drawn linear relations between the 0% and 100% speech reception scores (right-hand ordinate) for three classes of everyday speech. It is proposed that, for the sake of simplicity, hearing impairment for the different speech

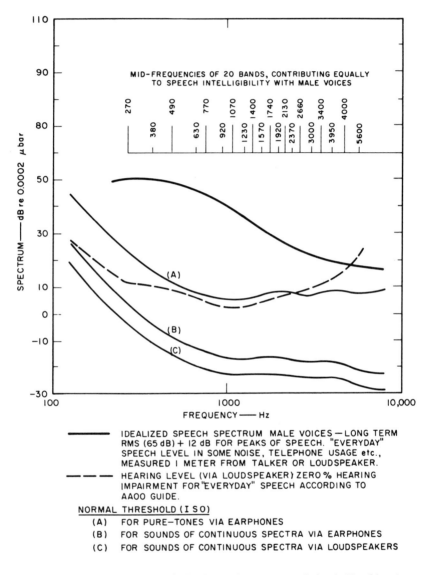

FIGURE 81.   Relations between idealized speech spectrum and thresholds of hearing.

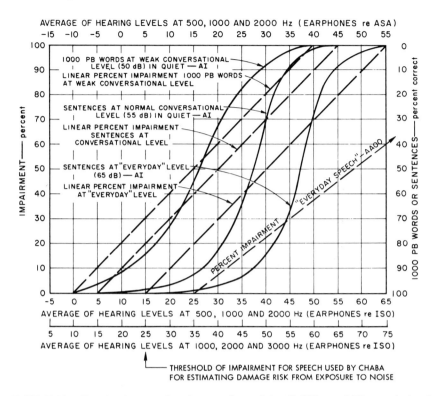

FIGURE 82. Relation between impairment of speech intelligibility and HL, as calculated by AI and as proposed by AA00.

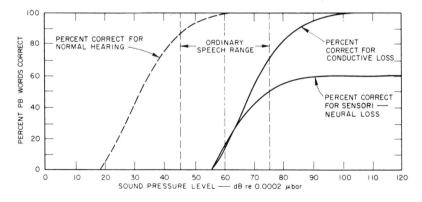

FIGURE 83. Speech intelligibility curves (percent PB words correct) at two different levels of amplified speech for normal hearing and for two types of impaired hearing. After Davis (169).

conditions be expressed in percent according to the left-hand ordinate of the figure and the straightline functions relating the left-hand ordinate to pure-tone hearing levels as expressed on the abscissa. This procedure of expressing impairment as some percentage loss linearly with dB increases in hearing level is also used by AAOO.

There are principally two practical speech conditions and two types of hearing impairment to be considered in relating HL for pure tones in dB to Percent Hearing Impairment for Speech.

1. Speech as typically heard in the relative quiet of home and office by persons with (a) conductive deafness, or (b) sensori-neural deafness, which includes noise-induced deafness.

2. Speech of amplified intensity (as made possible by the use of hearing aids, public address systems, to a limited extent by shouted speech, and by the listener getting within a foot or few inches of the talker), by persons with (a) conductive deafness, and (b) sensori-neural deafness.

It is proposed that a valid, as far as present research and clinical information permits, procedure for estimating hearing impairment for speech from pure-tone hearing levels is as follows.

1. The hearing levels re ISO of each ear at, preferably, 1000, 2000, and 3000 Hz are averaged and 10 dB is subtracted from the average; as a less desirable alternative the hearing levels of each ear at 500, 1000, and 2000 Hz are averaged.

2. For persons with sensori-neural deafness, for each decibel that the average hearing level exceeds the levels specified in the appropriate formula in Table 13, add 2 percentage points up to 100%.

3. For persons with conductive deafness, for each decibel that the average hearing level exceeds the level specified in the appropriate formula in Table 13, add 1 percentage point up to 100% for amplified speech, and 2 percentage points up to 100% for speech at conversational or everyday intensities. This difference in growth of impairment for these two groups is based on the functions shown in Fig. 83.

4. The smaller percentage (that for the ear with the better hearing) is multiplied by five, the result added to the percentage found for the other ear, and the total divided by six. The final percentage represents the evaluation of binaural hearing impairment.

Figure 84 shows the audiograms of two groups of men who have worked for a number of years in intense industrial noise. Table 14 illustrates the differences in percent of hearing impairment that would be calculated for the persons with the median, 50%, hearing levels shown on Fig. 84. It is seen in Table 14, that the

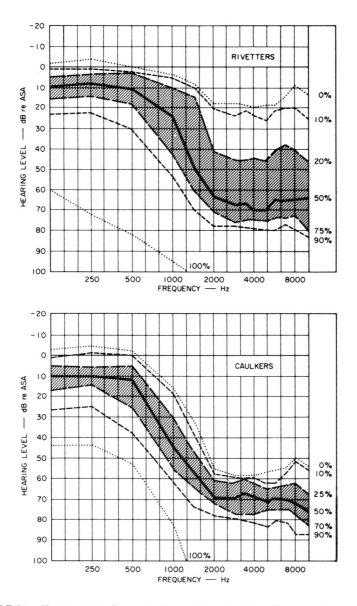

FIGURE 84.   Hearing level of some workers in heavy industry. From van Leeuwen (814).

AAOO method underestimates by a factor of about 2 the impairment, according to the Calculated Articulation Index, of these persons in their ability to understand sentences heard in the quiet when no more than one meter from a talker using a speech level that is 10 dB above that used for normal conversation in the quiet.

TABLE 13

Formula for Proposed New Method of Evaluating Percent Hearing Impairment for Speech for Persons with Noise-Induced or with Conductive Hearing Impairments

---

(1) $HLdB_{1,2,3, kHz} \cong \left(\dfrac{HL_{1\,kHz} + HL_{2\,kHz} + HL_{3kHz}}{3} \right)$

(2)   %THI (C) (Percent Hearing Impairment for total usable capacity for hearing amplified speech in persons with conductive deafness) = $HLdB_{1,2,3\,kc} - 10$.

Note for Formula 2: It is presumed, see Fig. 83, that with 50 dB of speech level amplification available, a range of about 100 dB in the intensity of speech signals is usable in a normal way by the person with conductive deafness.

(3)   %THI (N) (Percent Hearing Impairment for total usable capacity for hearing amplified speech by persons with noise-induced deafness) = $2(HLdB_{1,2,3\,kHz} - 10)$.

Note for Formula 3: For persons with noise-induced deafness it is presumed that the intelligibility of speech increases at about one-half the normal rate as its intensity is increased above his threshold and that a range in the intensities of speech signals of but 50 dB is usable in a normal way by these persons.

(4)   %EHI (Percent Hearing Impairment for "Everyday" unamplified speech in quiet by persons with either noise-induced or conductive deafness) = $2(HL_{1,2,3\,kHz} - 25)$.

(5)   %CHI (Percent Hearing Impairment for Conversational unamplified speech in quiet by persons with either noise-induced or conductive deafness) = $2(HL_{1,2,3\,kHz} - 15)$.

(6)   %SHI (Percent Hearing Impairment for Weak unamplified speech in quiet by persons with either noise-induced or conductive deafness) = $2(HL_{1,2,3\,kHz} - 10)$.

Note for Formula 4, 5, and 6: Impairment for unamplified speech, unlike amplified speech, will be approximately equal for persons with either noise-induced or conductive-type deafness of a given $HLdB_{1,2,3\,kH} - 10$ because the impairment in understanding will be primarily controlled by the inaudibility of the weaker speech components below the HL in both types of deafness and not by the result of some form of "overloading" or distortion that takes place in the ear with noise-induced deafness when presented with intense, amplified speech.

TABLE 14

Percent Hearing Impairment for Speech in Persons with Median (50%) Hearing
Levels of Fig. 84

	% Hearing Impairment for Speech	
	Riveters	Caulkers
AAOO Method	27%	40%
Proposed Method Fig. 82 and Table 13		
A. Everyday Speech	57	75
B. Normal Conversational Speech	77	95
C. Weak Conversational Speech	87	100

Chapter 5

Damage Risk from Exposure to Noise

Introduction

Noise-induced deafness is a significant health problem in most modern countries. Important among the reasons it has become a problem are:

1. When, as mentioned before, the person is in the noise environment, the loss in hearing represents no handicap to his performance.

2. It is often exceedingly costly, perhaps at times impractical, to reduce the noise or remove men from it; therefore, the economic pressures on industry and workers alike tend to keep men in dangerously noisy environments.

3. Techanical knowledge concerning the measurement of noise and the relations between the physical aspects of a noise environment and noise-induced hearing loss has only relatively recently reached proportions that permit more or less definite conclusions about relations between them.

In general, so-called damage risk criteria as used in industry and the government have not been aimed at specifying noise exposures that will not cause some hearing loss, but have been aimed at noise exposures that will cause, in no more than some percentage of the people, no worse hearing than is required to understand correctly undistorted speech heard in the quiet at a level of intensity which is normal for a talker in typical room noise. Later this chapter will discuss permanent damage to hearing as measured by threshold shifts to pure tones as the result of exposure to steady-state and impulsive noise, and procedures for predicting from physical measures of the noise the risk for temporary and permanent threshold shifts to hearing speech and pure-tones. First, however, it is in order to discuss possible relations between temporary and permanent, noise-induced threshold shift.

Temporary and Permanent Threshold Shift for Purposes of Evaluating Noise

It is customary to measure or estimate temporary threshold shifts (TTS) for pure-tones from exposure to noise two minutes after exposure to the noise (TTS_2) and to call the shifts as temporary provided the hearing of the subjects returns to preexposure levels within 16 hours after the exposure. Noise-induced permanent threshold shifts (NIPTS) are the audiometeric shifts re the pre-exposure threshold, or, if preexposure thresholds are not available, re audiometric zero minus a presbycusis or sociocusis factor related to the age of the person. The permanent threshold shifts are measured one month or so after exposure to the noise is stopped. It is estimated that NIPTS usually reaches its maximum, depending upon the intensity of the noise, following up to 20 years or so of near daily exposure to a given noise environment. These two measures — TTS_2 and NIPTS — represent the fundamental data on which opinions regarding risks to hearing from exposure to a given noise are based.

As far as noise being a hazard to the organ of hearing, studies of temporary threshold shifts are considered by some to be of academic interest because (a) no significant direct life-long tests have been (or probably can be) conducted with the same individual humans, and (b) susceptibility to TTS and NIPTS in some animals was not significantly correlated (Ward and Nelson [844]). (We disagree with this conclusion and will discuss these data further in a later section on Susceptibility to NIPTS and TTS.) Gravendeel and Plomp (315, 316) also question the possible close relationship between NIPTS and TTS on the fact that the average maximum permanent dip in the audiogram of one group of 228 soldiers was at 5900 Hz, whereas 36 twenty-year-old recruits on a gun-firing line showed an average maximum temporary dip at 4600 Hz. However, the small difference in the locus of the dip for the two groups could well be attributable to the small sample size of recruits, some natural selection factor in the older soldiers, and differences in the types of gun noise to which the two groups had been exposed. Gravendeel and Plomp (316) point out, on the other hand, that steady-state diesel noise seems to cause a similar pattern of TTS and PTS, as shown by Fig. 85. The power spectrum of the diesel noise is shown in Fig. 86; as we shall see later, this spectrum is much like the energy spectrum of military gun noise.

The presence of many influential factors within and without persons in a given noise environment obviously makes mandatory the need for a large number of subjects and a dependency upon statistical trends for getting answers to research questions in this problem area. Nevertheless, within the limitation of exposures up to about 8 hours per day, the following similarities between TTS and NIPTS up to 40 dB or so seem reasonably well established, as will be further illustrated later by research data:

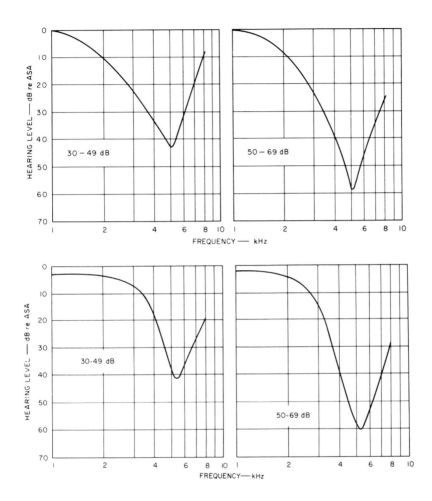

FIGURE 85. Average shape of the permanent hearing audiograms of a group of soldiers regularly exposed to gunfire (two upper graphs) and a group of workers regularly exposed to noise of diesel engines (two lower graphs). The curves represent the average for the audiograms of persons whose maximum permanent hearing level was between 30-49 and 50-69, as noted on the graphs. From Granvendeel and Plomp (315, 316). (By permission of the Archives of Otolaryngology.)

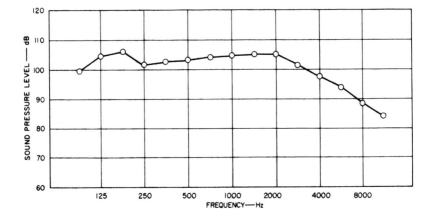

FIGURE 86.   One-third octave band analysis of diesel noise. From Gravendeel and Plomp
            (316). (By permission of the Archives of Otolaryngology.)

1. Exposure conditions that do not cause TTS_2 in persons with normal hearing can cause no NIPTS when NIPTS exposure is defined as 10-20 years of 8 hours or less of daily exposure. Somewhat contrary to this generalization are studies of Harris (342) and Trittipoe (806) in which evidence of "latent" fatigue not measured by TTS_2 was found; Ward (824), however, found no such latent effects.

2. Increasing the noise intensity above certain levels causes, within limits (up to no more than 40-50 dB), a roughly similar increase in TTS and NIPTS.

3. The greatest amount of threshold shift from a given noise band occurs within one octave above the frequency of the noise band for both TTS and NIPTS.

4. The frequency regions most susceptible to TTS are likewise most susceptible to NIPTS.

5. The locus of both temporary and permanent threshold fatigue or damage appears to be in the hair cells and their supporting cells on the basilar membrane.

The reasons for considering TTS data in evaluating permanent damage to hearing are practical:

1. There is always some uncertainty about the precise exposure conditions and other factors in NIPTS studies made after the fact in industry. These conditions can be more exactly controlled in laboratory studies of TTS.

2. The number of industrial noise environments for which there is available adequate measurements of the noise and the hearing of the people present in the noise has been somewhat limited.

3. There is a need to certify the potential hazards of noise environments for which there are no NIPTS data available.

4. Humanitarian considerations prevent the induction in humans of NIPTS.

5. Except possibly for data on traumatic exposures involving large (in excess of 40 dB or so) TTS values, TTS data on humans is probably of greater validity and usefulness in estimating most NIPTS conditions than is TTS or NIPTS data obtained on animals. This does not imply, of course, that the elucidation of basic physiological auditory mechanisms is not greatly enhanced, if not primarily dependent upon, auditory research on animals, but the setting of tolerable limits of noise exposure for humans is another matter.

The restriction that a period of at least 16 hours be allowed between regular daily exposures and that the longest continuous regular daily exposure be limited to about 8 hours is an important constraint on the generality of the prediction of NIPTS from TTS data. Mills, Gengel, Watson, and Miller (556a) found, for example, that TTS_2 (10 dB) did not increase as exposure duration was increased from 8 to 48 hours; however, the time required for recovery from these longer exposures required several days in the quiet.

Some Relations between TTS Data and Specific Industrial Studies of NIPTS

A key study of NIPTS from industrial noise was that conducted by Committee Z24-X2 of the American Standards Association under the chairmanship of Rosenblith (700). Various industries provided the Z24-X2 committee with audiometric records for hundreds of employees exposed to various noise environments. From the data made available, the general relations, called trend curves, between noise exposures and the hearing losses (hearing levels re ASA 1951) of the exposed employees were determined, as illustrated in Fig. 87. Table 15 shows that these trend curves predict reasonably well the apparent hearing losses found in persons exposed for extensive periods of time to six specific noise environments. The spectra of these environments are given in Fig. 88.

TABLE 15

Comparison of Mean Gross Hearing Losses (in dB) Measured in Six Noise Spectra with the Estimated Mean Hearing Losses That Are Predicted in the Trend Curves in Fig. 87

Continuous exposure to steady noise; not corrected for temporary threshold shift. From Rosenblith (700).

Noise Spectrum	No. of Subjects	Mean Age	Mean Exposure in Years	Spl in 300-600 Band	Hearing Loss at 1000 Hz		Hearing Loss at 2000 Hz		Spl in 1200-2400 Band	Hearing Loss at 4000 Hz	
					Meas.	Est.	Meas.	Est.		Meas.	Est.
A[*]	17M[†]	23	1[‡]	93	3.3	4	5.8	6	91	11.9	11
	16M	30	7		5.2	8	14.0	13		34.9	26[§]
	24M	40	13		7.6	12	18.5	19		45.6	39[§]
	19M	47	32		11.7	14	36.9	27		52.5	54[§]
B	6M	53	18	92	14	14	22.5	28	92	53.3	54[§]
	28W	41	2.2[‡]		9	9	11	11		18	18[§]
C[//]	46M	34	4	88	0	1	2.5	4.5	80	8.5	9
D[#]	20MW	28	1.5[‡]	93	4	5	5	7	95	16	14[§]
	16MW	28	2.3[‡]		2.5	5	7	9		20.5	16[§]
E[#]	20MW	23	1.5[‡]	86	0	2	2.5	3	84	9	8
F	21M	40	17	92	8.5	11	20	20	89	45	40[§]

[*] Spectra given in Fig. 88.
[†] M = Men; W = Women.
[‡] Estimated hearing loss is extrapolated when exposure time is less than 3 years.
[§] Extrapolated beyond sound pressure levels of trend curves in Fig. 87.
[//] Threshold shift after about one year's prior exposure.
[#] Threshold shift beginning with no exposure.

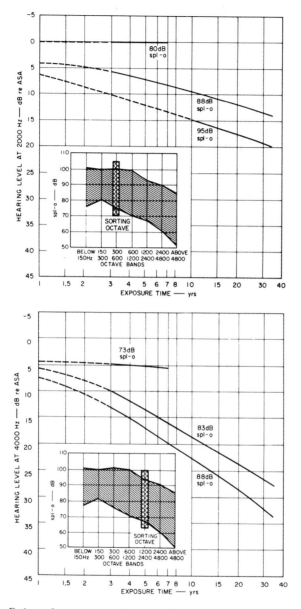

FIGURE 87.   Estimated average trend curves for net hearing loss after continuous
            exposure to steady noise, corrected for presbycusis, not corrected for
            temporary threshold shift. Each of these smoothed trend curves is identified
            by the sound pressure level in the sorting octave (spl-o) that most closely fits
            it. The shaded area of the inset figure represents the limits of the spectra on
            which these trend curves are based. From Rosenblith (700).

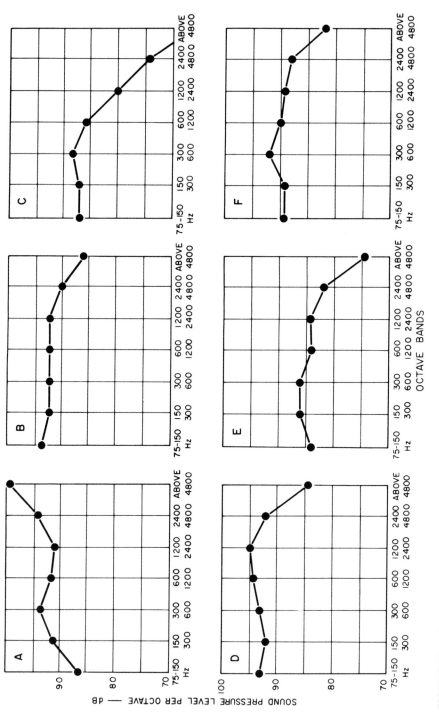

FIGURE 88. The six-noise spectra measured in the surveys reported in Table 15. These spectra were measured by different people and with different equipment. From Rosenblith (700).

We made an attempt (452) to demonstrate some relations between TTS and NIPTS using the results of Z24-X2 and more recent studies that provided sufficient detail and data to make possible direct comparisons among them. The results are given in Table 16 and the top graph of Fig. 89. Also shown on Fig. 89 (lower graph) are important data from a study conducted by Nixon and Glorig (585). These data show, among other things, that NIPTS from a given noise increases with exposure time (years on the job) up to about 20 years at which time it stablizes. However, as will be shown, it appears that the percentage of people in a group to develop a given amount of NIPTS increases as the group continues to work in the noise beyond 20 years. These two phenomena – stabilization of NIPTS within an individual after somewhere between 10-20 years exposure and on increase in the percentage of people in a group developing NIPTS with continued exposure – will be involved in the development of procedures for estimating damage risk to hearing in Chapter 6.

Similar patterns of TTS from various octave and broader bands of noise have been found by Ward *et al.* (848), Kylin (481), and Kryter (450), as illustrated in Fig. 90. It should be noted that Shoji *et al.* (739) recently also obtained TTS_2 from exposure to octave bands of noise that are in reasonably close agreement to those shown in Fig. 90.

Figures 91-95 exemplify the manner in which NIPTS develops over time. Because of individual differences and possible variability within an individual, the functions shown in Figs. 91-95 must be considered as statistical trends for relatively large groups of people. Comparison of the TTS and NIPTS audiograms on Figs. 84, 85, 87, and 89 to 95 shows a similar general pattern between TTS and NIPTS from roughly comparable noise spectra.

Shown in Fig. 96 is the general distribution of noise-induced hearing loss in industrial workers, based on the reports of Nixon and Glorig (586), Rudmose (709), and Kylin (482). Paschier-Vermeer (606) recently reanalyzed much of the published results of surveys (see Fig. 97) of hearing loss due to industrial noise. It is seen from Figs. 96 and 97 that when 50% of a population suffer a given degree of NIPTS, 25% of the group will have about 10 dB less and 25% about 20 dB more NIPTS, depending somewhat on the test frequency. This means, of course, that to protect 75% of a group from a given NIPTS rather than 50% (two damage risk criteria percentages that have been used in the past), the tolerable noise levels must be 5-10 dB less. Baughn (47) has reported similar results from an extensive analysis of over 6000 industrial audiograms which will be presented later in this chapter.

Workmen's Compensation for NIPTS

The question of workmen's compensation has been largely a legal problem involving adversary positions on the part of industry and labor (see Frazier

TABLE 16

Octave Band SPL's Required to Achieve Certain NIPTS and TTS$_2$ Values. From Kryter (452).

NIPTS 1000 Hz	Octave Band 300-600 Hz		NIPTS 2000 Hz	Octave Band 600-1200 Hz		NIPTS 4000 Hz	Octave Band 1200-2400 Hz		Study
	10-15 years	25-30 years		10-15 years	25-30 years		10-15 years	25-30 years	
10 dB	94 dB		15 dB	92 dB		20 dB	82 dB		Z24-X-2[700]
10 dB		93 dB	15 dB		91 dB	20 dB		80 dB	Z24-X-2
10 dB	83 dB		15 dB	81 dB		20 dB	82 dB		Rosenwinkel and Stewart
10 dB		83 dB	15 dB		80	20 dB		77 dB	Rosenwinkel and Stewart[703]
10 dB	89 dB		15 dB	94 dB		20 dB	79 dB		Kylin[481]
10 dB			15 dB	95 dB		20 dB	84 dB		Nixon and Glorig[585]
10 dB			15 dB		85 dB	20 dB		84 dB	Nixon and Glorig
Avg. 10 dB	88 dB	88 dB	15 dB	90 dB	85 dB	20 dB	82 dB	80 dB	

TTS$_2$ 1000 Hz	Octave Band 300-600 Hz 8 Hours (young, normal ears)	TTS$_2$ 2000 Hz	Octave Band 600-1200 Hz 8 Hours (young, normal ears)	TTS$_2$ 4000 Hz	Octave Band 1200-2400 Hz 8 Hours (young, normal ears)	Study
10 dB	--	15 dB	87 dB	20 dB	86 dB	Ward et al[848]
10 dB	--	15 dB	86 dB	20 dB	--	Kylin[481]
10 dB	89 dB	15 dB	86 dB	20 dB	84 dB	Kryter[450]
Avg. 10 dB	89 dB	15 dB	86 dB	20 dB	85 dB	

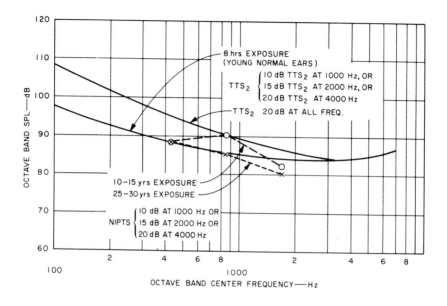

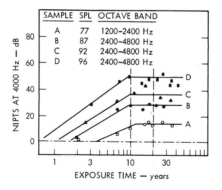

FIGURE 89.   Upper graph: Comparison of TTS$_2$ and NIPTS after exposure to noise in octave bands. After Kryter (452). Lower graph: Noise induced permanent threshold shift as a function of exposure time for four samples of employees. All values are corrected for age-effect. Note the change in growth between 10-20 years. After Glorig *et al.* (303). (With permission of the Controller of Her Britanic Majesty's Stationary Office.)

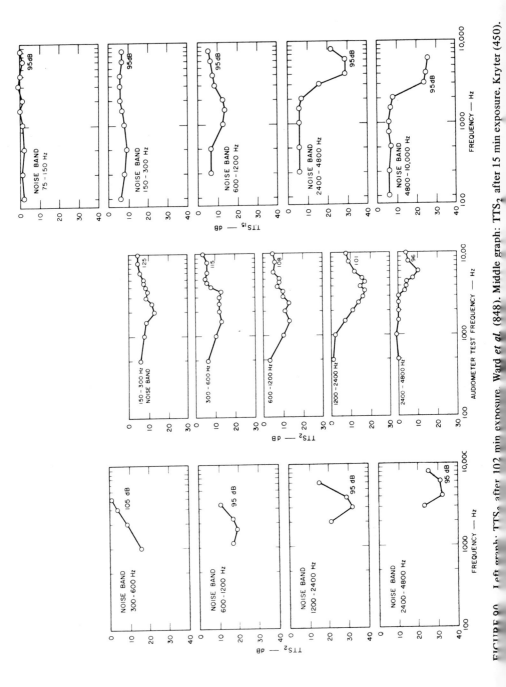

FIGURE 90. Left graph: TTS₂ after 102 min exposure. Ward *et al.* (848). Middle graph: TTS₂ after 15 min exposure. Kryter (450).

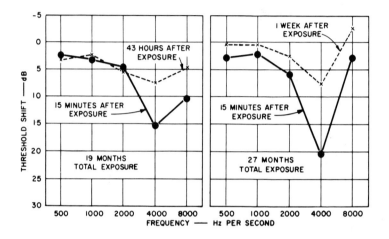

FIGURE 91. Threshold shifts at five frequencies as a function of the interval of time elapsed between the cessation of exposure and the measurement of hearing loss. Threshold shifts were measured in terms of a control group. The noise to which the people were exposed is shown in Fig. 88 (*d*). Twenty men and women were exposed to the noise for 19 months, 16 of them for 27 months. The mean age of the group was 28 years. From Rosenblith (700).

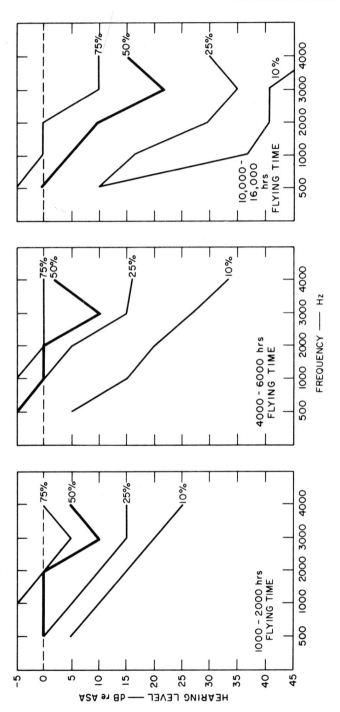

FIGURE 92.   The distribution of net hearing losses (corrected for presbycusis) of airline pilots, as a function of five test frequencies, for different ranges of flying time. The audiograms were taken several hours after exposure to the noise. From Rosenblith (700).

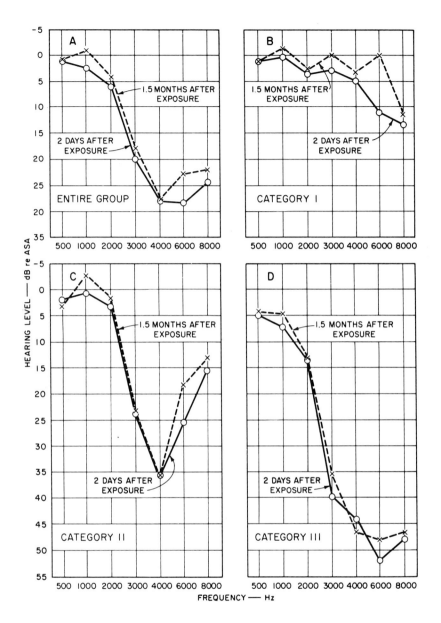

FIGURE 93. Recovery of seven frequencies as a function of the interval of time between the cessation of exposure and the measurement of hearing loss. The median age of the 36 persons was 31 years, and they had, on the average, been exposed to the noise for more than 10 years. The group was divided into three subgroups called Categories I, II, and III, on the basis of the amount of total hearing loss. There were 13 persons in Category I, 12 in Category II, and 10 in Category III. One person had hearing losses too large to be classified in Category III. From Rosenblith (700).

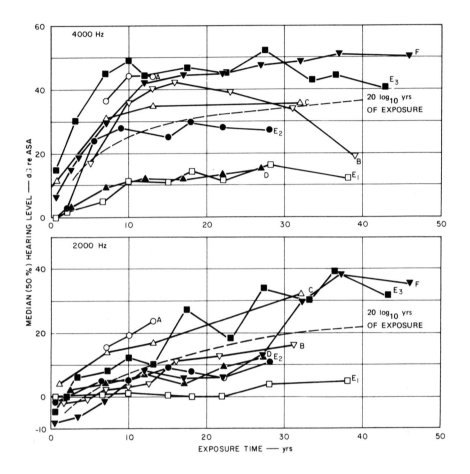

FIGURE 94.   Median noise-induced hearing loss as a function of exposure time. Data from
              the following studies: (A) Burns *et al.* (111); (B) Gallo *et al.* (270); (C)
              Rosenblith (700); (E) Nixon and Glorig (585); (F) Taylor *et al.* (793). From
              Passchier-Vermeer (606).

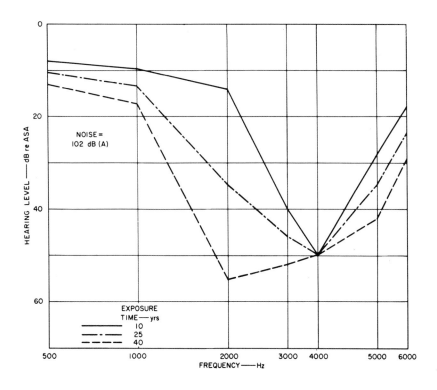

FIGURE 95. Median noise-induced hearing loss, as a function of frequency. After Passchier-Vermeer (606).

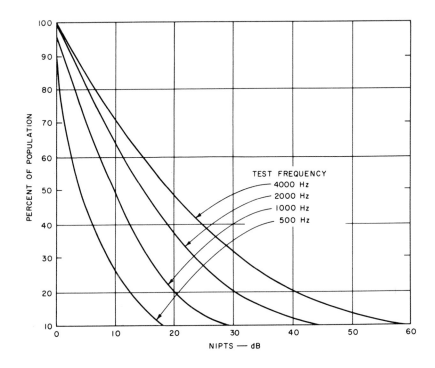

FIGURE 96.   Estimated percentage of people that will have as much or more presumed
              NIPTS than that indicated on the abscissa after more than 20 years of
              near-daily exposure to a given noise condition. Audiometric test frequency is
              the parameter. Data based on Nixon and Glorig (585), Rudmose (709), and
              Kylin (482).

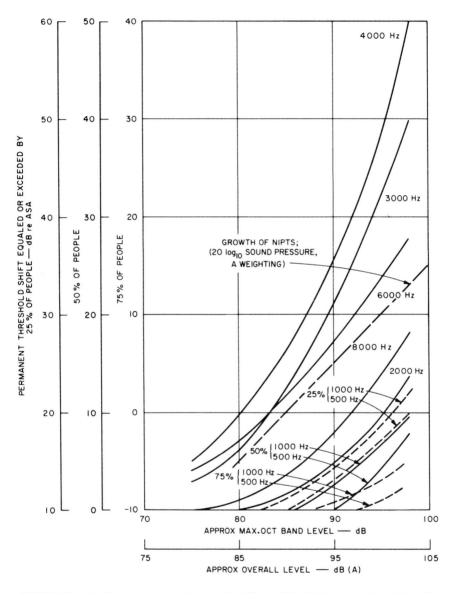

FIGURE 97. Median hearing loss (threshold shift), re ASA 1951, exceeded by 25%, 50%, and 75% of people exposed to noise for 10 years. The vertical ordinates for 75% and 50% of the exposed population are approximations. At 500 and 1000 Hz, separate curves are needed for 25, 50, and 75 percentiles. The dashed line, marked "20 $\log_{10}$," indicates that hearing losses of about 5-35 dB or so are linearly related, dB per dB, with sound pressure level. After Passchier-Vermeer (606).

[262-265], Loye [526], and Symons [788]). Scientific and medical data necessary to aid in determining appropriate safety standards and compensation for damage are still lacking with respect to certain details. The deduction of the percentage of people who will suffer hearing loss from exposure to noise in industry is complicated by the fact that industrial workers may often have less than normal hearing level when they enter a given occupation because of age and previous exposure to noise. Accordingly, the measurement of hearing loss in terms of the difference between a preoccupation and postoccupation hearing level will be less than the hearing loss found by comparing the hearing level of young adults upon entering any noisy industrial or military work and their hearing level after years of work. For example, Harris (347) reported NIPTS data for engine-crewmen in submarines which appeared to show much less threshold shift from the noise than was found in other studies of NIPTS due to industrial noise. However, an examination of these data reveals that the men studied had, on the average, significant amounts of presumably noise-induced hearing loss prior to the start of the noise exposures under question; at least the hearing acuity of the men as measured on a pure-tone audiometer was substantially less than normal, even relative to the ASA 1951 reference standard, at the start of the study.

It would seem reasonable that the measured hearing level, re ISO, of an individual who has worked in noise should be corrected for presbycusis in accordance with the curves of Fig. 78. However, some studies appear to show a greater amount of presbycusis or non-noise-induced hearing loss than that shown in Fig. 78. For example, data collected by Cohen *et al.* (144) suggests about a 5 dB greater reduction in hearing due to aging in a group from "nonnoisy" environments within industry than is shown by the general population involved in the data of Fig. 78. The definition of a "nonnoisy" industrial environment is obviously critical to this issue. It can be expected that persons concerned with the economics and operations of industry will take the position that industry should only not cause in its workers any more noise-induced deafness than is accrued in everyday living or in nonnoisy industry. However, from a hearing conservation point of view, this could well mean that many workers living in quiet environments would be insured of some degree of noise-induced hearing loss from their work, and that the possible reduction of other environmental noises in the future would be of little benefit to those working in industry.

Also important to the interpretation of the significance of noise-induced deafness in industry is the use of impairment as specified by AAOO (average 25 dB re ISO at 500 Hz, 1000 Hz, and 2000 Hz, and unlimited loss at higher frequencies) as a reference base for determining the incidence of hearing impairment due to industrial noise. This base is generally used in the U.S.A. as the hearing levels that must be reached in both ears of the worker before the worker may qualify for some amount of workmen's compensation.

Figure 98 from Cohen *et al.* (144) shows results (curves B and C) derived from industrial studies similar to those involved in the derivation of Fig. 97. Also shown are data points for several individual U.S. Public Health Surveys. It is seen that the percentage increase in incidence of compensable hearing impairment (as specified by AAOO) for noise at 80 dB(A) does not become significant until the age bracket of 36-45 years (presumably after about 10-20 years of noise exposure). It should be noted that these hearing level data are not corrected for presbycusis. Correction for this factor would undoubtedly reduce the percentage incidence of impairment given in Fig. 98 by a few points. One might deduce that 90 dB(A) is safe with respect to damage to ear because Botsford (83) found that only a few percent of the workers in nonnoisy locations in an industry have less average hearing loss at 500, 1000, and 2000 Hz than workers in a 90 dB(A) industrial environment. However, Figure 99 from Baugh (47) indicates that somewhere near 80 dB(A) is the level above which the noise in question starts to cause an increase in the percentage of people having more than an average hearing loss of 25 dB at 500, 1000, and 2000 Hz re ISO; in addition, again, it must be kept in mind that:

1. Levels of 80 dB(A) appear to increase hearing losses above presbycusis (which is itself probably a mixture of everyday noise-induced hearing loss and aging).

2. Hearing losses will occur, from long-term exposures to noise at that level, that are of a magnitude or at sound frequencies that are not considered compensable according to workmen's compensation laws. Indeed, to have a safe level in terms of complete hearing conservation, the levels apparently would need to be 10-15 dB less than 80 dB(A), and even then high-frequency noises damaging to the ear could be present.

Safety and Health Standards

The U.S. Department of Labor (223) has recently issued a notice of proposed rule-making which specifies that employees of government contractors shall not be exposed for 8 hours or more per day to noise at a steady level that exceeds 90 dB(A). Higher limits for individual octave bands and an increase of 5 dB for each halving of daily exposure duration (i.e., 95 dB[A] for 4 hours, 100 dB[A] for 2 hours, etc.) are specified. The apparent presumption of no hearing impairment with an average loss in hearing of 25 dB or less re ISO at 500, 1000, and 2000 Hz and unlimited losses at frequencies above 2000 Hz, and the lower limit of 90 dB(A) are perhaps controversial features of the proposed U.S. Department of Labor Standard.

The dB(A) is a reasonably appropriate weighting for damage risk up to frequencies of about 2000 Hz, but underestimates relative damage risk to higher

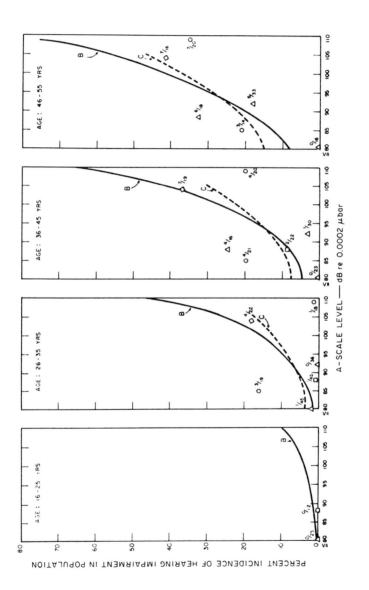

FIGURE 98. Proposed relationships between incidence of hearing impairment and exposure level to noise at work in different age groups. The hearing impairment plotted is that specified by AA00, i.e., greater than 20 dB at 500 Hz, 25 dB at 1000 Hz, and 30 dB at 2000 Hz re ISO and any level above 2000 Hz. Curve B is based on a survey made by Baughn (47), and curve C is based on a survey made by Botsford (83). The individual points represent incidence data obtained by Cohen *et al.* (144). From Cohen *et al.* (144).

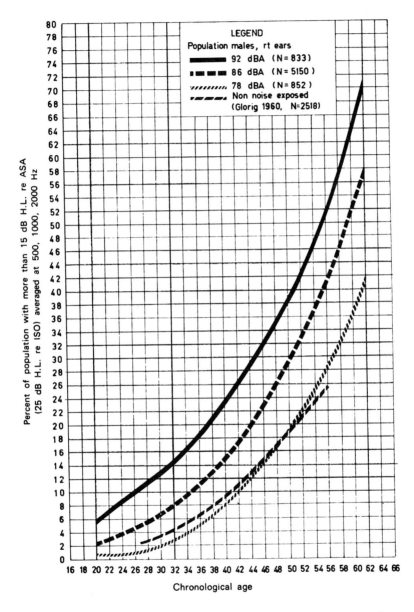

FIGURE 99. Shows the percentage of 6835 industrial workers having as much or more specified hearing impairment as a function of age and noise exposure level. From Baughn (47).

frequency bands. Since many industrial noises tend to be low frequency, and hearing losses above 2000 Hz are ignored in the formulation of damage risk in terms of workmen's compensation criteria, it is not surprising that the use of dB(A) is seemingly adequate for that purpose. Robinson (686) found that it was not possible to show any striking correlations between the spectral shape of a group of industrial noises and the locus of hearing loss suffered by workers; this perhaps unexpected result was probably due to the homogeniety of the spectra of the noises present for the majority of workers involved.

It might be noted, as will be discussed more fully in Chapter 6 that dB(D) and dB(A) rate sound frequencies up to about 1000 Hz about the same and would therefore evaluate damage risk from broadband industrial noise to hearing for frequencies up to about 2000 Hz equally well (see curves A and D_2 on Fig. 8). In addition, dB(D) would more properly rate than would dB(A) damage risk to hearing at frequencies above 2000 Hz to noise that had its predominate energy at frequencies above 1000 Hz.

Broadband vs. Narrow Band Noise and Pure Tones

If the critical band concept described in Chapter 1 is valid, it would follow that the TTS or NIPTS measured at a given test frequency should be dependent solely on the sound energy in a frequency band located at or somewhat lower than the test frequency; that is, energy outside this band (unless from the upward spread from a very intense lower frequency band) should not contribute to TTS or NIPTS at the test frequency region. That this condition more or less prevails is shown by Fig. 100 and Table 17 where it is seen that the TTS at a certain frequency is not influenced by the addition of energy to parts of the noise spectrum considerably below or above that of the test frequency.

A second consequence of the critical band mode of operation is that in order to predict the auditory fatigue effects of sound one must measure the distribution of sound energy in terms of the critical bandwidth of the ear. As seen in Table 1 and Fig. 14, each third octave band or each octave band between about 355 Hz to 10,000 Hz contains critical bands in about equal proportions. Therefore, for broadband noise, it is reasonably correct to measure the relative effects on TTS and NIPTS of different noises in terms of their one-third or full octave band spectra, although the relative effects of frequency bands below 355 Hz will be somewhat overestimated by these measurements. Miller (551) questioned the validity of the critical band notion when applied to TTS because correcting the overall SPL of white noise by the critical band (and threshold of hearing) at different TTS test frequencies did not provide equal TTS effects. However, applying the critical band corrections appropriate for the ear one-half to one octave below the test frequency (the sound frequencies presumably most

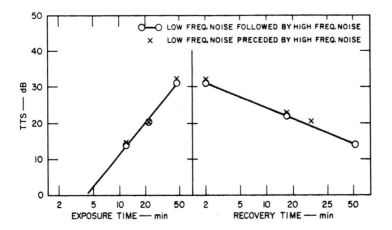

FIGURE 100.   Average TTS at 1500 and 2000 Hz (28 ears of 14 listeners). The left half of the graph shows the growth of TTS_2 (TTS 2 min after cessation of noise) upon exposure to a 600- to 1200-Hz band of noise at 110 dB SPL when preceded by silence (circles) and when preceded by exposure to 2400- to 4800-Hz noise at 100 dB SPL (crosses). The right half compares the course of recovery when the exposure is followed by quiet (crosses) or by the 2400 to 4800-Hz noise (circles). From Ward (825).

TABLE 17

TTS$_2$ After 15 Minute Exposure to Single Octave Band and
Wide Band Noises (450)

Note that octave bands presented singly had same SPL as octave
bands in wide-band noise.

Test Freq. Hz	Single Octave Band			Wide Band TTS$_2$
	Hz	SPL	TTS$_2$	TTS$_2$
1000	300–600	115 dB	10 dB	10 dB
1500	600–1200	107 dB	15 dB	16 dB
3500	1200–2400	98 dB	17 dB	14 dB
6000	2400–4800	85 dB	10 dB	10 dB
Average Difference Single Octave versus Wide Band ≤ 1 dB.				

responsible for the TTS) provides better consistency between Miller's test results and the critical band concept.

Sounds or noises containing tones or narrow (less than critical) bands of energy that exceed the energy in neighboring critical bands by more than 3 dB are not necessarily correctly measured, for purposes of estimating their auditory effects, by full octave band filters. This follows, of course, from the fact that, particularly in the frequency region above 355 Hz, the sound pressure in the critical band most influential on a particular test frequency can be 5 dB greater when the energy is confined to a single critical band than when the energy is distributed uniformly over the full octave band (in either case, the sound pressure as measured with a full octave band filter is the same). It is for this reason that the spectra of noises, above about 355 Hz, with strong pure-tone or narrow band components should be measured with band filters at least no wider than one-third octave or by adjusting full octave band measurements to make proper allowance for the presence of such pure-tone or narrow band components within a given octave band. It has been suggested that the octave band levels requiring such adjustment can be identified by the rule-of-thumb that any band that exceeds its neighbors by more than 3 dB should be considered as containing energy concentrated in pure tones or very narrow bands (Kryter *et al.* [478]).

Figure 101 shows some experimental results that substantiate the possible need to make allowance for the effects of pure tones or narrow bands of sound when octave bands are used to depict the spectra of a sound. The findings are in essential agreement with experiments conducted by Cohen and Bauman (142) except at the highest test frequency. Cohen and Bauman found that TTS was

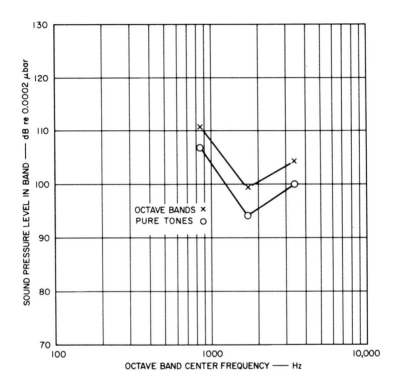

FIGURE 101.   Sound pressure levels, earphone listening, for octave bands and pure tones required to produce 15 dB TTS_2 from 10 min exposure. After Carter and Kryter (126).

proportional to octave band sound pressure levels regardless of whether or not a pure tone was present in bands above 2000 Hz, and they surmise that their results may have been influenced by action of the aural reflex.

Ward (828) suggests that the aural reflex, which causes attenuation of frequencies below 2000 Hz but which is more responsive to high frequency tones than lower frequency tones, makes the specification of a single pure-tone correction factor very unlikely. It appears probably that both the narrow band distribution of energy in a sound and the aural reflex activity must be considered in any detail understanding or explanation of the TTS and NIPTS from noise. Because of their complexity, the effects of the aural reflex are usually not considered in estimating damage risk to hearing from noise.

Susceptibility to NIPTS and TTS

It is reasonable to expect that the ear most susceptible to temporary auditory fatigue, other things being equal, is the most likely to suffer some permanent damage. Many tests of temporary auditory fatigue have been proposed with the view in mind of developing procedures for screening persons to be placed in noisy industrial environments. There have been several recent reviews of these efforts (Summerfield, et al. [787] and Ward [840]; see Table 18 after Ward). In general, while the relations between noise exposure and TTS, and noise exposure and NIPTS may be similar, it has been difficult, if not impossible, to demonstrate that the persons or animals most susceptible to TTS are likewise the most susceptible to NIPTS. One possible reason, as Griessen (327) and later Ward (840) found, is that, within certain limits, susceptibility to TTS within individuals from a given tone or band of noise is not too highly correlated with the TTS found from exposure to a different tone or band of noise. A possible, at least partial, explanation for this lack of correlation is that the HL at different frequencies could have been, for the different so-called normal subjects, the result of previous exposures to noise prior to the experiments involved, i.e., a certain amount of permanent threshold shift had already occurred at some selected frequency regions in different individuals due to the particular type of noise exposures they had had in their youth or due to minor otological disorders. Ward (840) in his study of susceptibility to auditory threshold shifts presumes, as do most investigators, that the subjects have no prior NIPTS if they have preexposure HLs that are within 25 dB re ISO normal. The fact that the standard deviation of the interobserver TTS difference was of the order of 8 dB, and the range of preexposure NIPTS could have been of the order of 35 dB (the range of "normals" from −10 to 25 HL re ISO, particularly at the higher frequencies where NIPTS usually first occurs) makes this assumption open to question.

TABLE 18

Proposed Susceptibility Tests Involving Temporary Threshold Shifts. From Ward (840).

Report	Exposure			Recovery Time (min)	Test Frequency (kHz)
	Stimulus (kHz)	Level* (dB)	Duration (min)		
Peyser (1940)	0.25	80 (HL)	0.5	0.5	0.25
Wilson (1943)	0.25	80 HL	5	1	octaves of 0.25
Peyser (1943)	1	100 (HL)	3	0.25	1
Theilgaard (1949)	0.5, 1, 2, & 4	100 HL	5	5	half-octave above exp.
Theilgaard (1951)	1	100 HL	5	5	1.5
Tanner (1955)	1	100 HL	5	"Immediately"	1
Theilgaard, according to Greisen**	1.5	100 HL	5	5	2
Wilson (1944)	2	80 HL	8	1	octaves of 0.25
Harris (1954)	2	97 SPL	5	Parameter	4
Palva (1958)	2	30 SL	3	2	2
van Dishoeck (1956)	2.5	100 (HL)	3	0.25	all (sweep)
Greisen (1951)	3	80 & 90 HL	5	5	4
Jerger and Carhart (1955)	3	105 SPL	1	Parameter	4
Jerger and Carhart (1956)	3	100 SPL	1	Parameter	4.5
Wheeler (1950)	Noise	105 SPL	30	Parameter	2, 4, 6
Gallegher and Goodwin (1952)	Noise	115 HF***	10	"Immediately"	2, 4, 6
Ruedi (1954)	Noise	Parameter	2	2	4
Falconnet et al. (1955)	Noise	100 SPL	3	Parameter	3
Christiansen (1956)	Noise	105 (HL)	3	0.5, 15	4
Ward (1967)	Noise, .7 - 5.6	120 SPL (Monaural)	1	2	1.7 to 5.6
	Noise, .7 - 5.6	106 SPL (Binaural)	15	2	1.7 to 5.6
Harris (1967)	Noise	110 SPL	1	2	1
	Noise	110 SPL	3	2	4
	Noise	110 SPL	10	2	4
	4	90 SPL	5	2	4
	4	90 SPL	25	2	4
	1	110 SPL	1	2	1

* SPL=Sound Pressure Level (dB re 0.0002 dyne/cm² rms pressure). SL=Sensation Level (dB above the individual listener's threshold). HL=Hearing Level (audiometer dial). Parentheses indicate that the article merely stated "dB" (no reference level given).
**Greisen says Theilgaard used 1500 Hz as a fatiguer, but Theilgaard's published reports indicate only 1000 Hz .
*** "115 dB above normal threshold.

Ward (840) and Harris (349) have recently completed rather extensive examinations of correlations of primarily TTS in subjects with normal hearing who had been exposed to a variety of tests for susceptibility to auditory fatigue and found there to be a number of measurable contributing factors. Ward concluded that a person could have different, in terms of TTS, susceptibilities to the following bands of noise: (*a*) 700-1400 Hz, (*b*) 1400-2800 Hz, and (*c*) 2800-5600 Hz. The susceptibilities could be measured, according to Ward, by sex and by TTS_2 at 1700, 2000, 2400, 2800, 4000, 5600, 6000, and 8000 Hz following a 3-minute monaural and a 15-minute binaural exposure to a broad-band noise encompassing the three frequency bands specified - 700 to 5600 Hz. Harris recommends, for testing susceptibility to auditory fatigue, the following: (*a*) a pure-tone of 4000 Hz at 90 dB for five minutes, and again for 25 minutes with TTS_2 measured at 4000 Hz, and a pure tone of 1000 Hz at 110 dB for one minute with TTS_2 measured at 1000 Hz; and (*b*) white noise at 110 dB for ten minutes with TTS_2 measured at 4000 Hz; white noise at 110 dB for one minute with TTS_2 measured at 1000 Hz; and white noise at 110 dB for three minutes with TTS_2 measured at 4000 Hz.

Ward and Harris found measures other than those outlined above, e.g., pre-exposure threshold at 1000 and 1400 Hz, $TTS_{0.5}$ at 4000 Hz from a band of noise of 1400-2800 Hz, to contribute only some small amount of unique information to total susceptibility to TTS.

Some human and animal studies on NIPTS suggest that susceptibility is possibly too influenced by varying general health factors within a subject to make strong relations between TTS and NIPTS likely. Indeed, patterns of NIPTS sometimes do not follow exactly precoursing patterns of TTS. For example, Miller *et al.* (554) found in cats that the frequency of the eventual maximum NIPTS occurred about one octave below the frequency of maximum TTS and, as mentioned earlier, Gravendeel and Plomp (316) suggest that somewhat similar findings occur in humans exposed to gun noise. However, the present studies on TTS and NIPTS in animals are probably not too helpful in answering this question. Ward and Nelson (844) and Miller *et al.* (554) created in animal subjects rather large degrees of NIPTS (ranging at some test frequencies from about 50 to 100 dB) in such relatively brief periods of time that the exposures can hardly be taken as sensitive indicators of differences in susceptibility to more moderate amounts of TTS and NIPTS. Just as a noise that creates no TTS is no measure of susceptibility to TTS or NIPTS, a noise that creates a TTS_2 of 40-50 dB or more is not necessarily a good measure of susceptibility to the smaller TTS_2 and NIPTS values of concern in most industrial noise environments. As Miller *et al.* note, such experiments would be completely pertinent only if the NIPTS was more than 30-40 dB and was achieved by long-term, daily exposures to the same noise exposure used to obtain TTS_2.

It is intuitively obvious, as various investigators point out, that a person with a high preexposure HL has less hearing to lose and will not show as much TTS as a person with good hearing; that this is generally true is shown in Fig. 102. Sataloff *et al.* (718), with a large and varied group of subjects with NIPTS, obtained results very similar to those shown in Fig. 102. Also, Ward (840) found that preexposure HL at some frequencies was negatively related to TTS.

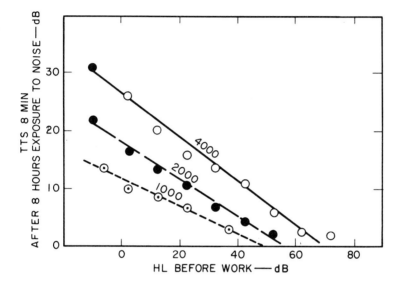

FIGURE 102.   Relation between TTS and preexposure hearing level (HL). From Glorig *et al.* (303). (With permission of the Controller of her Britanic Majesty's Stationery Office.)

Except for tests of ear overload distortion proposed by Lawrence and Blanchard (486), nearly all the tests of susceptibility have been some variation of the magnitude of threshold shift, or the time to recover from the shift. However, it would seem that the sum of the absolute Hearing Level of a subject two minutes after exposure to a noise, HL_2, plus TTS_2 (HL_2-HL preexposure) would logically be the best measure of susceptibility to NIPTS. A low value for these two measures could probably be taken to mean that the person had ears that had resisted NIPTS in the past (had a low HL_2) and would continue to resist NIPTS in the future (had a low TTS_2). A high postexposure HL and low TTS would imply that the person probably had susceptibility to NIPTS for that noise equal to a person with a somewhat lower HL but a larger TTS.

We were impressed to find, when measuring TTS to gun noise (Kryter and Garinther [463]), that some older, experienced soldiers with low HLs showed small TTS to gun noise that caused large TTS in some younger men and in some older soldiers with relatively high HLs. Parenthetically, it would appear from a study conducted by Loeb and Fletcher (521) that older persons are not necessarily more susceptible than younger persons to noise-induced threshold shift, as was suggested by Kryter (446).

It might be important to note that if an ear has differential susceptibility to NIPTS for different types of sounds, a difference could perhaps be expected for nonimpulsive steady-state vs. implusive sounds. In general, the variability of TTS to impulsive sounds for a given group of subjects appears to be about twice that for steady-state sounds (Carter and Kryter [128], Fletcher [240]), and it is probable that the aural reflex is more involved in the protection of the ear against implusive sounds than steady-state ones. Ward (827) found the variability of his subjects to impulsive sounds (clicks) to be so great that no meaningful conclusions could be reached regarding susceptibility. Hodge and McCommons (383) found that, while TTS for individuals from impulses was too unreliable to permit generalizations regarding impluse-noise effects, group (12 to 29 subjects) means and standard deviations varied only slightly with repeat tests of the same impluse noise condition; however, the impluse noise conditions for these tests did not cause a significant average amount of TTS and this could have prevented the observation of some underlying differences in susceptibility to more intense stimulation. The variability for TTS_2 as found by Ward for exposure to steady-state sounds is given in Table 19.

Stapedectomized Ears

Persons with stapedectomized ears (an operation in which part of the ossicular chain is replaced by wire or plastic and the oval windows of the cochlea cleared of otosclerotic bone) are not apparently more susceptible than normal hearing persons to TTS from, a least, high-frequency noise (Fletcher and King [248]).

TABLE 19

Unbiased Estimates of Standard Deviations of Differences in Threshold Shifts (Pre-minus Post-exposure Thresholds as Measured on the Same Day) and Shifted Thresholds (Average or Pre-exposed Thresholds Taken for a Number of Days Minus Post-exposure Measured on One Day) between Weeks 5 and 6 (Test-Retest). From Ward (840).

Test Frequency	Phone Exposure		Field Exposure	
	Threshold Shift	Shifted Threshold	Threshold Shift	Shifted Threshold
2 Hz	4.1 dB	2.7 dB	4.35 dB	3.35 dB
2.8	4.3	2.55	4.25	3.35
4	3.9	4.5	5.1	3.6
5.6	4.55	6.85	7.0	5.15

Vitamin A and Drugs

There has been some controversy as to the effect of the ingestion of Vitamin A upon the resistance of the ear to auditory fatigue as the result of exposure to noise. In carefully controlled studies it is usually found that the taking of Vitamin A does not of itself decrease susceptibility to noise-induced temporary threshold shifts (see Ward and Glorig [842] and Ward [839, 840]). It is clear, however, that severe oxygen deprivation and certain drugs such as quinine and ototoxic-mycins can cause sensory hearing losses, particularly at high frequencies (see Lenhardt [495]).

Summary of Discussion of Susceptibility

Proving a strong correlation between the results from susceptibility tests and eventual NIPTS in industry is probably an impossible task, if for no other reason than that the noise in industry may be but one of the noises to which men are exposed in their daily lives, thereby introducing some uncertainty and variability in the data. However, this does not mean that some persons do not have ears that are generally more resistant to NIPTS than other persons, or that under some circumstances testing and screening persons for this ability or lack thereof would not be worthwhile.

It would seem that the pure-tone and broadband noise tests of TTS_2 proposed by Ward and Harris, plus a TTS_2 test for impulsive sounds (if a practical one could be developed) would be appropriate for evaluating possible, it not probable, susceptibility to NIPTS. Further, it would seem logical to score these tests in terms of HL_2 plus TTS_2 as an index of susceptibility. The variability in an individual to repeat tests of TTS for a given noise suggests that susceptibility tests bear repeating under carefully controlled conditions and that both the tests and retest results be somehow combined.

It appears (at least as measured in persons with some possible small unknown degree of previous NIPTS) that the sensitivity in a person to develop a TTS from one frequency band of noise does not mean he will be equally sensitive to a different frequency band of noise. This does not imply, of course, that the pattern of TTS shown by a given ear to a given noise will not develop a similar pattern of NIPTS with long-term continued exposure to the same noise. Indeed the similarities between TTS and NIPTS indicate that, on the average, this is highly probable.

Finally, it follows from this inability to identify individual susceptibility to NIPTS, that the setting of tolerable limits for damage risk exposures to industrial and environmental noise must be based, within reason, on the general population statistics of NIPTS that have been gathered to date. The hypothesis that there might be some identifiable abnormal biological or temporary physiological weakness on the part of those persons developing NIPTS is without foundation at the present time.

Damage Risk Contours

Much of the TTS and NIPTS data cited above was used by a CHABA Working Group (478) to specify conditions for hazardous exposure to intermittent and steady-state noise. In those specifications special allowance was made for the protection of frequency regions important to speech reception. Also, some deference was made by the CHABA group to the AAOO procedures and the general Workmen's Compensation practices in the United States. The criterion of damage risk proposed by the CHABA group was that at least 50% of the people exposed nearly daily for 10 years to a noise environment should not suffer more NIPTS than 10 dB at or below 1000 Hz, 15 dB at 2000 Hz, and 20 dB at or above 3000 Hz. It was estimated that that criterion would be met as the result of exposure of groups of people to the sound conditions, called Damage Risk Contours (DRC), specified in Fig. 103.

Ward *et al.* (847-850) have developed rules for equinoxious intermittent noise exposures. These rules were applied, as shown in Fig. 104, to the CHABA Damage Risk Contours for Octave Bands of Fig. 103. It is important to note that

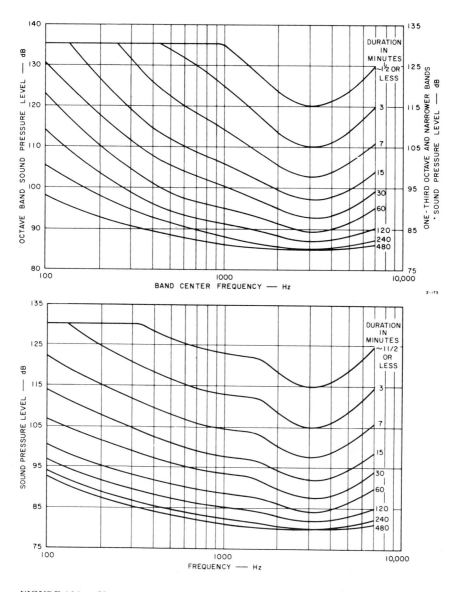

FIGURE 103. Upper graph: Damage-risk contours for one exposure per day to full octave (left-hand ordinate) and one-third octave or narrower (right-hand ordinate) bands of noise. This graph can be applied to the individual band levels present in broad-band noise. Lower graph: Damage-risk contours for one exposure per day to pure tones. From Kryter *et al.* (478).

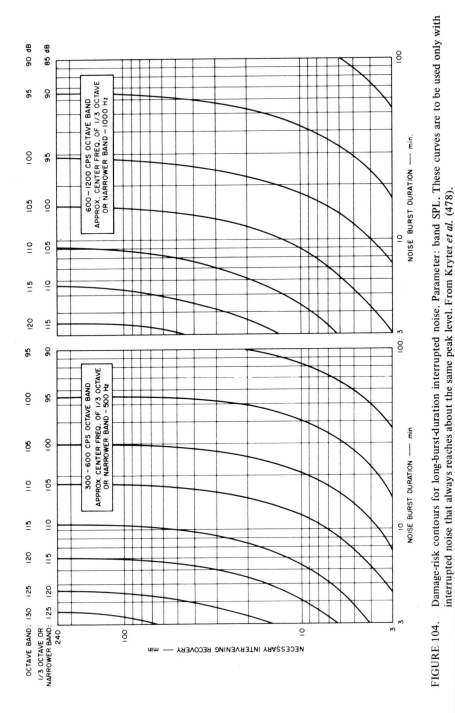

FIGURE 104.   Damage-risk contours for long-burst-duration interrupted noise. Parameter: band SPL. These curves are to be used only with interrupted noise that always reaches about the same peak level. From Kryter *et al.* (478).

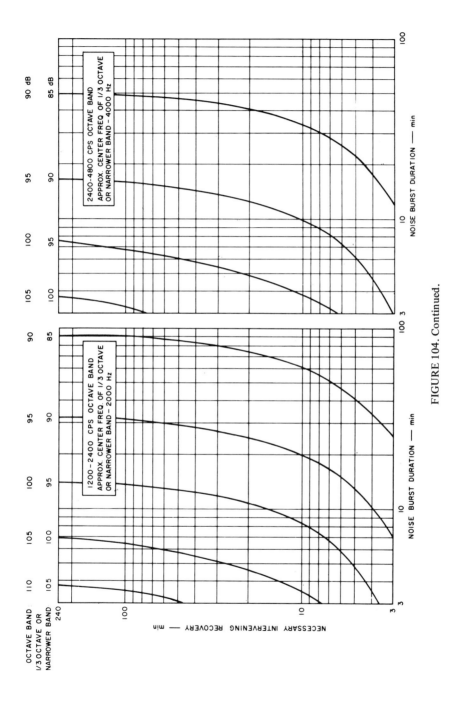

FIGURE 104. Continued.

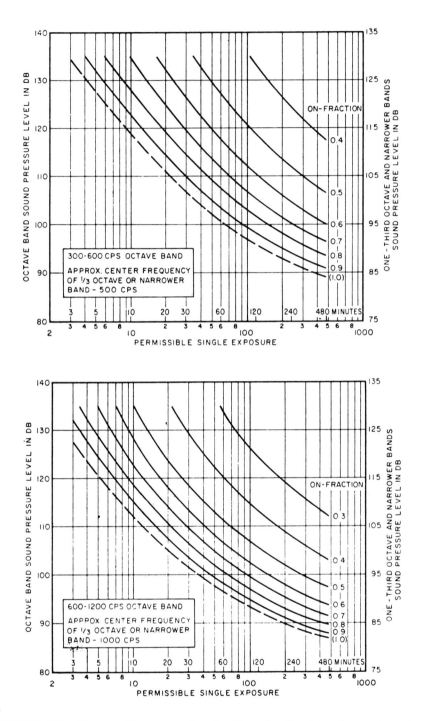

FIGURE 105. Damage risk contours for short-burst-duration intermitten noise – noise bursts 2 min or less in duration. From Kryter *et al.* (478).

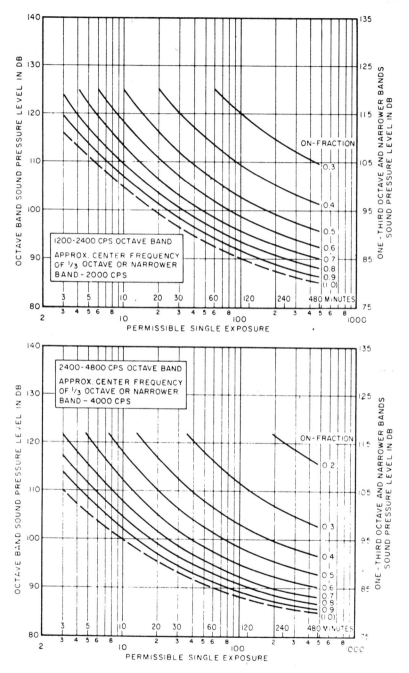

FIGURE 105. Continued.

the damage risk contours of Fig. 104 are applicable only when the individual noise bursts are of equal duration, do not exceed two minutes, and do not consist of pure tones, i.e., must be a band of more or less random noise. Figure 105 give the CHABA damage risk contour for short-burst duration noise (noise bursts less than two minutes long).

. Botsford (82) found that 80% of some 75 noises from various manufacturing industries (Karplus and Bonvallet [429]) had relative levels in dB(A) that were the same as the relative levels of damage risk to be found by comparing the octave band levels of each of the noises with the damage risk contours of Fig. 103. Botsford concluded that at least 80% of the time one would get the same estimation of damage risk to hearing from dB(A) measures as from octave band measures of typical industrial noises. Accordingly, he suggests, as had Flanagan and Guttman (229) previously, that overall dB(A) measures, rather than octave band spectra of noises, would often be adequate for estimating damage risk to hearing; this assumption makes possible the combination of Figs. 103 and 104 into Fig. 106. Botsford makes a valuable addition over Fig. 104 by extending the damage risk contours to on-off intervals of less than two minutes (damage risks to noise of less than two minute's duration were illustrated by a separate set of figures in the CHABA report).

The question of using dB(A) or some other overall measure such as dB(D), rather than octave band spectra, for evaluating damage risk from exposure to noise, assuming the CHABA damage risk contours are accurate, revolves to some extent around whether one is particularly concerned with a specific noise environment or with a description of the average damage risk of a number of noise conditions. The overall frequency measure of dB(A) underestimated the damage risk of 4% and overestimated the damage risk of 16% of the noises studied by Botsford. From a hearing conservation point of view, dB(A) could usually err, when it does, in overprotecting the hearing of the workers, a laudable goal. However, from the practical, economic point of view, the use of dB(A) or dB(D) as a yardstick could result in about 16% of the time in the application of possibly expensive and unnecessary noise control or hearing protective procedures. In addition, one cannot overlook (a) the 4% of noises for which the damage risk was underestimated, or (b) the possibilities of there occurring narrow band noises that are possibly more damaging to hearing than is estimated by overall dB(A) or dB(D).

During the past decade there have been many so-called "Damage Risk Criteria" or "Tolerable Noise Exposures" specified. Some have been somewhat ad hoc with no attempt to state fundamental assumptions followed in their derivations, and many have been influenced (as indeed to some extent were the CHABA recommendations) by practical considerations in deference to industry and the military services and workman's compensation practices rather than full hearing conservation. Somewhat similar sets of recommendations for tolerable

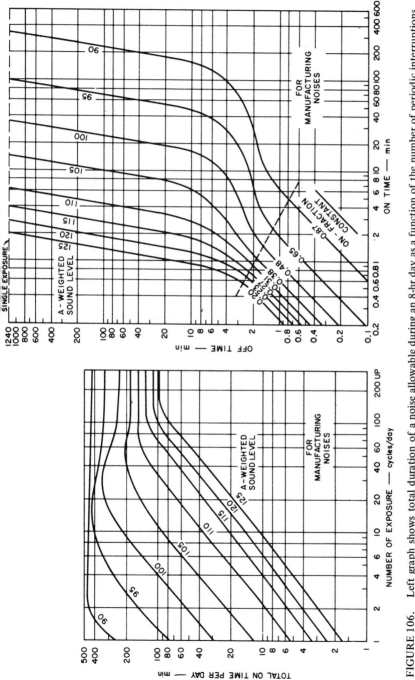

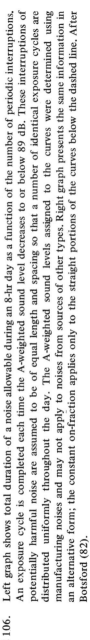

FIGURE 106. Left graph shows total duration of a noise allowable during an 8-hr day as a function of the number of periodic interruptions. An exposure cycle is completed each time the A-weighted sound level decreases to or below 89 dB. These interruptions of potentially harmful noise are assumed to be of equal length and spacing so that a number of identical exposure cycles are distributed uniformly throughout the day. The A-weighted sound levels assigned to the curves were determined using manufacturing noises and may not apply to noises from sources of other types. Right graph presents the same information in an alternative form; the constant on-fraction applies only to the straight portions of the curves below the dashed line. After Botsford (82).

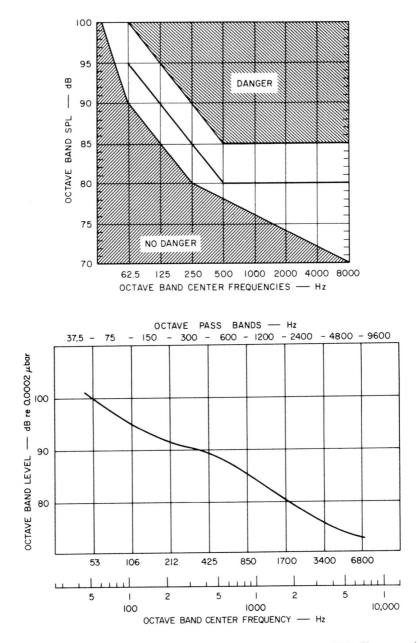

FIGURE 107.   Lower graph: Permissible noise levels. From Slavin (747). Upper graph: Permissible noise levels according to French Ministry of Health (508).

exposures have emerged in many countries, for example, Fig. 107 shows tolerable levels published by French (508) and Russian (747) investigators. The Russian contours appear to be somewhat higher than the French (more like the CHABA contours) but as Slavin, the Russian investigator, states, his contours are meant to be a practical balance between hearing conservation and necessary industrial noise conditions. The next chapter will return to the question of specifications of damage risk exposures to noise and will propose a set of new procedures for the evaluation of noise in this regard.

Implusive Noise

The CHABA Damage Risk Contours given above are specifically restricted to so-called steady-state noise. The specification of auditory damage risk from impulsive noise would be equally important and useful since NIPTS is common in persons exposed to gunfire and to the impulsive sounds to be found in industry.

A number of research studies of the effects of gun noise on TTS have been performed since the classical work of Murray and Reid (571, 572). It is difficult, however, to closely intercompare the results of these studies because of the differences in the number of rounds fired, the interval between last exposure and start of audiometric tests, and the preexposure hearing levels of the men tested. In addition, the relations between TTS and NIPTS have been hard to quantify because of (a) the often variable and unknown number of impulses to which people are exposed, (b) the problems of measuring impulse intensity, (c) the apparent great daily variability within an individual to TTS from impulses, and (d) the difficulty of simulating impulse noise in the laboratory for purposes of studying their effects on TTS. Nevertheless, several sets of tolerable limits have been proposed. In the immediate sections to follow, some presently proposed tolerable limits for impulses and related research data will be presented. From this information the basis for a new set of procedures for the evaluation of damage risk to impulse noise is derived.

U.S. Army Study

From some 178 subjects an extensive amount of TTS data were collected by Kryter and Garinther (463) under reasonably controlled conditions and with systematic variation in number and intensity of impulses from guns. Figures 108 and 109, and Table 20 summarize the results of this study. The waveforms of the four gun noises are shown in Fig. 110. It should be noted that Kryter and Garinther used HL_2 (HL two minutes postexposure) re ASA 1951 rather than

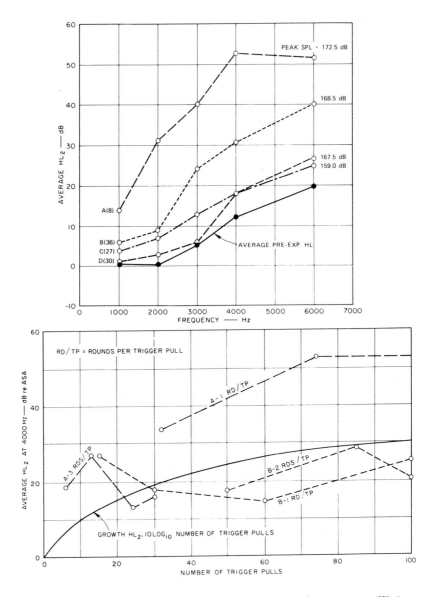

FIGURE 108.   Upper graph: Average hearing level 2 min after exposure (HL$_2$) as a function of the audiometric test frequency following 97-102 trigger pulls of each weapon. The average preexposure HL for all the ears exposed to these firing conditions is also shown. The numbers in parentheses indicate the number of ears averaged for each weapon. Lower graph: Average HL$_2$ at 4000 Hz following exposure to single and multiple rounds-per-trigger pull. Parameter is weapon and rounds-per-trigger pull. After Kryter and Garinther (463).

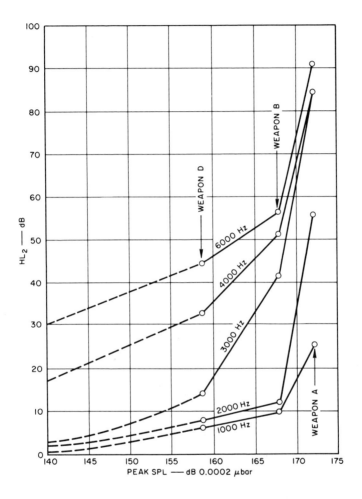

FIGURE 109.   HL$_2$ for Q$_3$ as a function of peak SPL. 100 trigger pulls, one round per pull, at the rate of one every 5 sec. Parameter is audiometric test frequency (see Table 20). From Kryter and Garinther (463).

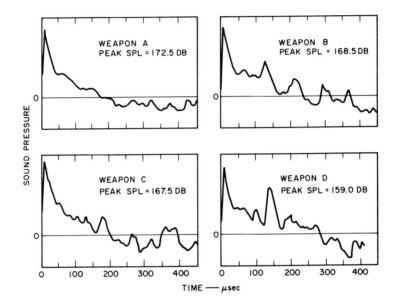

FIGURE 110.   Pressure waveforms and peak sound pressure level for the four types of weapons. The peak levels of the waveforms have been adjusted to have approximately equal peak amplitude. From Kryter and Garinther (463).

TABLE 20

Q_3 (75% Percentile) for HL_2 and Peak SPL for the Different Weapons, Grazing Incidence of Impulses to Ears. From Kryter and Garinther (463).

Weapon	Peak SPL (dB re 0.0002 μbar)	No. Trigger Pulls	Test Frequencies in Hz						Average at	
			500[a]	1000	2000	3000	4000	6000	500 1000 2000	1000 2000 3000
A	172.5	102	11[a]	26	56	84	84	91	31	55
		74	7	22	50	78	85	86	26	50
		32	0	10	16	27	60	57	9	18
					Grand Average				22	41
B	168.5	100	0	10	12	43	52	57	7	22
		60	0	8	10	22	25	29	6	13
		30	0	6	9	24	24	55	5	13
					Grand Average				6	16
C	167.5	97	0	7	10	12	14	25	6	10
		63	0	11	12	38	55	56	7	20
		23	0	12	18	51	65	73	10[b]	27[b]
					Grand Average				6	15
D	159.0	100	0	7	8	14	33	45	5	10

[a.] HL_2 for 500 Hz is estimated.

[b.] The data for weapon C are anomalous in that the fewer the number of impulses the greater was the amount of observed threshold shift. These results for weapon C are so inconsistent with the other findings of this as well as similar studies, they are not included in the further analyses of the data given in **Fig. 109.**

TTS_2 as their measure of TTS because the hearing level of the subjects (soldiers) had some small degree, on the average, of previously induced NIPTS prior to exposures to gun noise of the experimental tests (see top graph, Fig. 108).

From these data, and the assumption that TTS_2 in the normal ear will eventually lead to NIPTS, the following criteria of acceptability and tolerable limits can be suggested for about 100 rounds per day at about 5-sec intervals between rounds of rifle fire in the open field (see also Table 21):

Criteria No. 1 in terms of the AAOO criterion of an average HL (re ASA) of 15 dB or greater at 500, 1000, and 2000 Hz. A tolerable exposure for 90% of the people would be 100 rounds daily at a peak SPL of 160 dB, or at a level of 165 dB for 75% of the people.

Criteria No. 2 in terms of the criterion used by CHABA with steady-state noise of an average HL (re ASA) of 15 dB at 1000, 2000, and 3000 Hz. A tolerable exposure for 90% of the people would be 100 rounds at a peak SPL of 150 dB, or at a level of 160 dB for 75% of the people.

Criteria No. 3 in terms of good hearing for speech as well as other sounds, an average HL (re ASA) of 15 dB at 1000, 2000, 3000, and 4000 Hz. A tolerable exposure for 90% of the people would be 100 rounds daily at a peak SPL of 140 dB, or at a level of 150 dB for 75% of the people.

TABLE 21

Estimated Expected Permanent Hearing Level (ASA) Standard) to be Equaled or Exceeded in 50%, 25% and 10% of Ears Following Repeated Exposure to about 100 Rounds, at 5 Sec Intervals of the Noise from Shoulder Rifles.
From Kryter and Garinther (463).

Peak SPL's are specified at the listener's ears, grazing incidence.

Peak SPL	Test Frequency in Hz														
	1000			2000			3000			4000			6000		
	50%	25%	10%	50%	25%	10%	50%	25%	10%	50%	25%	10%	50%	25%	10%
170 dB	0	15	25	10	25	35	35	55	70	45	65	85	50	70	90
165 dB	0	9	16	0	10	20	12	32	42	25	45	60	47	52	67
160 dB	0	7	15	0	8	16	0	18	25	15	35	45	25	45	60
150 dB	0	3	10	0	4	15	0	8	15	10	25	35	20	40	50
140 dB	0	0	0	0	2	5	0	2	10	5	18	30	10	30	45

Review by Coles *et al.*

Recently Coles *et al.* (156) have reviewed TTS and NIPTS data about impulse noise. They recommend that impulse noises be divided into two types (see Fig. 111, left-hand graph) and that tolerable peak sound pressure levels for these two types would be those shown on Fig. 111, right-hand graph. By "tolerable," Coles *et al.* meant that no more than 25% of the people would have more NIPTS than 10 dB at 1000 Hz, 15 dB at 2000 Hz, and 20 dB at 3000 Hz (the CHABA criterion, except 25% rather than 50% of the people) if exposed to 50-200 impulses per day.

The use of A and B types (called A and B durations in the Coles *et al.* paper) is based on the observation by Coles *et al.* that guns fired in an enclosure or under some reverberent conditions cause more TTS or NIPTS than in free field. This is undoubtedly a valid observation but it is suggested that basic physical parameters controlling these auditory fatigue effects are not the duration types proposed but are best represented by the means of classifications proposed in Chapter 1 and illustrated in Figs. 12 and 13. In fact, the A-durations, see Fig. 111, have not been observed, to my knowledge, for gun impulses longer than a few msec or so.

It might be noted here, as will be demonstrated more clearly later, that

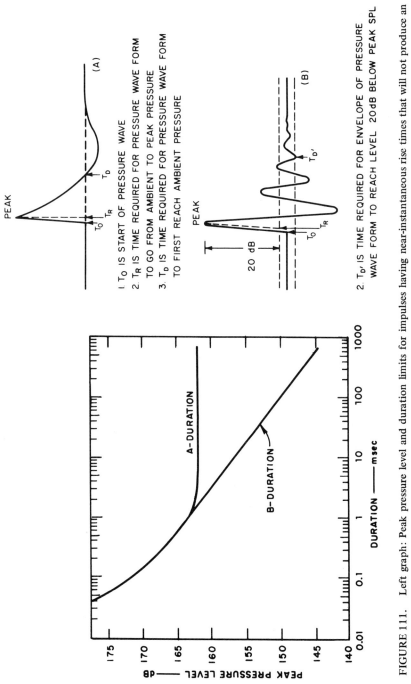
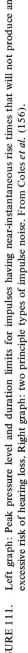

FIGURE 111.   Left graph: Peak pressure level and duration limits for impulses having near-instantaneous rise times that will not produce an excessive risk of hearing loss. Right graph: two principle types of impulse noise. From Coles *et al.* (156).

keeping the rise time and peak overpressure constant but making the impulse a form of damped sinusoid by means of reverberation, would have the effect of modifying the spectrum of the sound in various significant ways and increasing the amount of energy present at certain frequencies (see Fig. 13). It is possible that, at least for the range and type of impulse durations involved, spectra of the impulse will adequately describe their effect on TTS and NIPTS and that such a description will provide a general method for predicting the threshold shifting effects of sounds regardless of whether they are nonimplusive or impulsive.

Coles *et al.* also suggest that the ear is about 5 dB less tolerant to an impulse approaching the ear canal directly (at normal incidence) than at grazing incidence as would be the more typical case for a person firing a gun. This recommendation, according to Coles *et al.*, is made on the basis of: (*a*) measurements by Golden and Clare (306) showing that the pressure from a gunshot at the position of the eardrum in an artificial auditory canal is about 6 dB greater with normal than the grazing incidence of the wave to the opening to the canal, and (*b*) some TTS data obtained by Hodge *et al.* (385) that indicates that about 6 dB more TTS occurs from the normal than the grazing incident impluse from a gun.

Review by Ward

Subsequent to the review of Coles *et al.*, Ward (841) also proposed damage risk exposures to impulses (gunfire). His tolerable limits are very similar to those proposed by Coles *et al.* in Fig. 111 for the condition of 100 impulses per day, and 25% of the poeple experiencing NIPTS of 10 dB at 1000 Hz, 15 dB at 2000 Hz, and 20 dB at 3000 Hz. In addition, Ward proposed: (*a*) a correction of −3 dB for each doubling of impulses above 100 per day (a reduction in the tolerable levels), and an increase of 3 dB for each halving of the number of impulses below 100 per day; and (*b*) setting an upper limit of 179 dB for any type of impulse or condition of listening.

TTS as a Function of Number and Spacing of Impulses

Figure 112 from Murray and Reid, Fig. 113 from Ward *et al.*, and Fig. 108 from Kryter and Garinther show how TTS changes as the number of regularly-spaced impulses is changed. These data are interpreted to mean that TTS, at least for the frequencies from 1000 to 3000 Hz, grows linearly with $20 \log_{10}$ antilog of number of impulses or of exposure time.

The temporal spacing between the impulses has some influence on the threshold shift to be experienced. The tolerable exposures proposed for gun impulses usually presume, or are at least consistent with, intervals of 2-10 seconds

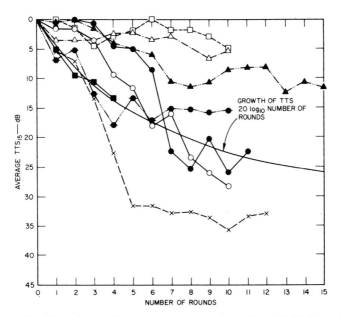

FIGURE 112.   Graph illustrating the effect of number of rounds fired on the average hearing loss (from 512 to 8192 Hz) for eight different ears in eight experiments. Firing at intervals of fifteen minutes. Peak SPL at ear, 180 dB. After Murray and Reid (571).

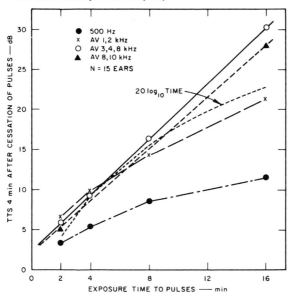

FIGURE 113.   Average growth of TTS from pulses (25 per min) as a function of exposure time. The dotted line shows the relation of TTS to $20 \log_{10}$ of exposure time. After Ward et al. (852).

TABLE 22

TTS_2 or HL_2 Average of 1, 2, and 3 kHz as Found in or Estimated
from Various Studies of Threshold Shift from Gun Noise. One – Ten Sec or so between Impulses

Study	Peak SPL Grazing to the Ear	Listening Condition	No. of Rounds	TTS_2 or HL_2 (Aver. 1, 2, and 3 kHz) Equaled or Exceeded by 25% of People		No. of Subjects
				Measured	Corrected to 100 Rounds by Adding $20 \log_{10}$ No. Rounds	
1. Coles and Rice[153]	160 dB	Open Field	10-50	7 dB	19 dB	20
2. Coles and Rice	160 dB	Reverberent	10-50	10	22	20
3. Coles and Rice	159 dB	Open Field	20-50	7*	19 dB	20
4. Coles and Rice	159 dB	Reverberent	20-50	19*	31 dB	20
5. Elwood et al.[219]	161 dB	Open Field	20	7**	20	12
6. Elwood et al.	173 dB	Open Field	1	7**	47 dB	12
7. Acton et al.[4]	138 dB	Open Field	100	0	0	19

8. Acton et al[4].	138 dB	Reverberent	100	5[***]	5	19
9. Murray and Reid[572]	159 dB	Open Field	100		20[****]	?
10. Murray and Reid	176	Open Field	10	5	27[*****]	1
11. Murray and Reid	181	Open Field		17	48	1
12. Smith and Goldstone[755]	158	Open Field	25	5 dB	17	30
13. Kryter and Garinther[463] Weapon D	159	Open Field	100	10	10	30
14. Kryter and Garinther Weapon B	168	Open Field	100	22	22	36
15. Kryter and Garinther Weapon A	173	Open Field	100	55	55	8

* TTS data were not given at 1 kHz; the Aver. for 1, 2, and 3 kHz was taken to be the TTS at 2 Hz.

** Data were given as TTS Aver. of 2–6 Hz, this Aver. multiplied by .3 was taken as TTS Aver. 1, 2, and 3 kHz.

*** Authors stated that "noise approached an auditory hazard for about 5% of people."

**** Estimated from authors' statement that 159 peak SPL "commonly" (taken to mean in 50% of people) caused 40 dB Peak TTS--after 100 rounds or more.

***** Aver. 512–8192 Hz TTS$_{15}$ corrected to TTS$_2$ by adding 10 dB (see Kryter) and then to Aver. TTS$_2$ at 1, 2, and 3 Hz by multiplying result by .5.

between rounds. It has been observed repeatedly that when the interval between impulses becomes less than one second, the TTS does not increase, and sometimes decreases, as shown in Fig. 114 from Reid, Fig. 115 from Smith and Goldstone, Fig. 116 from Ward *et al.* and, to some extent in Fig. 108 (Kryter and Garinther). Figure 117 summarizes some of the relevant data on this point. This phenomenon is undoubtedly related to action of the aural reflex and may possibly afford some protection against threshold shift from gun noise.

It is also clear that, when the interval between impulses increases beyond a certain point, the recovery from auditory fatigue between impulses permits a net decrease in TTS for a given total number of impulses. There is, however, conflicting data as to what is the exact interval at which recovery from threshold shift first starts. For example, Ward *et al.* (852) found in one study that impulses separated by slightly more than two seconds caused less TTS than when the interval was less than one second (see Fig. 116), but, in a second study with the same impulses, Ward (827) concludes that intervals between impulses as long as nine seconds do not influence TTS; see Fig. 118. The weight of the evidence suggests that perhaps only after about five seconds of relief from intense auditory stimulation does the ear start recovering from auditory fatigue.

TTS as a Function of Peak SPL of Impluses

Table 22 presents a summary of much of the data that can be used to show or estimate the average amount of threshold shift at 1000, 2000, and 3000 Hz for 100 gun noise impulses with 1-10 seconds between impulses. In order to interpret the data from the various studies it is necessary to convert, as best as can be done, the threshold shift data to a common set of exposure conditions; the rules for doing so, when required, are given in Table 22. Although the validity of such conversions may be questioned, their use permits, in my opinion a more meaningful and effective use of the data then is otherwise possible.

For gun noise, the relation of peak SPL to TTS, to a first approximation, can perhaps be represented by a straight line up to SPLs of about 170 dB at which point TTS appears to grow at a much accelerated rate. Such functions are shown in Fig. 119 along with data points taken from Table 22. The long solid curves drawn on Fig. 119 and the related rules in the caption appear to be a reasonable description, solely in terms of peak SPL, of apparent tolerable exposures to impulses.

Except for a few data points, the curves drawn in Fig. 119 tend to underestimate the threshold-shifting effects of the gun noise. Even so, the tolerable limits of peak SPL suggested by Coles *et al.* and by Ward are yet 8 dB higher than those derived herein for the same criterion, as shown in Fig. 119. Some of this difference may be due to the fact that although Coles *et al.* speak of PTS

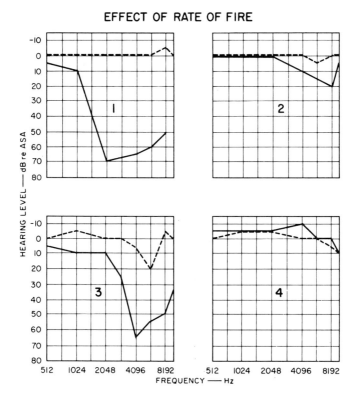

EFFECT OF RATE OF FIRE

FIGURE 114. Graphs illustrating the difference in the deafness produced after exposure to 28 rounds of Bren 0.303 fired rapidly (approximately 500/min) and after exposure to 28 rounds fired at intervals of from 10-20 sec. The former is shown by broken lines, and the latter by full lines. For graphs (1), (2), and (3) the test was made 10-20 min after completion of exposure and for graph (4), 2-7 min after.

(1) Subject D.G.–left ear.   (3) Subject Miss A.M.L.–right ear.
(2) Subject G.R.–left ear.   (4) Subject J.M.–right ear.

Preexposure thresholds are represented by the horizontal full lines through 0 dB. From Reid (667).

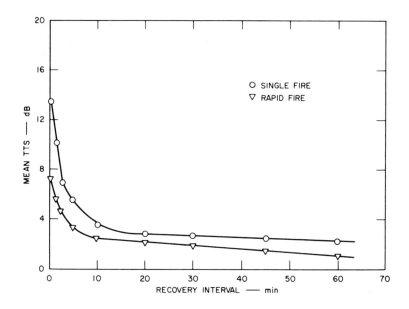

FIGURE 115.   Mean temporary threshold shift as a function of rate of impulse. From Smith and Goldstone (755).

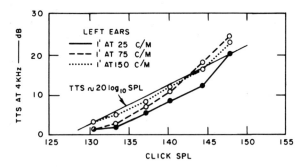

FIGURE 116.   Average curves showing growth of TTS 30 sec at 4000 Hz with three different rates of click per minute (C/M). After Ward *et al.* (852).

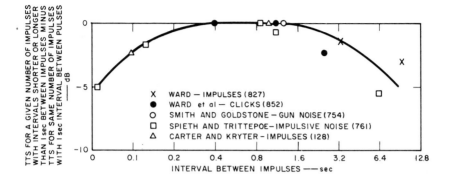

FIGURE 117. Showing that very short or relatively long intervals between impulses reduces, for a given number of impulses, the amount of TTS.

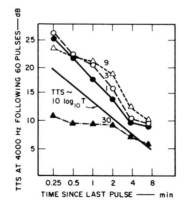

FIGURE 118. Recovery time of the TTS at 4000 Hz produced by 60 high-intensity pulses. Interpulse interval, in seconds, is the parameter. After Ward (827).

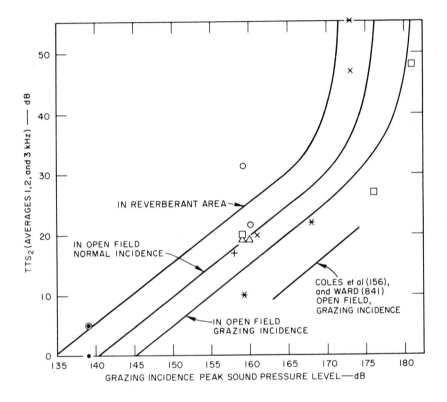

FIGURE 119.   Data points are TTS$_2$ (average at 1, 2, and 3 kHz) equaled or exceeded by
25% of people from 100 gun impulses per day spaced 1 sec or more apart
(see Table 22). For each doubling of numbered rounds (e.g. 100 to 200,
200 to 400, etc.) the peak SPL must be reduced by 6 dB to maintain TTS
relations shown. If more than one impulse occurs per second, the exposure
can be considered as only one impulse. The long solid lines represent
functions suggested herein for specifying tolerable exposure levels for a
given TTS$_2$ (average 1, 2, and 3 kHz). The short line gives tolerable limits
for open field, grazing incidence of 100 per day gun impulses as prescribed
by Coles *et al.* (156) and Ward (841).

from preexposure thresholds in presumably normal young ears, much of the data on which they base their conclusions may have been from TTS in military personnel with some degree of PTS due to previous exposure to intense noises. If so, the preexposure vs. postexposure audiograms (TTS) would underestimate the amount of threshold shift that would occur to the normal ear.

It should be fully understood that gun noise, because of its spectral characteristics, which equals but does not exceed the tolerable damage risk levels proposed by either Coles *et al.* by Ward, or by us in Fig. 119, will result in hearing losses in some people in excess of at least 25 dB at frequencies above 3000 Hz.

TTS and NIPTS for Impulses

It has been possible to demonstrate with steady-state noise some reasonable degree of substantiation for the hypothesis that TTS_2 in the normal ear can be used to predict NIPTS. There seems to be no good reason to presume that the same relations do not exist with impulses, at least for impulses that cause TTS_2 values up to about 40 dB, but uncertainty about previous exposures to gun and steady-state noises experienced by military personnel makes proof of this relation difficult. However, it would appear that the TTS_2-equals-NIPTS assumption, may lead to an underestimation of the degree of NITPS to be suffered from gun impulses, as is illustrated in Fig. 120. Again, a maximum 8-12 hr. daily exposure period is assumed.

Relation between Threshold Shifts and Spectrum of Impulses

The difficulties in the past of making a physical spectral analysis of impulses has impeded the determination of possible relations between the spectra of impulses and the degree of locus on the frequency scale of any threshold shifts resulting from exposure to the impulses. However, there are techniques now generally available for measuring, computing, or estimating the spectra of impulsive sounds.

Hecker and Kryter (359) reported the acoustic waveform and spectra of impluses (presented by means of loudspeaker and earphone) used in two different laboratory studies of TTS (see Fig. 121). The measured spectra on Fig. 121 were determined by rather complex means whereby the waveform, as photographed from an oscilloscope, was periodically scanned electronically and the resulting modulated wave then passed through an envelope detection device and band-pass filter. The procedures described in Chapter 1 were used for obtaining, from knowledge of the rise time, duration, and peak SPL of the pulses, an approximation of the spectra of the impulses, as is also shown in Fig.

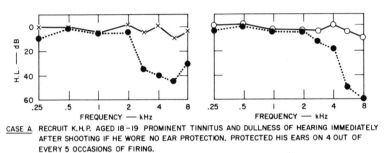

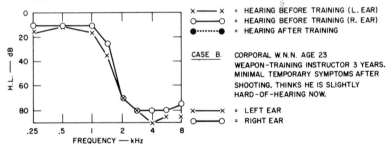

FIGURE 120.   Typical audiograms (hearing level, HL, is shown relative to British
Standard [2497] normal threshold for hearing of pure tones). From Coles
and Rice (153).

121. Kryter and Garinther also reported the spectra of the gun noises (see Fig. 110 for waveforms) used in their experiments, as shown in Fig. 122. These measured spectra were determined by the same means and equipment employed for the impulse spectra of Fig. 121. Also shown on Fig. 122 are the approximate spectra for these impulses as estimated by the use of Figs. 12 and 13 of Chapter 1. There appears to be reasonable agreement for these studies between the spectra obtained by actual measurement and by estimation from the graphs in Figs. 12 and 13.

It is interesting to note that the frequencies at which maximum threshold shift, relative to the normal threshold of hearing, occurred from the gun noises (at 4000-6000 Hz, see Fig. 108) are consistent with the observation that the maximum threshold shift for steady-state sounds occurs about one octave above the frequency band containing the greatest energy (around 2000 Hz, see Fig. 122). Also it would follow that since the impulses in the Kryter and Garinther study are probably representative of shoulder and hand guns with respect to rise time and duration, the general shape of the sound spectra and the threshold shifts, for a given peak SPL and number of firings, therefore should be similar for most guns; this is indeed the case as witnessed by Figs. 108, 114, and 120.

The importance of the spectral distribution of energy in impulses to threshold shift is well illustrated by a study performed by Fletcher and Loeb (253). The impulses, generated by an electrical spark gap, had the waveforms shown in Fig. 123. The spectra of these impulses were estimated, using the graphic procedure given in Chapter 1, as shown in Fig. 123. It is seen that the spectrum of the longer duration impulse peaks at about 5000 Hz, and at 13,000 Hz for the impulse for shorter duration. The pattern of TTS, shown in Fig. 124, is what one would predict—namely, the maximum threshold shift occurs at a higher frequency than the peak of the spectrum, and the greater the band levels the greater, in general, the amount of TTS. Some of the TTS data for impulses cited above, along with TTS to impulses as found by Ward *et al.* (852) and Carter and Kryter (128), will be used in the next chapter to show how well-measured TTS can be predicted from a knowledge of the spectral and temporal characteristics of a set of impulses.

Music

Perhaps an intermediate state of sound between impulsive and nonimpulsive would be music, particularly the highly amplified popular type sometimes called "rock." Figure 125 shows results of audiometric measurements made by Rice *et al.* (672a) of some young adults, not members of a "pop group," and young adult members of "pop groups" exposed to a program of rock music lasting an average of about 85 minutes.

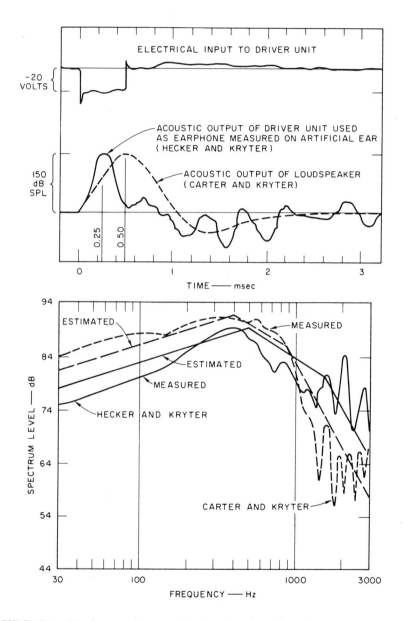

FIGURE 121.   Waveforms and measured and estimated (see Fig. 12) spectra of impulses
used in some studies of TTS from impulses. Carter and Kryter (128) and
Hecker and Kryter (359).

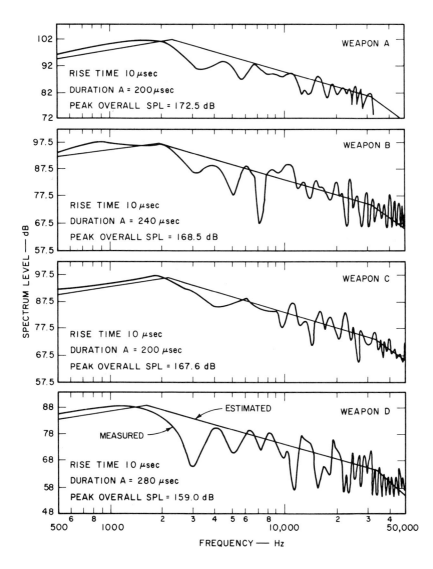

FIGURE 122.   Measured and estimated (see Fig. 12) spectra of gun impulses used by
Kryter and Garinther (463). See Fig. 110 for pressure-time waveforms.

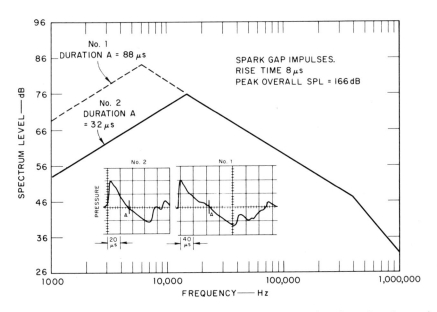

FIGURE 123.   Waveforms and estimated (see Fig. 12) spectra of spark gap impulses used
by Fletcher and Loeb (253).

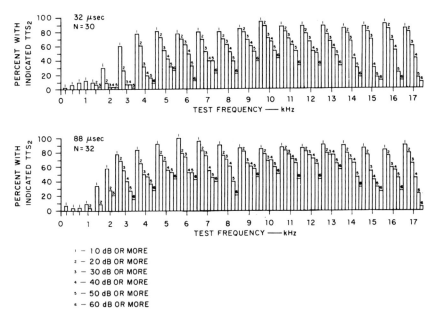

FIGURE 124.   TTS from spark gap impulses. The upper graph is for impulses with
duration of 32 μsec, the lower for impulses of 88 μsec (see Fig. 123 for
spectra). From Fletcher and Loeb (253).

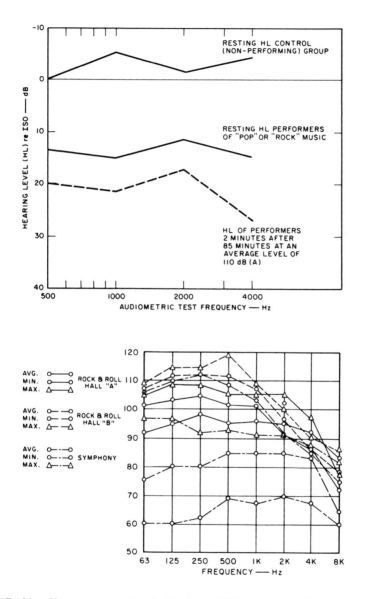

FIGURE 125.   Upper graph: Resting hearing levels (HL) of control and performer groups and HL of performers 2 min after performance. After Rice *et al.* (672a). Lower graph: Octave band sound pressure levels of "rock and roll" and loud, fortissimo symphonic music. From Lebo and Oliphant (488).

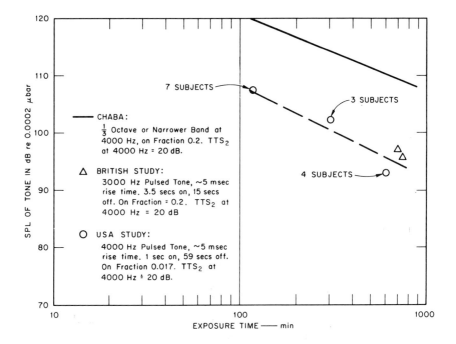

FIGURE 126.   Sound pressure levels of pulsed tones that give, for different exposure
times, TTS$_2$ of 20 dB at 4000 Hz according to CHABA contours (Fig. 104)
and as found in a British study (personal communication) and a study in
the United States. After Allen *et al.* (11).

It would appear likely, from the nature of the hearing levels of the pop groups, that the spectra of the music probably had levels of the order of 110 dB(A), and that the levels were equivalent to continuously present nonimpulsive sound. If this assumption of equivalent continuous level for rock music is true, it follows that the rock music greatly exceeds the tolerable limits to be set forth in this document and in the Federal Register (223).

Lebo *et al.* (489) report band spectra for rock music and deduce that such music exceeds proposed tolerable limits for the conservation of hearing. Lebo and Oliphant (488) made band spectral measurements for both rock and fortissimo symphonic music, and concluded that fortissimo symphonic music is below damage risk levels, whereas rock music is not; their findings are presented in Fig. 125. As was surmised from the hearing level data of Rice *et al.* (672a) damage risk from rock music extends over a wide range of frequencies, from at least 500 Hz to 4000 Hz.

Pulsed Tones

A rather unusual type of impulse is generated by rapidly turning on a pure tone, leaving it at steady-state for several seconds, and then turning it off rapidly. These tone bursts are found to create 10 to 15 dB more TTS than one would predict from TTS data obtained with bursts of random noise, as is shown in Fig. 126. There is no ready explanation, that we know of, for this apparent discrepancy, but it is clear that the damage risk from these pulsed tones is significantly greater than present procedures for evaluating damage risk would indicate. In the chapter to follow, new procedures for estimating damage risk to hearing are proposed which also underestimate by 10 to 15 dB the TTS data presented in Fig. 126. It would perhaps seem likely that the aural reflex is somehow involved differently for the tone than the noise stimulti, but, at the audiometric test frequencies involved, the aural reflex is supposed to afford little or no protection for either the pulsed tone or the noise.

Chapter 6

Proposed Procedures for Estimating Damage Risk to Hearing

Introduction

Below are described means of quantifying the damage risk of different conditions of noise exposure that are basically consistent with the CHABA contours but cover a broader range of conditions. In particular they differ from the CHABA contours in that these new proposed procedures:

1. Use as a criterion, protection for about 75% of the population against more than a 0 dB NIPTS for frequencies up to 2000 Hz and 10 dB for frequencies above 2000 Hz, whereas CHABA used as a criterion 50% of population, and 10 dB NIPTS at 1000 Hz, 15 dB at 2000 Hz, and 20 dB at 3000 Hz or above.

2. Simplify to some extent the shape of the frequency contours and level steps between the contours for different durations of exposure.

3. Attempt to handle, from a common set of definitions and assumptions, both impulsive (except for pulsed tones) and continuous and intermittent non-impulsive sounds (as defined in Chapter 1).

4. Permit the prediction of the average degree of threshold shift at specific sound frequencies as well as the degree of impairment to speech reception resulting from different gross patterns, on the frequency scale, of the threshold shifts.

In the CHABA and, implied in the AAOO, damage risk evaluation procedures, "negligible risk" was specified in terms of about 50% of the people and hearing losses that did not interfere appreciably with the understanding of speech in quiet. The present definition of negligible risk, while more stringent, still allows for some amount of hearing loss besides presbycusis (up to 25% of the people with average hearing losses of 10 dB for frequencies above 2000 Hz) that may be noise-induced, but is disallowed on the assumption that other than noise damaging stresses on the ear or some abnormal susceptibility to

207

noise-induced deafness occurs in some fraction of the people particularly at these higher frequencies. The matter of selecting a suitable criterion of handicap that would ensue from a given degree of impairment depends upon the sense of humanitarian, practical and economic values placed upon hearing by the person setting the criterion.

Before outlining these new procedures, TTS and NIPTS data which, along with Fig. 89, serve as their basis will be further developed.

TTS and NIPTS as a Function of Change in SPL

In Fig. 127 are plotted the readily available data that show that relation between TTS and SPL of a given noise. Shown on the upper graph of Fig. 128 are data obtained by Baughn showing, as a function of SPL of noise in dB(A), the HL (and presumably approximate NIPTS) of people exposed to industrial noises. Although Baughn's data are averages for different types of noise spectra and age groups, they are based on over 6000 audiograms and should represent about as clear a picture as is practically achievable of the relations shown. These data are also in agreement with the average, as found by Passhier-Vermeer, of other industrial studies of NIPTS as shown in Fig. 97.

Also drawn on each of Figs. 97, 116, 127, and 128 (upper graph) is a line representing what is a reasonable working approximation to all the data plotted, namely a monotonic growth dB for dB between TTS or NIPTS and sound pressure level of a noise, regardless of its frequency composition or temporal character. This relation is one to be used in the risk-to-hearing-contours to be presented later.

TTS and NIPTS as a Function of Exposure Duration

On the lower graph of Fig. 128 are plotted data from Baughn showing the growth of NIPTS as a function of years of exposure to given noise levels in dB(A) (see also Fig. 94). For purposes of comparison with the lower graph of Fig. 128, Fig. 129 shows all the data we could find that relates TTS to the duration of single exposures to various sounds.

It would appear that, on the average, a straight line such that a doubling of duration (or number of trigger pulls) results in about a 6 dB increase (20 $\log_{10}$ Time) in TTS (Figs. 108, 112, 113, and 129) or NIPTS (Figs. 94 and 128). This relation also forms an important part of the procedure to be described for estimating damage risk to hearing from a given noise condition.

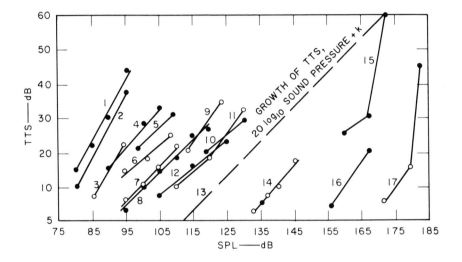

FIGURE 127.   Growth of TTS as a function of the sound intensity. The data points are from the following studies:

(1) Shoji *et al.* (739), 4 kHz, test freq. 2-4 kHz noise, 55 min duration. (2) Glorig *et al.* (303), 4 kHz, test freq. 1.2-2.4 kHz noise, 480 duration. (3) Lewis (personal communication), 4 kc pulsed tone, 60 min (2 sec on, 58 sec off). (4) Ward *et al.* (848), 4 kc, 1.2-2.4 kc, 47 min. (5) Spieth (762), 4 kc, WN, 20 min. (6) Allen *et al.* (11), pulsed tone. (7) Miller (551), 4.6 kc, WN, 12 min exp. (8) Davis *et al.* (180), 2-4 kc, 2 kc tone, 2 min. (9) Miller (552), 4 kc, WN, 3 min exp (run 2 min). (10) Davis *et al.* (180), 5-2 kc, 0.5 kc tone, 24 min. (11) Davis *et al.* (180), 4-8 kc, 4 kc tone, 2 min. (12) Ward (834), 3.4 kc, 1.4-2.8 kc, WN, 1 min. (13) Growth of TTS, 20 $\log_{10}$ sound pressure. (14) Ward *et al.* (852), 4 kc, clicks (75 min). (15) Kryter and Garinther (463), 1,2,3,4 kc, Gun 100 rounds. (16) Hecker and Kryter (359) 2,4 kc, 100 impulses. (17) Murray and Reid (571), 0.5-8 kc Gun 10 rounds.

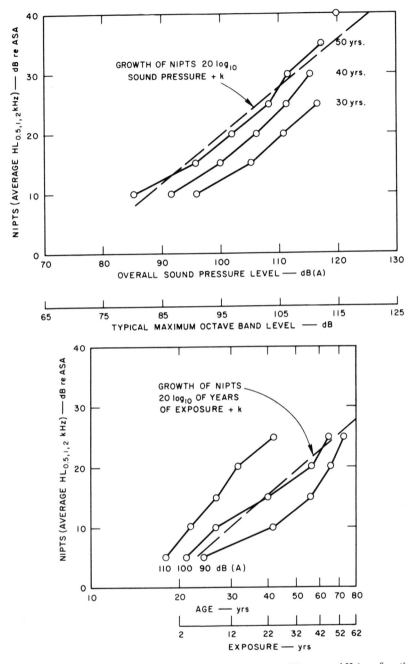

FIGURE 128.   Showing growth of median NIPTS (average (HL$_{0.5,1,2}$ kHz) as function of: SPL in dB(A) and maximum SPL in an octave band – parameter is age of workers in years (upper graph); and age and exposure of workers in years – parameter is SPL in dB(A) (lower graph). Based on Baughn (47).

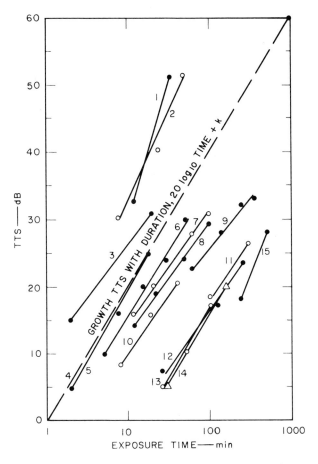

FIGURE 129. TTS as a function of the duration of the noise. The time following exposure before TTS varied among the studies, so that the absolute amount of TTS and the intercept with the abscissa are not constants for the different curves, although the slopes tend to be. The data points are from the following studies:

(1) Kryter and Garinther (463), Gun A, 4 kc, 1 per 5 sec. (2) Davis *et al.* (180), 2,4,8 kc, 120 dB, 2 kc tone. (3) Spieth and Trittipoe (762), 4 kc, 108 dB WN. (4) Growth TTS with duration, 20 $\log_{10}$ time. (5) Ward *et al.* (852), 4 kc, clicks (25 M). (6) Shoji *et al.* (739), 4 kc, 90 dB, 2-9 kc (N). (7) Ward *et al.* (848), 4 kc, 100 dB, 1.2-2.4 kc N, (1959a). (8) Glorig *et al.* (303), 4 kc, 100 dB, WN. (9) Davis *et. al.* (180), 0.5-2 kc, 120 dB, 0.5 kc tone. (10) Lewis, 4 kc, pulsed tones (personal communication). (11) Allen *et al.* (11), 4 kc, pulsed tone. (12) Kylin (482), 3,4,6 kc, 90 dB WN. (13) Ward *et al.* (847), 1 kc, 106 dB WN. (14) Murray and Reid, (572), 0.5-8 kc, Gun (1 per 15 min). (15) Kryter and Garinther (463), Gun B (1 TP per 5 sec).

Recovery from TTS

A third and crucial part of the procedure to be presented is a generalization of the average rate of recovery from a TTS or auditory fatigue. Recovery data from a large number of experiments are available as shown in Fig. 130., Fig. 131 from Spieth and Trittepoe, and Fig. 132 from Harris. Indicated on these figures are straight-line approximations to the general shape of the lines connecting actual data points. It is concluded that recovery, within the limits shown, occurs at the rate of 3 dB in TTS per doubling of recovery time, e.g., TTS reduction = $10 \log_{10}$ recovery time; or recovery in time following exposure from TTS occurs at one-half the rate of its growth in time during exposure. Although the evidence is meager, such as that shown in Fig. 117, it is proposed, for purposes of estimating damage risk from exposure to intermittent sounds, that recovery from TTS does not start until 5 seconds have elapsed subsequent to an exposure to noise.

Proposed DR Contours

Any sound of a given duration whose octave band (right-hand ordinate) or one-third octave band (left-hand ordinate) spectra does not exceed the DR contour shown in Fig. 133 for that given duration will cause (*a*) no more temporary threshold shift than 0 dB for frequencies up to 2000 Hz and 10 dB for frequencies above 2000 Hz, measured two minutes after initial exposure for the average normal ear, or (*b*) a like amount of permanent noise-induced threshold shift following 20 years of nearly 8 hours of daily exposure to noise in the hearing of no more than 25% of the population. Noise of this amount is said to contitute a negligible risk to hearing.

The proposal that 25% of the people could show some degree of permanent noise-induced threshold shift before damage risk is considered not negligible is based on the thought that some portion of the people may suffer some abnormal physiological condition that makes them seem abnormally susceptible to noise damage to the ear. As Hinchcliffe (372) has noted, apparently nonotological diseases and physiological conditions other than chronological age do affect hearing losses that simulate noise-induced deafness and it is perhaps not unreasonable to assume that the 25% of the people, at the extreme of the distribution of apparent noise-damaged ears, suffer some hearing losses from causes other than exposure to low intensity noise.

The duration of a sound is taken as the time in seconds that the nominal SPL of one or more of its band SPLs are above contour DR-O on Fig. 133. The nominal level of the bands of nonimpulsive sound is taken as the arithmetic average of each ten successive 0.5-sec rms levels of the sound.

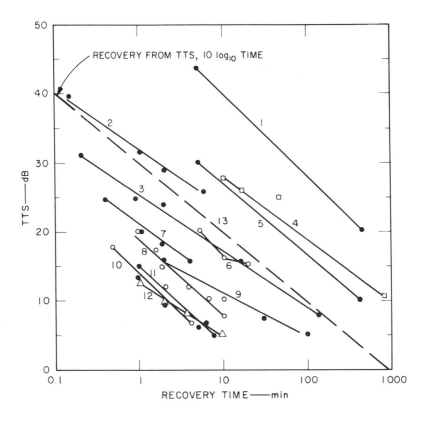

FIGURE 130. TTS as a function of postexposure time in minutes. The data points are taken from the following studies:

(1) Murray and Reid (572), #5 and 6, Fig. VI, Gun. (2) Miller (551), 4 kc, 3 min WN. (3) Spieth and Trittipoe (762), 108 dB, 20 min WN. (4) Lewis (personal communication), pulsed tones. (5) Murray and Reid (572), #2, Fig. VI, Gun. (6) Miller (551), 3 kc, 12 min WN. (7) Ward (834), 1.4-2.8 kc. (8) Cohen (139), WN. (9) Ward *et al.* (847), 106 dB, 1 kc. (10) Harris (349), 4 kc, Tone. (11) Harris (349), Impulse, 4 kc. (12) Smith and Goldstone (754), Gun, 4 kc. (13) Recovery from TTS, $10 \log_{10}$ Time.

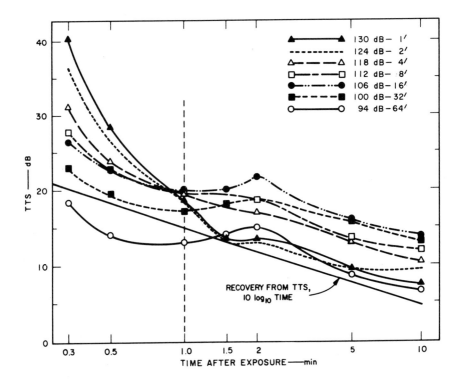

FIGURE 131.   Threshold shifts at 4000 Hz following exposure to each of the conditions listed. From Spieth and Trittipoe (762).

When the nominal level of the nonimpulsive sound or noise changes its level above a DR of 0 by more than 1 dB, or the level between any two successive 0.5-sec intervals changes more than 10 dB, a new noise is said to start (see Chapter 1).

When more than one impulsive sound occurs within a period of 1 sec, only the sound of maximum SPL is considered and the duration of the sound is taken as 0.5 sec. This is done to make allowance for the apparent protective action of the auditory reflex for impulsive sounds. It should be noted that this does not make any allowance for the attenuation of the second of two sounds that occur in some ears due to a first sound that is intense enough to elicit the aural reflex but does not of itself cause appreciable TTS_2; presumably under these conditions there will be less threshold shift than that to be predicted by Fig. 133. See Chapter 3 for discussion of the use of the aural reflex for hearing protection against impulses.

In addition to using information regarding the band spectra and duration of a sound, an estimate is made in Fig. 133 of the upward spread of auditory fatigue that occurs in the ear as the result of exposure to sound. The estimates are based on the nature of audiograms showing TTS and PTS and upward spread of masking that occurs in the ear as the result of exposure to sound.

Effective Damage Risk Level (EDRL)

The contours in Fig. 133 can be used as a basis for the calculation of a unit designated as Effective Damage Risk Level (EDRL) for hearing. To calculate the EDRL of a sound (see also Formula 1, Table 23):

1. Plot on Fig. 133 the actual or nominal, as defined above, sound pressure levels of the octave or one-third octave bands of a sound at the band center frequencies.

2. Draw from the plotted points, having the highest DR in sections A, B, and C, dashed lines (to be called "spread" lines) to the right that have the slopes specified in the graph on the lower left corner of Fig. 133. *Note*: With most broadband noises having band levels of less than 100 dB or so, (DR contour of about 24 or so), the spread-of-damage-risk from low to higher frequencies will usually be insignificant and no "spread" lines are required.

3. Note on section A, B, and C, as appropriate, the highest valued DR contour reached by a plotted point or transected by the dashed spread lines in each of those sections. Label these values as DR_A, DR_B, DR_C, as appropriate. Thus, for example, the value in section C, labeled DR_C, may be determined by the spread line from a point plotted on section B if that line transects a contour higher than any point plotted in C on the basis of actual band spectrum levels.

4. Subtract $20 \log_{10} (28,800/d)$ from the highest values found in step 3 where d is duration of the sound in seconds and 28,800 is number of seconds in 8 hours (see top graph, Fig. 134). As previously defined, the duration of the frequency bands of nonimpulsive sounds for these purposes is taken as the time that the nominal level of any one-third octave band of the sound is greater than that of the 0 DR contour shown in Fig. 133. The results are called the $EDRL_A$, $EDRL_B$, $EDRL_C$, respectively. Any values lower than zero are assigned the value zero.

5. The predicted threshold shifts at test frequencies of 1000 Hz or less are taken as the $EDRL_{(A \text{ or } B)}$, as appropriate, one-half octave below a given test frequency. At test frequencies above 1000 Hz, the predicted threshold shifts are taken as the $EDRL_C$ at a given test frequency. At test frequencies above 1000 Hz, the predicted threshold shifts are taken as the $EDRL_C$ at a given test frequency.

6. The average of the $EDRL_A$, $EDRL_B$, and $EDRL_C$ values is called EDRL and can be used to estimate average threshold shifts over the speech frequency range.

7. In the case of noise measured in units of Phons, PNdB, dB(A), and dB(D), the EDRL for the noise is taken as the measured value minus the value for that unit specified for DR-O on Fig. 133 minus $20 \log_{10} (28,800/d)$ (see Fig. 134).

Composite Damage Risk (CDR)

The results of the above steps (see Formulas 1 and 2, Table 23) are the Effective Damage Risk Levels (EDRLs) of one exposure of a sound. However, a common goal of noise measurement and control is the evaluation of a total noise environment with respect to its total damage risk to hearing. A unit called Composite Damage Risk (CDR) is proposed for this purpose (see Formula 3, Table 23). The CDR for some total period of time during which different sounds are present, or the same sound at different levels, or sounds interspersed with periods of quiet (sounds that fall below the DR contours on Fig. 133), can be determined by finding the EDRL for each sound and summing the EDRLs when corrected for auditory rest periods between sounds, as shown in Formula 4, Table 23. The correction for rest periods is based upon the recovery functions shown earlier for the TTS data—namely $10 \log_{10}$ of the time between sounds (see lower graph, Fig. 134). It is presumed that the total CDR is the maximum EDRL that occurs during a work day that is no longer than about eight hours, plus the other EDRLs, if any, corrected for recovery. It is also presumed that the temporal-intensity pattern of the noise is approximately the same on each work day.

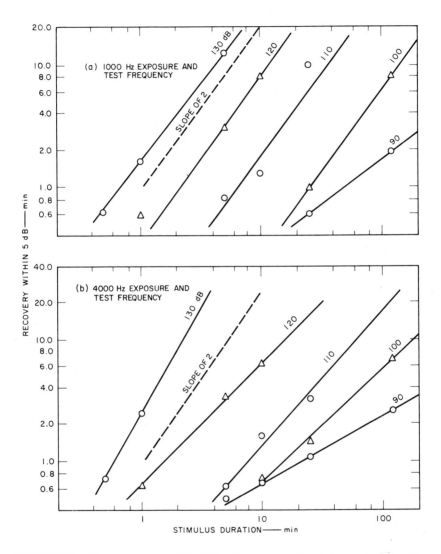

FIGURE 132. Time to recover within 5 dB of preexposure threshold as a function of log stimulus duration; parameter in SPL. The dashed line shows hypothetical growth of TTS with stimulus duration at rate twice that of recovery rate. After Harris (349).

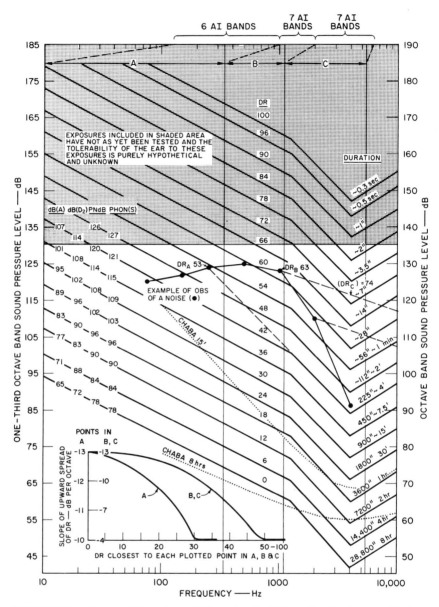

FIGURE 133.   Damage risk contours (DR) used as basis for estimating damage risk from exposure to continuous or intermittent impulsive or steady-state sounds. Dotted lines are CHABA DRCs adjusted to present criterion of no or negligible threshold shift in at least 75% of people. Dashed lines on example noise (octave band spectrum, OBS, plotted as solid dots) show effective spectrum due to upward spread of sounds of higher intensities according to functions shown on insert in lower left of figure. In the insert the vertical ordinate and curve marked A gives slope of upward spread line for points in Section A and those marked B and C give the slope of the upward-spread line for points in Section B and C. The values of dB(A), dB(D_2), PNdB, and Phons (Stevens) are based primarily on EDRLs for 7 representative noises (see Tables 7 and 75).

TABLE 23

General Formulae and Assumptions for the Calculation of the Effective Damage Risk Level
(EDRL) and the Calculation of the Composite Damage Risk (CDR) for a Series
of Daily Noises, and Their Relation to Hearing Level (HL)

(1)　*Formula 1:*　$EDRL_A, EDRL_B,$ or $EDRL_C$ = Highest $DR_{(A, B, \text{ or } C)} - 20 \log_{10}$
$28{,}800/d$, where d is the duration in seconds of a noise: e.g., if first noise starts at
8:00 A.M. and ends at 8:30 A.M., d_1 = 1800 secs., and if noise #2 starts, say, at
9:00 A.M. and ends at 9:10 A.M., d_2 = 600 secs. The first noise of a daily cycle is
said to start when the nominal SPL of one of its bands exceeds the DR-O Contour
on Figure 133.

(2)　*Formula 2:*　$EDRL = \dfrac{EDRL_A + EDRL_B + EDRL_C}{3}$

　　　Note: $EDRL_{A, B, \text{ or } C}$ less than zero are set at zero for this calculation.

(3)　For continuous noises – *Formula 3:*

　　　$CDR_{A, B, \text{ or } C} = EDRL_{A, B, \text{ or } C}$ and $CDR = EDRL.$

(4)　For intermittent noises – *Formula 4:*

$$CDR = EDRL_{max} \; |+| \; \left[\left[EDRL_1 - 10 \log_{10} (\frac{t_{S2} - t_{E1}}{5}) \right] \right.$$

$$|+| \; \left[EDRL_2 - 10 \log_{10} (\frac{t_{S3} - t_{E2}}{5}) \right] \ldots |+| \; \left[EDRL_{N-1} - 10 \log_{10} \right.$$

$$\left. (\frac{t_{SN} - t_{EN-1}}{5}) \right] \; |+| \; \left[EDRL_N - 14 \right],$$

$$\left. |-| \; \left[EDRL_{max} - 10 \log_{10} (\frac{t_{SN_{max} + 1} - t_{EN_{max}}}{5}) \right] \right]$$

where $EDRL_{max}$ is the largest valued EDRL for any noise occurring during the total
daily noise exposure cycle; $EDRL_1$ is that of the first noise in the cycle, $EDRL_2$,
the second, etc; $|+|, |-|$ is 20 $\log_{10}$ antilog addition or subtraction; and t_{EN} is clock
time and end of total daily noise exposure cycle.

Note 1:　$t_{SN} - t_{EN-1}$ is the time in seconds between the end of one noise (t_{EN-1})
and the start (t_{SN}) of the next succeeding noise of a daily noise exposure
cycle, e.g., if t_{EN-1} is 10:30 A.M., and t_{SN} is 11:00 A.M., ($t_{SN} - t_{EN-1}$)
= 1800. t_{EN} is not to exceed start of first noise plus 8 clock hours.
These periods serve as the basis for estimation of recovery from threshold
shift.

Note 2:　$t_{SN} - t_{E-1}$ that is less than 5 is set to 5. An interval of silence or a decrease
in the noise to a level below DR of 0 for a period of 5 seconds or less is
considered as no interruption to a sound, i.e., no effective recovery from
auditory fatigue is said to occur during this interval (see Fig. 134 bottom
graph, left-hand ordinate).

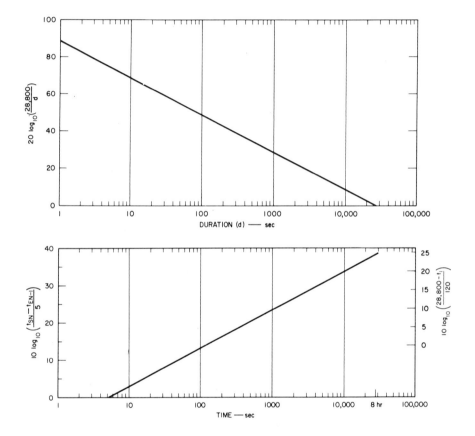

FIGURE 134.   Upper graph: Used for finding the growth factor, $20 \log_{10} (28,800/d)$, to be applied to DR based on band SPL (see Table 23). Lower graph: left-hand ordinate: used for finding recovery factor, $10 \log_{10} (^{t}SN - ^{t}EN-1/5)$ (see Formula 4, Table 23); right-hand ordinate: used for finding recovery factor, $10 \log_{10} (28,800-t_1/120)$ (see Table 25).

Table 24 is designed as a work sheet for the calculation of EDRLs and CDRs; also shown in Table 24 is an example of such calculations for exposure to an impulsive noise. Figures 15, 16, 17, and 134 provide graphic aids that can be used with Table 24 for the calculation of EDRLs and CDRs.

Damage Risk from Career Exposures of Less than 20 Years

It may be presumed that most workers in industry may accumulate at least twenty years of exposure to noise environments that exceed a damage risk value of zero. However, many persons can be expected to be exposed for appreciably less than twenty years to a noise environment that would be maximally harmful only after twenty years or so, and not be exposed subsequently to more harmful noise. When this assumption is a reasonable one, it is proposed that the EDRL for the exposure conditions be reduced by the subtraction of 6 dB for each halving of total anticipated exposure below twenty years as shown in the top graph on Fig. 135. It is suggested that the EDRL or CDR for these intense, but relatively brief, noise exposure careers be marked by a subscript that indicates the total elapsed length of anticipated exposure. For example, according to the top graph of Fig. 135, CDR_2 mo. is equal to CDR_{20} yrs plus 21, i.e., the amount of tolerable noise is 21 dB greater than it is if nearly 20 years or more of additional equivalent exposure to noise are anticipated. In this document EDRLs and CDRs without a subscript will presume noise exposure careers of 20 or more years.

As noted in the previous chapter (see Fig. 89), whereas NIPTS within an individual may not change appreciably after 20 years of exposure, the percentage of people within a group suffering NIPTS continues to increase. The definition of negligible risk given for EDRL or CDR, to repeat, is 25% of the people having a given NIPTS after 20 years of near daily exposure. The approximate percentage of people who would have a given or larger NIPTS (average HL for 500, 1000, and 2000 Hz) for any given number of years of exposure can be estimated from the bottom graph on Fig. 135; this graph is extrapolated from the data shown in Figs. 97 and 99.

EDRL and CDR from dB(A), dB(D), PNdB, Phons

Although predictors of loudness and perceived noisiness such as dB(A), dB(D), PNdB, and Phons (to be discussed in Part II) are not necessarily appropriate for the evaluation of damage risk from noise, these measures are sometimes used, as mentioned previously, for this purpose. For this reason, levels are indicated on Fig. 133 that would be measured for broadband noises by PNdB,

TABLE 24

Work Sheet for Calculation of EDRL and CDR and Example of Its Use

NOISE OCCURRENCE	1 SPL, dB Max 1/3 Oct or Oct Band — A	B	C	(A)	(B)	(C)	2 DR (From Fig.133) A	B	C	3 Duration (d) in sec.	4 $20 \log_{10}(28{,}800/d)$ (See Fig. 134)	5 $EDRL_{A,B,C}$ Col. 2 − Col. 4 — $A^†$	$B^†$	$C^†$	6 EDRL $\frac{A+B+C}{3}$	6 Clock Time Start	End	7 Diff. in Secs of Start of Following Noise Minus End Preceding Noise $(t_{SN}-t_{EN-1})$	8 $10 \log_{10}\frac{(t_{SN}-t_{SN-1})}{5}$ See Fig.134	9 Col. 5 $\oplus$ Σ Col. 9 — $A^†$	$B^†$	$C^†$
1	84	90	104	84	90	104	15	30	58	720	32	−17	−2	26	8.7	8 AM	8:15	180	16	−33	−18	10
2	84	90	104	84	90	104	15	30	58	720	32	−17	−2	26	8.7	8:12	8:27	120	14	−31	−16	(12)*
												Max $EDRL_C 26$			Max EDRL 8.7					Σ−33	−18	10
1.																						
2.																						
3.																						
etc.																						

Max $EDRL_{A,B}$, or C ———— Max EDRL ——

† Negative values are recorded and used in calculations.

* Negative values of $EDRL_A$, B, or C are set at 0 when calculating EDRL.

** CDR_A, B, or C set on 0 when Max Col. 5 $\oplus$ Col. 9 is negative and when calculating CDR.

\# $\oplus$ and Σ of Col. 9 are additions on a $20 \log_{10}$ antilog basis, + is arithmetic addition. The largest values in Col. 9 for a $EDRL_{\text{Max A, B, and C}}$ are excluded from summation of Col. 9 values.

$$CDR_{A,B,C} = Max_{A,B,C}\ Col.\ 5 \oplus \Sigma\ Col.\ 9$$

$$CDR = \frac{CDR_A + CDR_B + CDR_C}{3}$$

$CDR_A = 0$

$CDR_B = 0$

$CDR_C = 27^{**}$

$CDR = 9$

* Largest Col. 9 values for $EDRL_{\text{Max A, B, and C}}$ are excluded from summations.

** $CDR_C = Max\ EDRL_C\ (26) \oplus Col.\ 9\ (10) = 27$ (See Fig. 16).

Example: Impulse from mechanical "clicker," see Fig. 140 and Tables 27 and 28

Phons, dB(A) and dB(D). Thus, after substituting for these respective units the DR values on Fig. 133 (or by subtracting from a sound measured in dB[A], dB[D], PNdB, or Phon the value given on DR-O for the respective units of measurement), estimates of EDRLs and CDRs can be calculated and interpreted in terms of the effects of the noise or noise environment on average hearing levels and impairment to hearing speech. For example, a noise of 90 dB(A) has an equivalent DR level of 25; with this latter number and knowing the durations of the noise, one proceeds to calculate EDRLs and CDRs by means of the formulae given in Tables 23 and 24.

The use of these overall units of measurement can lead to some overestimation, as well as underestimation, of the damage risk to hearing for particular noises. In general, the use of these units will be reasonably conservative, compared to damage risk calculated from noise band spectra, except dB(A) (but not dB[D], PNdB, or Phons) will underestimate damage risk to the frequency regions above 2000 Hz from noises having significant portions of their energy above 1000 Hz.

Simplified Procedures for Estimating EDRL and CDR of Nonimpulsive Noise

Table 25 can be used as an alternate means for estimating EDRL, CDR, and the hearing impairment to be expected from 20 years of near daily exposures to noise environments of a given EDRL and CDR. As with Figure 133 and Tables 23 and 24, the EDRLs and CDRs found with Table 25 can be increased to allow for less than 20 years of near daily exposures by subtracting 6 from the EDRLs and CDRs for each halving of total exposure below 20 years (see top graph, Figure 135).

Table 25 is designed to be used as follows:

1. If a person works more or less continuously 8 hours or less per day in a single noise environment, enter Table 25 at the proper exposure duration, and noise level present during that period (lunch and normal rest breaks are not counted as interruptions to the noise). Read from column 5 on the right the CDR on the same line as the noise entry.

2. If a person works during an 8-, or less, hour day in noise at different levels, proceed as follows:

    Step 1. Enter Table 25 at the duration and level of each noise occurrence and record the highest EDRL found. Exclude the noise occurrence giving the highest EDRL from further consideration in Steps 2-5 below.

    Step 2. Excluding the noise, if any, having the highest EDRL above -10, find the total length of time in seconds that he is in each noise level and subtract each sum from 28,800.

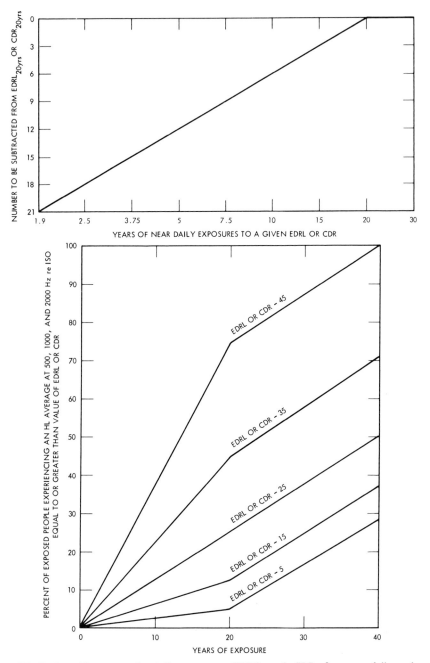

FIGURE 135.   Upper graph: Adjustment to EDRL and CDR for near daily noise exposures to a common total number of career exposure years. Lower graph: Percentage of people having an average HL re ISO at 500, 1000, and 2000 Hz equal to or greater than a given EDRL or CDR as a function of years of exposure.

Step 3. Enter the above table with each noise level condition as found in Step (2) and find the EDRL from Column 4 for each noise level condition.

Step 4. Subtract from each EDRL the recovery factor of:

$$10 \log_{10} \left(\frac{28,800 - t_i}{120} \right)$$

as appropriate for that EDRL (see lower graph, right-hand ordinate, Figure 134).

Step 5. Add the EDRL values found in Step (4) on a $20 \log_{10}$ antilog basis (see Figures 16 and 17).

Step 6. Add the sum found in Step (5) to the EDRL, if any, of Step (1) on a $20 \log_{10}$ basis (see Figure 16). The result is the Composite Damage Risk (CDR) for the total work day.

Step 7. If appropriate, subtract from EDRL or CDR an amount read from Figure 135, upper graph.

3. The impairment or damage risk to hearing is read from the column of one's choice on the right that is on the same line as the EDRL equal to the CDR found in Step (6) above, or, if appropriate, Step (7) above.

Correction for Presbycusis

For each five years above the age of forty that a person is exposed to noise, a certain number of dB, taken from the insert tables in Fig. 78, are to be subtracted from the HLs in columns 4 and 5 of Table 25, or the HLs deduced from CDRs found with Table 24. The resulting HLs and associated percent impairments for speech are presumed to be the best estimate of hearing impairment attributable exclusively to noise-induced deafness.

Relation between EDRL, CDR, HL, and Impariment for Speech

Hearing Impairment in dB (HIdB) or percent Hearing Impairment (%HI) to the reception of speech, EDRL for a given sound, and CDR for a total daily exposure to sounds are believed to be related to hearing level for pure tones in "dB-like" units as shown in Tables 12, 13 and Fig. 82. The relation between CDR, TTS_2 and HL can be summarized as: (a) CDR $\approx TTS_{2\,(0.5,1,2\ kHz)}$, in normal ears, and $\approx HL_{0.5,1,2\ kHz\ re\ ISO}$, after 20 years exposure; (b) CDR $\approx TTS_{2\,(1,2,3\ kHz)}$, in normal ears, and $\approx HL_{1,2,3\ kHz\ re\ ISO}$, after 20 years exposure.

TABLE 25

Simplified Procedure for Estimating EDRL, CDR and Impairment to Hearing from Exposure to Noise

Note: SPL in dB(A). Subtract 6 from dB(D), and 13 from PNdB or Phon to Convert to Equivalent dB(A) Values for Typical Noises.

From Column 5 Find EDRL for Each Noise Level at Each Different Duration During a Work Day (' = hours, " = minutes, '" = seconds.)

1 — Normal Speech Level in Quiet, % Hearing Impairment for Speech (Measured by Tests or Estimated by AI in Quiet; Listener 1 meter from Talker)	2 — Everyday Speech Level (Normal +10 dB), %	3 — Recommended by AAOO for Everyday Speech in Quiet, % Hearing Impairment for Pure Tones (NIPTS in dB Equalled or Exceeded in 25% of People)	4 — Aver. HL 1000, 2000, and 3000 Hz re ISO, dB	5 — EDRL, CDR and Aver. HL 500, 1000, and 2000 Hz re ISO, dB	8	7'20"	6'40"	6'10"	5'40"	5'10"	4'40"	4'10"	3'40"	3'15"	2'55"	2'35"	2'20"	2'5"	1'50"	1'40"	1'30"	1'20"	1'12"	56"	50"	45"	40"	36"	32"	28"	25"	22"	20"	18"
			0	(-10)*	55	56	57	58	59	60	61	62	63	64	65	66	67	68	69	70	71	72	73	74	75	76	77	78	79	80	81	82	83	84
			5	(-5)*	60	61	62	63	64	65	66	67	68	69	70	71	72	73	74	75	76	77	78	79	80	81	82	83	84	85	86	87	88	89
			10	0	65	66	67	68	69	70	71	72	73	74	75	76	77	78	79	80	81	82	83	84	85	86	87	88	89	90	91	92	93	94
0			15	5	70	71	72	73	74	75	76	77	78	79	80	81	82	83	84	85	86	87	88	89	90	91	92	93	94	95	96	97	98	99
10			20	10	75	76	77	78	79	80	81	82	83	84	85	86	87	88	89	90	91	92	93	94	95	96	97	98	99	100	101	102	103	104
20	0		25	15	80	81	82	83	84	85	86	87	88	89	90	91	92	93	94	95	96	97	98	99	100	101	102	103	104	105	106	107	108	109
30	10		30	20	85	86	87	88	89	90	91	92	93	94	95	96	97	98	99	100	101	102	103	104	105	106	107	108	109	110	111	112	113	114
40	20	0	35	25	90	91	92	93	94	95	96	97	98	99	100	101	102	103	104	105	106	107	108	109	110	111	112	113	114	115	116	117	118	119
	30	7.5	40	30	95	96	97	98	99	100	101	102	103	104	105	106	107	108	109	110	111	112	113	114	115	116	117	118	119	120	121	122	123	124
	40	15.0	45	35	100	101	102	103	104	105	106	107	108	109	110	111	112	113	114	115	116	117	118	119	120	121	122	123	124	125	126	127	128	129

Notes (right side of table):

Note 1: EDRL's read to closest dB, i.e., a noise on for 8 hrs at a level of 86 dB(A) has an EDRL of 21.

Note 2: CDR equals maximum EDRL for any noise plus sum of EDRL's for all other noises minus factor for recovery time.

* (-10) and (-5) used only for EDRL's.

Top header row (EDRL values): 105 106 107 108 109 110 111 112 113 114 115 116 117 118 119 120 121 122 123 124 125 126 127 128 129 130

Main table (staircase values by size):

Size	Values
34″	114 119 124 129
38″	113 118 123 128
43″	112 117 122 127
48″	111 116 121 126
54″	110 115 120 125 130
1″	109 114 119 124 129
1 1/8″	108 113 118 123 128
1 1/16″	107 112 117 122 127
1 1/25″	106 111 116 121 126
1 1/36″	105 110 115 120 125 130
1 1/48″	104 109 114 119 124 129
2″	103 108 113 118 123 128
2 1/12″	102 107 112 117 122 127
2 1/30″	101 106 111 116 121 126
2 1/48″	100 105 110 115 120 125 130
3 1/12″	99 104 109 114 119 124 129
3 1/36″	98 103 108 113 118 123 128
4″	97 102 107 112 117 122 127
4 1/30″	96 101 106 111 116 121 126
5 1/4″	95 100 105 110 115 120 125 130
5 1/40″	94 99 104 109 114 119 124 129
6 1/18″	93 98 103 108 113 118 123 128
7 1/6″	92 97 102 107 112 117 122 127
8″	91 96 101 106 111 116 121 126
9″	90 95 100 105 110 115 120 125 130
10″	89 94 99 104 109 114 119 124 129
11″	88 93 98 103 108 113 118 123 128
12″	87 92 97 102 107 112 117 122 127
14″	86 91 96 101 106 111 116 121 126
15″	85 90 95 100 105 110 115 120 125 130

Left-hand conversion scales:

%	%	%	dB	dB
70	50	22.5	(-10)*	0
80	60	30.0	(-5)*	5
90	70	37.5	0	10
100	80	45.0	5	15
	90	52.5	10	20
	100	60.0	15	25
			20	30
			25	35
			30	40
			35	45

%	%
	0
	10
0	20
10	30
20	40
30	50
40	60

Theoretical Model

The formulae for taking into account the threshold shift effects of continuous and intermittent noise are based on the concepts that: (*a*) on each portion
of the basilar membrane are a large population of receptor or response units that
have a different threshold of stimulation, each cell unit being, in this regard,
essentially independent of the others; i.e., fatiguing, as the result of constant
stimulation of a very sensitive response unit, does not make a less sensitive unit
more sensitive or susceptible; (*b*) the threshold of audibility is the result of a
certain minimum number of receptor units being stimulated simultaneously or
nearly so, and the loudness of the sound is a joint function of the particular
units being stimulated and the number of stimulations per unit, per a unit of
time; (*c*) the degree of TTS to be experienced by the ear is a function of the
number of receptor units (and/or related neural elements) having a given sensitivity threshold that are in the process of recovering from fatigue; (*d*) the degree
of PTS is a function of the number of receptor units (and/or related neural
elements) having a given sensitivity threshold that are damaged or have otherwise
lost their ability to recover from stimulation. Accordingly, subjecting an ear with
noise capable of causing a TTS_2 of 20 dB at a moment in time when the ear is
suffering a 10 dB TTS should not cause, at its termination, a 30 dB TTS, but a
20 dB TTS plus a few dB equivalent to the fatigue that would result from a
somewhat prolonged exposure of the sound. Ward *et al.* (850, p.793), has
described this phenomena in terms of an "exposure-equivalent duration" for
exposures to noise of varying intensity: "the TTS existing at the beginning of a
particular exposure can be regarded as additional time of exposure to the noise
concerned."

It might also be noted that this hypothetical description of auditory fatigue
and the functioning of the ear takes cognizance of the phenomenon of loudness
recruitment that is observed in ears with either TTS or NIPTS because the less
sensitive receptor units, presumably those contributing the most to loudness, are
unfatigued and fully operational. It should again be noted, however, that in spite
of the recruitment of loudness, the person with a sensori-neural threshold shift
loses some of his ability to discriminate among sounds that are audible. It would
appear, if the simple model of auditory receptors described in the preceding
paragraph is at all valid, that the receptor units involved in the detection of weak
signals at threshold levels are also involved, perhaps as part of more complex
nerve networks, in the discrimination of complex frequency-intensity-time patterns of suprathreshold auditory signals.

Validity of Proposed DRCs, EDRL, and CDR

The proposed damage risk contours of Fig. 133 below 60 or so should, on the average, be valid within the definitions provided because they represent a rather concise general picture of the sense of most of the TTS and much of the NIPTS data presented earlier. The allowable intensities, for brief duration exposures and particularly for the lower frequencies, seem perhaps inordinately high compared to previous damage risk contours. There appears to be no data available on the effect, on the normal and unprotected ear, of sounds above 20 Hz having band spectra levels above about 130 dB; accordingly, the tolerability of the open ear to exposures covered by the shaded area in Fig. 133 is hypothetical and must be tested before given much credence. On the other hand, the human ear can tolerate pressures at low frequencies of a few Hz up to 180-190 dB, at which time rupture of the ear (without any necessary damage to the sensitivity of the inner ear) is likely to occur in some ears.

Continuous Nonimpulsive Sounds

Figure 136 from Mohr *et al.* (561) summarizes a series of tests in which five adults were exposed to a number of intense, predominately low-frequency, pure tones and bands of random noise. For most of the exposures, the subjects wore earplugs and/or muffs which probably afforded about 30 dB attenuation for the sound frequencies down to 100 Hz or so. The exposure durations were typically 1 to 2 minutes long.

In addition to some clinical observations of extra-aural effects, the investigators noted that there were no measurable TTSs. The hearing tests were sometimes made, however, one hour after the exposures. That no significant TTS was to be expected from the exposures cited in Fig. 136 is substantiated by free-field tests (Kryter and Pearsons, unpublished data) in which two subjects listened without ear protection for a period of one hour to each of two pure tones, one of 3 Hz, and one of 23 Hz at an SPL of 130 dB. Pre- and two-minute post-exposure audiograms revealed no threshold shifts at any test frequency.

Reference to Fig. 133 reveals that none of the exposures reported by Mohr *et al.* should result in any TTS_2, and the one-hour exposure to a 23 Hz tone of 130 dB should possibly cause a 15 dB TTS_2 in a restricted frequency region around 100 Hz. Thus, the contours in Fig. 133 would seem to be ultra conservative at the very low frequencies and intensities.

Figure 137 from Davis *et al.* (180) illustrates the general pattern of threshold shifts obtained from exposures to intense pure tones. Figure 138 shows how average measured threshold shifts (solid lines visually fitted to data points) for various pure tones at various SPL compare with TTS to be expected (the dashed

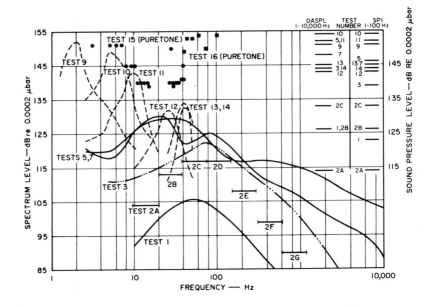

FIGURE 136.   Summary of test environments. Summary analysis of representative noise
                exposures for Tests 1-16. Random noise exposures are plotted in spectrum
                level (left ordinate) with overall sound pressure levels indicated on right
                ordinate. From Mohr *et al.* (561).

lines) on the basis of Fig. 133. To determine the expected TTS_2 values, Fig. 133 was entered with the one-third octave band SPL (which would be the same as that for a pure tone at the same center frequency) and the DR intersected by that level and center frequency noted. The sum of that DR and the growth duration factor for the exposure in question represents the amount of TTS_2 to be expected at frequencies one-half to two octaves above the frequency of the exposure tone.

It is seen in Fig. 138 that, except for the 4000 Hz tone or for TTS_2 values above 30-40 dB, predicted TTS_2 is about equal or somewhat greater than measured. As mentioned earlier, 40 dB is the probable maximum one can expect to estimate at all closely on the basis of the generalizations of the growth of TTS, particularly with respect to SPL, represented by Fig. 133. Even so, there is approximate agreement between the predicted and obtained threshold shifts.

Figure 133 also shows how the newly proposed and representative CHABA Damage Risk Contours (DRCs) compare. The CHABA DRCs were adjusted (1 dB reduction in SPL for 1 dB reduction in threshold shift) to provide equal (0 dB) threshold shift at all frequencies and 10 dB to accommodate 75% of the population, rather than the 50% used for the CHABA contours, in order to meet the criterion used in developing Fig. 133.

The general shape of the CHABA and presently proposed contours for continuous single exposures to steady-state sound are only roughly similar. We believe, in retrospect, that the flattening of the CHABA contours for four- and eight-hour exposures in the frequency region around 2000-4000 Hz is probably not justified; in particular, it is seen in Fig. 139 that a somewhat better extrapolation of the TTS data for exposures longer than one hour than the original extrapolation would lower the tolerable levels for eight-hour exposure by 5-10 dB. Other than that, the primary justification for the new contours is that they permit a relatively simple procedure or formula for estimating either temporary or permanent damage risk for a given frequency region or for the impairment of speech reception that can be applied within broad limits to any noise exposure regardless of temporal or spectral pattern.

A specific example of good agreement between a CDR based on Fig. 133 and/or Table 25 is to be found in the NIPTS data reported by Baughn in Fig. 99. A noise having a dB(A) level of 90 would cause, according to Fig. 133 and Table 25, an average HL at 500, 1000, and 2000 Hz of 25 dB re ISO in about 25% of the people exposed to it for eight hours a day for about twenty years. It is seen Fig. 99 that about 25% of the men, age forty-two years, who worked in noise of 92 dB(A), have average HLs of this amount or greater.

Intermittent Nonimpulsive Sounds

The research data on TTS and the interpretation of those data by Ward and his colleagues represent a significant portion of the available knowledge

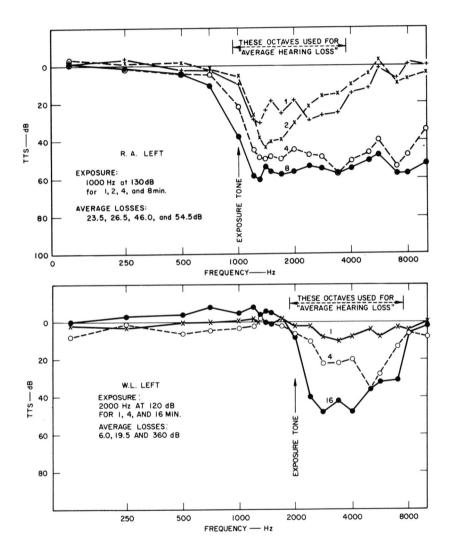

FIGURE 137.   TTS from exposure to pure tones. From Davis *et al.* (180).

concerning temporary threshold shifts due to exposure to noise. Their formulations (847-850) of the functional relations between TTS, spectrum, and intermittency of steady-state sounds are as follows:

Exposure Band	Test Frequency	$TTS_2 - dB$
600-1200 Hz	1500 Hz	$0.53r(S\text{-}71)(\log_{10}T\text{-}0.44)\text{-}3$
600-1200 Hz	2000 Hz	$0.41r(S\text{-}68)(\log_{10}T\text{+}0.15)\text{-}8$
1200-2400 Hz	3000 Hz	$0.58r(S\text{-}65)(\log_{10}T\text{+}0.55)\text{-}13.5$
1200-2400 Hz	4000 Hz	$0.61r(S\text{-}70)(\log_{10}T\text{+}0.33)\text{-}9.5$
2400-4800 Hz	4000 Hz	$0.91r(S\text{-}75)(\log_{10}T\text{+}0.19)\text{-}8$
2400-4800 Hz	6000 Hz	$0.51r(S\text{-}68)(\log_{10}T\text{+}1.80)\text{-}22$
Broadband	4000 Hz	$1.06r(S\text{-}85)(\log_{10}T/1.7)$

Where r is "on" fraction ($\frac{\text{on time}}{\text{on + off time}}$), S is SPL, and T is the total duration of the exposure.

Ward (personal communication), on the basis of additional tests, believes that the equation for the 600-1200 Hz band significantly underestimates the threshold shift for exposure durations less than about 20 minutes (see also data on Fig. 139 for that band).

Using these and related formulae, Ward *et al.* have predicted the TTS_2 to be expected from a number of noise exposures for which TTS_2 data are available. In Table 26, TTS_2, as predicted by the procedures of Ward *et al.* and as predicted by EDRL, are compared with TTS_2 as measured by Ward *et al.* It is seen that EDRLs do a reasonably accurate job of predicting TTS, although the Ward *et al.* procedures are usually somewhat more accurate. This might be expected since their prediction equations were based largely on most of the TTS data involved.

Impulsive Sounds

The procedures illustrated in Fig. 133 and in Table 24 can be validly used with most impulses. It is necessary to convert the time-pressure waveform of impulses first into energy spectrum level (which can be done by the use of Fig. 12 or 13 depending on the type of impulse) and then into a one-third octave band spectra (which can be done through the use of Fig. 9). The general ability of the procedures, outlined in Fig. 133 and Tables 23 and 24 for estimating damage risk to hearing, are illustrated in Table 27 for impulses from guns, spark gaps, mechanized "crickets," earphones, and loudspeakers.

The spectra of various impulses, as estimated from procedures outlined in Figs. 12 and 13 and as measured by analog filters or by calculation of the Fourier transform, are illustrated in Figs. 121-124 and Figs. 140-142. It might be noted that most of these impulses are distinctly either the single-spike or the exponentially damped-sinusoid type of impulses. The impulse shown in Fig. 141

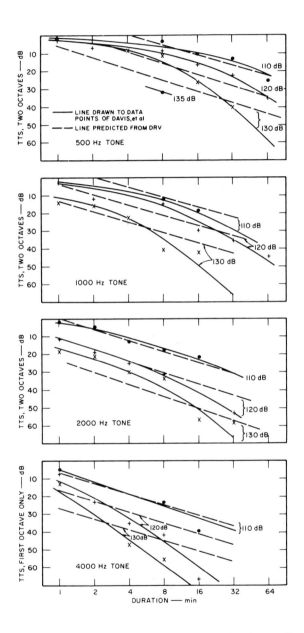

FIGURE 138.   TTS from exposure to pure tone as measured and as predicted by EDRL. Data from Davis *et al.* (180).

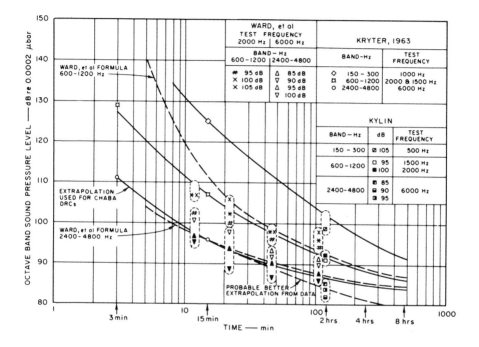

FIGURE 139. Obtained and estimated TTS$_2$ values from exposure to certain octave bands of noise. From Kryter (452).

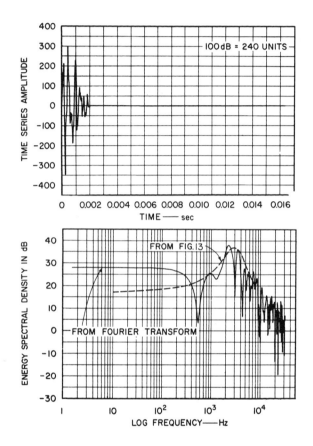

FIGURE 140.   Time series and energy spectral density records of mechanical clicker or "cricket" impulse used by Ward *et al.* (852).

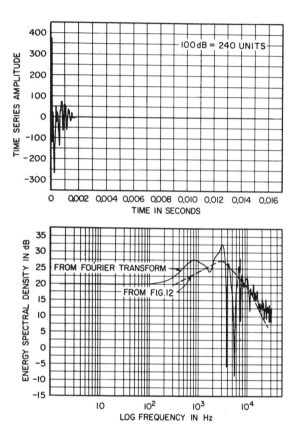

FIGURE 141. Time series and energy spectral density records of impulse from Altec 288B Loudspeaker used by Ward *et al.* (852).

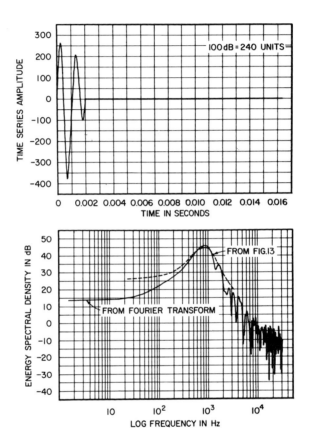

FIGURE 142.   Time series and energy spectral density records of impulse from Altec 20801 Loudspeaker used by Ward *et al.* (852).

contains some characteristics of both types but is classified as being predominately single-spike.

Using these spectra, and the procedures outlined in Fig. 133 and Tables 23 and 24, estimates were made of the amount of TTS_2 to be expected from various exposures to these impulses. The predicted and measured TTS_2 values at 4000 Hz are given in Table 27. Part of Table 24 illustrates the calculation of EDRL and CDR with data for one of the impulses, and Table 28 shows the steps taken to predict, by EDRL and CDR, the TTS_2 at 4000 Hz from two of the impulses.

While there are some differences, the predicted and measured TTS_2 values are perhaps of surprisingly similar magnitude. It is clear that peak overpressure and number of impulses are in and of themselves inadequate indicators of the damage risk to hearing, but that, as with nonimpulsive sounds, spectra, exposure durations, and recovery periods between exposures apparently provide an adequate base for estimating from physical measures the damage risk of exposures to acoustic impulses.

Comparison of Schemes for Predicting NIPTS from TTS Data

The functions, or closely similar ones, proposed by Ward *et al.* for the rate of growth of TTS, when the ear is exposed to noise, and the rate of recovery therefrom, when the ear is in subsequent quiet, have been incorporated into an electronic device by Botsford and Laks (83a). Botsford and Laks have proposed that the results of noise measurements made with this device, or for that matter from the application by calculation of these functions to noise measures, be used as a means of predicting eventual NIPTS, based on the same general reasoning used to develop the procedure we have just proposed. There is, however, a possibly significant difference between our procedure and that proposed by Botsford and Laks.

Botsford and Laks propose that after 20 or more years of almost daily exposure to continuous or intermittent noise, NIPTS will equal the residual TTS that would be present 2 minutes after the end of the eight-hour workday, as the result of the auditory fatiguing and recovery actions from exposures to noises occurring during the workday; whereas we propose that:

Continuous Noise. In an ear with normal pre-exposure hearing following exposure, up to eight-hour's duration within a 24-hour period, to a continuous noise, TTS_2 will equal NIPTS after 20 or more years of near daily exposure to this noise.

Intermittent Noise. After 20 or more years of near daily exposure to intermittent noise, NIPTS will equal (1) the TTS_2 in an ear with normal pre-exposure hearing after exposure to the eight-hour workday noise that is capable of causing the greatest amount of TTS_2 from a single exposure, plus (2) the residul TTS that would be present 2 minutes after the end of the eight-hour workday, as the result of the auditory fatiguing and recovery actions from exposures to the other There are not sufficient data available to demonstrate which of these two

TABLE 26

Comparison of Predicted and Measured TTS for Continuous and Intermittent
Steady-State Noise

Random noise, 106 dB OA SPL (Oct. bands 75 to 4800 Hz) at 97 dB). TTS_2 at 4000 Hz

Noise 12 min on, 18 min off	Minutes of Exposure				
	12	42	62	122	172
Measured – Ward et al.[850]	19	21	23.5	23.5	25
Predicted – Ward et al.	18.5	22	23.5	23.5	23.5
Predicted – CDR_C	19	20	20	21	22

Noise on continuously	Minutes of Exposure			
	12	22	47	102
Measured – Ward et al.[847]	19.1	25.1	32.1	40.6
Predicted – Ward et al.	19	24.5	33	40
Predicted – CDR_C	19	25	31	39

Noise 30 sec on, 30 sec off	Minutes of Exposure			
	12	22	47	102
Measured – Ward et al.[847]	9.5	12.8	16.3	19.9
Predicted – Ward et al.	9	12	16	20
Predicted CDR_C	8	13	19	25

Noise 30 sec on, 60 sec off	Minutes of Exposure			
	12	22	47	102
Measured – Ward et al.[847]	6.5	8.3	10.5	13.1
Predicted – Ward et al.	6	8	10	13
Predicted – CDR_C	5	10	16	22

Noise 60 sec on, 30 sec off	Minutes of Exposure			
	12	22	47	102
Measured – Ward et al.[847]	12.2	17.5	20.8	26.8
Predicted – Ward et al.	12	17	21	27
Predicted – CDR_C	14	19	25	31

TABLE 26 (Continued)

100 Min Uninterrupted Exposure to Diesel Noise

			2 kHz	3 kHz	4 kHz	6 kHz
Noise			TTS_2	TTS_4	TTS_2	TTS_4
Octave Band Levels						
300- 600 Hz	94 dB	Measured - Ward et al[848]	16	19.5	17.5	7.5
600-1200	96	Predicted - Ward et al	17	22	19.5	11
1200-2400	90	Predicted CDR_C	18	19	22	9
2400-4800	84					

200 Min Uninterrupted Exposure to Sheet Metal Noise

			2 kHz	3 kHz	4 kHz	6 kHz
Noise			TTS_4*	TTS_2	TTS_2	TTS_4
Octave Band Levels						
300- 600 Hz	95 dB	Measured - Ward et al '59a	28	38	43	36
600-1200	95.5	Predicted - Ward et al '59a	27	41	43	33
1200-2400	97	Predicted CDR_C	28	40	40	38
2400-4800	97					

* CDR nominally predicts TTS_2 values. TTS_4 values predicted by CDR obtained by subtracting 3 dB from the TTS_2 values that were predicted from application of basic CDR procedures.

procedures is the more valid, and for many noise conditions they would give essentially the same answers. Furthermore, the sum of the two factors in our scheme for predicting the effects of intermittent noise can, of course, be no larger than 3 dB greater than the larger factor.

On the other hand, there are possible noise conditions in which the predicted NIPTS would be considerably different for the two prediction schemes. According to the Botsford and Laks procedure, for example, a noise exposure condition that starts a workday at a very intense level and then declines as the workday progresses would be rated as less damaging than the same noise that starts out at a low level and then increases during the workday. However, by our method the two noise conditions would be rated essentially equivalent in damage risk.

It might be noted that the Botsford and Laks scheme implies that the recovery process in the ear is complete at the moment predicted TTS-minus-recovery reaches 0; however, it cannot be deduced from present data that this is

TABLE 27

Comparison of Predicted and Measured TTS_2 for Impulses

All measured TTS data for other than 2 minutes post-exposure corrected to TTS_2 on the assumption that recovery occurs at a rate equivalent to 3 dB per doubling time.

SINGLE-SPIKE IMPULSES. FIG. 12 USED TO ESTIMATE SPECTRA.

	Peak OA SPL	No. of min. exposure; pulses per min.	Measured TTS_2 at 4000 Hz	TTS_2 at 4000 Hz predicted by CDR_C
1. Gun Noise, Weapon D Kryter and Garinther[463] (see Fig. 122)	159 dB	10 min; 12/min	32	34
2. Spark Gap Fletcher and Loeb[254] (see Fig. 123, Duration 32 μsec)	156	≈10 min; Aver. 40 min	7	8
3. 288B Loudspeaker Ward et al[852] (Primarily Single-Spike plus low-level damped sinusoid) (see Fig. 141)	145	12 min, 3 min break, 12 min; 75/min	11	19
4. Loudspeaker Carter and Kryter[128] (see Fig. 121)	163	2 min; 60 min	20	22
5. Earphone Hecker and Kryter[358] (see Fig. 121)	163	1 min 40 sec; 60/min	20	21

EXPONENTIALLY DAMPED SINUSOIDS. FIG. 13 USED TO ESTIMATE SPECTRA.

1. Mechanical Clicker. Called "Cricket" by Ward et al[852] (see Fig. 140)	145	12 min, 3 min break 12 min; 75/min	22	27
2. 20801 Loudspeaker Ward et al[852] (see Fig. 142)	145	1 min; 75/min	14	9

indeed the case. Also, the empirical basis for the TTS_2-NIPTS relation is more directly related to a TTS_2 value generated by a noise that occurs once every 24 hours, during workdays, than it is to TTS_2 present 2 minutes after the end of the eight-hour workday. It is primarily for these reasons that we felt obliged to suggest the two-step procedure outlined in this section.

TABLE 28

Examples of the Calculation of EDRLs and CDRs of Impulses [*]

(1) Gun Noise, Weapon D, Fig. 122.

    (a)   From Fig. 12, Spectrum Peak Level at 1800 Hz = 86 dB; level at 4000 Hz = 78 dB

    (b)   From Fig. 9, 1/3 Oct Band Level = 112 at 1800 Hz (26 + 86) and 108 at 4000 Hz (30 + 78)

    (c)   From Fig. 133

$$DR_c = 67 \quad \text{based on level at 4000 Hz}$$
(Enter on Column 2, Table 24)

    (d)   From Fig. 134, Durational Factor = 33 (Enter Column 4, Table 24)

    (e)   From Column 5, Table 24, $EDRL_C$ and CDR_C = Column 2-4, Table 24 = 67 −33 = 34

    (f)   EDRL and CDR = 11.3

(2) Mechanical Clicker, OA SPL 145 dB (See Fig. 140 for Spectrum, OA SPL of 103 dB. Also see bottom Table 24.)

    (a)   From Fig. 8 Spectrum Level at 3000 Hz = 75 dB

    (b)   1/3 Oct Band Level = 104 dB

    (c) DR_c = 58

    (d)   Durational Factor = 32 (1st and 2nd 12 min exp.)

    (e) $EDRL_C$ (2nd 12 min exp.) = 58 −32 = 26

    (f)   Recovery Factor 3 min rest = 16

    (g) $EDRL_C$ (1st 12 min exp.) = 10

    (h) CDR_C = 27

    (i)   CDR = 9

[*]Also see Table 27.

PART II

SUBJECTIVE RESPONSES TO NOISE

Introduction

The ability of the ear to detect information in the presence of noise, and to become fatigued and damaged from the noise, was the primary subject matter of Part I of the book. Part II is concerned with the equally important, but more subjective, psychological attributes of sound or noise. These attributes are usually described in terms of the relations between the physical characteristics of a noise stimulus presented to a person and his verbal response to questions asked him about his auditory experience.

Two of the most popular questions asked in the past have had to do with the pitch (subjective "height") of a sound when its physical frequency content was changed, and the loudness (subjective intensity) of a sound as its physical intensity was varied. However, Laird and Coye (483), Thomas (800), and others have asked the subject different questions — how "annoying," "big," "sharp," etc., did the subjects judge one sound to be relative to another as the spectrum or intensity of the sound stimulus was varied. Peters (616) asked the subjects to judge merely how far, psychologically, a large number of sounds, taken two-at-a-time, were from each other; an analysis of the results showed that three major factors were operative: one was correlated with the spectral frequency, one was the intensity, and the third was a joint function of these two. For the most part, these basic attributes of sound and noise have been studied with stimuli that are as meaningless as possible to the listeners or with stimuli that have more or less equal meaning for the individual listeners. The two attributes of most general interest with respect to noise will be discussed in the next two chapters — loudness and perceived noisiness, or as sometimes called, annoyance.

Chapter 7

Loudness

There are three basic psychological-physical relations, ignoring temporal factors, that depict the perceptual attribute of sound called loudness:

1. The intensity levels required to make each tone, or critical band of frequencies, in the audible frequency range appear to be subjectively equally loud to each other.

2. The growth of loudness as the bandwidth of the sound spectrum is widened.

3. The growth of loudness upon some numerical scale as the physical intensity of a given sound is increased.

The Dependence of Loudness on Frequency

Fletcher and Steinberg (239) and Fletcher and Munson (237) appear to have made the first major attempts to define and measure loudness. Fletcher and Munson defined loudness as the "magnitude" of a sound, and conjectured that the loudness was proportional to the number of impulses leaving the cochlea upon stimulation. Fletcher and Munson specified a 1000 Hz tone as the standard sound against which other tones would be judged for loudness. Stevens (773) suggested that the unit of loudness be called the sone, and that one sone be ascribed to a 1000 Hz tone set at a sound pressure level of 40 db. The sone scale, which will be discussed more fully later, is such that a sound twice as loud as a sound of 1 sone is given a value of 2 sones, four times as loud is called 4 sones, etc.

Equal Loudness Contours

Fletcher and Munson found the sound pressure levels at which pure tones taken over an extended range of frequencies were judged equal in loudness to a 1000 Hz reference tone set at a fixed sound pressure level; the results were called equal loudness contours for pure tones. A number of other investigators have also determined equal loudness contours for pure tones as well as bands of noise, using a tone or band of noise centered at 1000 Hz as a reference sound against which other sounds are judged. The loudness contours found by various investigators have their differences and their similarities as shown in Fig. 143. Robinson and Whittle (692) recently proposed an averaging of the band contours obtained by Stevens (774), Cremer, *et al.* (165), Robinson and Whittle (692), and a set of contours calculated according to a method recently proposed by Zwicker (900, 901). The result is shown in Fig. 144. Pollack (635) had subjects judge the loudness of some octave bands of noise using a reference sound — broadband white noise from 100 to 10,000 Hz. Pollack's contours tend to be somewhat "flatter" than the contours found when the reference sound is a tone or band of noise centered at 1000 Hz. Robinson *et al.* (691, 695) also determined the difference between equal loudness contours for free-field frontal incident sound and diffuse sound as would be heard in a typical room. Their findings are shown in Fig. 145.

Stevens' Methods for Calculating Loudness

Equal loudness contours, whether for pure tones or bands of noise, are of somewhat academic interest unless they can somehow be used for evaluating the loudness of complex noises and sounds found in real life. Fletcher and Munson (237) proposed a procedure for calculating from physical measurements the loudness of a complex sound consisting of a number of tones. Their method, however, was not used much because of its complexity, and because the sounds of greatest practical interest tend to be broad-spectra sounds and not pure tones.

Gates (Churcher and King [136]) and later Beranek *et al.* (61) proposed that a simple summation of the loudness in sones of octave bands of sound would give a reasonable approximation to the perceived loudness of a complex sound consisting of one or more octave bands of random noise. It was assumed for these purposes that an octave band of random noise having the same overall SPL as a pure tone of the same center frequency would be equally loud. In addition to the equal loudness contours for octave bands of random noise, Stevens (774, 776, 779) also published new procedures to be used for evaluating the total loudness of broad, continuous spectra sounds. Stevens demonstrated that his method was more accurate in predicting the judged loudness of complex

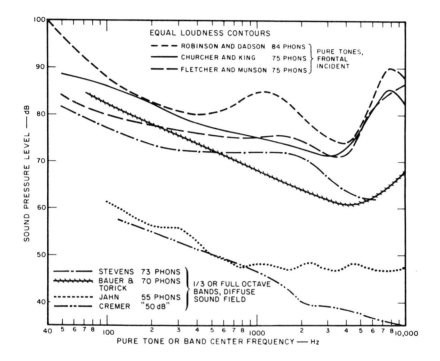

FIGURE 143. Comparison of equal loudness contours for pure tones and bands of noise.

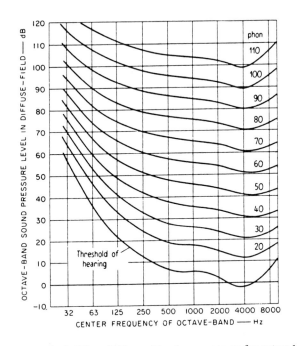

FIGURE 144.   Smoothed diffuse field equal loudness contours for octave bands of noise. From Robinson and Whittle (692).

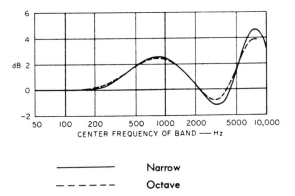

FIGURE 145.   Difference between sound pressure levels of frontally-incident and diffuse sound fields at equal loudness. Thus, at 1000 Hz a frontally incident field must be about 2dB greater than a diffuse field for equal loudness. From Robinson and Whittle (692).

sounds consisting of bands of random noise than the method of simply adding together the sone values of individual bands.

Stevens' general formula is to add to the sone value of the loudest band a fractional portion of the sum of the sone values of the remainder of the bands:

$$\text{Loudness} = S_m + f\,(\Sigma s - s_m)$$

where Σs = sones in all bands, S_m = maximum number of sones in any one band, and f = fractional portion dependent on bandwidth. Stevens derived the fractional portion to be applied when the spectra of the sound was measured in either full ($f = 0.3$), one-half ($f = 0.2$), or one-third ($f = 0.15$) octave bands.

Stevens (779) slightly modified his earlier method (776) of calculating loudness and named this new method Mark VI. Mark VI has been adopted by the U.S.A. Standards Institute as the procedure to be used for the calculation of loudness of noise measured in either octave, one-half octave, or one-third octave bands (30). The procedures and formulae for the calculation of loudness, Mark VI, are the same as that in the Stevens' 1957 article, except that individual band values of loudness are found from a graph depicting loudness index (I) contours (see Fig. 146) and somewhat different than equal loudness (sone) contours. For example, in the original Stevens' procedure for calculating the loudness of bands of noise, the octave band 600-1200 Hz at a sound pressure level of about 38.5 dB has a loudness of 1 sone; in the Mark VI modification, the same band at 34.5 dB is given a loudness index of "1."

It has become practice, however, to express the loudness of a given sound in terms of the sound pressure level in dB of the reference sound when it is as loud as the given sound, rather than in units of loudness, or sones. The result is called loudness level in phons. The unit phon can be calculated from psychological units, sones, but not directly from physical measurements of sound pressure because the relation between sound pressure level and loudness varies as a function of frequency differently at different levels of intensity.

The International Organization for Standardization (ISO) has recommended Mark VI as the method to be used for calculating the loudness of sounds measured with octave band filters, and Zwicker's method, to be described below, when the sounds are measured with one-third octave band filters (30).

Zwicker's Method for Calculating Loudness

As previously mentioned, Fletcher and Munson suggested that loudness is proportional to the number of nerve impulses per second reaching the brain from the auditory nerve fibers. Further, they noted that two tones competing for the attention of a single nerve fiber would interfere with simple loudness

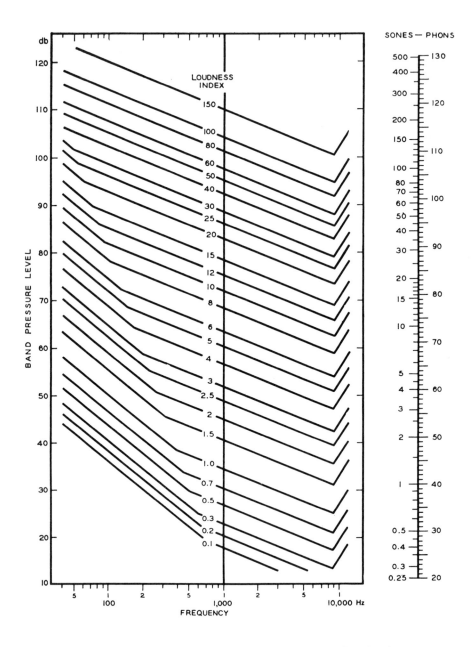

FIGURE 146.   Contours of equal-loudness index. From Stevens (779) and Ref. 30.

summation and that it must be necessary to group together all components within a certain frequency band and treat them as a single component. The width of these "grouping together" bands was estimated by Fletcher and Munson to be 100 Hz for frequencies below 2000 Hz, 200 Hz for frequencies between 2000 and 4000 Hz and, 400 Hz for frequencies between 4000 and 8000 Hz. From subjective tests of loudness and masking, Zwicker *et al.* (907) also determined the frequency groupings, "frequenzgruppen," that take place in the cochlea of the ear (see Table 1). Frequenzgruppen are sometimes referred to as critical bands.

Zwicker (900) determined the spread of masking for narrow bands of noise, the threshold of audibility of pure tones, and the change in level of a 1000 Hz tone to obtain a doubling (or halving) of loudness. His results on the growth of loudness are similar to those found by Stevens (775) and Robinson (682). His data for spread of masking for narrow bands of noise are more or less, as far as can be determined from his published results, like the spread-of-masking data obtained by Egan and Hake (206), Ehmer (211), and Carter and Kryter (127) (see Chapter 2). Zwicker's assumption that there is a functional correspondence between masking and loudness is well substantiated by data on the critical bandwidth of the ear.

Zwicker (901), on the basis of these concepts, developed a graphic method for depicting and calculating the loudness of a complex sound. For calculation purposes he prepared ten graphs (covering both diffuse and free-field conditions, see Fig. 147 for an example) in which the horizontal ordinates are marked off in equal frequenzgruppen (approximated for practical purposes by one-third octave steps above 280 Hz), and the vertical divisions for each frequenzgruppen, in loudness units, are proportional to sones. The short-dashed curves show the area covered by the upward spread of masking.

Plotting a sound spectrum on Zwicker's graph and drawing in the lines for spread of masking are supposed to show, in essence, what proportion of available "nerve impulse units" are made operative as the result of exposure of the ear to a given sound; accordingly, this area on the graph is proportional to total loudness. A planimeter is used for measuring the area encompassed by a given sound, as plotted on one of Zwicker's graphs, although the area can also be estimated with reasonable accuracy by visual inspection.

Zwicker defines as one sone the area encompassed on his graph by a one-third octave band of noise centered at 1000 Hz at a sound pressure level of 40 dB, including the additional area encompassed by the dashed curve that takes into account the upward spread of loudness and masking.

It should be noted that, in Stevens' Mark VI method, either a one-third octave or full octave band of noise, centered at 1000 Hz at a level of 34.5 dB, would have a loudness index of 1.0. Because of this and other differences between the Stevens and Zwicker methods, the loudness levels calculated by the two procedures for the same sound often differ by 3 to 5 phons.

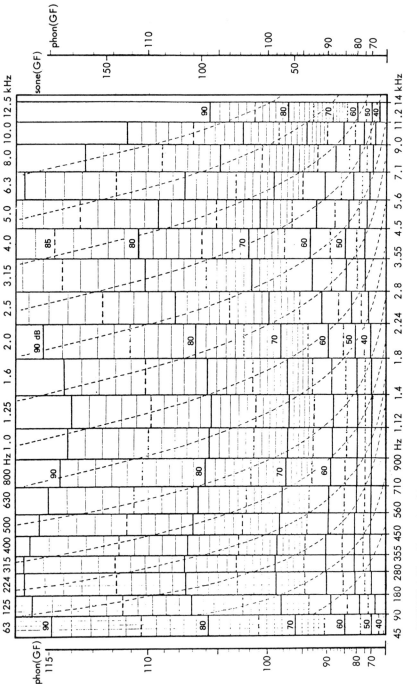

FIGURE 147. Example of loudness computation graph; the range is 70-115 Phons. From Ref. 30, after Zwicker (901).

Munson's Method

Munson (567) has proposed a modification of the equivalent-tone-sone summation method suggested by Churcher and King (136) and Beranek *et al.* (61) to take into account the spread-of-masking and loudness effects that are acknowledged in Stevens' and Zwicker's schemes for calculating loudness. Munson's procedure, as he states, is not based on any published theoretical model and, perhaps to some extent for that reason, has not been widely used.

Dependence of Loudness on Intensity (Growth of Loudness)

The studies of loudness discussed to this point have been concerned primarily with (*a*) the loudness of individual pure tones or narrow bands of noise of different frequency relative to the loudness of a standard, usually, 1000 Hz tone and (*b*) the loudness of several pure tones or bands of noise heard together, i.e., the effect of variations in total bandwidth of a complex sound upon judged loudness. Although there are differences in equal loudness contours found by various investigators, their shapes are in reasonable agreement (see Fig. 143). On the other hand, scaling the growth of loudness of a sound as a function of changes in its intensity into steps that are subjectively equal in size has been a much more controversial problem. Reviews of the work in this area have been made by Stevens (777) and Gzhesik *et al.* (332).

There have been three general methods used for scaling the growth of loudness of a sound, usually a 1000 Hz tone, as a function of changes in sound pressure level.

1. Monaural vs. binaural loudness.

2. Magnitude and ratio estimation.

3. Equal section or equal interval.

Monaural vs. Binaural Loudness

The argument for the method used by Fletcher and Munson (237), which followed from their assumption that loudness was proportional to the number of auditory nerve impulses reaching the brain, was that the same sound delivered to the two ears should appear to be twice as loud as when presented only to one ear. Fletcher and Munson found that the level of the monaurally presented tone had to be set about 10 dB higher in level than the level of an equally loud binaurally presented tone. Thus, they concluded that, over at least the middle range of loudness levels, subjective loudness about doubles for each 10 dB increase in the sound pressure level of a sound.

Reynolds and Stevens (670) found that the loudness scale for monaural listening was somewhat different than the loudness scale for binaural listening, indicating that the Fletcher and Munson assumption about the summation of loudness from the two ears appeared less than perfect, at least at some intensity levels. However, Hellman and Zwislocki (364) later found nearly perfect, within experimental error, interaural summation of loudness, as shown in Fig. 148.

Magnitude and Ratio Estimation

The monaural vs. binaural equal loudness scale is very similar to the average of those developed on the basis of magnitude estimations of the loudness of sound presented only monaurally or only binaurally. In this method, the subjects assign a number, say 100, to a tone at, say, 100 dB SPL; they are then asked to assign the number 50 to the tone when it sounds half as loud as it did at 100 dB. Another method is that of estimating loudness ratios or fractions; here the subjects may adjust the level of a tone until it is one-half or one-tenth, etc., as loud as a standard or reference level.

Results of studies by various investigators using the magnitude estimation and ratio judgment methods differ rather widely. Garner (288, 290) believes that the differences among the results of experiments on judgments of loudness fractions are due in part to context effects. That is, a subject will give different judgments about what appears half as loud when he knows the total range of levels available to him for judgment than when he does not. In most studies of loudness fractionation, the minimum or zero loudness is assumed to be threshold of hearing, a rather inexact and individualistic value that would change the general context of level range available to different listeners.

Garner (290) was able to train different groups of subjects (a training period plus 600 experimental trials) to state that the half-loudness of a 90 dB tone was either 60, 70, or 80 dB depending on the range of intensities available to each group as a choice for half-loudness. A second factor is that different people apparently have different rules they follow when making ratio or fraction judgments. Evidence of this variability in individual loudness function was found by Garner (287) by the method of fractionation (one-half), shown in Fig. 149.

A third factor, probably related to the second above, that has caused some variability in loudness estimation, is that numbers apparently have semantic meaning beyond their strict arithmetic character. Hellman and Zwislocki (363), using the method of magnitude estimation, obtained results that suggest that the number one, for example, was appropriate for the loudness of a 1000 Hz tone at 40 dB, and the number ten for a level of about 70 dB, as indicated in Fig. 150. Figure 151 shows that different loudness scales are found when the number ten is assigned by the experimenter to different reference sound pressure levels, and

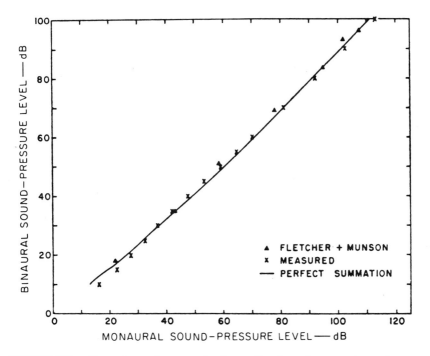

FIGURE 148.   Binaural sound-pressure level as a function of monaural sound-pressure
level at equal loudness. From Hellman and Zwislocki (364).

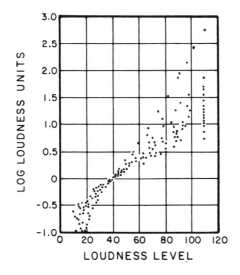

FIGURE 149.   Results from loudness tests with 18 observers based on fractionation data.
From Garner (287).

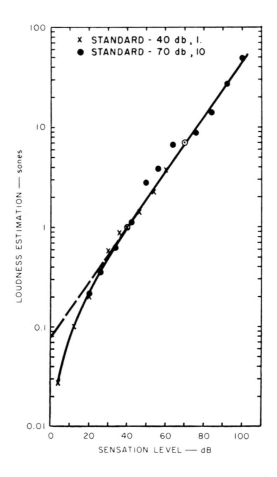

FIGURE 150.   Median loudness estimates for two reference standards normalized to the
40 dB reference standard. From Hellman and Zwislocki (363).

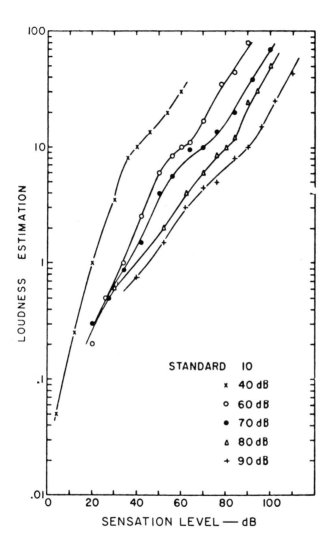

FIGURE 151. Median loudness estimates as a function of sensation level (SL) obtained with a reference number 10 assigned to give reference SL's. From Hellman and Zwislocki (363).

the subjects then estimate the subjective number they feel should be assigned to the tone presented at different sensation levels.

Stevens (777), in reviewing loudness scaling procedures, makes the point that although obtaining a loudness scale from a listener is a difficult problem, it is a function that must be determined if the concept of loudness is to have any practical utility. Stevens suggests that the best method (called magnitude production) is to allow each subject to use whatever number scheme he wishes, and then to average results across subjects after normalizing the results for individual differences in the choice of numbers used.

Equisection Loudness Scale (Equal Intervals)

In addition to the one-ear vs. two-ear, and the methods of magnitude and ratio estimation, a method of equal intervals or equisections has been suggested as a suitable method for deriving a scale of loudness. In this method, the subject hears a tone presented at, in the simplest case, two different levels of intensity; he is then told to adjust the third level of the same tone such that the difference in loudness between the second and third levels is equal to that between the first and second levels. Using this method, Wolsk (882), Kwiek (479), and Garner (287) measured equal intervals over various ranges of intensity of a 1000 Hz tone.

Unlike the magnitude and ratio estimation methods, the results obtained by various investigators using the method of equal intervals are in close agreement with each other. However, there is no real knowledge obtained from the equal interval method as to what changes in level are required in order that the listener report a subjective sensation of the doubling, or halving, or some other fraction, in the loudness of a sound.

Garner concluded that loudness scales, based on ratio judgments and magnitude estimations, are too inconsistent among different subjects to be meaningful. Instead, Garner derived a loudness scale from judgments of equal loudness intervals found by the equisection procedure. The results of a series of equal-interval tests are shown in Fig. 152. Although words like "one-half," or "twice," or a numbering scheme are not included in the instructions to the subjects, the method of equal intervals or equisection is in the last analysis a special case of magnitude estimation where the subject is presented with a very restricted range of intensities he is asked to bisect. The repeatability of the experimental findings of various investigators may be as much due to this restricted range of levels involved in any one set of judgments as it is to the unambiguousness of the task assigned to the listeners.

Since the loudness scale, derived by the equal-interval method, is so different from the scales derived by other methods (see Fig. 153), we must choose one or

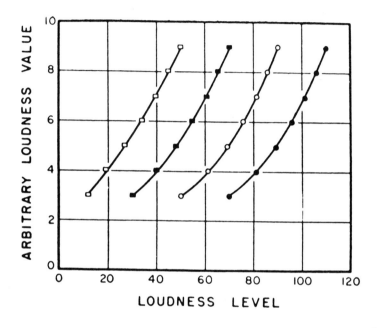

FIGURE 152.   An illustrative set of data obtained from an equisection procedure for loudness judgments. From Garner (287).

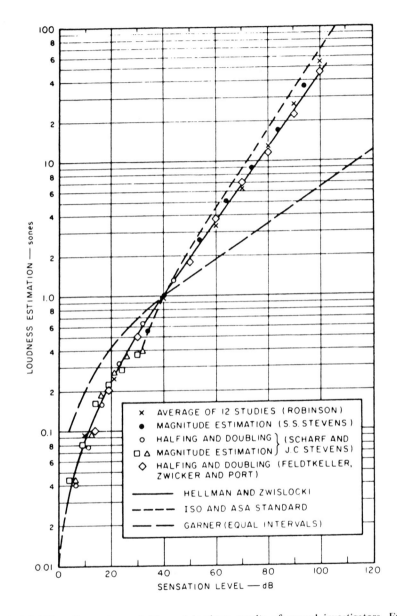

FIGURE 153.   Comparison of binaural loudness results of several investigators. From
              Kryter (454).

the other for practical use. It would seem reasonable to decide which of these forms of loudness functions is the most appropriate on the basis of how the loudness scale is to be used. If, for example, it is intended to say that sound "A" is twice (or some portion) as loud as sound "B," then we are obliged to use a loudness scale based on ratio or magnitude judgments. On the other hand, if we want to decide whether the difference in loudness between sounds A and B is equal to the difference in loudness between B and C, then the Garner-Kwiek loudness scale would be more meaningful. If, and we would assume that such is the case, the general interest in loudness judgments in real-life situations is more in terms of apparent magnitude or relative loudnesses than in terms of equal intervals, it would seem that we must accept the loudness scale based on magnitude estimation as being the more appropriate for general use.

Changes in Loudness with Time

For the most part, loudness judgments have been made only of sounds having durations of fractions from one second to several seconds long. According to Miller (544) and Garner (284), loudness presumably remains more or less constant after the first 100-500 milliseconds of duration of a sound (see Fig. 5, 18, 19 and 154). Continued exposure to a steady-state sound produced another change in loudness that normally goes unnoticed by the listener. It is most striking when one ear is exposed and the other ear is not exposed to an intense sound. When both ears are then subsequently exposed to the same sound, the loudness in the previously unexposed ear is greater than in the previously exposed ear. The effect has been called perstimulatory fatigue and an example of its effect on loudness is shown in Fig. 155. It is not clear whether the effect is due to receptor fatigue or to a purely perceptual loudness adaptation, or to both.

Loudness Predicted by Sound Level Meters

Although the loudness of a complex sound may best be estimated on the basis of its band spectrum, the sound level meter that integrates acoustic energy over the audible spectrum to achieve a single overall value is widely used for this purpose.

The present standardized sound level can be operated in four modes (see Fig. 8):

1. with a network that causes all frequency components within a sound to be weighted equally — the flat scale;

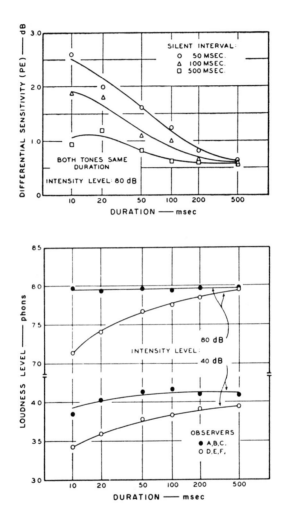

FIGURE 154.   Upper graph: The effect of tone duration on differential sensitivity for intensity when both the standard and comparison tones have the same duration. Lower graph: The effect of tone duration on loudness, at two intensity levels, for two different groups of observers. Each plotted point is the average of data obtained with three different silent intervals between tones. From Garner (285).

2. with a network that more or less weights the intensity value of the frequency components in a sound in accordance with the shape of the Fletcher-Munson equal loudness contour at the level of 100 phons — the C scale;

3. with a network that weights the frequency components more or less in accordance with the 70 phon contour — the B scale; and

4. with a network that weights the frequency components more or less in accordance with the 40 phon contour — the A scale.

The validity and use of the sound level meter with weighting networks for the evaluation of noises will be discussed later, especially in Chapter 9.

Meters involving a set of octave band filters and various other electronic circuits that will automatically give loudness level in phons, as would be found by the Stevens method of calculating loudness, have been developed by Anderson (12) and Bauer and Torick (45). Blaesser (70) described a meter and spectral analysis display that provides the loudness level in phons of a sound according to the procedures proposed for this purpose by Zwicker.

Impulse Noise

The measurements and methods of estimating loudness described above are applicable to sounds that are more or less steady-state in time. Some attention has been given to the judged loudness of impulse noise and the design of meters that would give readings that are correlated with the loudness of such sounds. In particular, the work of Niese (577-579) and Port (652) should be referred to. Niese proposed that, for impulse noise, a sound level meter with A-weighting but a 23-msec time constant (which would make it a quasipeak meter) be used. Port found that a sound level meter with a time constant of 70 msec does a reasonable job of estimating the loudness of impulse or impact sounds.

Because the ear is sensitive to the spectral content of the impulse, the time constants chosen by Niese and Port may provide fortuitously appropriate indications of the effective level of some impulsive sounds. This question of how to measure the sound pressure level and spectrum of impulse sounds for estimating auditory threshold shift effects was discussed in Part I, and the method for estimating perceived noisiness will be discussed in later chapters.

Interrupted Noise

The value 200 msec has usually been interpreted as a time constant of the auditory system required to integrate energy and maintain a steady level of response. As Garner (282) noted, this time constant also describes the critical

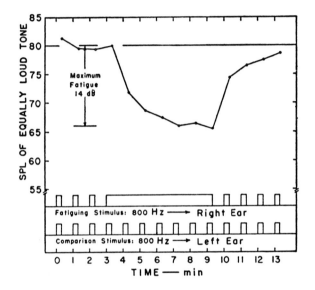

FIGURE 155.   Sound pressure level of an equally loud comparison tone as a function of
the time at which the loudness balance was made. The intensity of the
fatiguing stimulus was always 80 dB, and the intensity of the comparison
stimulus was adjusted by the listener. As shown by the on-off markers near
the bottom of the graph, the perstimulatory fatiguing period began at the
third minute and lasted for 6 min. From Egan (205).

bandwidth of the ear as analogous to a filter: the wider the filter, the shorter is the time required for a signal of a given bandwidth to reach a steady level at its output. Since the critical bandwidth of the ear varies as a function of frequency, the growth of loudness in time of brief sounds should also, according to this reasoning, be a function of its center frequency.

Garner (284) found that a series of repeated short tones can be louder than a steady tone of the same peak intensity. He pointed out that these results are understandable in view of the shape of equal loudness contours and the complex spectra of interrupted tones. Pollack (634) later avoided the change in spectra that occurs when tones are interrupted by using a random white noise as his signal.

Pollack (642) found that when a noise burst was one-tenth as long or longer than the interval between bursts plus the duration of the burst, the loudness was as though the noise was on continuously. For shorter burst-time fractions, the loudness level in phons was directly proportional to a change in log burst-time fraction. Pollack presents formulae whereby one can calculate the loudness of interrupted noise. Pollack reported that a noise, interrupted at a rate of 2-10 per second, was louder than a continuous noise of the same energy. He interprets this finding in terms of the ability of nerve fibers to fire best when stimulated at certain rates and also notes that the so-called electrical "alpha rhythm" of the central nervous system is about 10 Hz; it is conceivable that aural reflex action along with reduced auditory fatigue may somehow also be involved in this phenomenon.

Carter (124) compared the judged and calculated loudness (Stevens' and Zwicker's methods) of transients having a triangular waveform (0.5 msec rise and decay times) presented at a rate of 10 per sec with that of white noise interrupted 10 times per sec. He found that the calculated and judged loudnesses differed by as much as 8 phons, but that there was a constant 40 dB difference between peak-to-peak SPL of impulse and rms level of interrupted noise (i.e., 120 dB peak-to-peak pulse was about as loud as 80 dB interrupted noise). Because of uncertainty about the actual spectra of the sounds, it is not possible to draw firm conclusions from this study.

Figure 156 shows the results of various loudness tests made with impulsive-type sounds; these data show that, as the rise time of any impulse is shortened, keeping peak overpressure constant and duration nearly constant, the loudness increases. This finding is to be expected since the intensity of the higher frequency components also increases (see Figs. 12 and 13). However, the earlier tests conducted by Steudel (767) and Bürck et al. (104) show inexplicably large changes in loudness as rise time is shortened, whereas the results of Zepler and Harel (899) appear to be much more reasonable. Some additional judgment tests have been made of the loudness and the perceived noisiness of impulsive sounds (sonic boom). The results of these tests will be discussed in Chapters 8 and 9.

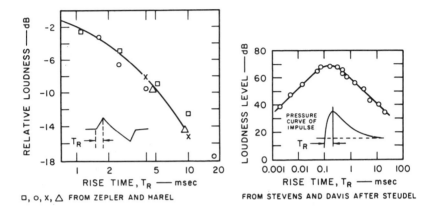

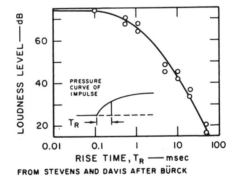

FIGURE 156.   Change in loudness of impulses as a function of their rise times, as found by various investigators: Zepler and Harel (899), Steudel (767), and Burck *et al.* (104), as reported by Stevens and Davis (781).

Chapter 8

Perceived Noisiness (Annoyance)

Introduction

It would appear that a model having the following four points is implicit to most of the quantitative approaches that have been made to the evaluation of annoyance due to environmental noise:

1. Since individuals in a community live in somewhat similar and daily repetitive ways, the average amount of annoyance from physical measures of the noise should be somewhat predictable in a statistical sense.

2. The effects of spectral and temporal changes in expected sounds upon judged annoyance, keeping meaning and cognitive aspects as constant as possible, can be used as a basis for noise measurement and control purposes.

3. Various psychological and sociological factors present in individuals and a community influence the annoyance felt and the behavior expressed by people in response to the annoyance caused by noise.

4. Because of differences in work or living requirements, different tolerable limits of noise exposure must be set for different rooms and community environments. Special cognitive meanings associated with given sources (fear of injury, etc.) may on occasion require different tolerable limits for different sources.

The next two chapters are concerned with the basis and general validity of this four-point model. The present chapter is concerned primarily with basic concepts and research relating physical measurements of individual noise occurrences with psychological judgments of the noise. Chapter 9 discusses noises as found in our general living, some work, and some transportation environments, and methods for evaluating the total complex of noises to which people are exposed during their 24-hour day.

269

Perceived Noisiness

The subjective impression of the unwantedness of a not unexpected, nonpain or fear-provoking sound as part of one's environment is defined as the attribute of perceived noisiness. The measurement or estimation of this subjective attribute or quantity is of central importance to the evaluation of environmental sounds or noises with regard to its physical content. For this reason, this topic will be discussed in considerable detail.

Confusion sometimes results in the use of the word noise as a name for unwanted sound because there are two general classes of "unwantedness." The first category is that in which the sound signifies or carries information about the source of the sound that the listener has learned to associate with some unpleasantness not due to the sound per se, but due to some other attribute of the source — the sound of the fingernail on the blackboard suggests perhaps an unpleasant feeling in tissues under the fingernail; a baby's cry causes anguish in a mother; the squeak of a floorboard is frightening as indicating the presence of a prowler; a sonic boom is disturbing because it is an unfamiliar sound, etc. In these cases it is not the sound that is unwanted (although for other reasons it may also be unwanted) but the information it conveys to the listener that is unwanted. This information is strongly influenced by the past experiences of each individual; because these effects cannot be quantitatively related to the physical characteristics of the sounds, they are rejected from the concept of perceived noisiness. After all, the engineer, attempting to control the noise from a given source, must shape the characteristics of the noise in as effective a way as possible for the majority of the people and the most typical of circumstances; those legislating or adjudicating the amounts of noise to be considered tolerable must also have a quantitative yardstick that is relatable to groups of people and typical circumstances.

Psychological judgment tests have demonstrated that people will fairly consistently judge among themselves the "unwantedness," "unacceptableness," "objectionableness," or "noisiness" of sounds that vary in their spectral and temporal nature provided that the sounds do not differ significantly in their emotional meaning and are equally expected. Presumably this consistency is present because men learn through normal experience the relations between the characteristics of sounds and their basic perceptual effects: masking, loudness, noisiness, and, for impulses, startle. This is a basic premise of the concept of perceived noisiness and of the word noise as unwanted sound. Although noise evaluation procedures — specific to individual effects such as speech masking, loudness, and auditory fatigue — are available, a single number rating for the average unacceptability or perceived noisiness of normal environmental noises appears to be adequate for community noise control and management from a physical standpoint.

It is also an hypothesis of the concept of perceived noisiness that even though the absolute level of noisiness or unacceptability of the sound from a given source, say the buzz of insects, may be vastly different than the sound of another source, say an automobile, because of differences in the meaning to the auditor of the two sounds, the relative effects of variation in the frequency content, duration, and spectral complexity upon perceived noisiness will be similar for noises of near similar meanings. This is what will be meant in this text when it is said that a proper unit of physical measurement for estimating perceived noisiness is independent of the source of the noise.

Perceived noisiness is obviously synonymous with what is often implied by the word annoyance. However, the word annoyance is commonly used to signify one's reaction to sound that is based both on what we have attempted to delimit for perceived noisiness and also on the emotional content and novelty (which are excluded from perceived noisiness) the sound may have for the particular individual. The phrase "perceived noisiness," although somewhat redundant, was chosen in an attempt to avoid the ambiguity possible from the word annoyance when speaking of the attribute with which we are concerned.

Psychological-sociological factors can usually be reconciled with the general attribute of sound called perceived noisiness. For example, Borsky (80) and Cederlof *et al.* (131) found that propaganda, stressing the importance of military aviation to the people and the plans of the government to control and lessen the noise, reduce the willingness of citizens near military airports to complain about the aviation noise; the reduction was equivalent to the effect that would have been obtained by lowering the noise levels by 6 dB or so. At the same time, the concept of perceived noisiness would maintain that reduction of the actual noise level should further reduce the willingness of the average person to complain about the noise, regardless of his particular absolute willingness at a given moment, and that this amount of average reduction in complaints would be a function of how cleverly, and compatible, to the attribute of perceived noisiness, the noise spectrum and its duration were tailored. Hawel (353) has proposed to obtain the quantitative relations between some of these psychological, sociological, and attitudinal factors and noise exposure. According to this concept, one could apply correlations or adjustments during the calculation or measurement of noise exposures to take these factors into account. Although the evaluation of the relative contribution of the physical aspects of sounds to their perceived noisiness should in no way interfere with or diminish the manipulation of psychological and sociological factors in the control of environmental noise, basic aspects of perceived noisiness probably set certain fundamental limits, as will be discussed later, on the tolerability of noise.

Loudness vs. Noisiness

Loudness of sounds is often assumed to be an adequate indicator of the unwantedness, for general noise control purposes, of sounds. Experiments have shown, however, that for many sounds there are differences between some physical aspects of sounds, and judgments of loudness compared to judgments of perceived noisiness. The difference between loudness and perceived noisiness in terms of spectral content per se (the equal loudness vs. equal noisiness contours) is insignificantly small for broadband sounds, as shown in Figs. 157 and 158. On the other hand, the differential effects of duration and spectral complexity upon these two attributes, as will be shown, are rather large.

The fact that loudness is apparently not influenced by duration and spectral complexity features of a sound would seem to disqualify loudness as an appropriate attribute for the estimation of the unacceptability of environmental noises. Although loudness and perceived noisiness differ in some respects, an assumption of the concept of the perceived noisiness of nonimpulsive noises is that, as the intensity of a noise changes, keeping other factors constant, the subjective magnitude of loudness and noisiness change to a like degree; e.g., a 10 dB increase in the physical intensity of nonimpulsive sounds causes a doubling of the subjective magnitude of its loudness and its noisiness. There is some experimental proof of this common relation between this subjective scale of noisiness and loudness, but, as with loudness, the scale found is somewhat dependent on the experimental methods used and sounds judged (99, 601, 592a).

Instructions to Subjects

The words used in the instructions to the subjects for judgment tests of the acceptability of sounds have some influence upon their rating of sounds, as illustrated in Fig. 159. It is difficult and probably academic to fathom what is the basis for the range of differences shown in Fig. 159, such as whether the words used really mean different things to different people. In any event, there is no apparent reason why listeners should not be asked to rate directly sounds in terms of their unwantedness, unacceptability, annoyance, or noisiness, as synonyms, rather than to rate their loudness in the expectation that the latter is an indirect clue to the noisiness or unwantedness of the sounds.

Following are parts of the instructions that have been given to subjects who were asked to make subjective judgment tests of the noisiness of sounds. *"Instructions, Method of Paired-Comparison, for Judgments of Noisiness.* You will hear one sound followed immediately by a second sound. You are to judge which of the two sounds you think would be the most disturbing or unacceptable if heard regularly, as a matter of course 20 to 30 times per day in

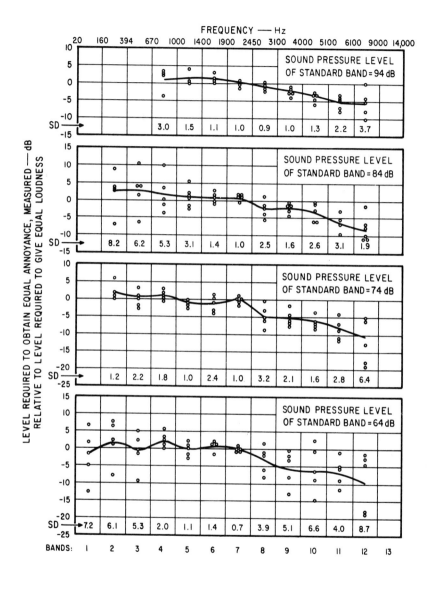

FIGURE 157.   Equal annoyance contours for bands of noise. Band 7 was reference band. Individual subjects are shown by circles. After Reese *et al.* (663).

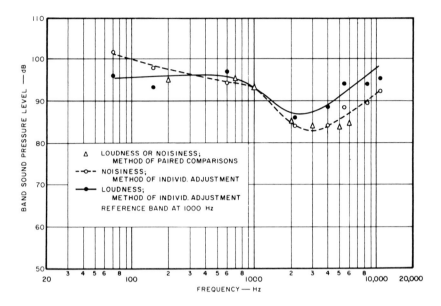

FIGURE 158. Equal loudness and equal noisiness judgments. From Kryter and Pearsons
(466).

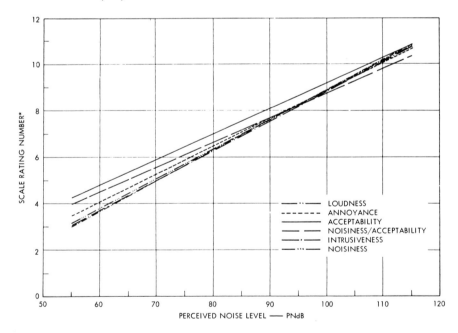

FIGURE 159. Mean response rating of all noise stimuli for all laboratory test sessions.
From Pearsons and Horonjeff (612).

your home. Remember, your job is to judge the second of each pair of sounds with respect to the first sound of that pair. You may think that neither of the two sounds is objectionable or that both are objectionable; what we would like you to do is judge whether the second sound would be more disturbing or less disturbing than the first sound if heard in your home periodically 20 to 30 times during the day and night." The purpose of including in the instructions to the listeners a number of terms in rating the noisiness or unwantedness of expected sounds is to try to reduce possible differences in how different subjects might interpret the purpose or intent of the judgments when only one term such as "disturbing" or "annoyance" is used.

Although it might be thought that the term loudness would be an unambiguous instruction for getting subjects to rate the annoyance values of sounds, such is not the case even though loudness is clearly a significant determinant of annoyance. Some subjects will surmise that the experimenter is concerned with the annoyance value of the sounds when he is asked to judge loudness (and for this reason the results of the tests for these two attributes will be fortuitously similar); some subjects, however, think of loudness as the correlate of the intensity of a sound and whenever two sounds they are asked to judge are (a) of equal peak intensity, but different durations, (b) of varying intensity, or (c) contain spectral-complexities, these subjects sometimes ask the experimenter for further definitions of what is meant by the term loudness. In this case the experimenter must usually reveal that the real intent, if true, of the instructions is to get at annoyance, unwantedness, or, in short, perceived noisiness. Not only is confusion avoided by asking the subjects directly to rate the unwantedness of sounds rather than loudness when the former attribute is of interest, it also leaves unsullied the term loudness for the attribute of the peak or steady subjective intensity of a sound.

The importance of defining the terms used in the instructions to the subjects is illustrated by the findings of Kerrick et al. (430). The subjects were asked to rate a number of sounds on a scale marked "loudness," a scale marked "noisiness," and a scale marked "unacceptable." No instructions or definitions of the meaning of those terms were given to the subjects. As seen in the left-hand graph of Fig. 160, the sounds were rated about the same on the scales of loudness and noisiness. However, as seen in the right-hand graph of Fig. 160, the sounds were rated differently on a scale of "unacceptable" than on a scale marked "noisy." Experiments in which single, undefined adjectives are used, without further instruction, as bases for rating scales of sounds, particularly when a number of such adjectives are used as separate instructions, are perhaps more related to questions of semantics than to the elucidation of the attribute of perceived noisiness.

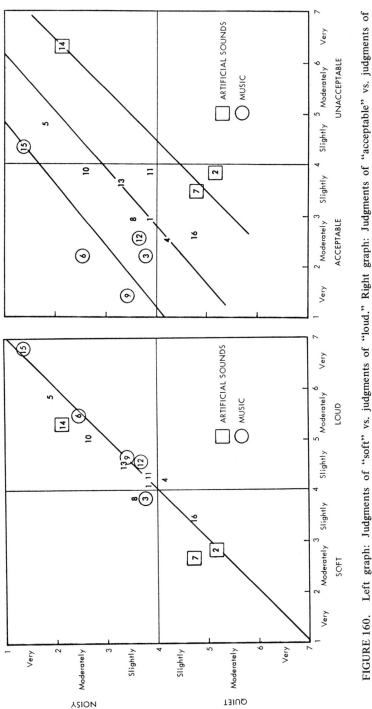

FIGURE 160.   Left graph: Judgments of "soft" vs. judgments of "loud." Right graph: Judgments of "acceptable" vs. judgments of "unacceptable." Sound stimuli: (1) DC8 flyover, (2) octave band centered at 1000 Hz, (3) Bernstein (jazz), (4) motorcycle passby, (5) helicopter flyover (plus 20 dB), (6) popular music, (7) shaped synthetic broadband noise, (8) auto passby, (9) folk music, (10) 720B flyover, (11) helicopter flyover, (12) Vivaldi (classical), (13) truck passby, (14) tone complex, (15) popular music (+20dB), (16) rain. From Kerrick *et al.* (430).

Influence of Cognitive Values

Perhaps the attribute of loudness has often been proposed as an appropriate basis for rating the relative unacceptability of sounds on the hope that judgments of loudness would be unaffected by the meanings or cognitive values that different sounds have for different people. However, it appears that cognitive factors, at least in some experiments, influence both loudness and noisiness ratings. For example, as seen in the right-hand graph of Fig. 160, Kerrick *et al.* found that music was in general rated more acceptable than recordings of transportation noise, and both were rated more acceptable than "artificial sounds."

The findings of Kerrick *et al.* are not inconsistent with, but are rather irrelevant to, the attribute of perceived noisiness as defined, in that the sounds that were intercompared — music vs. transportation noises for example — were of different cognitive meaning. The concept of perceived noisiness would say that, within each cognitive category of sounds, the relative effects of spectrum, duration, and complexity should similarly affect perceived noisiness; it is interesting to note in Fig. 160 that the sounds within each class do appear to be rated along a single continuum, which is consistent with this concept.

Five Physical Aspects

So much for the general concept of the perceived noisiness of individual sounds. For practical purposes the measurable physical aspects of a sound that are most likely to control its perceived noisiness must be determined. To date, five significant features have been identified or suggested — (1) spectrum content and level; (2) spectrum complexity (concentration of energy in pure-tone or narrow frequency bands within a broadband spectrum); (3) duration of the total sound; (4) duration of the increase in level prior to the maximum level of nonimpulsive sounds; and (5) the increase in level, within an interval of 0.5 sec, of impulsive sounds. Some physical aspects that might seem important — for example Doppler shift (the change in the frequency and sometimes noted pitch of a sound as a sound source moves towards and away from the listener, Nixon *et al.* [584] and Ollerhead [592]) and modulation of pure tones (Pearsons [611]) — appear to be very secondary in their effects on people compared to the five physical characteristics mentioned above.

Historical Résumé

Next is a brief historical resume of the development of procedures for measuring or estimating perceived noisiness as influenced by the five aspects of a

TABLE 29

Chronology of Data and Procedures Related to Calculation of Perceived Noise Level and Composite Noise Rating

									Presently Suggested
FREQUENCY									
Weighted Band Spectrum	'29 L & C^{483}	'44 R, K & S^{663}				'63 K & p^{466}	'67 w868	'68 O592	'63, K & p^{466} Adjusted < 355 Hz
Band Summation								'56 S774	'59, K^{444}, PNdB; '69, K^{462}, PNdB
OA-Weighted Spectrum				'58 Y896 dB(A)	'59 K^{444} dB(A), dB(N)		'68 K458 dB(N)	'68 Y & p898 dB(RC)	'69, K^{462} Adjusted dB(D) < 355 Hz
Spectral Complexity					'61 L513	'62 W & B869	'63 K & p^{466}	'68 S758	'63, K & p^{466}; '68, S758
TIME									
Total Energy — Single Occurrences						'63 K & p^{466} Estimated Effective		'68 K458 Effective Threshold	'69, K^{462} Effective 8-sec Reference Duration
Total Energy — Multiple Occurrences over 24 hrs			'53 R & S^{702} NR*	'57 S & P769 CNR*	'63 Mc536 NNI	'65 B, et al105 Q̄	'68 L & S485 TNI	'69 R686a NPL	'68, K458 CNR in EPNL
Onset Duration								'68 N, vG&R584	'69, K^{462} EPNL
Impulse Level reached in unit of time						'60 N577		'63 p652	'69, K^{462} EPNL

*Original NR and CNR, unlike NNI and Q̄, incorporated adjustments for time of day and for socio-economic status of a neighborhood. See Chapter 3

sound just mentioned. This resume will perhaps help show the rather uneven development of noise measurement procedures into a system that is seemingly comprehensive. Table 29 traces this history.

Noisiness Contours and Band Summation

Some experiments were performed in 1943 at the Harvard Psychoacoustics Laboratory under the direction of Professor S.S. Stevens to pursue the earlier work of Laird and Coye (483) on the annoyance values of sounds containing different frequencies. The data from these studies, reported by Reese *et al.* (663) and Kryter (441), showed that the higher frequencies tended to be more annoying than the lower frequencies even though they were equally loud.

In the 1943 study, some curves were obtained relating the intensity of bands of random noise to their band center frequencies when the bands were judged equally "annoying." Although the curves were based on a rather small amount of data, they were renamed equal noisiness contours and were used by Kryter (444), in conjunction with formulae and contours developed by Stevens for calculating the loudness of complex sounds, in an attempt to predict the results of subjective judgment tests of the unwantedness of aircraft noise. To distinguish noisiness from loudness, it was proposed that the subjective unit of noisiness be called the "noy" in parallel to the "sone" for loudness. A sound of 2 noy was said to be subjectively twice as noisy as a sound of 1 noy; 4 noy was assigned to the sound four times as noisy as a sound of 1 noy, etc. "PNdB" was coined as the name of the unit of perceived noise level (PNL) as calculated for a sound. The PNdB unit is a translation of the subjective noy scale to a dB-like scale; an increase of 10 PNdB in a sound is equivalent to a doubling of its noy value.

Kryter and Pearsons (466, 467) later obtained further data on equal noisiness contours (see Figs. 161 and 162) which they proposed be used in place of the contours suggested earlier in 1959. These contours were obtained with bands of random noise in the middle to higher levels (60-100 dB or so) of intensity; the contours for the very low and highest levels were extrapolated from the lowest and highest experimentally found contours. Wells (868) obtained a set of equal noisiness contours by having subjects adjust the octave band levels within a broadband random noise until the noise seemed to be subjectively equal in noisiness to the standard band centered at 1000 Hz. Ollerhead (592) obtained equal noisiness contours with one-third and full-octave bands of noise. S.S. Stevens compiled composite equal loudness and noisiness contours from various published loudness and noisiness contours, and labeled the contours Perceived Level, Mark VII. These contours and the contours Wells and Ollerhead obtained are similar in general shape to those obtained by Kryter and Pearsons, as shown in Fig. 162.

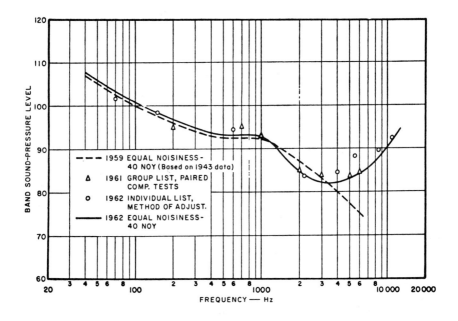

FIGURE 161.   Comparison of equal-noisiness contours – 1959 and 1962. Kryter and
Pearsons (466).

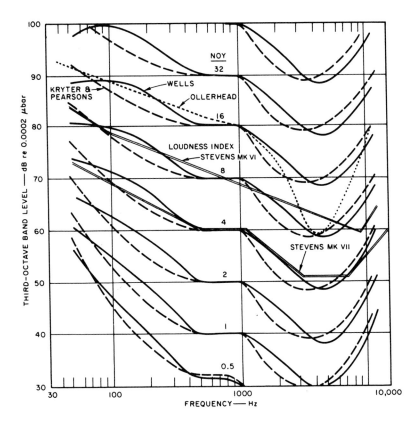

FIGURE 162.   Equal-noisiness contours as found by Kryter and Pearsons (466), Wells (868), and Ollerhead (602); loudness index, MK VI, Stevens (779); and MK VII, Stevens (780a,b).

Kryter (462) proposed and demonstrated that the SPL in frequency bands below 355 Hz should be combined in a way that would reflect the energy in critical bands of the ear (in a manner similar to that proposed earlier by Zwicker in his procedure for calculating loudness) prior to the calculation of the perceived noisiness of a sound. It was tentatively proposed (462) that the calculation of PNdB be simplified by discarding the band summation procedures adopted from Stevens' method for calculating loudness.

OA Frequency Weightings

It was noted (896, 444) that a sound level meter with the A-frequency weighting could be used as a means of rating noises, and in 1960 it was recommended (458) that the converse of the 40 noy contour, called N, be used in preference to dB (A). It was suggested by the International Technical Commission (L. Batchelder, Technical Note, *J. Acoust. Soc. Am.* **44**, 1159, 1968) that this first proposed N-weighting be designated as D and is herein designated as D_1 (see Fig. 8). Kryter (462) proposed that a new frequency weighting network, herein called D_2, be used as a means for estimating with a sound level meter the perceived noise level of broadband sounds. Young and Peterson (898) suggested that a frequency weighting, herein called D_3, requiring only a particular single electronic resistance-capacitance (RC) combination be used. This weighting function provides a straight monotonic approximation to the D_2-weighting, leaving out the sharp inflection of the D_2-weighting at 1000 Hz.

Spectral Complexity

Little (513), Wells and Blazier (869), and Kryter and Pearsons (466) proposed procedures for increasing above normal the value of the PNdB of a noise that contained strong pure-tone components. Sperry (758) quantified the general method suggested by Little for purposes of aircraft noise evaluation.

"Energy" Summation

In 1953, Rosenblith and K.N. Stevens (702) (see also refs. 76, 625, 626, 769, and 771) proposed that some frequency-weighted measure of noises summed over a 24-hour period on an estimated "energy" basis serve as a Noise Rating (later called Composite Noise Rating, CNR) of the noise environment of a given neighborhood. A somewhat similar proposal, suggesting a unit called "Noise and Number Index" (NNI), was made later in Great Britain by McKennell (536). In Germany, Bürck *et al.* (105) proposed that the noise energy be truly integrated over time, giving a value they called "$\bar{Q}$." Kryter (458) also proposed that an integration procedure (0.5 sec basic time unit) be used to achieve an Effective Perceived Noise Level (EPNL) for a noise, and EPNL's integrated over 24 hr. to provide a CNR. In 1968 Langdon and Scholes (485) proposed a "Traffic Noise Index" (TNI), and in 1969 Robinson (686a) suggested a modified version of TNI which was called "Noise Pollution Level" (NPL). TNI and NPL summate noise measurements over a 24-hour period somewhat differently than the other units cited above, but more importantly, TNI and NPL also include in their calculation a weighting factor related to the range of noise levels found in a given environment.

Onset Duration and Impulse Level Corrections

A correction for the effects of the length of time occupied by nonimpulsive sounds as they are increasing in intensity (onset time) is proposed later in this chapter to account for the unwantedness of an approaching sound source, as described first by Nixon *et al.* (584). It is also suggested later in this chapter that certain adjustments be made to the sound pressure levels of impulsive sounds to provide the EPNL of nonimpulsive sounds as indicators of their relative subjective noisiness.

The remainder of this chapter is concerned with some discussion of the basis and details of the experiments, concepts, and units mentioned in the preceding historical review, and an analysis of the statistical reliability of subjective judgments of loudness and perceived noisiness. CNR, NNI, $\bar{Q}$, TNI and NPL will be discussed more fully in following chapters concerned with environmental noises.

Judged Perceived Noisiness and Perceived Noise Level

In order to determine the functional relations among various physical aspects of sound with respect to perceived noisiness, the spectrum shape, intensity, and duration of a standard reference sound should be defined against which other sounds can be judged or described. There has been no single precisely defined reference sound used in tests of loudness or perceived noisiness. A standard reference sound that should be appropriate for perceived noisiness is defined in Chapter 11 as an octave of random pink noise extending from 710 to 1400 Hz that maintains a steady, maximum level for 2 sec and has an onset and decline rate of 2.5 dB per 0.5 sec. It is not practical to specify a standard reference sound that is but one critical bandwidth wide and 0.5 sec in duration, a bandwidth and duration apparently commensurate, as mentioned previously, with the basic functional characteristics of the auditory system. For one thing, the audible spectrum of a sound changes as its rise time and duration are made shorter than about 0.5 sec, and a band of random noise that is too narrow takes on a tonal character that in itself contributes to perceived noisiness, as will be discussed. At the same time, it is important that our physical knowledge of a noise be grained fine enough so that it will reflect the sensitivity of the auditory system to changes that occur in the spectrum and duration of sounds. Accordingly, the contribution to judged perceived noisiness of the energy in the standard reference sound at a maximum sound level of 40 dB, for one 0.5-sec interval of time, is designated, for purposes of scaling the attribute of perceived noisiness, as one noy. As will be described later, various temporal characteristics of a noise also influence its perceived noisiness, but this need not alter the relations found for spectrum per unit (0.5 sec) of time.

The total judged noisiness (JPN) of the standard reference sound is presumed to be depictable by the following formula:

$$JPN = 10^4 \Sigma_i N_i^{3.3249}$$

where i is successive 0.5-sec intervals between the times the standard reference sound is 15 dB or less below its maximum level, and N_i is the number of noy of the standard reference sound in ith interval. The 15 dB range between the maximum and 15 dB down level specified for judged PN is selected partly for the psychological reason that the sound energy considerably below its maximum level contributes but little to judged noisiness, as will be discussed later with respect to the threshold of noisiness. A second reason is related to problems of making physical measurements and magnetic tape recordings of real-life broadband sounds. Real-life sounds at their maximum level often have a 10-12 dB peak factor (peak instantaneous levels above rms pressure) and a spectrum with frequency components that are 30-40 dB below the level of other frequency components. Accordingly, faithfully recording or measuring with standard electronic instruments such sounds at rms levels more than 15 dB or so below their maximum rms level requires a dynamic range in the instruments (60-70 dB) or special filtering procedures and gain adjustment not usually available.

An alternative way of expressing the JPN of the standard reference sound is to recite the integrated sound pressure level in dB of the standard reference sound. The result is called the Judged Perceived Noise Level and can be represented by the following formula:

$$\text{Judged Perceived Noise Level (JPNL)} = 10 \log_{10} [\Sigma_i \log_{10}^{-1} (SPL_i/10)]$$

where i is successive 0.5-sec intervals between the times the standard reference sound is 15 dB or less below its maximum level. A comparison sound judged equally to the standard reference sound is said to have a JPNL in dB equal to that found from the formula above.

Methods of Predicting Perceived Noise Level from Physical Measures and Calculations

The perceived noisiness and perceived noise level of a given sound, as would be determined by subjective tests, can be estimated approximately from certain physical measurements, or from certain calculations or operations performed on physical measures made of the sound. Perceived noisiness values and perceived noise levels obtained by means of subjective judgment tests should be described, as was done above, as "Judged PN" and "Judged PNL," to distinguish them from PNs and PNLs estimated from physical measurements. Henceforth in this document, PN and PNL will be used to refer only to calculated PN or PNL unless otherwise specified.

The fundamental unit for calculated PN, as for JPN, is the noy, but for calculated PNL the fundamental unit is PNdB, not dB. It is common practice to report the PNL rather than the PN of a sound. It will be recognized that a difference between judged and predicted PNL is that in the case of the former the sound pressure level of the standard reference sound serves as the measure of equivalent judged perceived noisiness (the spectrum or sound pressure level of a comparison or given sound need not even be measured), whereas for predicted PNL, in PNdB, the sound pressure levels of spectral bands of a comparison or given sound are converted into noy values summed in certain ways and then converted into an equivalent dB level of the standard reference sound. For certain frequency and intensity ranges, these summations and conversions are not linear with sound pressure level changes, and for this reason, as well as for identification purposes, the term PNdB seems justified.

The general formulae for PNdB and PN, which are the same in form as those developed by Stevens for loudness level and loudness, are as follows (see also Chapter 11).

$$\text{PNdB}_i = 40 + \frac{10}{0.30103} (\log_{10} \text{PN}_i)$$

where i is a 0.5-sec interval of time, and

$$\text{PN}_i = N_{i_m} + f(\Sigma N_i - N_{i_m})$$

where i is a 0.5-sec interval of time and ΣN_i are noys in all bands, N_{i_m} is maximum number of noys in any one band, and f is a fractional portion dependent on bandwidth, 0.3 for octave bands, and 0.15 for one-third octave bands. Band SPL's below 355 Hz should be combined in certain ways.

As noted earlier, and as will be described in detail later, it has been tentatively proposed that in the future, if additional verification of the appropriateness of the procedure is forthcoming, PNdB be taken as the sum, on a power basis, of the band SPLs of a sound adjusted according to the equal noisiness contours.

Other objective units than PNdB that have been developed or might be used for predicting the PNL present during the occurrence of a sound (even though some of the units were developed for other purposes) are the following:

PNL in Phons (Stevens) $+k$, or Phons (Zwicker) $+k$,

PNL in dB(D) $+k$, and

PNL in dB(A) $+k$, where k is a constant the size of which depends on the unit of measurement and to some extent, the spectral content of the noise.

Peak and Maximum Perceived Noise Level

Two general practices have been followed for reporting the perceived noisiness of a sound: one is to measure and refer only to the Peak or Maximum (Max) PNL reached by a sound during its occurrence, and the second procedure is to somehow sum the perceived noisiness over the entire occurrence of the sound.

The peak and maximum perceived noise levels are calculated by the following procedures:

1. *Peak PNdB.* The highest sound pressure level as reached in each spectral band for any 0.5-sec interval of time during the occurrence of the sound is used for the calculation of this perceived noise level.

2. *Max PNdB.* The perceived noise level is calculated from the spectral band levels for each successive 0.5-sec interval in time during the occurrence of a sound. The highest level for any 0.5-sec interval is designated as Max PNdB to distinguish it from Peak PNdB.

For stationary and most moving sources of sound, the peak or maximum perceived noise levels will be found to be the same. However, for some moving sound sources such as jet aircraft (because of the geometry of the vehicle and engine and its mounting on the vehicle), different portions of the sound spectra reach their maxima at a given point in space at slightly different times; this may cause the Max PNL for some aircraft sounds to be 1 to 3 PNdB lower than Peak

PNL calculated by the first method cited. The basis for the difference between these two procedures is illustrated in Fig. 163.

There appear to be no psychological judgment data available that clearly demonstrate which of these two methods, whether Peak or Max PNdB when they differ, predict most accurately judged perceived noise level. However, an argument can be made that the higher value Peak PNdB would be a better indicator than Max PNdB of the judged PNL on the basis that the overall effect of sound of given energy is probably less when the energy tends to be concentrated at a single moment in time than when the energy in the sound is distributed more irregularly in time. Also, the peak sound pressure levels required for the calculation of Peak PNdB are usually more easily obtained. However, as will be discussed later, neither Peak nor Max PNL will give good estimates of the judged noisiness of sounds having widely different durations.

Nonimpulsive Sound in Max dB(D), dB(A)

In order to provide with a simple sound level meter the best approximation possible to the judged PNL of different spectra, or types of sounds, different valued constants (k) can often be added to the sound level measures that have been weighted according to A or D. For example, it is typically found that the dB(D) weightings given in Fig. 8 are such that, for many noises, k on the average equals 6 dB; dB(D) levels plus 6 (see Chapter 11) are called herein dB(D') and are taken as approximate estimations of PNdB. Adding 13 dB to dB(A) values is also occasionally done as a means of estimating PNdB and is called herein as dB(A'). Actually somewhat different k values are appropriate for A and D with different types of noises (see Chapter 11 and Table 77).

As mentioned earlier, it was proposed (458) that the converse of the 40 noy contour (weighting D_1 in Fig. 8) be used in conjunction with a sound level meter for estimating perceived noise level. The 40 noy contour weighting is strictly appropriate, as a frequency weighting for a sound level meter, for use only with broadband sounds having their energy predominately above 355 Hz or for very narrow bands of sounds (one critical band or less wide) regardless of its center frequency. The D_2 weighting given in Fig. 8 is the 40 noy weighting adjusted to take into account relativly fewer number of critical bands in broadband sounds at frequencies below 355 Hz than above. (462).

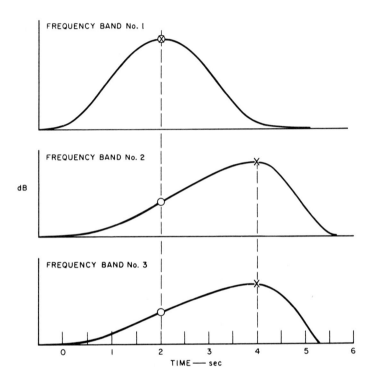

FIGURE 163.   Levels in different frequency bands from a moving sound source can reach their maximum level at different moments in time. Peak PNdB is calculated from the highest levels that occur anytime during noise cycle (levels marked X), and Max PNdB from levels that occur at the 0.5-sec interval in time when the PNdB calculated for the successive 0.5-sec interval is at a maximum (levels marked 0 at time 2 sec).

Tone Corrections

The perceived noisiness is greater for sounds that contain, within a broadband spectrum, relatively high concentrations of energy in narrow bands (one-third octave or less wide) than is the perceived noisiness of broadband sounds of equal overall energy but without these frequency regions of more highly concentrated energy. This phenomenon has been amply demonstrated in laboratory tests with steady-state and modulated single and multiple pure tones. The arguments for bands of sound one-third octave or narrower as being equivalent in perceived auditory effect as a tone of the same center frequency are primarily that: (a) bands of this width correspond fairly closely, except at frequencies below 355 Hz or so, to the so-called critical bandwidth of the ear as determined by tests of loudness and auditory fatigue; and (b) critical bands of random noise have pitch or tonal quality to them suggesting that they are rather closely related to tones in terms of some qualitative effects upon the ear.

The judged perceived noisiness of the sounds that have energy concentrated in tones or in narrow bands that exceed adjacent band levels can be estimated approximately through corrections applied to the measured sound pressure levels normally used in calculating perceived noisiness. Correction factors developed by Kryter and Pearsons for this purpose are given in Fig. 164. A somewhat simplified version of Fig. 164 is recommended for general use in Chapter 11.

It is to be noted in these figures that corrections are applied to those bands that exceed the energy of adjacent bands by 3 or more dB. It may also be noted that the accuracy, with which one can define and correct for the presence of tones or narrow bands of concentrated sound energy, is better for spectra determined with narrower band filters. With any set of fixed frequency filters, unless they are one-sixth octave or less wide, care must be taken to insure that the presence of a pure tone or very narrow band of concentrated energy in a broader band is not overlooked because the center frequency of the tone or narrow band of sound happens to fall in the center, or thereabouts, of the crossover frequencies between two adjacent filter bands. Following the correction, if any, to the sound pressure levels of each band, PN and PNdB are calculated in accordance with the usual procedures and tables. The results of these calculations have been designated in the past as PNL in PN_t or $PNdB_t$ respectively to distinguish these units from nontone-corrected PNs or PNdBs. However, it is suggested in Chapter 11 that PN or PNdB calculated by the procedures described be designated without the subscript even though the sounds involved required corrections. In this text, unless noted otherwise, all PN and PNdB units are determined by the procedures given in Chapter 11.

Little (513) suggested another method for correcting calculated PNdBs of sounds that contain pure tones. In Little's method one calculates, according to the usual procedures, the PNdB of a sound for a given 0.5-sec interval of time.

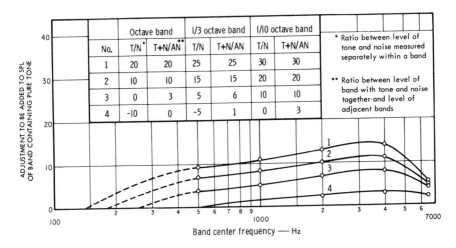

FIGURE 164.   Correction in dB to be applied to a frequency band containing a pure tone
or very narrow band of energy. From Kryter and Pearsons (468).

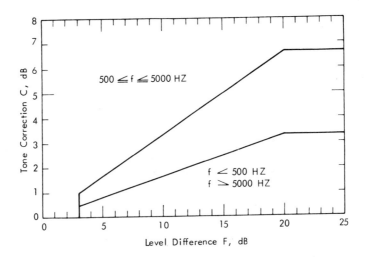

FIGURE 165.   Showing the tone correction procedure proposed by Little and developed
for use by U.S. Federal Aviation Administration. The dB value of the tone
correction on the vertical ordinate is added to PNdB calculated from band
spectral levels. The level difference is the amount, in dB, a spectral band
exceeds the bands adjacent to it. Parameter is bandcenter frequency. From
Sperry (758).

The band spectra of the sound for that interval are then examined to determine the presence of strong pure-tone components, which are specified in terms of a tone-to-background noise ratio. If this ratio, called level difference, exceeds a certain value, a correction, in dB, is added to the PNdB for the sound (see Fig. 165 from Sperry [758]). The magnitude of the correction is a function of the tone-to-noise ratio and frequency of the tone. However, unlike the procedure outlined in the preceding paragraph, only one correction is added to a sound even though more than one pure-tone component is present, and the magnitude of the correction is independent of the absolute intensity of the tone and background noise. For example, with respect to the latter point, the same dB correction would be applied in the case of a sound with a 5 dB tone-to-noise ratio regardless of whether that tone and noise band were very weak or very strong relative to the sound pressure levels in other bands within a sound.

Wells and Blazier (869) have proposed a method for computing the subjective reaction to complex sounds that attempts to account for the effect of pure-tone components on judged noisiness. In the Wells and Blazier approach, the value of one of a family of frequency-weighted contours (that are tangentially closest to a given sound spectrum) is assigned to the actual spectrum of the sound in question (upper graph, Fig. 166). This value is, however, corrected according to the number of one-third octave bands within 5 dB of the highest contour tangent to the sound spectrum (lower graph, Fig. 166). Wells and Blazier also provided graphs for use with octave band spectra that are not shown here.

Validation of Tone-Correction Procedures

Most of the experiments conducted to date with artificial sounds have shown that, when but one tone is present in a background of broadband noise, tone-corrected PNdB correlated better with the judgments of perceived noisiness than did PNdB not tone-corrected.

Perhaps the most exhaustive tests on this question are those recently carried out by Pearsons (611) with single and multiple steady-state and modulated tones. Some of his findings are shown in Figs. 167 and 168, and Fig. 169 illustrates the general spectra of the standard or reference sound and the comparison sound. It is seen in Figs. 167 and 168 that tone-corrections gave PNdBs for the comparison noises that were comparable to the PNdBs of the reference standard when they were judged to be equally noisy, whereas some PNdBs without tone corrections differed considerably from each other. The results of some earlier judgment tests with modulated and multiple pure tones are somewhat inconsistent with these findings (611a).

Several recent studies of the noisiness of aircraft sounds, some of which contain pure-tone components, are described in Chapter 9. It might be

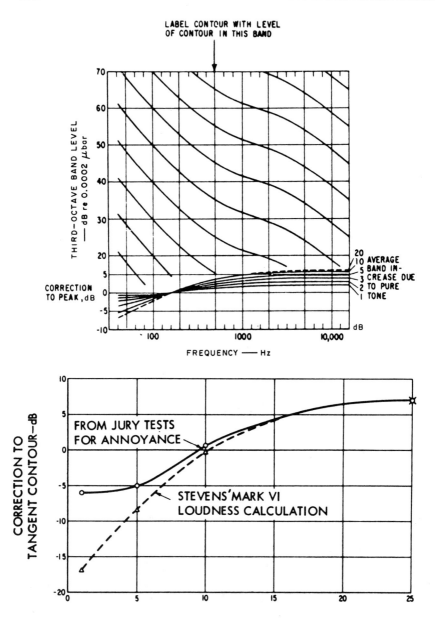

FIGURE 166.   Equal annoyance contours and pure-tone correction curves for the third
octave bands (upper graph), and correction for effect of spectrum shape on
annoyance (lower graph). From Wells and Blazier (869).

mentioned here that in these studies of aircraft noise, tone-correction did not always appear to be necessary for obtaining the best estimates of judged perceived noisiness. Further research study of this problem is obviously required.

Integrated Perceived Noisiness (IPNL)

Up to this point, the discussion has been concerned primarily with the calculation of the peak or maximum level of sounds of presumably equal durations. It is often a matter of importance to compare the noisiness of sounds of different durations. As mentioned before, the human auditory system has a temporal-intensity integrating characteristic, apparently of about 0.1 to 0.5 sec (see Figs. 5, 18, 19, and 154); also 0.5 sec represents the approximate time-constant to be found on a standard level meter set on slow meter action, and 0.2 sec represents one set on fast meter action. As a practical matter, it seems likely (as was also discussed in Chapter 1) that the slow-meter action setting will be used when sound level measurements of most noises are being made. It is proposed that, for the continuous measurement of the perceived noisiness of a sound, a sound measuring device having the characteristics corresponding to the slow-meter now specified for the standard sound level meter be used. It is also suggested that measuring the sound at 0.5-sec intervals would provide an adequate approximation to the continuous time-intensity history of steady-state sounds.

It appears that man's auditory system can combine into a perceived entity of noisiness, the distribution of energy present in the sound spectrum at any point in time; it also appears that man perceptually usually integrates successive intervals of noisiness into an entity of perceived noisiness for the total duration of an identifiable sound. To perform physical integrations of sound as a means of estimating judged perceived noisiness, it is first necessary to decide, among other things, at what threshold level of intensity this integration process must be started. In this regard, it is interesting to note that:

1. A broadband sound at PNL of 70 (PNdB, dB[D'], or dB[A']) during a 0.5-sec interval of time generated and measured outdoors would provide a level of about 50 inside a typical house because of the attenuation of the sound by the house structures.

2. People apparently require a noise environment within their home that is 20 PNL or so lower than that which they find to be acceptable when heard outdoors. Robinson et al. (673), Bishop (67), Bowsher et al. (85), and Kryter et al. (473-474) found this indoor-outdoor difference in acceptability of aircraft noise in experiments where people located inside and outside of houses were exposed to the sound from actual aircraft (see Fig. 170). This difference in

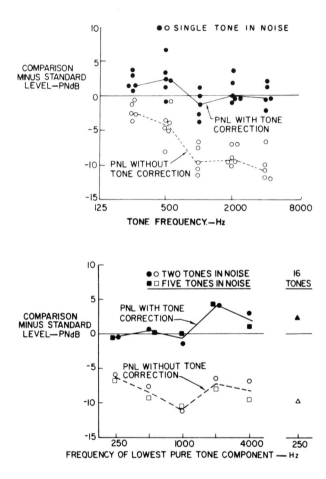

FIGURE 167.   Difference in PNL when comparison and standard are judged to be equal.
Standard noise without tones, comparison with single and multiple tones.
Tone corrections according to the method of Kryter and Pearsons (468).
From Pearsons (611).

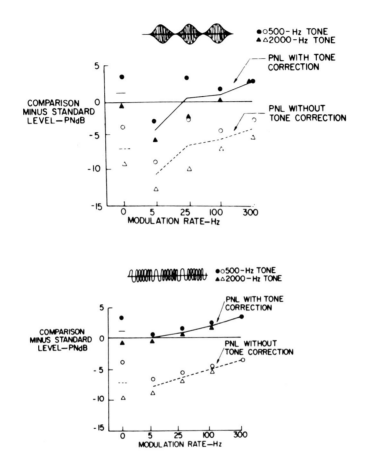

FIGURE 168.   Difference in PNL when comparison and standard are judged to be equal. Amplitude modulated tones (lower graph), and frequency modulated tones (upper graph). Tone corrections according to method of Kryter and Pearsons (468). From Pearsons (611).

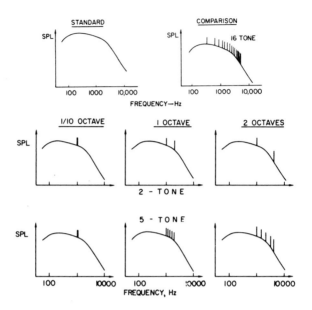

FIGURE 169.   Spectrum of standard and 16-tone comparison stimuli (upper graphs) and samples of spectra of 2- and 5-tone comparison stimuli (lower graphs). From Pearsons (611).

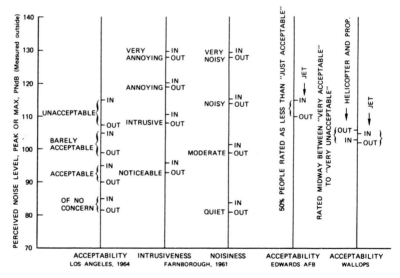

FIGURE 170.   Comparison between perceived noise level of aircraft flyovers and category scales of acceptability, intrusiveness, and noisiness. Subjects were from civilian communities except those at Edwards who were residents of a military Air Force Base. After Bishop (67), Robinson *et al.* (693), and Kryter *et al.* (473,474).

indoor vs. outdoor tolerability to noise may actually be somewhat different than that deduced on the assumption that the house attenuation, which was not actually measured in these studies, was equivalent to 20 PNdB, as will be discussed in Chapter 9. The attenuation by a house of outdoor sound may be 0.10 PNdB greater, depending on the house and the particular sound, than 20 PNdB.

Although it has been suggested that this phenomenon could be a subjective projection of the indoor sound as to how it would appear if audited outdoors, it seems likely that this apparent difference in the outdoor vs. indoor threshold of acceptability of sounds is due to the fact that activities which can be disrupted by noise, such as talking, listening, and mental concentration, are usually somewhat more demanding and important to people indoors than outdoors. In any event, it turns out that, to a first approximation, a single threshold level can be used for predicting the subjective reaction to noise heard indoors or outdoors provided the noise source is located outdoors and the measurements of the noise are made outdoors.

It is suggested, for present purposes that a sound, heard and measured indoors, is said to start occurring when: (*a*) its perceived noise level for any 0.5-sec interval exceeds a PNL of 40, and to stop when its perceived noise level falls below 40; or (*b*) its perceived noise level exceeds a PNL 15 PNdB, dB(D'), or dB(A') below Max PNL reached by that sound, and to stop when its perceived; noise level falls more than 15 below the Max PNL reached by that sound, provided the Max PNL is equal to or greater than 55.

A sound heard indoors or outdoors, but generated and measured outdoors, is said to start occurring when: (*a*) its perceived noise level for any 0.5-sec interval of time exceeds 60, and to stop when its perceived noise level falls below 60; or (*b*) its perceived noise level exceeds a PNL 15 below the Max PNL reached by that sound, and to stop when its perceived noise level falls more than 15 below the Max PNL reached by that sound, provided the Max PNL is equal to or greater than 75.

The threshold of noisiness of 40 PNdB, dB(D'), or dB(A') indoors, and 60 outdoors is for the more sensitive people; it is estimated that the threshold for the average person is about 10-20 dB higher (see Fig. 238, Chapter 11). For example, in a recent British study of the attitude and behavior of people to everyday outdoor sounds, it was found that sounds that have Peak PNdBs of 80 or less did not appear to add, on the average, to the annoyance or general noisiness of the environment.

The second alternative definition of effective duration given above—the sound within 15 dB of Max PNL—follows from some subjective judgment data which showed that, for sounds which increased to a maximum in a period of several seconds and then decreased from the maximum level at about the same rate as the sound had increased, the energy in the sound below a level of about 10-15 dB lower than the maximum level did not appear to contribute significantly to the perceived noisiness of the sound (Kryter and Pearsons [466]). In reality there will often be little difference between the magnitudes of Integrated Perceived Noise Levels achieved when the threshold-of-noisiness rule, and when the Max PNL −15 dB rule is used to establish the levels at which the integration process is started; the reason, of course, is that the weaker portions of the sound environment contribute but a fraction of a dB, on a physical basis, to the Integrated Perceived Noise Level. The level of 15 PNdB, dB(D'), or dB(A') below the Max PNL of a sound will be called the "practical threshold" of perceived noisiness. The Max PNL −15 dB rule should not, of course, be applied in the evaluation of an environment where noise that is more than 15 PNL units below the Max PNL levels, but still appreciably above the "real" threshold values for perceived noisiness specified above for considerable periods of time.

It is suggested that the Integrated Perceived Noise Level (IPNL) of an occurrence of a sound be represented by the following formula:

$$\text{IPNL} = 10 \log_{10} [\Sigma_i \log_{10}^{-10} (\text{PNL}_i/10)]$$

where i is successive 0.5-sec intervals of time.

Integrating continuous PNL as a function of time on a $10 \log_{10}$ basis is justified because this relation fits certain judgment data (see Fig. 171) reasonably well, particularly over durations from about 5 to 30 seconds.

Correction to IPNL for Duration of Onset of Nonimpulsive Sound

It seems, on the basis of everyday observation, that the longer the duration of noise, the less wanted it is. Perhaps more subtle is the apparent fact that the longer the duration in the build-up of the intensity of a noise, the more unacceptable it is, even though the total duration remains the same. Nixon et al. (584) reported that a sound that increases slowly to a given peak level and then decreases rapidly is much more objectionable than one of the same total duration and maximum intensity that increases rapidly and then decreases slowly in intensity (see Fig. 172). Comparison of the results for the first pair in Fig. 172 (an intensity but no frequency shift) vs. those for the second pair (intensity as well as frequency shift) reveals that a shift in frequency, such as would be present with an actual moving sound source—the so-called Doppler shift—does not appear as was mentioned earlier, to have a significant effect on the results. Nixon et al. found that, to be judged equally acceptable, the level of signal A of Fig. 172 had to be about 7 dB less than signal B. These investigators suggest that, as long as a sound is increasing in intensity, the listeners presume that the source of the sound is approaching and may come dangerously close. Therefore, the onset portion of the sound is judged noisier than the portion that

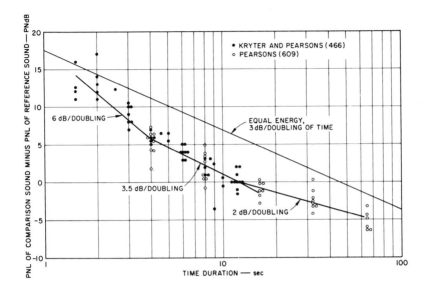

FIGURE 171.   Relative effect upon PNL of changing duration of a noise relative to a duration of 12 sec. After Pearsons (609).

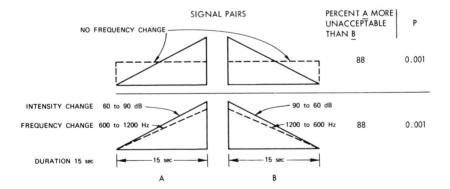

FIGURE 172.   Temporal and frequency pattern of pure-tone signals used in judgment tests and test result. Note that a "Doppler" frequency change did not increase the percentage of people who judged signal A to be more unacceptable than signal B. From Nixon *et al.* (584).

is decreasing in level, even when these two portions are of equal duration.

This trend in the judgements of these noises is consistent with the so-called "time-error" of subjective judgments. That term is used to describe the phenomenon that the more recent of two physically equal, but intense, stimuli is subjectively judged as being the more intense. Accordingly, the sound whose peak level occurred closest in time to the end of a sound might be judged the more intense by an amount equivalent to 1 to 2 dB, other things being equal. However, the hypotheses of a fear of an oncoming source, or, and particularly, the longer uncertainty felt by the listener as to just how intense an increasing noise may become, seem more reasonable explanations for the phenomenon found by Nixon *et al*. In addition, the effect is generally more than could usually be explained by the subjective time error. As with the effects on perceived noisiness of other physical variables of the noise, the factor of onset duration is presumed to have about equal meaning to the average listener for sounds not foreign to his environment.

Tone-corrected and integrated PNL values were used to predict the results of a series of judgement tests of the relative noisiness of a variety of aircraft noises conducted at Wallops, Virginia (474). This will be discussed in some detail in the next chapter. It was found in these tests that the aircraft sounds that had relatively long onset durations were judged noisier than aircraft sounds where the onset portion of the noise was relatively brief, even though the two aircraft sounds had the same integrated or effective values. In order to account for this difference, a correction function was estimated using the data from Nixon *et al.* to establish its approximate slope, as shown in Fig. 173, and then was applied to the judgment results that suggested its necessity. This correction procedure improved the prediction of the subjective judgments of some of the aircraft noise. It might be noted that in these aircraft noise tests the listeners were reasonably well adapted to the noises, as the result of repeated exposures, and that the aircraft were performing operations normal when over a community near an airport.

Whether the relation shown in Fig. 173 can be applied to other types of sounds remains to be demonstrated. Because of the meager amount of data available on this phenomenon, the correction to be applied in the calculation of IPNL must be considered as tentative.

Effective Perceived Noise Level of Nonimpulsive Sounds (EPNL)

It is suggested that integrated PNL values be referenced to a period of 8 seconds, and that the result be called an Effective Perceived Noise Level (EPNL). The formula for calculating EPNL of nonimpulsive sound is as follows:

$$EPNL = IPNL - 12 + oc \text{ (onset correction)}$$

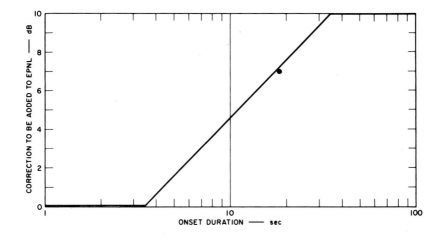

FIGURE 173. Correction to EPNL for contribution to perceived noisiness of onset duration of nonimpulsive sounds. The data point, from Nixon *et al.* (584), is plotted against a suggested standard onset duration of 3.5 sec.

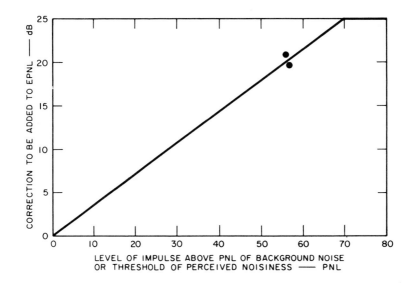

FIGURE 174. Correction to EPNL for contribution to perceived noisiness of startle to expected impulsive sounds. The level of the impulse is taken as the amount, in PNL, that the impulse exceeds the PNL of the background noise or the threshold of perceived noisiness, whichever is higher. The plotted points are from judgment tests (Kryter *et. al.* [473]) of the unacceptability of sonic booms vs. the noise from subsonic jet aircraft.

On occasion, EPNL can be referenced to some duration other than that of 8 sec. For example, if a reference of 0.5 sec were used, EPNL would equal IPNL; in these cases the duration should be indicated by a subscript, e.g., $E_{0.5}$PNL for a reference duration of 0.5 sec. If a reference duration of 16 sec is deemed appropriate, the subscript 16 would be used, e.g., E_{16}PNL. EPNL without such a subscript is reserved in this text for a reference duration of 8 seconds. The use of different reference durations will change the magnitude of the constant (-12 for an 8-sec reference duration) to be subtracted from IPNL (see Chapter 11). The presumption is being made here that a sound of a given EPNL value will be as effective as, or equivalent to (in its effect on people) the standard reference band having an EPNL numerical equal to the EPNL of the sound in question. The thought behind the terms "effective" or "equivalent" is the same as that proposed by Pietrasanta and K.N. Stevens (625) for the evaluation of aircraft noise, and that applied by Eldred *et al.* (212) to the evaluation of sound with regard to its fatiguing effects upon the ear.

The use of the reference duration specified, although somewhat arbitrary, seems appropriate because (*a*) it is that of the reference standard sound proposed for judgment tests, and (*b*) it is near the average duration of a number of common important noises. Aircraft noises will often be somewhat longer, whereas other common noises, such as from cars or trucks, will be somewhat shorter in duration.

Background Noise for Judgment Tests

The attempt to quantify, in terms of the physical stimulus, the "onset duration" factor in the calculation of EPNL has certain implications for the definition and specification of the temporal intensity pattern of a standard reference sound and the relation of that sound to a standard reference background noise. Clearly, if the onset duration is a factor in judged perceived noisiness, then the degree to which the standard reference at its maximum level exceeds its initial level will have an influence upon its onset duration and noisiness. For this reason, it is specified in Chapter 11 that, as far as possible and particularly in laboratory testing with the standard reference sound, a standard background noise be continuously present during the judgment tests of the standard reference sound and any comparison sounds. The level of the background noise should be at least 15 dB below the maximum level at all frequencies of the standard reference and comparison sounds.

Background Noise in Real Life

There is some data and much anecdotal evidence that a sound, such as that heard from an aircraft flying overhead, is not as noticeable in a high background noise as it is in a quiet environment. If one assumes that background environmental noise becomes the natural or adapted-to-environment of the average person, then it would follow that increasing the background noise level would make the occasional noises that exceed this level more acceptable. This question will be more fully discussed in Chapter 9.

An alternative interpretation is that the environment with the high level of background noise is judged relatively unwanted, and the intruding higher level noises are able to add some, but only a relativley small amount, additional annoyance. In any event, the function of Fig. 173 provides some probably appropriate corrections beyond that gained from the integration of PNLs over 0.5-sec intervals of time, because the higher the background noise level the shorter will be the onset of an intruding noise regardless of its final absolute level, and therefore the smaller is the correction added to the EPNL.

Estimated Effective Perceived Noise Level

Just as relatively simple overall physical measurements, such as dB(D) or dB(A), can be used to estimate the effects of variations in frequency spectra upon the subjective noisiness of some sounds, it is possible to estimate the effects of varying duration upon the subjective noisiness of sounds by simpler procedures than those outlined above for EPNL. The procedure most commonly used is as follows.

Step 1. Determine the Peak or Max PNL present during the occurrence of a sound.

Step 2. Determine the time in seconds (*a*) when the sound is above 40 PNL for sounds heard and measured indoors, (*b*) when the sound is above 60 PNL for sounds heard either indoors or outdoors but generated and measured outdoors, or (*c*) between the moment the sound first reaches a level 15 PNL below Max PNL and the moment it declines in intensity to a value 15 PNL below the max level. The PNL levels required for this step are typically found by the use of sound level meter with a D- or A-weighting network.

Step 3. Add to the Peak or Max PNL value the $10 \log_{10}$ of the duration in seconds (see Fig. 17 for a graphic aid) found in Step 2 above.

Step 4. Subtract 12 from results of Step 3 (ref. duration of 8 sec).

Step 5. Add to the results of Step 3, an onset correction from Fig. 173. The result is called the Estimated Effective Perceived Noise Level (EEPNL).

When using EEPNL, one assumes that there is a rectangular distribution in time of the PNLs of the sound (i.e., that a sound comes on fairly abruptly, stays at a steady level for some period, and then declines abruptly), whereas the EPNL takes into account the effects of the actual variations every 0.5 sec in the level of the sound as a function of time. The maximum difference between EEPNL and EPNL for a given sound can conceivably be rather large, but for typical sounds the difference is usually of the order of 2 or 3 PNdB, dB(D), or dB(A); for example, in the case of the sound made by an aircraft flying overhead, where the sound rather slowly rises to and falls from a peak level, the EPNL typically is from 2 to 4 less in value than the EEPNL.

Correction for Impulsive Level

Common observation and experience indicate that when the PNL of a sound increases faster than at a certain rate, this rate of change contributes to an unwanted startle effect, even when the sound is expected. Because rate and duration of onset are clearly not independent, one would perhaps expect this effect to be opposite that plotted in Fig. 173, i.e., shortening the onset duration of a sound would cause an increase in subjective noisiness rather than a decrease as shown in Fig. 173.

Nevertheless, it is believed that, beyond a certain rate in change of level, a sound takes on an impulsive characteristic that in-and of-itself contributes to the unwantedness or perceived noisiness of a sound. Figure 174 depicts a correction value that can ge applied to the PNL of impulsive segments of sound to take into account this effect. The function shown in Fig. 174 is drawn on the basis of three points, one from the presumed threshold level, and two from an experiment in which subjects judged the relative noisiness outdoors of expected sonic booms vs. the sound of a subsonic aircraft (see Chapter 9). Whether this proposed correction has general validity and, if so, whether it (a) represents some nonlinear behavior in the fluid or mechanical parts of the peripheral ear when the energy flux of a sound is increased above some amount, or (b) whether it is a reflection of some central perceptual response to a startle or suddenness attribute to sound, are open questions. The recommendation that sounds be measured over 0.5-sec intervals would seem to be inappropriate for closely occurring impulses that are so short that more than one can occur within a half second, such as the beginning and end of a sonic boom. However, this measurement time interval appears to be of practical accuracy even here since the loudness and annoyance of two pulses within a half second, according to Shepherd (personal communication), is controlled by the one with the highest peak level, i.e., the presence of the other impulse is not significant.

As is within keeping of the previous definitions of perceived noisiness, it is presumed that the impulsive sounds to be evaluated in regard to their perceived noisiness are familiar to the listeners and are an expected part of their noise environment. This was the case for the judgment tests of sonic booms and the noise from subsonic aircraft.

EPNL for Impulsive and Nonimpulsive Sounds

Obviously, studies of possible startle effects of impulses with longer rise times than those common to such impulses as sonic booms must be conducted before the generality of the function given in Fig. 174 can be established. The impulse correction procedure is tentatively proposed as a practical matter to provide a completely general procedure for the evaluation of the perceived noisiness of any type of noise, impulsive or nonimpulsive. The formula for the effective perceived noisiness of nonimpulsive and impulsive sounds is:

$$EPNL = PNL - 12 + oc + ic \text{ (impulse correction)}.$$

The value of oc is zero for impulsive sounds. The value of ic is zero for nonimpulsive sounds.

Impulsive Sounds in dB(D') and dB(A')

To some extent, measurements of single impulses for a frequency-weighted sound level meter set on slow meter action should be an appropriate estimate of the perceived noise level, loudness level, or damage risk to hearing. For one thing, the spectra of the impulses is broadband, and secondly, all the frequency components important to these auditory functions should be registered by the meter. The frequency weighting would appropriately weight all frequency components contributing to the auditory functions mentioned.

It is perhaps even more appropriate to use dB(D') or dB(A') for measuring single impulse than nonimpulsive sounds because impulse spectra tend to be less complex than nonimpulsive sounds in terms of pure tones or concentrations of energy in bands less than a critical bandwidth wide with relatively little energy in adjacent critical bands. There appears to be no reason why standard sound level meters could not be used for evaluating the auditory effects of impulsive sounds with the addition of certain constants. The constants shown in Table 30 are appropriate for nonimpulsive and impulsive sounds in order to approximate the PNdB values that would be calculated from the band energy spectra of these sounds. The values for the impulsive sounds are derived from sound level meter, with A- and D-weightings, readings of sonic booms, and the values for nonimpulsive sounds are based on a variety of data, some of which will be presented later (Table 78, Chapter 10).

TABLE 30

Estimated Constant to be Added to SLM Values, Slow Meter Action, to Approximate PNdB

| | Non-impulsive Sounds Maximum Energy in Frequency Region | | | Impulsive Sounds | | | |
| | | | | Rise Times Shorter than 0.1 msec | | | |
	Below 400 Hz	400-1200 Hz	Above 1200 Hz	Longer than 4.0 msec	0.4 to 4.0 msec (Duration)	Shorter than 0.4 msec	Average
dB(D)	6	6	6	7	6	6	6
dB(A)	12	11	15	13	11	15	13

Note: The values of the constants given in this table are for typical broadband sounds. For accuracy, particularly with sounds containing concentrations of energy in narrow frequency bands or for different classes of sounds, specific constants for dB(D) or dB(A) should be determined by means of calculations or measurements that permit comparison between dB(D) or dB(A) and PNdB.

Although it may seem incongruous to use the sound level meter as described for measuring single impulse sounds, sight should not be lost of the fact that the purpose of these particular measures is to predict the response of man's auditory system to sound. Measuring the acoustic energy in sounds that fall beyond the frequency limits of the ear can only tend to confuse the meaning of the measurements, as will be shown in Chapter 9. The fact that impulsive sounds have a greater psychological effect may only, in our opinion, be fortuitously related to the fact that the true physical peak level reached by an impulse over all frequencies is greater than the peak level as read on a sound level meter with limited frequency bandwidth and slow meter action. Perhaps the more legitimate procedure here is to measure the acoustic energy as does the ear (in this case about like a frequency-weighted sound level meter set on slow) and then to add a factor, as shown in Fig. 174, proportional to some additional psychological factor, presumably, for impulses, that of startle.

Effect of Differences in Sources, Subjects, and Test Conditions

Later in the text, the results of laboratory and field tests of the judged noisiness and loudness of noises having a wide variety of spectra and temporal characteristics will be given. Before doing so, the results of a study concerned with finding the general effects of some nonacoustic variables are presented.

Different Sources

Figure 175 shows that while there were some apparent systematic differences in the ratings of noisiness given sounds of the same EEPNdB from different sources, there were no striking dissimilarities that could be attributed to the type of source from which the sounds came. The range of differences were equivalent to about 5-10 EEPNdB. It might be noted that this range would possibly have been smaller had EPNdB been available and applied to the physical measurements.

Type of Subjects

In general it has been found, as will be shown in Chapter 9, that age, sex, and occupation have been minor variables in the experiments of judged noisiness or loudness. This is shown in Fig. 176 with subjects selected from the general adult community and from a population of college students.

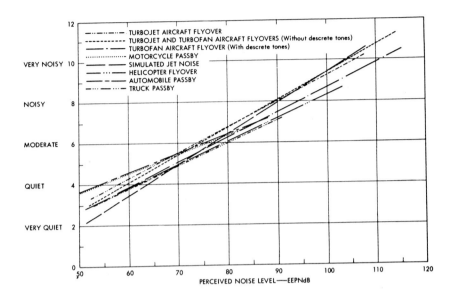

FIGURE 175.   Mean noisiness rating vs. perceived noise level of noise-stimulus groups for all laboratory test sessions. From Pearsons and Horonjeff (612).

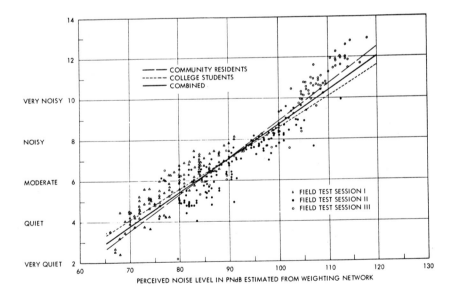

FIGURE 176.   Mean noisiness rating for all stimuli during all field test sessions. From Pearsons and Horonjeff (612).

Laboratory vs. Field Test Conditions

As will be also described more fully in Chapter 9, attempts have been made to conduct paired-comparison and absolute rating tests of perceived noisiness with the noise sources (usually transportation vehicles) operated in the vicinity of listeners under semireal-life conditions. The results have been perhaps surprisingly consistent with the results of similar tests conducted under carefully controlled laboratory conditions with recorded noise signals. Figure 177 gives the results of a study in which subjects judged the noises "live" in the field, and another group of subjects judged recordings of similar noises in the laboratory.

Reliability of Subjective Judgments

For noise-control purposes, specific levels of acceptance and tolerability are specified as a matter of practical necessity. There is, unfortunately, no realistic way of qualifying a boundary level in deference to the variability or reliability of human judgments; for example, a boundary level of 100 EPNdB plus or minus 5 dB is tantamount to setting the boundary at 105 EPNdB. In any event, the results of a number of studies in the literature in which people made equal loudness or equal noisiness judgments of various sounds have been gathered together for the purpose of examining their reliabilities.

Pure Tones and Bands of Random Noise

Reese *et al.* (663) obtained some data in which subjects judged both the loudness and annoyance of narrow bands of filtered white noise and reported the results for the individual subjects with respect to the annoyance judgments. In this study the subjects adjusted the intensity level of a comparison sound until it was judged by the listener to be either as annoying, or, depending upon the instructions, as loud as a reference sound; this experimental procedure is called the method of individual adjustment. Figure 157 shows the standard deviations for these subjects at each band for the judgments of equal annoyance. The subjects in the Reese *et al.* study made two judgments, separated by several days, of the loudness and annoyance of each of the narrow bands of noise; analysis of the test-retest results reveals that the standard deviation between the first and second loudness judgments was 1.65 dB and was 2.23 dB between the first and second annoyance judgments.

Figure 157 reveals two interesting points: (*a*) the variability appeared to be greater when the sounds were judged at the weaker intensities than when they were presented at higher sound pressure levels (average of 5 dB vs. 2 dB), and (*b*)

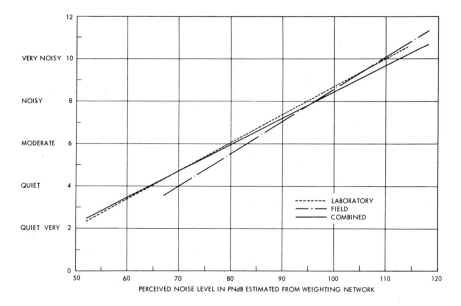

FIGURE 177.　Mean noisiness rating for all noise stimuli during laboratory and field test sessions. From Pearsons and Horonjeff (612).

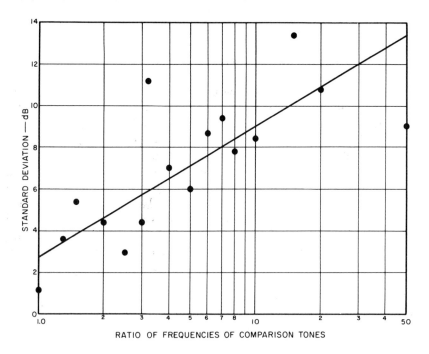

FIGURE 178.　Standard deviation in equal-loudness tests as a function of the frequency separation of the comparison tones. After Robinson and Dadson (690).

the greater the distance along the frequency scale between the standard band and the band being compared to it, the greater was the variability amongst the subjects; that is, the highest frequency band, No. 12, and the lowest frequency band, No. 1, showed the greatest variability.

Robinsons and Dadson (689, 690) found equal loudness contours for pure tones presented in a free-field (anechoic) chamber. Figure 178 shows the variation in standard deviations that they found as a function of the differences in frequencies between two tones being equated to each other in loudness. It is seen that there is a marked increase in the variability of the judgments as this difference is increased. The intensity levels used for some of the judgment data shown in Fig. 178 were of low, or near-threshold levels, which accounts for the generally large size of the standard deviations.

Stevens (774), also using the method of individual adjustment, obtained equal loudness contours for bands of noise. He reports that the average standard deviation of the judgments for the different bands was 4.5 dB, ranging from 1.9 to 9.0 dB, and that, as Reese *et al.* found, the standard deviations were less at the high intensity levels than at the more moderate levels. Cremer *et al.* (165) report an average standard deviation of about 2.5 dB for the equal loudness contours that they obtained with octave bands of noise, whereas Robinson and Whittle had standard deviations of about 8 dB for equal loudness contours of octave bands of noise at low intensity levels and about 6 dB at the higher intensity levels.

Figure 179 shows the distribution of subject responses when judging the subjective noisiness or acceptability of bands of noise relative to a standard noise extending from 600-1200 Hz. As with Figs. 157 and 178, we see that the greater the difference, in terms of frequency, between the standard and the comparison noise, the greater the variability of the subject.

It appears reasonable to conclude that the reason the high frequency and low frequency bands gave the highest dispersions among the subjects was because they were psychologically the most different from the standard or reference band against which they were judged, and not because they were high or low frequencies per se. This contention is perhaps borne out by Fig. 180 which shows that there is no general greater variability in thresholds for different frequencies except possibly that due to high frequency hearing loss in the older age groups.

In addition to obtaining data by the method of individual adjustment, as shown in Fig. 179, Kryter and Pearsons (466) used the method of paired-comparisons in which the bands of random noise were recorded in pairs and presented to groups of subjects via loudspeakers. From these data, as shown in Fig. 181, one is able to estimate approximately the standard deviation units, assuming that the distribution from which it was taken is normal or nearly so. The standard deviations thus estimated for paired-comparison data ranged from

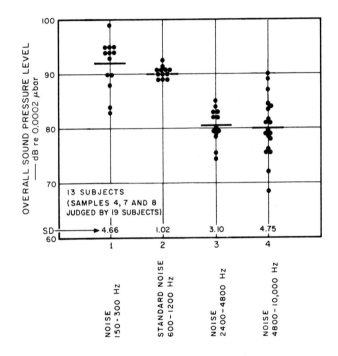

FIGURE 179.   Sound pressure level of sounds judged equally acceptable or noisy. Method
of individual adjustment. After Kryter and Pearsons (466).

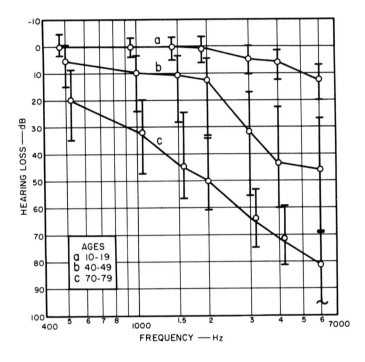

FIGURE 180.   Median and Q_1-Q_3 quartile range of hearing of men in total sample tested (left ear only). After Glorig *et al.* (304).

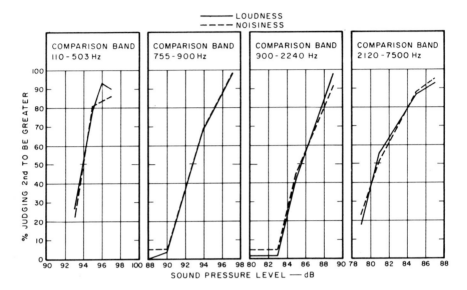

FIGURE 181.   Percentage of subjects judging the second (comparison) of a pair of sounds
to be louder or noisier than the first (standard) of the pair. Group (n=43)
judged loudness on the first day and noisiness one week later. The standard
was a band from 755-900 Hz at 94 dB. From Kryter and Pearsons (466).

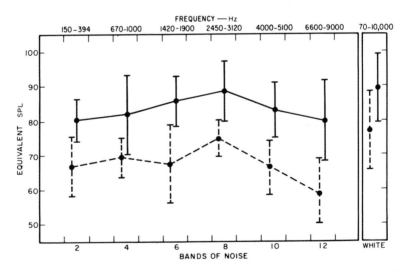

FIGURE 182.   The upper and lower light lines represent the mean judgment of threshold
of annoyance of the noise workers and quiet workers of the noise bands,
respectively. The vertical bars represent the 95% confidence limits of the
means. From Spieth (759).

about 1 to 3 dB. There was essentially no difference found in this study between the results of the paired-comparison tests for judged loudness or for noisiness.

Figure 182 shows the 95% confidence limits of the means found in an experiment by Spieth (759) in which 112 individuals made judgments, using the method of individual adjustment, of the threshold of annoyance for bands of filtered white noise. The average standard deviation of the various distributions shown for the different bands in Fig. 182 is about 6 dB. Spieth's experiment was conducted in an Air Force office building under more or less real-life, as distinct from laboratory, conditons and the data, for this reason, may be expected to be somewhat more variable than the data from previously mentioned laboratory studies.

A number of investigators have reported data showing the variations among subjects judging either the loudness or the noisiness of various complex sounds when heard live in the field in contrast to the laboratory studies in which recorded or simulated real-life noises were used. Table 31 summarizes these data along with, for comparison purposes, data on loudness and noisiness judgment tests conducted with filtered white noise.

There is a limiting accuracy of prediction that is imposed by (*a*) the reliability with which people can make or repeat their judgments of the difference in perceived noisiness of two sounds, and (*b*) the individual differences among people in their judgments of the perceived noisiness of two sounds. It appears that a person is able to repeat his judgment about the equal perceived noisiness of two reasonably similar sounds with a standard deviation of about 1.5 dB in their intensities (see Table 32). The variability among people is apparently about twice that found with repeated judgments by an individual, a standard deviation of about 3 dB for sounds that are somewhat similar (see Table 32).

All together, the data indicate that the standard deviation of the distribution of the levels of a comparison sound, when judged to be as noisy as a standard reference sound at a fixed level of intensity, would be probably no greater than 3 dB provided that the sounds were of a somewhat similar type and 10 or more subjects were used. If we assume that 50 is a practical maximum number of people to use for testing a given set of noises and listening conditions in a single experiment, it is possible to deduce approximately how precise the average results of the subjective judgment test are likely to be. This result, which is developed in the hypothetical example in Table 33, sets the approximate maximum accuracy with which any physical measure could possibly predict or estimate the average of the subjective judgments of a group of people. It appears likely, as shown in Table 33, that the limiting statistical and practical significant accuracy with which a given physical measure can be shown by experiment to predict average perceived noisiness as judged by a group of 50 people is equivalent to between 0.25 and 0.5 dB for a group of sounds not too dissimilar from each other.

TABLE 31

Standard Deviations Among Subjects Judging the Loudness and Noisiness of Sounds

Values are representative averages for the different studies. Estimated (est) values are based on assumption that distributions of data obtained were statistically normal. From Kryter (454).

Judgments of Equality

Study	Type of Sound	Judgment	Tests Lab.	Field
Quietzsch[655]	Variety Real-Life Noises	Loudness	2.5 dB	
Rademacher[661]	Motor Vehicles	Loudness	3.0	
Kryter[444]	Subsonic Aircraft	Noisiness	3.0	
Copeland et al[159]	Subsonic Aircraft	Loudness + Noisiness	4.4	
Williams (unpublished)	Subsonic Aircraft	Threshold of Noisiness	7.0	
Kryter et al[465]	Subsonic Aircraft	Noisiness	2.2	
Kryter et al[466]	Variety	Noisiness	4.0	
Pearsons et al[613]	Sonic Boom vs. Subsonic Aircraft	Noisiness	5.5	
Kryter et al[473]	Sonid Boom vs. Subsonic Aircraft	Noisiness		8.0 (Est)
Pearsons et al[613]	Sonic Boom vs. Sonic Boom	Noisiness	2.5	
Kryter et al[473]	Sonic Boom vs. Sonic Boom	Noisiness		5.0 (Est)
Kryter et al[473]	Subsonic Aircraft	Noisiness		5.0 (Est)
Lubcke et al[527]	Machine	Loudness	2.4	
Broadbent et al[99]	Sonic Boom vs. Regular Aircraft	Noisiness	10.0 (Est)	
Reese et al[663]	Bands of Noise	Loudness + Noisiness	4.0	
Pollack[635]	Bands of Noise	Loudness	4.0	
Stevens[774]	Bands of Noise	Loudness	4.5	
Spieth[760]	Bands of Noise	Threshold of Annoyance	6.0	
Robinson et al[692]	Bands of Noise	Loudness	6.0	
Kryter et al[466]	Bands of Noise	Loudness + Noisiness	4.0	
Cremer et al[165]	Bands of Noise	Loudness	2.5	
Zwicker[900]	Bands of Noise, Tones	Loudness	4.3	

In the above studies, the judgments, when obtained, of the loudness or noisiness of the same sound against itself show Standard Deviations of 1 to 2 dB.

Judgments of Differences - Rating and Magnitude Scaling

Study	Type of Sound	Judgment	Tests Lab.	Field
Stevens[773]	Tones	Loudness	3.0 dB	
Garner[287]	Tones	Loudness	5.0 dB	
Poulton et al[654]	White Noise	Loudness	4.2 dB	
Mills et al[558]	Live Motor Vehicles	Noisiness		4.0 dB
Robinson et al[693]	Live Aircraft	Noisiness		4.5 dB
Bishop[67]	Live Aircraft	Noisiness		6.5 dB

TABLE 32

Difference in Sound Pressure Levels, C Scale, between Flyover Noise of Super-Constellation
at 400 Ft Altitude and 707 at 960 Ft Altitude When the Recorded Sounds
Were Judged to be Equal in Noisiness. From Kryter (444).

Subject	1	2	3	Test 4	5	6	7	Mean	S.D.
HM	14	16	14	15	17	15	16	15	1.1
RN	15	11	14	14	13	13	12	13	1.5
KP	7	9	4	7	8	10	9	8	2.0
NM	13	13	14	14	12	11	12	13	1.7
CG	14	10	9	9	9	11	10	10	1.8
Mean	12	12	11	12	12	12	12	Average S.D. = 1.6 dB	
Standard Deviation (S.D.)	3.3	2.5	4.5	3.7	3.5	2.0	2.7		

Average S.D. = 3.2 dB

Summary of Reliability Data

There are several major sources of variability present in subjective judgments
made of the loudness and noisiness of sounds.

1. *Intrasubject variability.* A person, other factors being constant, will repeat
his judgment of the loudness or noisiness of one sound at suprathreshold levels,
relative to the same or very similar sounds with a standard deviation of 1-2 dB.

2. *Intersubject variability in judging equality.* The standard deviation of
judgments of equal loudness or noisiness is about 2-4 dB for a group of people
judging sounds at suprathreshold levels that are of a similar "class"—bands of
noise, aircraft, motor vehicles, sonic booms, etc. This includes intrasubject
variability.

3. *Intersubject variability in judging subjective absolute magnitudes (as
distinct from judgments of relative magnitudes).* The standard deviation of
judgments of the absolute magnitude of loudness or noisiness on a numeral or
verbal scale is about 4-7 dB. This includes intrasubject variability.

4. Judgments of either equality or magnitude of loudness, or noisiness at
threshold have a standard deviation of about 7 dB.

5. A physical measure cannot be expected to predict with statistical
significance the average of judgments made by a group of 25-50 subjects of
perceived noisiness of typical sounds under good test conditions with an
accuracy greater than an amount equivalent to about 0.25 dB to 0.5 dB in
intensity level of the sounds.

TABLE 33

Illustrative Statistical Analysis of Precision of a Group of Individuals in Judging the Equal Perceived Noisiness of Two Sounds

$$\delta_{X_1} = \sqrt{\Sigma\,(X_1 - M_1)^2 / N\text{-}1}$$

$$\delta_{MX_1} = \delta_{X_1} / \sqrt{N}$$

$$\text{Diff} = M_1 - M_2$$

$$\delta\,\text{Diff} = \sqrt{\delta_{MX_1}{}^2 + \delta_{MX_2}{}^2 - 2r\delta_{MX_1}\,\delta_{MX_2}}$$

where X is the difference between levels of two sounds judged to be equally noisy when measured by a given physical unit such as dB(D), PNdB, etc. (designated by subscript 1, 2, etc.). M is the average difference between the measured physical levels of the two sounds judged by a number of people to be subjectively equally noisy. N is the number of people, and r is the product moment coefficient of correlation between the values for each physical measure when the two sounds were judged by the people to be equally noisy.

Example:

say, $M_1 = 1.5$ dB(A)

$\delta_{X_1} = 3.0,$ and $N = 50$

$\delta_{MX_1} = 0.43$

and,

say, $M_2 = 1.0$ dB(D)

$\delta_{X_2} = 3.0,$ and $N = 50$

$\delta_{MX_2} = 0.43$

$\text{Diff} = M_1 - M_2 = 0.5$ dB

$\delta\,\text{Diff} = 0.28$ (assumed r of 0.85)

$$\frac{\text{Diff}}{\delta\,\text{Diff}} = 2.0$$

t test shows 95% confidence level of probability that difference $M_1 - M_2$ would be between ±0.5 dB.

Relative Accuracy of Physical Units for Predicting Judged Perceived Noisiness

Two standard statistical techniques for evaluating the accuracy with which the physical units of normal measurement predict judged perceived noisiness have been generally used in the past: (*a*) product moment coefficients of correlation (see Tables 34 and 35) between the physical measures and judgments; and (*b*) a rank ordering of the average differences and average or standard deviation of these differences between the physical measures of noises that were judged to be subjectively equal (see Table 36). For a single experiment, in which a variety of noises are judged to be equal to a single reference noise, the average of the differences between the reference and comparison noises according to a given unit of physical measurement is not necessarily a good indicator of the overall predictive accuracy of a given unit. For example, if the reference noise were the only noise of the group not properly evaluated by a given unit, the average of the differences between it and the comparison noises would be larger than had a different noise been chosen from the group to serve as the reference noise. For this reason, the standard deviation of the differences is usually a better measure than the average difference of the relative accuracy with which units of physical measurement can predict judged perceived noisiness.

However, the typical statistical tests of the significance between two units having different standard deviations cannot, to our knowledge, be legitimately applied to these standard deviation statistics. The reasons being, of course, (*a*) that the various units of PNL tend to be highly correlated, e.g., they all increase and decrease in value as the sound pressure level of a given noise increases or

TABLE 34

Coefficients of Correlation between Peak Physical Measurements and Subjective Ratings of the Sound from Various Vehicles. From Cohen and Scherger (143).

Physical Measures	Pearson Product Moment Coefficient (r)	Spearman Rank Order Coefficient (rho)
phons (Zwicker)	.96	.98
phons (Stevens)	.91	.92
PNdB[*]	.90	.92
dB(A)	.83	.72
dB(C)	.75	.68

[*]Calculated by procedures given in ref. 466

TABLE 35

Rank Listing of Several Peak Noise Rating Methods Studied and the Percentage Probability That a Significant Difference Exists Between Any Two. From Ollerhead (592).

RANK	2	5	7	9	12	15	23	28
	PNdB* (Tone-Corrected)	Z Phon (Zwicker)	PNdB	dB(D)	Phon (Stevens)	dB(A)	dB(B)	dB(C)
	66%							
	69	53%						
	70	54	51%					
	72	57	53	53%				
	78	63	60	59	56%			
	93	85	83	83	81	76%		
	99	98	98	98	98	97	89%	
	99	99	99	99	99	99	99	99%

*Calculated in accordance with ref. 466

decreases, and all the units give more weight to the mid-to-higher than to the lower frequencies, and (*b*) standard deviation measures are not necessarily normally distributed. The application of the *t* test to the average differences of prediction is, of course, possible when the correlation between the measures are provided.

Rather than attempt to use these statistical tests of significance of the relative average accuracy and variability in accuracy of the different units of measurement in their prediction of judged perceived noisiness, P.J. Johnson and I developed the following argument and procedure. Let us presume that a person has taken physical measures of pairs of sounds chosen at random from those evaluated by paired-comparison tests, and that these measures are, in turn, converted (or are made directly) into one of the units of physical measurement. One obvious question to be asked is which units will be in closest agreement and how often, with respect to the judged perceived noisiness of the sounds.

To answer this question, a table is made of the percentage of time that the value of each of the units of measurement would be within ±2 and ±4 "dB" units of the judged perceived noise level for any noises chosen at random from those tested. This percentage is the normal probability to be expected according to the number or portion of standard deviations of a given unit found between the average difference for that unit and the criterion of ±2 or ±4 units from exact agreement with subjective judgments. The general concept is illustrated in Fig. 183.

In an attempt to determine which of various units of measurement best predicts judged perceived noisiness of, primarily, aircraft noise, a somewhat detailed examination is made below of judgment data that can be related to Peak and Max PNL units, and data that permit the use of both Peak, Max, and Effective PNL units.

In order to clearly indicate the character of the various units to be evaluated, the following symbols will be used in the remainder of this chapter:

Phon—Stevens, Mark VI (776)

PNdB—calculated in accordance with reference 466 and 467

PNdB-M—calculated in accordance with Chapter 11 (same as PNdB except bands below 355 Hz combined in certain ways)

Subscript t_1 refers to tone corrections in accordance with Kryter and Pearsons (468), and t_2 in accordance with Sperry (758)

dB(A), (C), (D_1), (D_2), (D_3) are those based on frequency weightings given in Fig. 8

The prefixes E and EE refer to temporal integrations of units between the 10dB downpoints, as defined in earlier paragraphs.

TABLE 36

Differences between Objective Measurements [dB(C), dB(A), PHONS (S), PHONS (Z), and
PNdB] of a Band of Noise Centered at 1000 Hz, or a 1000 Hz Tone, and Recordings
of Various Machinery, Motor Vehicle, Auto Horns, Aircraft, etc., Noises When
the Tone or Band of Noise Was Judged to Be Just as Loud [or Noisy,
According to Kryter and Pearsons] as the Recorded Noises
From Kryter (454).

Investigator(s)	Peak dB(C) 1	2	Peak dB(A) 1	2	Peak Phons (S) 1	2	Peak Phons (Z) 1	2	Peak PNdB** 1	2
Quietzsch	-11.2	5.2	-13.7	3.9	-1.7	3.3	+2.9	3.2	-1.4	3.2
Rademacher	- 1.8	2.8	-10.9	2.3	+3.1	1.2	+6.8	1.1	+2.6	1.2
Niese '57 63 Phon	- 3.7	2.1	-10.5	2.9	+0.3	1.9	+5.8	2.1	-1.6	0.9
Niese '57 60 Phon	- 6.5	2.8	-11.0	2.0	-0.6	1.1	+6.1	1.4	-2.3	2.0
Niese '59 80 Phon	- 9.5	3.1	-12.6	3.2	-2.2	2.3	+4.5	2.4	-0.6	3.0
Niese '60 64 Phon	- 9.5	5.4	-13.5	4.7	-1.6	1.9	+2.3	1.4	-3.7	2.0
Niese '60 85 Phon	- 8.0	3.9	-11.4	4.9	-3.0	2.1	+2.4	1.9	-1.8	2.3
Lubcke et al. (Berlin)	-12.8	1.4	-15.5	1.5	-3.4*	1.6*	+1.8	1.3	-2.5	1.5
Lubcke et al. (Stuttgart)	-14.4	2.1	-16.9	1.6	-6.4*	1.5*	-0.1	1.1	-5.3	1.6
Kryter + Pearsons	- 7.4	3.4	-10.6	1.0	-3.3*	1.3*	-3.8	2.8	-2.8	1.4
Average	- 8.5	3.2	-12.8	2.8	-1.9	1.8	+2.9	1.9	-1.9	1.9

Column 1 - Average difference between subjective and objective values of various noises.
Column 2 - Deviation, regardless of sign, of data about average difference.

*Phons (S) were calculated by the Mark VI method. Phons (S) in the other studies were
calculated by Mark II.

**PNdB values are calculated in accordance with ref. 466.

Note: The objective measures for this table were not always provided in the original
articles referred to in the table. In those cases the necessary calculations were
made on the basis of octave or 1/3 octave band data included in the articles or
kirdly sent to us by the authors. In some cases, octave band spectra were converted
(by subtracting 5 dB) to 1/3 octave band spectra in order to calculate phons (Z).

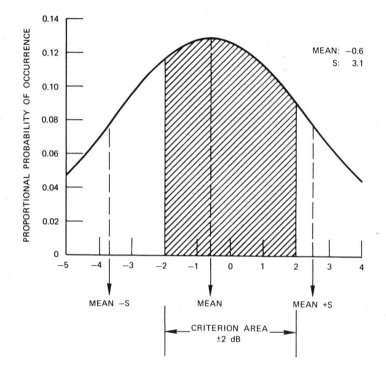

FIGURE 183.   Schematic diagram showing statistical method used for evaluating accuracy
of units of measurement for estimating judged perceived noisiness. 0 on the
scale is the "true" subjective rating given by listeners; minus 1 indicates
that the physical measurement (PNdB, dB(D), Phon, etc.) underestimates
the judged noisiness by a 1 dB unit; +1 indicates an overestimation. The
curve is the statistical normal probability of the accuracy of a hypothetical
unit of measurement that has been found by test to have a given average
difference from judged noisiness of -0.6 and a given standard deviation of
those differences of 3.1. The area under the curve for a criterion of ±2 dB
units is shaded. This area is 47% of the total area under the curve, i.e., 47%
of the time the noisiness of sounds as estimated by the hypothetical unit
would be within ±2 dB of the judged noisiness of the sounds.

Peak and Max PNL

In Table 37 are presented all the paired-comparison, equal noisiness data we could find that permit a comparison between judged perceived noisiness and Peak and Max PNL. The judgment data were obtained in the various studies, as indicated, and the objective perceived noise levels were calculated at Stanford Research Institute from one-third or one octave band spectra of the noises by means of computer routines.

The bottom rows of Table 37 can be interpreted as showing the percentage of times a given unit will have an accuracy in predicting judged perceived noisiness with ±2 or ±4 units of measurement. Other difference criteria can, of course, be calculated. For example, about the same percentage of the time (53%) that Max dB(A) is within ±4 dB units, Max dB(D_2) is within about ±2.5 dB units; or, in other terms, Max dB(D_2) will predict the judged perceived noisiness of about 50% of the aircraft noises within a range of 5 dB units, and Max dB(A) within a range of 8 dB units. As with all statistics, the practical significance of the differences in the summary percentage figures, as well as the standard deviation values, are a matter of judgment and the circumstances in which noise evaluations are to be made. However, an improvement of but ±0.5 (a range of 1) in dB units from a practical point of view would probably be significant. As discussed earlier, the average judgments of groups of 50 or more people about the relative perceived noisiness of two noises usually has a test-retest reliability such that a difference of usually less than 0.5 dB in noise level is perceived with a statistical level of confidence exceeding 95%.

Peak, Max, and Effective PNL, Including Tone and Onset Duration Corrections

It is unfortunate that in most of these published studies of the judged perceived noisiness or loudness of real-life noise, measurements of the band spectra that were present preceding and following the Peak or Max levels were not usually made or, if made, not reported. Judgment tests of aircraft noise conducted at Wallops (474) provide the only extensive field tests for which are available the acoustical data that permit a comparative evaluation of the relative accuracy of Peak, Max, and Effective units of PNL in predicting judged perceived noisiness. Tables 38 and 39 summarize the data obtained at Wallops.

If one accepts the criteria that (a) a change of 0.5 dB in nominal sound pressure level is a matter of practical physical importance, and (b) a criterion that there must be agreement between judged and predicted perceived noisiness some given percentage of the time, it follows that a difference in predictive accuracy between two units of about four percent of the time is significant. This follows from the function of Fig. 184 where the change in percentage of time that the average unit of prediction will be within a given range of accuracy is

TABLE 37

Showing the Average Difference between Reference and Comparison Noises (Col. D) and Standard Deviation (Col. S) of the Differences for Each Unit of Max PNL, 136 Aircraft, 1 Diesel Train, and 6 Filtered Random Noises

Experiment	N	Max $dB(D_2)$		Max PNdB-M		Max Phons		Max PNdB		Max $PNdB_{t_1}$		Max $PNdB_{t_2}$		Max $dB(D_1)$		Max $PNdB_{t_1}$		Max $PNdB_{t_2}$		Max $dB(D_3)$		Max $dB(A)$		Max $dB(C)$	
		(D)	(S)	(D)	(S)	(D)	(S)	(D)	(S)	(D)	(S)	(D)	(S)	(D)	(S)	(D)	(S)	(D)	(S)	(D)	(S)	(D)	(S)	(D)	(S)
Copeland et al[159]	3	1.0	1.1	1.1	1.8	1.2	2.8	1.3	2.5	1.1	1.8	2.1	1.1	1.9	2.2	1.3	2.5	-1.8	0.8	-6.7	1.2	0.6	1.3	4.0	8.1
Robinson and Bowsher[687]	5	0†	1.6	0	1.5	0	1.5	0	1.4	0	1.2	0	2.6	0	1.7	0	1.0	0	2.5	0	2.1	0	2.0	0	3.2
Pearsons, Helio-S[610]	8	0.4	1.5	-0.4	1.6	-0.8	1.7	-0.9	1.7	-0.9	1.5	0.1	1.8	0.1	2.2	-1.3	1.7	-0.4	1.8	1.6	1.4	1.0	1.8	-0.7	3.7
Pearsons, Helio-D[610]	8	1.5	2.6	1.1	2.8	0.4	2.5	0.6	2.8	0.7	2.8	0.1	3.0	0.9	3.1	0.2	2.7	-0.4	3.0	2.6	2.5	2.3	2.8	0.4	4.4
Hinterkeuser, et al[373]	12	2.6	3.7	2.0	3.9	1.4	3.4	1.9	4.0	2.0	3.9	2.4	4.0	1.4	4.1	1.9	4.0	2.2	4.2	3.4	3.5	3.7	3.8	-0.7	5.3
Ollerhead[592]	35	1.9	3.5	-0.6	3.4	-1.8	3.1	-2.1	3.5	-1.1	3.4	-1.1	3.8	0.6	3.3	-2.5	3.5	2.5	3.9	7.0	3.7	4.9	3.5	-4.4	5.0
Kryter - Indoor[444]	10	4.0	2.8	3.9	2.2	3.9	2.2	3.7	1.0	3.9	2.2	3.6	2.3	4.9	3.3	4.4	2.5	4.0	2.7	1.2	1.7	3.4	2.9	8.9	6.0
Kryter - Outdoor[444]	5	3.5	0.6	3.7	0.7	4.0	1.0	3.7	1.0	3.7	0.7	4.0	1.7	3.9	1.0	3.7	1.0	4.0	2.0	2.7	1.2	5.1	1.4	9.0	4.8
Kryter and Pearsons[465] Tbl.1A	4	-1.6	1.9	-1.3	3.0	-2.0	2.8	-1.5	3.1	-1.3	3.0	-2.2	2.2	-1.8	2.3	-1.5	3.1	-2.4	2.3	0.2	1.5	0.6	3.3	-1.9	7.0
Kryter and Pearsons[465] Tbl.1B	4	-1.0	2.5	-1.0	1.9	0.2	2.2	-0.9	1.8	-1.0	1.9	-1.0	3.2	-1.3	2.6	-1.3	2.0	-1.3	3.2	0.2	2.2	0.2	1.5	-1.8	5.4
Kryter and Pearsons[465] Tbl.2A	4	5.7	0.9	-7.4	1.1	7.2	1.0	7.4	1.2	7.4	1.1	5.9	1.5	6.6	0.9	7.4	1.2	5.9	1.6	4.6	1.1	7.8	1.8	14.8	2.8
Kryter and Pearsons[465] Tbl.2B	4	1.6	2.5	2.7	1.6	3.9	1.2	3.6	1.4	2.7	1.6	2.9	2.9	3.1	2.4	3.6	1.4	3.8	2.7	-0.3	2.2	2.8	1.5	11.8	0.7
Kryter and Pearsons[466]	8*	4.6	2.4	2.5	2.5	4.3	2.0	2.2	2.6	2.0	3.2	4.5	1.7	4.2	2.4	1.6	3.2	4.2	1.5	9.3	2.3	10.7	2.0	8.7	6.1
Hecker and Kryter[361] Tbl.XIV	11	-0.8	2.1	-3.2	1.6	-3.0	1.8	-3.8	1.6	-4.5	1.9	1.7	1.1	-0.9	2.0	-5.0	1.7	1.2	1.1	2.1	2.1	3.3	2.3	2.2	3.6
Kryter, Johnson and Young Edwards[473]	4	1.9	1.9	1.5	1.7	2.3	1.9	1.5	1.7	-3.9	3.0	-1.9	1.7	2.1	2.1	-3.7	2.3	-1.9	1.2	1.8	1.3	4.0	2.4	5.4	5.1
Kryter, Johnson and Young Wallops (880 reference)[474]	12	0.1	4.0	2.9	3.9	1.3	4.2	1.8	4.0	2.9	4.3	2.7	3.3	1.0	3.8	3.2	4.4	2.5	3.3	-0.7	3.6	-1.2	3.3	1.9	5.6
Kryter, Johnson and Young Wallops (1049G reference)[474]	6	-0.9	2.9	2.6	3.0	-5.8	2.6	-5.6	2.6	-4.4	3.8	-4.5	3.7	-5.3	2.7	-5.4	3.4	-4.8	3.3	-3.0	2.1	-6.1	2.0	-8.1	5.5
	N = 143																								
Aver. Diff. (D) Stand. Dev. (S)		1.6	3.3	0.6	3.6	0.3	3.8	0.1	4.0	0.2	4.0	2.0	3.7	1.1	3.7	-0.2	4.3	1.6	3.7	3.1	4.2	3.1	4.3	1.1	7.3
Percentage of time a unit will predict PNL within ±2 " " " " ±4		41 73		41 73		40 70		38 68		38 68		36 66		39 65		35 65		37 61		28 53		28 53		20 42	

† One jet aircraft, one diesel train, and six filtered random noises.

* In the Robinson and Bowsher study each of the five noises was judged against each of the other noises and the average difference is therefore set at 0.

plotted for average standard deviation and differences found in the Wallops study. It is seen that the slope of this latter function is such that a change of 0.5 dB range of accuracy in prediction is equal to a change of four percentage points. Because the value of the average, as well as the standard deviations, interact to give somewhat different rank orderings for different criteria of accuracy, the ranks in Table 37 and 38 are based on the average of the percentages for the ±2 and ±4 dB criteria.

Pearsons and Bennett (613a) recently reported the results of paired-comparison judgment tests of the relative perceived noisiness of a wide variety of recorded real and simulated aircraft noises. The tests were administered in an anechoic chamber to 20 subjects. The results of these tests, presented in Fig. 185, are seen to be in essential agreement, insofar as the prediction measures calculated by Pearsons and Bennett permit, with those obtained in the field at Wallops (Tables 38 and 39).

It should be kept in mind that there was possibly present in all the studies reported a certain amount of unavoidable experimental error due to subject variability, variability between acoustic spectra that was presumed by the experimenter to be the same (room or even outdoor acoustic conditions cause variation in a sound as heard by subjects seated at different locations) and errors in sound measurement or analysis. Because of these errors and the somewhat small differences in the predictive accuracy of the different units, some inconsistancies are to be found among the results of judgment tests. In addition, the virtues of tone and duration corrections are not, for these reasons among others, demonstrable when the noises being judged do not differ markedly with respect to tonal content and/or durations.

Summary of Accuracy Data

On the basis of auditory theory and the results of judgment tests, it is concluded that:

1. The perceived noisiness of broadband sounds when measured at the position of the listener can be best predicted by a combination of one-third octave band sound levels below 355 Hz in certain ways prior to the calculation of the unit PNdB (PNdB-M), or by the power summation of sound frequency weighted according to the 40 noy contour adjusted at low frequencies for the critical bandwidth of the ear, dB(D), (dB[D_2]). dB(A) and particularly dB(C) do less well than dB(D).

2. Effective, time-integrated, measures of sound are significantly better predictors of judged perceived noisiness than are so-called Max or Peak sound measures.

3. Corrections should be applied to (a) the EPNLs or nonimpulsive sounds of different tonal content and onset durations, and (b) the impulsive sound.

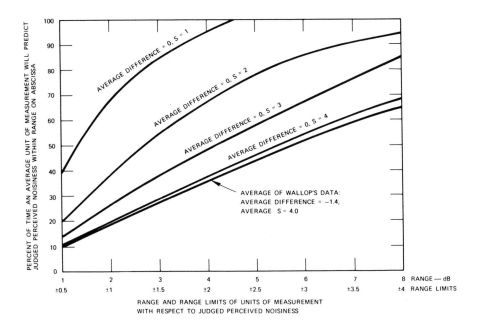

FIGURE 184.   Percentage of time that the average units of measurement predicted judged
perceived noisiness of a variety of aircraft tested at Wallops Island, Virginia
(474) within the range of unit accuracy given on the abscissa, and the
accuracy relations for idealized data having various degrees of variability.

TABLE 38

Average Differences and Standard Deviations of Physical Noise Measurements of Reference
and All Comparison Aircraft Noises When Judged Equally Noisy
or Unacceptable. From Kryter *et. al.* (474).

Judgments and physical measurements made outdoors. Thirty-five listeners, 18 comparison aircraft. Also shown are percentage of time the various units of noise measurement would agree with ±2 and ±4 units used to predict judged equal perceived noisiness.

Col. 1 Rank (see Col. 5)	Col. 2 Measure	Col. 3 % Times Between -2 and +2	Col. 4 % Times Between -4 and +4	Col. 5 Average of Percentages (Col. 3 and 4)	Col. 6 Average Difference	Col. 7 Standard Deviation
1	$EPNdB_{t_1}M_o$	52	84	68	-0.8	2.7
2	$EdB(D_2)_o$	46	83	64.5	-2.1	2.0
3	$EPNdB\text{-}M$	48	80	64	-1.3	2.8
4	$EPNdB_{t_2}M_o$	48	80	64	-1.3	2.8
5	$EdB(D_2)$	47	81	64	-1.9	2.3
6	$EPNdB_{t_1}M$	47	79	63	-0.6	3.1
7	$EPNdB\text{-}M_o$	49	77	63	-1.6	2.5
8	$EPNdB_{t_2}M$	46	78	62	-0.9	3.1
9	$EPNdB_{t_1}o$	45	78	61.5	-1.1	3.1
10	$EEPNdB_{t_1}M$	45	77	61	-0.9	3.2
11	$Max\ dB(D_3)$	42	74	58	-1.4	3.3
12	$EPNdB_{t_1}$	41	71	56	-0.8	3.7
13	$Peak\ dB(D_2)$	40	71	55.5	-0.9	3.7
14	$EdB(D_1)_o$	40	71	55.5	-2.2	2.9
15	$EdB(D_1)$	39	70	54.5	-1.9	3.3
16	$EPNdB$	38	69	53.5	-1.6	3.6
17	$EPNdB_{t_2}$	38	68	53	-1.2	3.8
18	$Max\ dB(D_2)$	37	67	52	-1.1	4.0
19	$EEPNdB_{t_1}$	37	67	52	-1.1	4.0
20	$EEPNdB_{t_2}$	36	64	50	-1.3	4.2
21	$EdB(D_3)$	30	69	49.5	-3.0	2.0
22	$Max\ PNdB_{t_2}M$	34	63	48.5	0.1	4.5
23	$Max\ dB(D_1)$	33	61	47	-1.1	4.6
24	$EEPNdB$	33	61	47	-1.9	4.2
25	$Max\ PNdB\text{-}M$	32	60	46	-1.4	4.5
26	$Max\ PNdB_{t_2}$	33	59	46	0.1	4.8

TABLE 38 (continued)

Col. 1 Rank (see Col. 5)	Col. 2 Measure	Col. 3 % Times Between -2 and +2	Col. 4 % Times Between -4 and +4	Col. 5 Average of Percentages (Col. 3 and 4)	Col. 6 Average Difference	Col. 7 Standard Deviation
27	Peak PNdB	32	60	46	-0.3	4.8
28	Max dB(A)	31	60	45.5	-2.8	3.7
29	Peak Phons	32	58	45	-0.7	4.9
30	$EdB(D_3)_o$	28	61	44.5	-3.3	2.4
31	Max PNdB	31	58	44.5	-0.7	5.0
32	Max Phons	31	57	44	-1.0	5.0
33	EdB(A)	29	57	43	-3.4	3.1
34	Max PNdB$_{t_1}$ M	29	54	41.5	0.4	5.4
35	$EdB(A)_o$	25	54	39.5	-3.7	2.8
36	Max PNdB$_{t_1}$	27	51	39	0.3	5.8
37	Max dB(B)	24	46	35	-2.4	6.3
38	Max dB(C)	21	41	31	-1.4	7.3

4. Tone-correction procedures, as outlined by Kryter and Pearsons (468) and Sperry (758) appear to be about equally effective when applied to subsonic aircraft noise.

5. The fact that $EdB(D_2)$ as predictive accuracy EPNdB-M suggests the possibility that the band summation method proposed by Stevens for the calculation of loudness, and adopted for use with PNdB could be discarded. In its place one could use the simpler procedure for summing on a power basis the sound pressure levels of band spectra that have been weighted according to the noy contours and the critical bandwidths of the ear. Procedural steps for this are outlined in Chapter 11, and the resulting unit is designated as PNdB'. We believe that the basic unit of PNdB' may turn out to have desirable features from a physical measurement standpoint, and to be the most general unit to use for estimating perceived noisiness.

6. Although effective (time integrated) PNLs are not generally available for other than aircraft noises, there is no reason to believe that the above conclusions are not applicable to the evaluation of all types of environmental noises, such as automobiles, trucks, trains, industry, etc.

TABLE 39

Showing Relation between Results with Phons (Stevens) and PNdB and Average Effect of Various Modifications and Corrections to PNdB and Overall Frequency Weightings

All score values are percentage of time a given unit of measurement would, for the 18 aircraft noises tested, fall within ±4 units of judged equal-perceived noisiness; 35 listeners outdoors. From Kryter et al. (474).

Units Calculated from 1/3-Octave Band Spectra

Max Phons	57%	Max PNdB	58%	Max PNdB$_{t_1}$	51%	EPNdB	69%	EPNdB$_{t_1}$	71%	EPNdB$_o$	(69)%*	EPNdB$_{t_1 o}$	78%	EEPNdB	61%	EEPNdB$_{t_1}$ 67%
Peak Phons	58			Max PNdB$_{t_2}$	59			EPNdB$_{t_2}$	68			EPNdB$_{t_2 o}$	(78)*			EEPNdB$_{t_2}$ 64
Aver.	58															

Max PNdB	58	Max PNdB-M	60	Max PNdB$_{t_1}$M	54	EPNdB-M	80	EPNdB$_{t_1}$M	79	EPNdB-M$_o$	77	EPNdB$_{t_1}$M$_o$	84	EEPNdB-M	(70)*	EEPNdB$_{t_1}$M 77
Peak PNdB	60			Max PNdB$_{t_2}$M	63			EPNdB$_{t_2}$M	78			EPNdB$_{t_2}$M$_o$	80			EEPNdB$_{t_2}$M (74)*
Aver.	59															

*Estimated, not calculated

Units Calculated from Overall Frequency Weightings

dB(D$_1$)	dB(D$_2$)	dB(D$_3$)	dB(A)
Max dB(D$_1$) 61	Max dB(D$_2$) 66	Max dB(D$_3$) 67	Max dB(A) 74
EdB(D$_1$) 70	EdB(D$_2$) 81	EdB(D$_3$) 69	EdB(A) 69
EdB(D$_1$)$_o$ 71	EdB(D$_2$)$_o$ 83	EdB(D$_3$)$_o$ 61	EdB(A)$_o$ 61
Aver. 67	Aver. 77	Aver. 77	Aver. 68

Average Effect of Frequency Modification for Critical Bandwidth of the Ear (M, D$_2$)

All PNdBs	66	All dB(D$_1$)	67
All PNdB-Ms	73	All dB(D$_2$)	77
Aver. Improvement	7% pts	Aver. Improvement	10% pts

Average Effect of Summation over Frequency Range (Freq. Weighting plus Stevens' Band Summation vs. Freq. Weighting plus Sound Energy Summation)

All PNdBs and PNdB-Ms except for tone-corrected units	68
All PNdB$_t$s and PNdB$_t$Ms	71
All dB(D$_1$)s and dB(D$_2$)s	72
Aver. Improvement dB(D$_1$)s and dB(D$_2$)s vs. PNdBs and PNdB-Ms	4% pts
vs. PNdB$_t$s and PNdB$_t$Ms	1% pts

Average Effect of Duration (Max vs. Effective (E) and Estimated Effective (EE))

All Max PNdBs and PNdB-Ms	58	All Max dB(D$_1$) and dB(D$_2$)	64
All EEPNdBs and PNdB-Ms	69	All EdB(D$_1$) and dB(D$_2$)	76
All EPNdBs and PNdB-Ms	76	Aver. Improvement	12% pts
Aver. Improvement Re/Max: EE 11% pts; E 18% pts			

Average Effect of Onset Duration Correction (o)

All EPNdBs and PNdB-Ms no onset correction	74	EdB(D$_1$) and dB(D$_2$) no onset correction	76
All EPNdBs and PNdB-Ms with onset correction	78	EdB(D$_1$) and dB(D$_2$) with onset correction	77
Aver. Improvement	4% pts	Aver. Improvement	1% pt

Average Effect of Tone Corrections

All PNdBs - no tone corrections	68
All PNdB$_{t_1}$ - tone-corrected	70
All PNdB$_{t_2}$ - tone-corrected	71
Aver. Improvement, t_1	2% pt
Aver. Improvement, t_2	3% pts

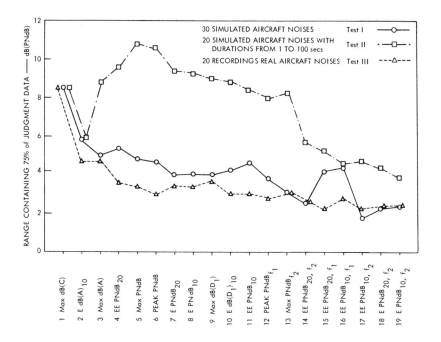

FIGURE 185. Comparison of the three judgment tests for all prediction measures. The unit designations on the abscissa are the same as those in Tables 37 and 38 except that the subscripts 10 and 20 refer, respectively, to integrations between 10 dB and 20 dB downpoints. From Pearsons and Bennett (613a).

Chapter 9

Environmental Noise and Its Evaluation

Introduction

The question of noise pollution is a matter of concern to industry, science and local, national, and international levels of government. There is available much information about and some quantitative methods for measuring noise pollution in man's environment and his behavior in response to that pollution. Research and engineering data about environmental noise and methods for relating or predicting the average behavior of individuals and communities to long-term noise pollution are divided and presented in the following parts:

1. Office and room noise
2. Noise surveys in the community
3. Motor vehicle noise
4. Aircraft noise
5. The sonic boom
6. Regulatory codes for community noise
7. Composite Noise Rating Schemes

Office and Room Noise

As noted earlier, Rosenblith and K.N. Stevens (702) proposed a method for the evaluation of the general bothersomeness of outdoor noise in a community. According to the original form of this method, one finds the highest equal loudness contour for octave bands that is tangent at least at one point to the octave band levels of a sound being evaluated. The authors ascribed "level ranks" to these contours that signified the general acceptability of a noise reaching a given level rank. Work sheets for plotting octave band spectra and finding the highest rank contour tangent to or touched by the octave band spectra of a noise were developed. Beranek (55) recommended that a similar method be used for rating office noise, primarily with regard to speech communication problems, and proposed a set of contours closely like those suggested by Rosenblith and K.N. Stevens. Beranek called these contours SC, for speech communication.

In 1957 Beranek (58) slightly modified the SC contours, relabeled them as NC (Noise Criteria) contours, and assigned to each contour the sound pressure level of the 1200-2400 Hz band, as shown on the left-hand curves in Fig. 186. The number of the highest NC contour just reached by any octave band of a noise plotted on Fig. 186 is used as a means of rating a noise. Beranek also proposed a set of NCA contours (see the right-hand curves of Fig. 186) to be used for office spaces where noise control could be achieved only with great difficulty and expense. A committee of ISO has proposed changing the number of NC contours to make them correspond to the SPL of the octave band centered at 1000 Hz; it was proposed that these contours (see Fig. 187) be called NR, for noise rating.

Beranek published articles (55-58) that have served as the validation of the NC method for the evaluation of noise in buildings. The data in these articles came from (a) questionnaires (concerning ease of speech communications, general bothersomeness, etc.) administered to general office workers and executives, (b) an experiment in which noise was introduced by loudspeakers into a frequently used conference room and ratings were made by conferees of the disturbance caused by the noise, and (c) case histories of situations where remedial acoustical help had been requested because of complaints about noise. On the basis of some of this information, the effects of sounds having various NC ratings were derived. The effects associated with the NC spectra are determined by reference to Tables 40 and 41. It should be noted that the conclusions reached by Beranek are in good agreement with criteria for room noise put forth earlier by Knudson and Harris (434) in terms of dB(A) levels.

At the time the NC method was proposed, Stevens' and Zwicker's procedures for calculating loudness and the procedure for the calculation of perceived noise level were not available. It would appear, in view of the probably greater accuracy with which these latter procedures reflect the hearing process, that the substitution of any of them for the closest tangent-octave-band method (the Level Rank, SC, NC, NCA, or NR method) would improve the general accuracy of evaluating office and room noises in terms of the NC criteria. As a matter of fact, overall dB(A) or dB(D) are apparently more accurate, as will be shown, for any purpose or type of noise evaluation than is Level Rank, SC, NC, NCA, or NR procedure. Wells and Blazier found, for example, that certain noises having peaked spectra may be judged as being 18 dB or so less objectionable than broad spectra noises having the same NC value.

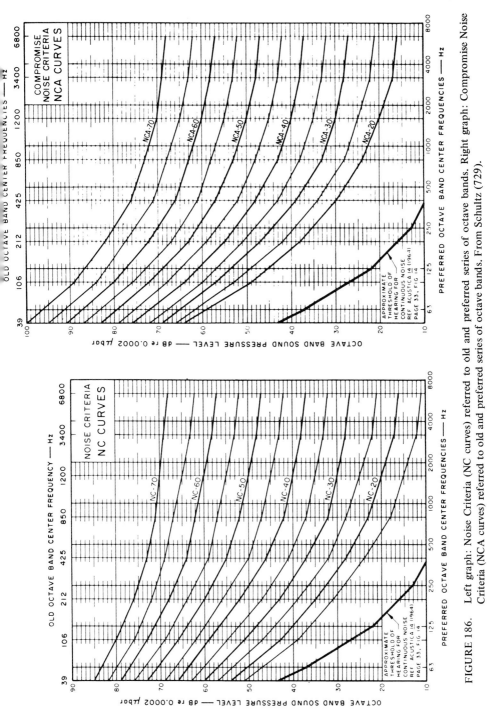

FIGURE 186. Left graph: Noise Criteria (NC curves) referred to old and preferred series of octave bands. Right graph: Compromise Noise Criteria (NCA curves) referred to old and preferred series of octave bands. From Schultz (729).

Although NC values are a relatively poor means of rating certain noises relative to each other, the NC procedure of plotting octave band spectra on graph paper marked with more or less equal loudness contours provides the engineer with insight into what portions of the noise in question are contributing the most to its loudness or perceived noisiness, and it is probably this feature of the NC (tangent contour) method that contributed most to its rather widespread use. The major shortcoming of the NC method is, of course, that the rating it assigns to a complex sound is determined by the level in only one octave band — the octave band closest or tangent to the highest contour. The contribution of the remainder of the spectrum to the audibility of the noise is ignored.

There is little reason, however, to doubt that the criteria in Tables 40 and 41 are accurate for noises with broadband spectra reasonably similar to the contours in Fig. 185. We have, accordingly, taken the liberty of calculating the approximate Max PNL, EPNL (for continuous noise) in PNdB, dB(D), and dB(A), and CNR values of the NC curves, and included these values in Tables 40 and 41. Beranek (58) also calculated dB(A) values for the NC curves with regard to noise criteria for rooms but recommended they not be used for specification purposes. The suggestion here is that instead of using the closest-tangent-octave-band NC number, one finds PNdB, dB(D), or dB(A) values for a room noise. Continuous noise whose Max PNL, EPNL, or CNR values equal those specified in Tables 40 and 41 would be expected to result in the criterion behavior described.

The behavior criteria to be used in conjunction with the NC method, or the PNdB, dB(D), or dB(A) values as herein proposed, are oriented towards the effects of noise on speech communication. However, there is nothing inherent in either the NC, PNdB, dB(D), or dB(A) values obtained for a noise that guarantees that any of these measures will reflect the interference effect of that noise with speech. A noise that was strongly peaked in either the very high or low frequency region could have an NC, PNdB, dB(D), or dB(A) level all out of proportion to its masking of speech (see Chapter 2).

Speech Interference Level (SIL) was proposed by Beranek (55) as a simple way to rate the speech masking effectiveness of a noise. SIL is usually the arithmetic average of the sound pressure levels in the octave bands 600-1200, 1200-2400, 2400-4800 Hz (see Chapter 2). Beranek (57, 58) found, in studies of the rating of office noises with respect to interference with work and speech communications, that SIL did not predict the obtained ratings as well as did loudness levels in phons (Stevens). Although Beranek felt that this reflected the contributions to the rating of the general loudness of a noise over and above its speech interference, it is probable in view of the strong low frequency components in some of the noises present in the offices studied, that SIL was a poorer measure of speech masking of the particular noise in question than was

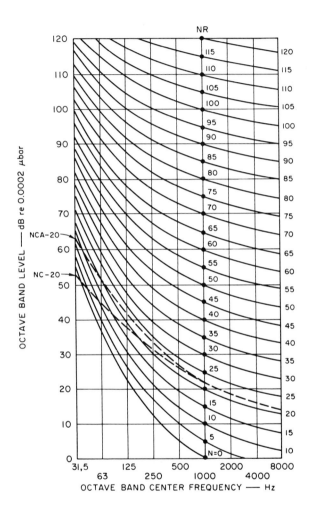

FIGURE 187. Curves and noise rating numbers (NR) proposed by ISO for rating acceptability of noises (Kosten and Van Os [436]). NC-20 and NCA-20 noise
rating curves are shown for purposes of comparison. (With permission of
the Controller of Her Britanic Majesty's Stationery Office.)

TABLE 40

Tolerable Limits in Various Rooms for Noise Continuously Present 7 AM to 10 PM

Tolerable maximum levels or exposures in various rooms for more or less continuous noise from 7 AM to 10 PM. Equal max PNLs for different noises are comparable to each other only when the noise has a broadband spectrum approximately similar in shape to the 40 noy contour and does not contain any strong puretone or line spectrum components. After Beranek (58).

Noises or noise environments of equal EPNL or equal CNR values are presumably equal in their effects on people regardless of the spectral or temporal complexities of the noises or noise environments they represent.

Type of Space	NC	Max PNL			PNdB	EPNL EPNdB EdB(D') EdB(A')	CNR
		dB(A)	dB(D)				
Broadcast studios	18	28	35		41	78	66
Concert halls	18	28	35		41	78	66
Legitimate theaters (500 seats, no amplification)	23	33	40		46	83	71
Music rooms	25	35	42		48	85	73
Schoolrooms (no amplification)	35	35	42		48	85	73
Apartments and hotels	28	38	45		51	88	76
Assembly halls	28	38	45		51	88	76
Homes	30	40	47		53	90	78
Motion picture theaters	30	40	47		53	90	78
Hospitals	30	40	47		53	90	78
Churches	30	40	47		53	90	78
Courtrooms	30	40	47		53	90	78
Libraries	30	40	47		53	90	78
Offices – Executive	25	35	42		48	85	73
– Secretarial (Mostly typing)	40	50	57		63	100	88
– Drafting	35	45	52		58	95	83
Meeting rooms (sound amplification)	35	45	52		58	95	83
Retail stores	37	47	64		60	97	85
Restaurants	45	55	62		68	105	93

Note 1: The noise levels outdoors from sources located outdoors (aircraft, road traffic, etc.) would be typically about 20 dB greater for the average house and 30 dB for masonry or well sound-insulated buildings than the levels given in the above table.

Note 2: dB(A') −13 = dB(A); dB(D') −6 = dB(D).

TABLE 41

Recommended Noise Limits for Offices

Noise measurements made for the purpose of judging the satisfactoriness of the noise in an office by comparison with these levels should be performed with the office in normal operation, but with no one talking at the particular desk or conference table where speech communication is desired (i.e., where the measurement is being made). Background noise with office unoccupied should be lower, say by 5-10 units. (See also Table 40.) From Beranek (57).

Communication Environment	Typical Applications	NC	Max PNL			EPNL	CNR
			dB(A)	dB(D)	PNdB	EPNdB EdB(D') EdB(A')	
1. Very quiet office-telephone use satisfactory-suitable for large conferences.	Executive offices and conference rooms for 50 people.	25	35	42	48	85	73
2. "Quiet" office; satisfactory for conferences at 15-ft table; normal voice 10 to 30 ft; telephone use satisfactory.	Private or semi-private offices, reception rooms, and small conference rooms for 20 people.	33	43	50	56	93	80
3. Satisfactory for conferences at a 6- to 8-ft table; telephone use satisfactory; normal voice 6 to 12 ft.	Medium-sized offices and industrial business offices.	38	48	55	61	98	86
4. Satisfactory for conferences at a 4- to 5-ft table; telephone use occasionally slightly difficult; normal voice 3 to 6 ft; raised voice 6 to 12 ft.	Large engineering and drafting rooms, etc.	45	55	62	68	105	93
5. Unsatisfactory for conferences of more than two or three people; telephone use slightly difficult; normal voice 1 to 2 ft; raised voice 3 to 6 ft.	Secretarial areas (typing), accounting areas (business machines), blueprint rooms, etc.	53	63	70	76	113	101
6. "Very noisy"; office environment unsatisfactory; telephone use difficult.	Not recommended for any type of office.	55	65	72	78	115	103

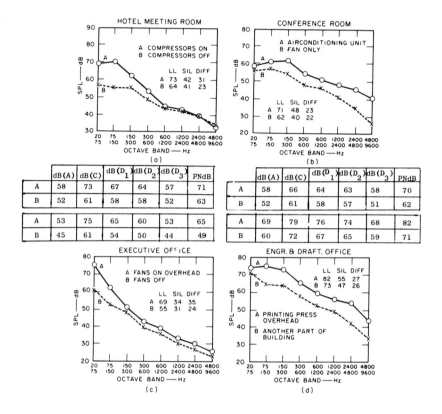

	dB(A)	dB(C)	dB(D_1)	dB(D_2)	dB(D_3)	PNdB
A	58	73	67	64	57	71
B	52	61	58	58	52	63
A	53	75	65	60	53	65
B	45	61	54	50	44	49

	dB(A)	dB(C)	dB(D_1)	dB(D_2)	dB(D_3)	PNdB
A	58	66	64	63	58	70
B	52	61	58	57	51	62
A	69	79	76	74	68	82
B	60	72	67	65	59	71

FIGURE 188.   Sound-pressure levels vs. octave-band number for office and rooms (a) through (h) after changes in the noise. The calculated loudness levels (LLs) and SILs, and the differences, are given on the graphs. When finally corrected, the noise levels of the offending machines were reduced below the levels given by curve B for each office. After Beranek (57).

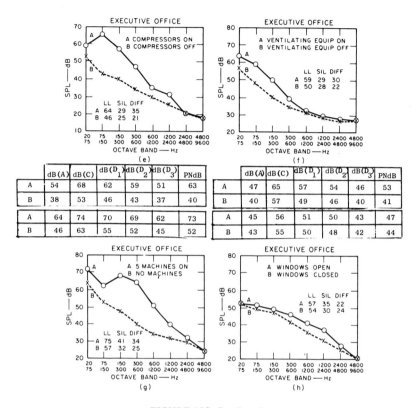

	dB(A)	dB(C)	dB(D_1)	dB(D_2)	dB(D_3)	PNdB
A	54	68	62	59	51	63
B	38	53	46	43	37	40
A	64	74	70	69	62	73
B	46	63	55	52	45	52

	dB(A)	dB(C)	dB(D_1)	dB(D_2)	dB(D_3)	PNdB
A	47	65	57	54	46	53
B	40	57	49	46	40	41
A	45	56	51	50	43	47
B	43	55	50	48	42	44

FIGURE 188. Continued.

loudness level. Figure 188 shows octave band spectra in some offices where noise was considered to be a problem.

Embleton *et al.* (220) obtained questionnaire data with regard to the effect of noise in offices containing business machines. While his results in general agree with the NC criteria for business machine offices, he believes that the levels can be increased by 5 to 10 dB above those now specified for this type of office, provided the room is made nonreverberant and the width-to-length ratio is 1.6 to 1. Young (897) found the correlations for Embleton's noises between dB(A) and loudness levels in Phons (Stevens), SIL, and NC to be 0.97, 0.99, and 0.98 respectively. Young suggests that this demonstrates that dB(A) is as meaningful and valid a measure for rating office noise as loudness level, SIL, and NC. While his conclusions may be true, the correlations primarily prove that the spectrum shape of the noises present in the offices studied by Embleton *et al.* were very similar to each other, as was the case.

Speech Privacy

Cavanaugh *et al.* (129) suggest that one of the most bothersome sounds in offices, particularly private offices, is the presence of speech that intrudes through the walls from adjacent rooms. When such speech is present, the occupants feel the privacy of their own speech is lacking. These investigators further believe that it is the degree to which the intruding speech can be understood, rather than its intensity level or loudness, that destroys the feeling of office privacy. These authors developed a nomograph (see Fig. 189) whereby, from a knowledge of the sound attenuation properties of the walls of a room, one could estimate the Articulation Index (AI) present in a room from speech uttered in an adjacent room. Essentially, this is accomplished by plotting on Fig. 189 the attenuation in each one-third octave band afforded by the walls between the rooms, and then counting the number of dots lying above the attenuation curve. The number of such dots divided by 100 gives the approximate AI of speech from the adjacent room. Cavanaugh *et al.* found that an AI of greater than 0.05 gave rise to some expressions of dissatisfaction with respect to ratings of the privacy of an office space. This need for a sense of privacy provides the interesting situation where the introduction of some speech masking noise, sometimes referred to as "acoustic perfume," increases feelings of satisfaction about unusually quiet offices that are not isolated from intruding speech signals.

Noise Inside the Home

Common noises found in the home are shown in Figs. 190 and 191. Many of these noises are, of course, intermittent. We will return later to a discussion of the generally tolerable levels of noise in the home.

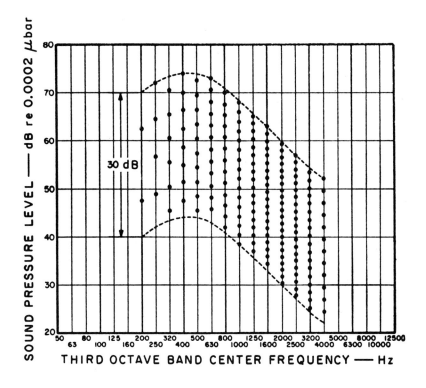

FIGURE 189.   Graphical representation of normal speech levels. Number of dots in each third-octave band signifies relative contribution to articulation index. From Cavanaugh *et al.* (129).

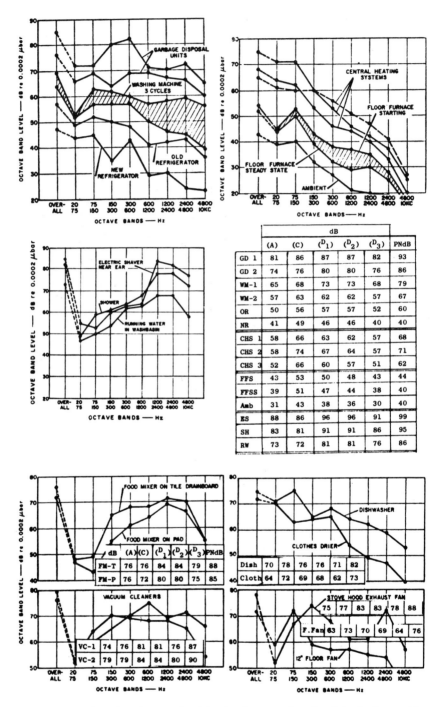

FIGURE 190. Overall and octave-band levels of some noises found in the home. After Mikeska (541).

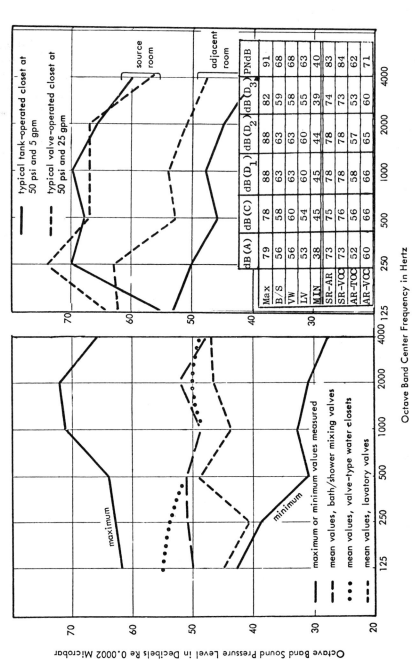

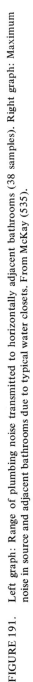

Octave Band Center Frequency in Hertz

FIGURE 191.   Left graph: Range of plumbing noise transmitted to horizontally adjacent bathrooms (38 samples). Right graph: Maximum noise in source and adjacent bathrooms due to typical water closets. From McKay (535).

Noise Inside Aircraft and Motor Vehicles

A special room noise is that found inside aircraft and motor vehicles. The internal noise in commercial aircraft has been extensively studied but no fixed criteria of tolerable levels have been established. Lippert and Miller (512) developed an "acoustical comfort index" for aircraft noise based on questionnaires given to passsengers and crew members of commercial airliners. Figure 192 shows the borderline levels, as best could be determined from the limited types of aircraft noise studied, between intolerable and ideal quiet.

Sternfeld *et al.* (766) studied the effects of the internal noise in military helicopters on pilot ratings of interference with speech communications, disturbance of muscle coordination, feelings of fatigue, hearing loss, etc. The octave band sound pressure levels present at the ears of the pilots were correlated against the ratings obtained as shown in the lower graph of Fig. 193. It is seen that the higher frequencies — those that would interfer most with speech communication, auditory fatigue and annoyance — correlated highest with the ratings given.

Figure 194 shows noise levels as found in some typical vehicles used for transportation. It might be noted that near-daily exposures to some of the aircraft noises illustrated could, after a few years, result in some degree of permanent high-frequency hearing loss according to the damage risk conditions as described in Chapter 5. Indeed, such hearing losses are found in airline stewards and hostesses (627).

Noise Inside Space Vehicles

Figure 195 shows the sound pressure levels just outside the nose cone of a space vehicle during launch and during reentry. The maximum octave band spectra of the sound during launch is given in Fig. 196. It would appear that persons inside the space vehicle would not be exposed during any portions of launch and flight to intolerable levels because of the attenuation (probably in excess of 50 dB in the speech frequency region) of the sound that would be provided by the cabin of the vehicle, and the protective helmets that would be worn, and the short duration of the more intense levels during launch.

Noise Surveys in the Community

A number of studies and surveys of environmental sounds that bother people have been made. The sounds measured include, to name a few, those from rain,

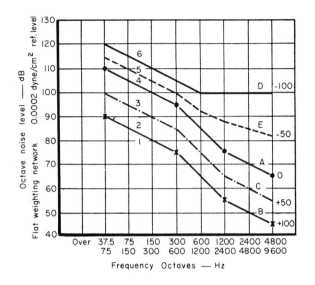

			dB			
		(A)	(C)	(D_1) (D_2) (D_3)		PNdB

			(A)	(C)	(D_1)	(D_2)	(D_3)	PNdB
1.	Ideally quiet	B	77	91	84	82	75	89
2.	Comfortable	C	87	101	94	92	85	100
3.	Quasi comfortable	A	97	111	104	102	95	109
4.	Uncomfortable	E	110	116	108	107	101	116
5.	Very uncomfortable	D	117	121	110	116	120	124

FIGURE 192. Graphic classification of acoustical comfort for transport aircraft. Curve 3(A) should be considered as defining the upper limit of comfort for an airplane noise spectrum according to Lippert and Miller (512).

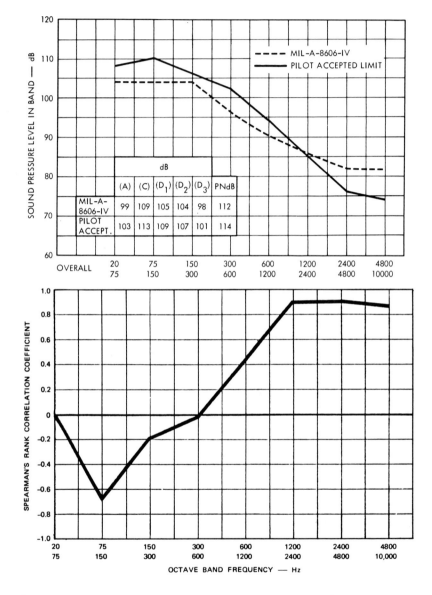

FIGURE 193.   Upper graph: Comparison of noise limits for aircraft by octave band according to judgment tests and U.S. Army specification for maximum internal noise in military aircraft (MIL-A-8606-IV). Lower graph: Correlation of
rank pilot opinion of noise with ranked sound pressure level in octave
band. From Sternfeld *et al.* (766).

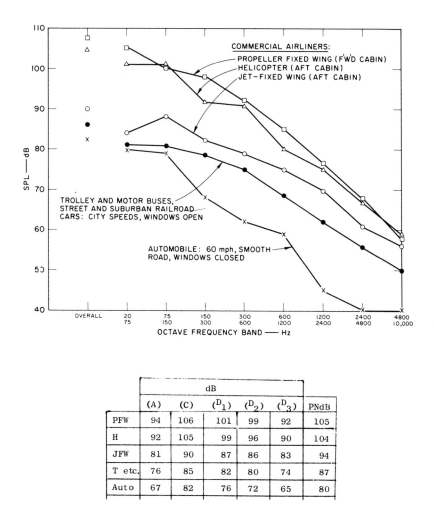

	dB					
	(A)	(C)	(D_1)	(D_2)	(D_3)	PNdB
PFW	94	106	101	99	92	105
H	92	105	99	96	90	104
JFW	81	90	87	86	83	94
T etc.	76	85	82	80	74	87
Auto	67	82	76	72	65	80

FIGURE 194.   Overall and octave-band levels of noise inside various transportation vehicles. Aircraft, after Miller and Beranek (555); trolley, buses and railroad cars, after Bonvallet (78); and automobile, after Wiener (875).

thunder, barking dogs, birds, factories, autos, trains, and aircraft (68, 78, 79, 114, 115, 148, 220, 230, 273, 429, 488, 490, 541, 555-557, 712, 726, 799, 816, 875, 876, 878, 880).

Figure 197, from a survey made by Veneklasen (816), illustrates the overall levels and octave band spectra of some of the sounds found in residential, business, and industrial areas in or near a city. Many of these noises are, of course, intermittent. Table 42 shows, in terms of band spectra and dB(A), other noises found outside and/or inside some common vehicles. Noises from outdoor sources as found in the home are shown in Fig. 198.

A major analysis of the reaction of people to noise in the community was recently undertaken by a committee under the auspices of the British government (536, 880). Some of the results of a general questionnaire survey made at 450 points spaced over 36 square miles of Central London are given in Table 43 and in Fig. 199. The results presented in Table 43 were more or less similar for all neighborhoods except those near airports, where the aircraft noise became the predominant noise that disturbed people. Some of the more specific findings of motor vehicle and aircraft noise studies will be presented later.

The area enclosed by the solid and dashed lines in Fig. 199 includes about 70% of the opinions obtained and shows the wide variations in people's judgments. It must be remembered in interpreting the "street noise" contour in Fig. 199 that the results represent more or less peak noise levels found out of doors – their frequency, duration, and time of occurrence for any one dB(A) level were not specified.

It is generally recognized that the sound heard on the ground from present-day commercial aircraft passing overhead and during engine "run-up" while the aircraft is on the ground is a major source of annoyance in neighborhoods near airports. In the United States, lawsuits have been brought against airlines and airport operators as creators of "intolerable" noise (Tondel [803]). Some understanding of why this occurs can be gained from Fig. 200. It is seen in Fig. 200 that aircraft, particularly the present-day jet, represents a significant increase, for single exposures, in noise over that produced by surface transportation vehicles. In addition, the duration of flyover noise is, in some localities near airports, surprisingly long, as shown in Table 44 from Cohen and Ayer (141). The problem of human response to external aircraft noise will be discussed in a later section devoted exclusively to research studies concerned solely with aircraft noise.

The engines of vehicles for flight into outer space create considerable external noise when they are being launched, even at distances of several miles from the launch site (see Fig. 201 from Regier et al. [668]). Fortunately, as the sound pressure level increases, the predominate sound frequency becomes lower, as shown in Fig. 202. From what little is known about the effects of low frequency sounds (frequencies below 50 Hz or so) it appears that man, particularly as far as

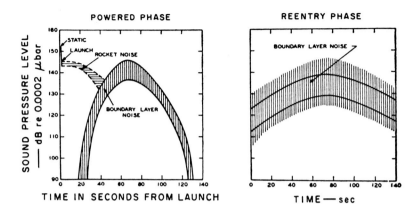

FIGURE 195.   Predicted overall sound-pressure levels just outside the nose cone are plotted here for both powered and re-entry phases of a typical space craft. From von Gierke (818).

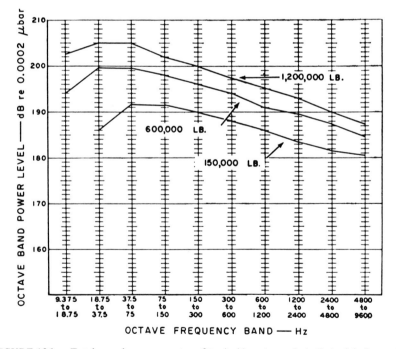

FIGURE 196.   Total sound-power spectra of typical booster rockets that might be used in various combinations for this space craft during launch. From von Gierke (818).

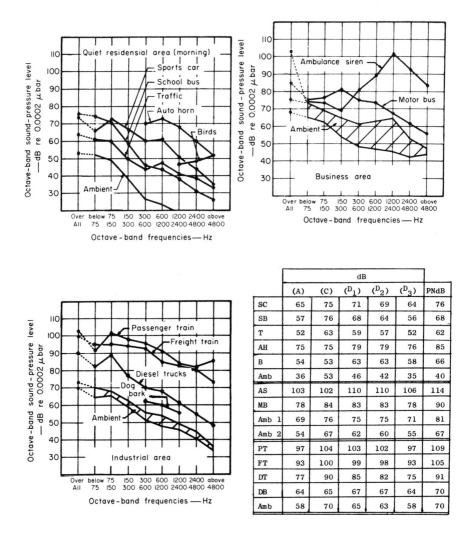

	dB					
	(A)	(C)	(D_1)	(D_2)	(D_3)	PNdB
SC	65	75	71	69	64	76
SB	57	76	68	64	56	68
T	52	63	59	57	52	62
AH	75	75	79	79	76	85
B	54	53	63	63	58	66
Amb	36	53	46	42	35	40
AS	103	102	110	110	106	114
MB	78	84	83	83	78	90
Amb 1	69	76	75	75	71	81
Amb 2	54	67	62	60	55	67
PT	97	104	103	102	97	109
FT	93	100	99	98	93	105
DT	77	90	85	82	75	91
DB	64	65	67	67	64	70
Amb	58	70	65	63	58	70

FIGURE 197.   Noise in residential, business, and industrial areas. From Venaklasen (816).

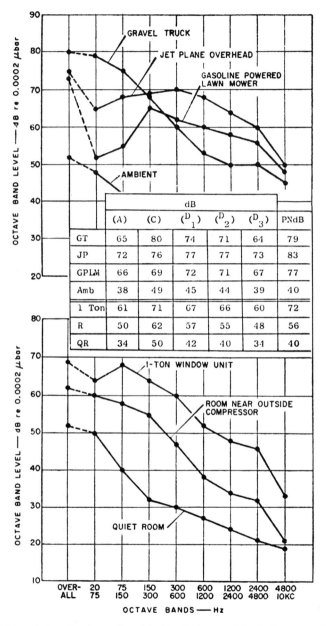

FIGURE 198.   External noises as heard inside the house with windows open in the summer (upper graph). Noise levels from air-conditioning equipment, averaged throughout the room with equipment in full operation (lower graph). From Mikeska (541).

TABLE 42

Noise Levels in dB(A), Inside and or Outside Some Common Transportation Vehicles

Data collected and furnished by Botsford (Personal Communication).

The data on public transportation noise was obtained by C.R. Bragdon and that on power boats by R.A. Campbell.

Public Transportation Noise (Philadelphia)	Below Ground	Above Ground
Subway-elevated train		
inside cars	82-95 dB(A)	78-90 dB(A)
on boarding platform	93-98	83-93
at cashier's booth	90-93	82-88
Trolley car		
inside cars	74-87	65-75
on boarding platform	84-100	80-85
at cashier's booth	83-84	None
Power Boats (at seat nearest motor)		
Cruising speed	83-104 dB(A)	
Full speed	85-90	

Maximum Noise at Operator's Ear	
Cranes	85-113 dB(A)
Outboard Motor	85
Street Sweeper	96
Buses	82-96
Trucks	81-92
Tractors	83-113
Road Graders	97-100
Self-Propelled Camper	92
Diesel Tractor-trailer trucks	
Engine room	101-112 dB(A)
Shop, steering room	94-98
Other rooms	73-78
River Barge Tow Boat (919 gross tons)	
Idle (400-700 rpm)	68-79 dB(A)
Low rpm (1000-1500)	75-87
High rpm (2000-2500)	82-92

TABLE 43

Relation of Noise to Other Factors (Top) and
Noises Which Disturb People at Home, Outdoors and at Work (Bottom)
From Mc Kennell (536). (With Permission of the Controller
of Her Britanic Majesty's Stationery Office.)

The one thing that people most wanted to change (i)	The percentage of people who wanted to change it (ii)
Noise	11
Slums/dirt/smoke	10
Type of people	11
Public facilities/transport/council	14
Amount of traffic	11
Other facilities/shopping/entertainment ...	7
Other answers...	1
No answer, or vague reply...	5
Would change nothing	30

Description of noise (i)	Number of people disturbed, per 100 questioned		
	when at home (ii)	when outdoors (iii)	when at work (iv)
Road traffic	36	20	7
Aircraft	9	4	1
Trains	5	1	--
Industry/Construction works	7	3	10
Domestic/Light appliances	4	--	4
Neighbors' impact noise (knocking, walking, etc.)	6	--	--
Children	9	3	--
Adult voices	10	2	2
Wireless/T.V.	7	1	1
Bells/Alarms	3	1	1
Pets	3	--	--
Other noise	--	--	--

his ear is concerned, becomes less and less affected by sound as its spectral
frequency is lowered. However, it has been found that certain structures of the
body will start vibrating with very intense sound at certain frequencies below 20
Hz or so. Figure 203 (Cole and Powell, 147) gives preliminary estimates of
maximum allowable sound pressure levels from space vehicles with regard to
ground-based personnel near the vehicles at time of launch.

Motor Vehicle Noise

It appears that, next to aircraft, automobiles and trucks are responsible for
much of the unwanted sound in our environment. The results of judgment tests

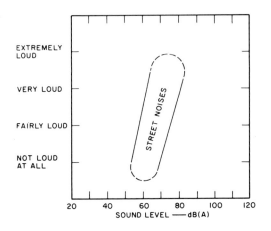

FIGURE 199. Judgments of street noise obtained in social survey in London. From Wilson (880). (With permission of the Controller of Her Britanic Majesty's Stationery Office.)

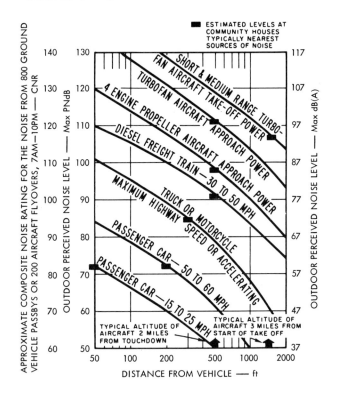

FIGURE 200. Max PNL in PNdB and dB(A) of noise from various transportation vehicles [dB(A) typically equals PNdB -13] and approximate CNR for certain specified noise exposures. For these approximate CNRs the duration, to the 10 dB downpoints, of the noise of the ground vehicle passby was taken as 4 sec and, for the aircraft flyover noise, as 16 sec. (Copyright 1966 by the American Association for the Advancement of Science.)

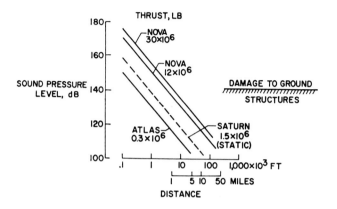

FIGURE 201. Maximum overall sound-pressure levels as a function of distance for some large rocket-powered vehicles during launching. From Regier *et al.* (668).

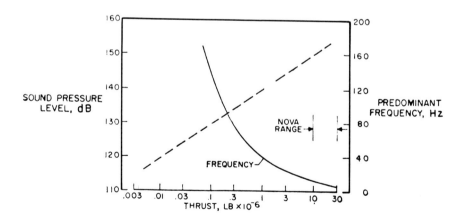

FIGURE 202. Overall sound-pressure levels and predominant frequencies at a distance of 1000 feet from several rocket engines. From Regier *et al.* (668).

TABLE 44

Length of Time That Flyover Noise Outdoors Exceeded SIL of 65 dB
From Cohen and Ayer (141).

	Jet			Propeller	
Type	Altitude (feet)	Time (seconds)	Type	Altitude (feet)	Time (seconds)
707	1800	30	DC-7	800	15
DC-8	2000	30	Electra	1000	8
DC-8	2000	25	DC-6	1200	1
DC-8	2000	38	DC-6	1200	22
707	2500	40	Electra	1500	16
DC-8	2500	35	Viscount	1500	16
DC-8	2600	60	DC-7	1500	22
DC-8	3000	25	DC-6	1650	14
DC-8	3000	48	Viscount	1700	12
720	3100	56	DC-6	1800	6
DC-8	3300	25	DC-7	2000	10
DC-8	3300	35	DC-7	2000	17
DC-8	3300	60	Electra	2000	22
DC-8	3400	40	DC-7	3000	6
DC-8	3500	18	Electra	3000	8
DC-8	4000	25			
Caravelle	4500	35		Median	13.5'
	Median	35			

of the loudness of motor vehicles conducted by Rademacher (661), Niese (578), Quietszch (655), and Lubke et al. (527) are mentioned in Table 36. In addition, Calloway and Hall (116) found a correlation of 0.83 with the dB(A) levels as the noises judged. Cederlof et al. (130) studied the relative annoyance reaction to recordings of car, trucks, and motorcycles; they found that car noise was judged the least annoying, even when its overall noise level on dB(C) was some 8 dB greater than the dB(C) level of the motorcycles.

Hillquist (370) conducted an extensive study of the ratings of the subjective preferences that jurors gave to 100 recorded noises from moving trucks (46 gasoline-engined and 54 diesel-powered). The distribution of the octave band spectra of the noises is shown in Fig. 204 and coefficients of correlations found between various measures of the noise and the subjective ratings are given in Table 45.

Andrews and Finch (14), in the United States, had subjects rate the recorded truck noise on a ten-point scale ranging from "quite inoffensive" to "quite annoying." Robinson et al. (694), Mills (557), and Mills and Robinson (558) conducted experiments in Great Britain in which subjects rated a variety of motor vehicles operated outdoors on a scale ranging from "quiet" to "excessively noisy." A somewhat similar study using various types of motor vehicles was conducted in Switzerland (18).

TABLE 45

Results of Judgment Tests of 100 Truck Noises

The product moment correlation coefficients PNdB (1969) in Chapter 11 of this book and D_1, D_2, and D_3 were calculated subsequent to publication of this paper, Hillquist (370).

		Product Moment Correlation Coefficient
1.	ARF (SAE J672) Loudness Level	.96
2.	Phons, Stevens Mk VI	.95
3.	Phons, Stevens Mk II	.93
4.	PNdB (1963)	.95
5.	PNdB (1969)	.94
6.	A-Weighted Sound Level, dB(A)	.95
7.	B-Weighted Sound Level, dB(B)	.92
8.	C-Weighted Sound Level, dB(C)	.86
9.	D_1-Weighted Sound Level (40 Noy Contour), dB(D_1)	.94
10.	D_2-Weighted Sound Level (1969), dB(D_2)	.94
11.	D_3-Weighted Sound Level, dB(D_3)	.94
12.	DIN 3-Weighted Sound Level	.94
13.	0.5/1/2 kHz Octave Band Level	.93

Figure 205 summarizes the results of several studies conducted on this problem. The area in Fig. 205 is divided between dB(A) levels that were considered acceptable and those that were noisy or offensive. The difference between the Andrews and Finch results and the other studies is possibly due to the fact that the former study was conducted in the laboratory with recordings of the motor vehicle sounds, while the latter were performed in open air with actual vehicles. There is no ready explanation for the difference between the British and Swiss studies other than systematic differences in the interpretation by the subjects of the meaning of the words used on the various rating scales and, of course, possible differences in the actual average tolerance that the different groups of subjects had to noises.

Jonsson et al. (427) also found possible nationalistic differences in tolerance to road noise. These investigators report that the annoyance reaction to road

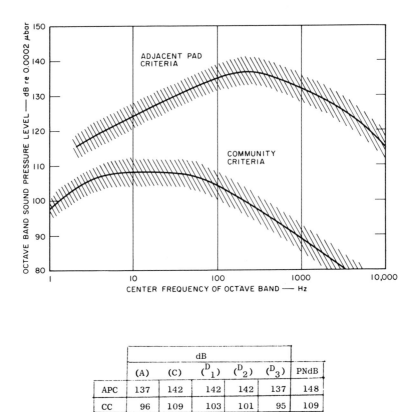

	dB					
	(A)	(C)	(D_1)	(D_2)	(D_3)	PNdB
APC	137	142	142	142	137	148
CC	96	109	103	101	95	109

FIGURE 203.   Proposed maximum octave-band levels of noise from space vehicles at the pad adjacent to a pad from which another vehicle is launched and at any community. From Cole and Powell (147).

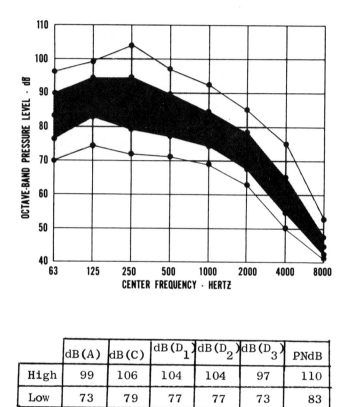

	dB(A)	dB(C)	dB(D₁)	dB(D₂)	dB(D₃)	PNdB
High	99	106	104	104	97	110
Low	73	79	77	77	73	83

FIGURE 204. Distribution of the octave-band levels measured for the 100-truck sample. Mean values are represented by the connected points; the extreme levels are shown by the outer data points. The intermediate points are one-standard-error points of the distribution. From Hillquist (370).

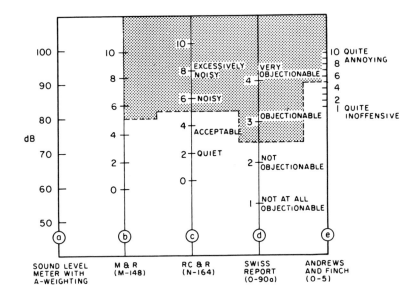

FIGURE 205.   Comparison of subjective and sound-level scales for motor vehicle noise. After Robinson *et al.* (694) and Mills and Robinson (558).

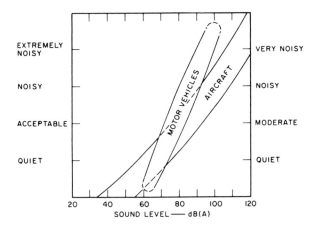

FIGURE 206.   Comparison of ratings of noise from motor vehicles and from aircraft. From Wilson (880). (With permission of the Controller of Her Britanic Majesty's Stationery Office.)

traffic noise was greater in Stockholm, Sweden than in Tarrara, Italy. For example, 61% of selected populations in Stockholm, but only 49% in Tarrara, said they were "disturbed" by motor vehicle noise of comparable levels and amount; however, the difference of numbers of people "greatly disturbed" was not statistically significant — 23% in Stockholm, and 21% in Tarrara.

Mills and Robinson found that there were no apparent significant differences in the results of the judgment tests due to the age or sex of the subjects. Similar conclusions, as mentioned earlier and as will be discussed later, have also been observed when aircraft noise was judged.

Method for Measuring Motor Vehicle Noise

It has become general practice to measure the sound from motor vehicles with a sound level meter set on A scale. Andrews and Finch, as did Hillquist, found a higher correlation between the judged noisiness of truck noise and loudness in sones than with dB(A), and Galloway (271), using the Mills and Robinson data, found that for diesel motor vehicles, PNdB (calculated according to reference 466) and Phons (Stevens) correlated better with judged noisiness than did dB(A). However, Galloway found that dB(A) values correlated slightly higher with the results of judgment tests of gasoline-engined vehicles than did loudness level (Phons, Stevens) or perceived noise level (PNdB, calculated according to reference 466).

It should be noted, however, that in all the above-mentioned experiments with the noise from motor vehicles, the spectra of the noises from various vehicles are somewhat similar (the sound of energy is predominately below about 500 Hz), and for this reason a reasonably high correlation could be expected with any frequency weighting function or band weighting procedure that weights the lower frequencies relatively less than the higher frequencies.

Relation to Judgments of Aircraft Noise

It is interesting to note in Fig. 206 that, at certain peak levels, aircraft noise was rated as more acceptable than automobile noise; Robinson et al. (693, 694) suggest that the subjects expect aircraft noise to be more intense than motor vehicle noise and therefore find it more tolerable at higher levels. This expectation or set for different noises on the part of the subject is undoubtedly a factor behind these results; another at least partial possible explanation is that the dB(A) values do not properly reflect the subjective noisiness of the several classes of motor vehicles and aircraft noises involved in these studies. Robinson et al. point out that the aircraft and motor vehicle noises varied greatly and irregularly with respect to number of occurrences, duration, rate of onset, and line spectra. For these reasons it is impossible to do more than speculate about the general validity of Max dB(A), PNdB, or Phons for predicting the subjective noisiness of the aircraft and motor vehicle noise in question. It is possible that the use of Effective PNL would have provided a general rating procedure such that the difference noted in Fig. 206 between aircraft and automobile noise would disappear or be reduced.

Noise from Subsonic Aircraft

The studies of individual and community response to aircraft noise represent a broad approach to the noise problem, going from the rather precise laboratory situation, to the ostensibly more valid conditions of the field tests, and finally to community behavior. A rather consistent and relatable pattern of findings emerges from the laboratory, field, and community studies of human response to aircraft noise. The field and laboratory experiments will be described next, and the community studies (primarily attitude surveys) wil be presented afterwards.

Laboratory and Field Studies

In most of the studies of actual, recorded, or simulated aircraft noise, paired-comparison tests were performed in which the subjects rated, with respect to subjective noisiness or unacceptability, the noise from one aircraft relative to a reference noise or the noise from another aircraft. Tables 37 and 38 summarized most of the readily available data of paired-comparison tests of the subjective relative noisiness or unacceptability of the external sound from aircraft.

Most of the studies summarized in Table 37 were conducted in the laboratory with recorded noises being presented to the subjects via loudspeakers. It is perhaps in order to briefly describe the studies conducted in the field with the noise coming "live" from the aircraft. In particular, the field situation is thought to be sufficiently similar in certain respects to real-life so that more meaningful, absolute ratings of the degree of unacceptability of the noises (as distinct from the relative ratings found in the paired-comparison tests) are obtainable.

Edwards Air Force Base and Wallops Island Studies

One hundred adult subjects were located inside and outside typical residential houses located at Edwards AF Base (473). In addition to making judgments of

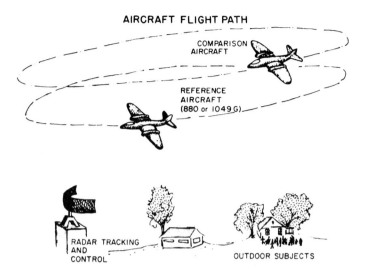

AIRCRAFT FLIGHT PATH

COMPARISON
AIRCRAFT

REFERENCE
AIRCRAFT
(880 or 1049 G)

RADAR TRACKING
AND
CONTROL

OUTDOOR SUBJECTS

FIGURE 207.   Schematic diagram of aircraft flight paths and location of houses and subjects used for aircraft noise judgment tests at Wallops Island, Virginia.
From Kryter *et al.* (474).

FIGURE 208.   Photograph showing outdoor subjects for aircraft noise judgment tests at
Wallops Island, Virginia. From Kryter *et al.* (474).

sonic booms (see later sections of this chapter), the subjects judged the acceptability of the noise from a fanjet subsonic aircraft vs. the noise from a turbojet subsonic aircraft that was flown over the test houses. Tests similar to those conducted at Edwards Air Force Base were performed at Wallops Island, Virginia (474), but a much larger number of subsonic aircraft were involved. Figure 207 illustrates the general setup and operation, and Fig. 208 is a photograph of some subjects. Figure 209 gives representative one-third octave band spectra of the aircraft flyover noises. The noise spectra of many of the present-day commercial jet aircraft will be found on Fig. 209.

In addition to the paired-comparison judgments, the subjects rated on a scale, ranging from very acceptable to very unacceptable, each aircraft flyover noise. Of particular interest is the fact that the indoor subjects rated a noise about the same as did the outdoor subjects (see Fig. 170) even though the noise was reduced by an average of about 20 dB as the result of attenuation of the sound by the houses (see Fig. 210). However, it is apparent that the ratings by the subjects indoors, relative to the ratings by the subjects outdoors, were lower (less acceptable) for the lower frequency noise from primarily the propeller-driven aircraft than for the higher frequency noise from primarily jet aircraft (see right-hand bar graphs on Fig. 170). This is to be expected because of the lesser attenuation by the house of the lower than the higher frequency components in the noises.

Farnborough Study

Robinson et al. (693) had over 60 subjects, placed near Farnborough Airport during the period of an air show, rate the "intrusiveness" and the "noisiness" of aircraft flying overhead. Sometimes the subjects were indoors and sometimes outdoors. The noise levels were expressed in both Max PNdB and dB(A). Some of their results are shown in Fig. 211. Of particular interest, in Fig. 211 and as discussed earlier (see Fig. 170), is the fact that the subjects were less tolerant, by about 18-20 PNdB, of the aircraft noises heard indoors than they were of the noises heard outdoors.

Los Angeles Study

In a study performed at Los Angeles and reported by Bishop (67), 55 adults selected from neighborhoods at various distances from a major commercial airport were placed in groups of 2 and 3 in rented apartment rooms under the flight path of aircraft taking off and landing. They were asked to rate the sounds made by aircraft flying overhead on the basis of the categories given on the

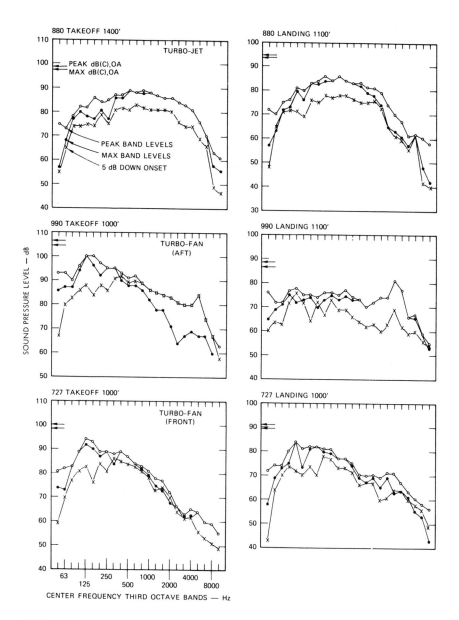

FIGURE 209. Representative third octave band spectra for aircraft used in experiment at Wallops Island, Virginia. Make of aircraft, altitude, and operation (with normal power setting) is given above each graph; engine type and overall sound pressure level in dB(C) is given in the graphs. From Kryter *et al.* (474).

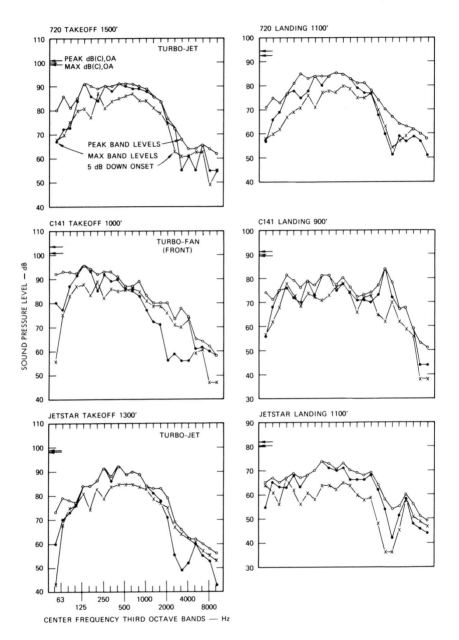

FIGURE 209. Continued.

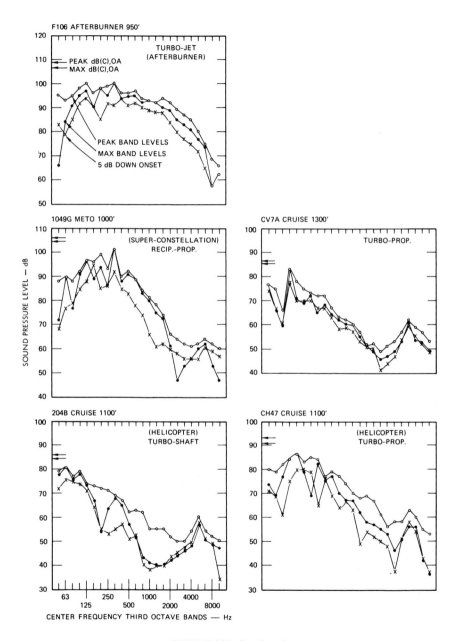

FIGURE 209. Continued.

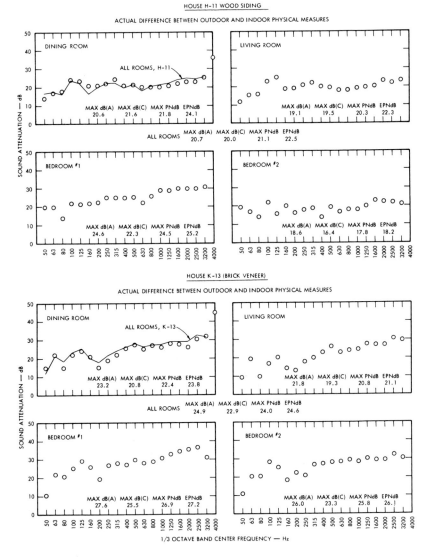

FIGURE 210. Sound attenuation (level outdoors minus level indoors) characteristics of four rooms in a wood-sided house (upper graphs) and brick-veneer wood frame house (lower graphs). Windows and doors closed. Based on an average of four aircraft flyover noises. From Young (893).

vertical ordinate in Fig. 212. The average results given in Fig. 212 are also to be found in Fig. 170. The present study, as is consistent with studies involving motor vehicle noises, showed that, for practical purposes, ratings on an absolute scale of annoyance are similar for men and women.

One feature of this study was a series of tests in which some of the subjects were presented, via a loudspeaker placed in the room, a recorded aircraft flyover noise and asked to assign the number 100 to it. The next live aircraft noise they heard was to be assigned the number 50 if they thought it one-half as noisy, 200 if twice as noisy, or whatever number expressed the proper fractional relation between their subjective ratings of the recorded and actual flyover noises. They found that a change of 16 PNdB was required to achieve a doubling of noisiness as rated by this numbering scheme. This was somewhat unexpected since the perceived noisiness scale holds that a 10 PNdB change would cause a doubling of the noisiness. Broadbent and Robinson (99) found that 13 PNdB resulted in a doubling of the perceived noisiness of the noise from subsonic aircraft, but Parnell *et al.* and Ollerhead (529a) found that an average of 10 PNdB was required to achieve a doubling or halving of perceived noisiness when the subjects rated sounds on a fractional scale. Parnell *et al.* found, on the other hand, that if they allowed the subjects to assign a number indicating magnitude to a sound, a change in the intensity of the sound of about 30 dB was required to obtain a doubling of the size of the number assigned to the sound. These results are in some disagreement with other results of either fractionation or magnitude estimation of loudness or noisiness.

It appears, overall, that the present scale (a 10 PNdB increase in level is equal to a doubling of subjective noisiness) is perhaps correct for nonimpulsive sounds, but that, as we shall see later in this chapter, the noisiness of sonic booms appears to grow at a somewhat faster rate as its intensity is increased than does the noisiness of the more common nonimpulsive noise from subsonic aircraft. Green *et al.* (323) point out that the scaling ratio per se has essentially no effect on the relative rank order ratings that are to be earned by different sounds.

V/STOL and STOL Aircraft

Vertical and steep takeoff and landing aircraft (V/STOL and STOL) may become an important means of transportation over densely populated areas. Because of their operational characteristics they tend to reduce some noise problems (they can reach relatively high altitudes while still within the boundaries of an airport before passing over populated areas) and to create new problems (they may fly close to occupied buildings, and when landing or taking off they tend to create noise for longer periods of time than do other types of aircraft).

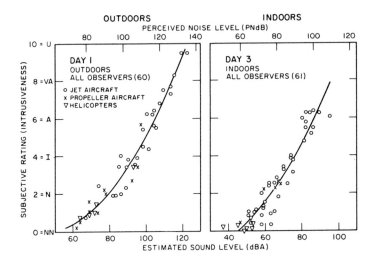

FIGURE 211.   Outdoor (left graph) and indoor (right graph) judgments of the category
scale of intrusiveness plotted against sound level, dB(A), and perceived
noise level, PNdB. PNdB calculated or estimated in accordance with ref.
466. From Robinson *et al.* (693).

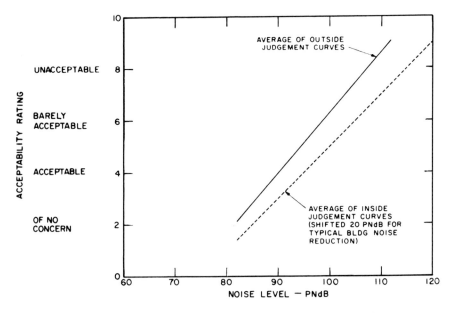

FIGURE 212.   Comparison of average outdoor and indoor acceptability-judgment curves.
Combined takeoff- and approach-noise judgments. PNdB calculated in
accordance with ref. 466. From Bishop (67).

The noise from helicopters, a common type of V/STOL aircraft, were judged by the subjects in the Farnborough experiments (693) and the results, in terms of PNdB (calculated according to reference 466) and dB(A) do not appear to differ appreciably from the results for other types of aircraft. Cross *et al.* (166) attempted to synthesize external helicopter noise by modulating a recording of jet aircraft noise following takeoff at rates of 4, 8, and 12 times per second, with an 8 dB difference between the peaks and troughs. They found no difference in the judgments of the noisiness for the three modulation rates used, and concluded that the peak PNdB (calculated according to reference 466), rather than some average level, should be used for estimating the noisiness of helicopters.

Pearsons (610) had subjects seated in an anechoic chamber judge the subjective noisiness of the recorded flyover sound from various helicopters relative to that of the sound, both simulated and from an actual recording, from a fixed wing commercial jet aircraft. Typical results of these tests are shown in Fig. 213. Interestingly, peak PNdB closely predicted the results of the judgment tests suggesting, among other possibilities, either that (*a*) the differences in duration among these particular noises were unimportant to their perceived noisiness, or (*b*) there were factors in other aspects within these sounds that somehow compensated for the durational differences.

Hinterkeuser and Sternfeld (373) prepared recordings of the noise to be expected from various new or experimental types of V/STOL aircraft. In their tests they found that durational corrections to peak PNdB level improved predictions of the results of the judgment tests, but that tone corrections did not as consistently improve the correlation between the physical and psychological measures. The results of the tests are shown in Fig. 214, and Fig. 215 illustrates variations in the spectra of some of these noises. The right-hand graph on Fig. 215 is perhaps particularly interesting in that it shows the complexity of the noise from various sources contributing to the overall noise of a particular V/STOL aircraft, a so-called tilt-wing, during terminal operations.

As mentioned earlier, some of these aircraft employed for the Wallops Station tests were standard fixed-wing commercial-type aircraft, and others were V/STOL aircraft. Although the judgments of some types of aircraft noises are apparently more predictable from the physical measures than are other types, the subjective judgments of noises from V/STOL aircraft appear to be as predictable by present measurement techniques as are the noises from other types of aircraft, provided that durational factors are taken into account. Table 46 shows, for one case at least, the decided improvement in the prediction by the physical measures of the subjective judgments when the noisiness during the duration of the sound was summed (EPNdB of -3.5) compared to maximum PNL (Max PNdB of -6.5), and the further improvement when the onset duration correction was used (EPNdB$_{OC}$ of +1.0).

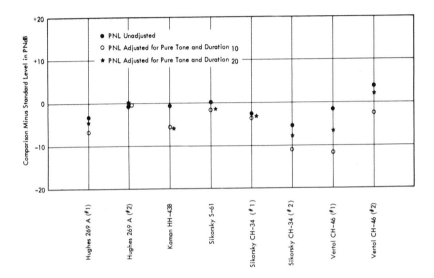

FIGURE 213.   Equal noisiness judgments of helicopter flyovers in terms of PNL using a
DC-8 jet flyover as a standard. PNLs calculated in accordance with procedures of Kryter and Pearsons (466). Durations 10 and 20 refer to the time
the flyover noise was, respectively, between 10 dB and 20 dB below Max
PNL. From Pearsons (610).

TABLE 46

Differences between Certain Physical Measures When Noise
from 204B Helicopter Was Judged as Acceptable as Noise from
Fixed-Winged Aircraft

$EPNdB_{oc}$ is corrected for onset duration. From Kryter *et
al.* (474).

Max PNdB	EPNdB	EPNdB-oc
−6.5	−3.5	+1.0

Masking of Speech by Aircraft Noise

It has been proposed from time to time that the annoyance of aircraft sounds
should be evaluated in terms of their speech-masking effectiveness rather than in
terms of overall loudness or perceived noisiness level.

This notion has not been adopted or implemented for several reasons:

1. Some complaints about the general noisiness and bothersomeness of
aircraft sounds do not appear to be concerned with the masking of speech.

2. High frequency, narrow band, or impulsive noises, or noises with strong
pure tones are not necessarily effective maskers of speech but are generally
perceived as being very noisy or annoying.

3. A simple and proven method of quantitatively measuring, or indirectly
inferring, on the basis of some calculation procedure, the masking characteristics
of the time-varying sound from different aircraft has not been proposed.

A study was performed by Kryter and Williams (471) to show how the sound
reaching a point on the ground from aircraft during engine run-up, shortly after
takeoff, and during approach to landing operations reduces the percentage of
test words correctly heard by a crew of trained listeners. For these tests,
magnetic tape recordings were used of the noise from (*a*) takeoff and landing
operations that started and terminated at 15 dB below peak level, and (*b*) the
run-up operation that started and terminated at about 3 dB below peak level.
These recordings were made into loops that could be played continuously with
no pause between repeated cycles of the noises.

Recorded Modified Rhyme Tests were mixed electronically with the recorded
aircraft noise and presented to the listeners via earphones. Since aircraft noises

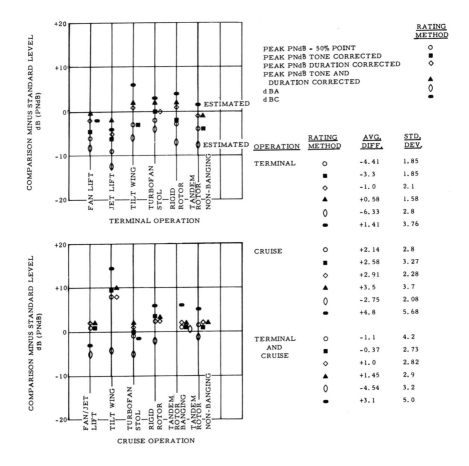

FIGURE 214.   Comparison of methods for estimating the judged noisiness of the sound
from V/STOL aircraft. Peak PNdB tone and duration corrected in accordance with Kryter and Pearsons (466). From Hinterkeuser and Sternfeld
(373).

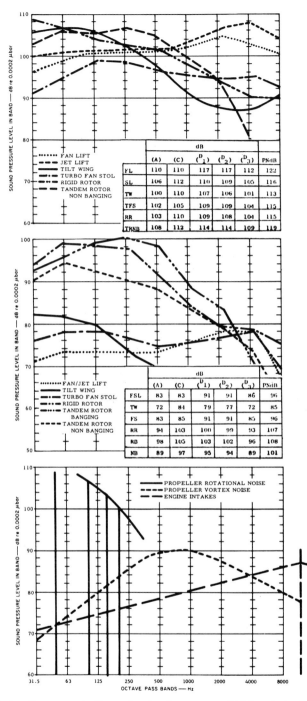

FIGURE 215.   Lower graph: Octave-band noise spectra at 500 feet during V/STOL mode. Middle graph: Octave-band noise spectra at 2000 feet during cruise mode. Upper graph: Tilt wing terminal spectrum. From Hinterkeuser and Sternfeld (373).

TABLE 47

Number of Words That Would Be Masked with Normal Rate Speech (Estimated to Be 140 Words Per Minute at a Level of About 5-20 dB Above Conservation Level) Per Flight Operation for Each of Four Aircraft

From Kryter and Williams (471)

	Duration between 15 dB downpoints (secs)	Percent words masked during test		Number of words that would be masked with normal rate of talking	
		Indoor 69 dB speech	Outdoor 84 dB speech	Indoor 69 dB speech	Outdoor 84 dB speech
TAKEOFF					
707-120	35	46%	48%	37.5	39.1
720B	23	39	42	20.9	22.5
727	25	19	34	11.1	19.8
SC	24	9	13	5.0	7.3
LANDING					
707-120 (1500 ft)	34.5	25%	40%	20.1	32.2
720B	23	29	42	15.5	22.5
727	32	31	41	23.1	30.6
SC	12	13	34	3.6	9.5
707-120 (610 ft)	9.5	55	61	12.2	13.5

are heard indoors as well as outdoors, half of the intelligibility tests were administered with the noise filtered to achieve indoor spectra and levels. To do this, the noise signal from the test loops was passed through filters that were designed to provide the attenuation that would be imposed upon a sound passing from out-of-doors into a typical one-family frame house.

During the outdoor test condition, the listeners heard each aircraft noise at a peak sound pressure level of 100 dB (as would be measured on a sound level meter set on C scale, fast meter action). During the indoor test condition, the "indoor" filters reduced the noise level to an average peak sound pressure level of 85 dB for all except the ground run-ups which were reduced to 87 dB. Some of the results are presented in Table 47.

Williams *et al.* (878) later conducted a somewhat similar study with comparable results with respect to the masking of speech by aircraft noise. Details of the relations found in these two studies between the masking of speech by aircraft noise and various physical measures of the noise, such as the AI, PNdB, dB(A), SIL, etc., were presented and discussed in Chapter 2. In addition to measuring the masking of speech, Williams *et al.* also had the subjects rate the "acceptability" of the aircraft noise. The results, Figs. 216 and 217, are in close agreement with previous findings that levels of about 85 Peak PNdB are barely acceptable; at this level, speech at a level of 77 dB gave a correct MRT score of about 80%.

Community Reactions to Noise from Subsonic Aircraft

Elwell conducted the first experiment, to the best of our knowledge, on neighborhood reaction to different types of aircraft noise. A small aircraft, sometimes without and sometimes with noise reducing engine exhaust, propeller, and gears, was flown over 10 communities near Boston at an altitude of 500 feet. There was a significant reduction in the number of complaints when the noise-quieted aircraft was flown, as compared to when the regular aircraft was flown. Sound levels were measured with an SLM set on C scale and no particular relation was noted between dB(C) levels and complaints.

A detailed study of aircraft noise was made in 1957-68 for the Port of New York Authority by Bolt Beranek and Newman Inc. (556). The purpose of the study was threefold: (*a*) to determine the noise exposure from aircraft being then experienced in communities surrounding New York airports in terms of octave band spectra of the noise on the ground at various distances from the airport, and the average number and times of occurrences per day of the aircraft operations; (*b*) to apply, and develop as necessary, methods for relating these physical data to the reactions of people to the noises from propeller-driven, reciprocating-engined, aircraft; and (*c*) to apply the same methods of evaluation used in depicting the existing aircraft noise environment to the jet aircraft noise as might be present in future operations.

With such information available, it was possible to estimate what the impact would be, relative to the present-day noise exposure conditions, for specified operating procedures of the jet aircraft. One could also specify what the perceived noise levels and numbers of operations of jet aircraft must be if the noise environment on the ground was not to exceed that presently being experienced. In brief, this method allowed a comparison to be made between existing and anticipated future noise conditions. The procedure was not a direct measure of tolerability of the aircraft noise.

Nature of Complaints Against Aircraft Noise

Examination of data collected at a center established in New York City (60) for receiving complaints about aircraft noise, and the testimony given by witnesses at a trial brought by citizens of Newark, New Jersey, in 1958, against the airlines for creating a noise nuisance, revealed that the greatest single complaint was concerned with the interference of the aircraft noise with talking and listening; the second complaint (in number, but not in intensity of feeling) was concerned with the disturbance of sleep and rest, and the third, was with the fear of crashes. This pattern is in general agreement with the results of other similar studies.

Beranek *et al.* (60) have been able to demonstrate in a quantitative way how certain aircraft and environmental factors influence complaint activity about aircraft operations. Figure 218 shows, for example, that complaint activity varies with the season of the year; it is highest in the warm months when windows and doors are open. Also, it was found that only after several successive days of similar exposure to flyover noise (which varied for a given neighborhood because of particular flight patterns as dictated by wind direction) did complaints about the noise become numerous.

Figure 219 shows, among other things, that the ratio between complaints and aircraft activity is the greatest from 11:00 P.M. to 12:00 midnight. Presumably the noise interferes with falling asleep, but drops perceptibly after midnight and reaches a minimum after 1:00 A.M. to 2:00 A.M. It is possible that there is a physiological explanation of this reduced complaint behavior. Kryter and Williams (unpublished data, 1960) measured, in a human subject, changes in sleep activity following exposure to aircraft and other sounds at various intensity levels. This experiment showed that during deep stages of sleep, the intensity of an aircraft flyover noise (obtained from a tape recording) had to be increased 85 dB or so above the level required during drowsing to cause a response. This elevation in response threshold appeared to be independent

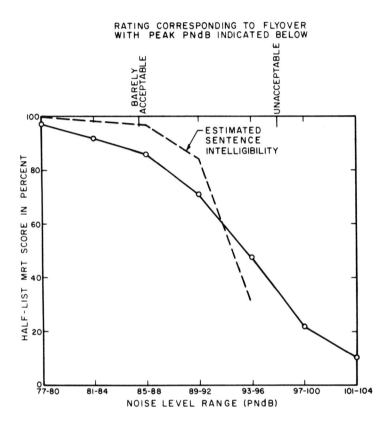

FIGURE 216.   Noise level range and corresponding intelligibility score when listeners judged aircraft noises to be barely acceptable and unacceptable. Speech presented at level of 77 dB, 12 dB above normal conversational level 1 from the talker. From Williams *et al.* (878).

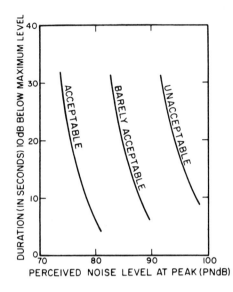

FIGURE 217.   Contours of equal noisiness based on judgments of acceptability of aircraft noises having different durations. From Williams *et al.* (878).

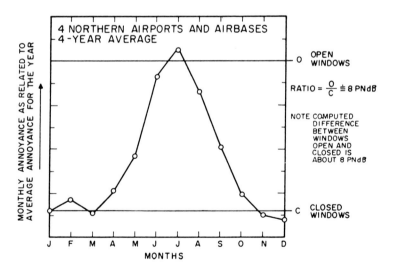

FIGURE 218.   Total monthly annoyance varies with the season of the year. This curve is a four-year coverage for all communities near four airports in the north-eastern quarter of the United States. From Beranek *et al.* (60).

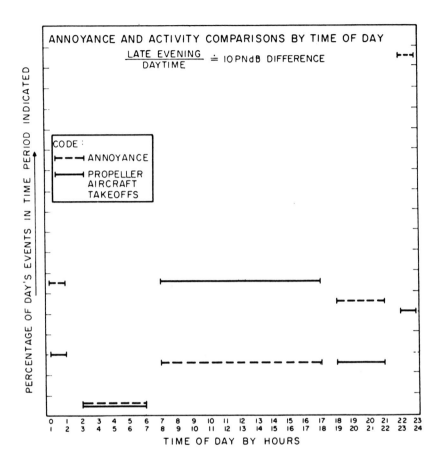

FIGURE 219. Annoyance and takeoff activity around one airport are plotted here by the hour for a 24-hour period. Data for nine months were averaged. Activity is expressed as the percentage of the day's events occurring in the hour indicated. Note that the ratio of annoyance to activity increases in the evening and particularly in the period when people retire for the night. From Beranek *et al.* (60).

of the type of sound, be it a pure tone, band of random noise, or an aircraft noise (see also Lukas and Kryter [528] and Chapter 12).

The Relations between Noise Environment and Opinions Obtained through Interview Surveys

The more or less empirically derived relations shown between complaint activity and noise exposure have proven to be useful in predicting, in broad terms, human behavior. These relations, nonetheless, are not very precise and do not lead to an understanding of the exact nature or basis for the human behavior recorded in complaints. It is obvious, for example, that telephone and letter complaints of citizens are not necessarily representative of the feelings and thoughts of all persons in a neighborhood. Deliberate surveys of peoples' attitudes should be helpful in this regard.

European Surveys of Attitudes Toward Civilian Aircraft Noise

A social attitude survey (536) conducted in the vicinity of Heathrow Airport near London reveals in a quantitative way how people feel about and react to aircraft noise heard in and near their homes. The British study was conducted in September 1961 and included interviews with 2000 people taken from residential districts within 10 miles of the London Heathrow Airport. In addition to the social survey data, physical measurements of aircraft noises were made at 85 locations in the area covered. The physical measures were converted into Max PNL values. Tables 48, 49, and 50 and Fig. 220 represent some of the results of the British social survey study.

It is possible to conclude from the data obtained that:

1. As previously mentioned, aircraft noise was, on the average, not significantly annoying when the level was below 80 PNdB.

2. The greater the number of flight operations, the greater was the annoyance — the relation between Max PNL and number of aircraft and annoyance score was best predicted by a Noise and Number Index (NNI) which consists of the sum of the average Max PNL plus $15 \log_{10}$ of the number of aircraft operations. The relation means that doubling the number of aircraft operations is equivalent in terms of annoyance scores to increasing the noise level by 4.5 PNdB. (It is interesting, but no doubt fortuitous, that this is the same relation Kryter and Pearsons [466] found between exposure level and durations typical for aircraft flyover, 15 seconds or so; see Fig. 171.) However, this is different than the intuitively derived equal energy exposure method proposed by Pietrasante and Stevens (625, 769) and the principle also

TABLE 48

The Number of People with Various Annoyance Ratings Classified by Noise Level and Number of Aircraft Per Day From McKennell (536).
(With permission of the Controller of Her Britanic Majesty's Stationery Office.)

(i) Noise level in PNdB	(ii) Average number of aircraft per day	(iii) Annoyance Score						(iv) Average annoyance score	(v) Number people in stratum
		0	1	2	3	4	5		
84–90	5.75	230	128	113	5	5	31	1.1	512
	22.5	45	33	26	17	12	22	1.9	155
	81	5	7	2	7	10	7	2.8	38
91–96	5.75	51	41	28	17	11	10	1.5	158
	22.5	90	64	55	45	35	32	1.9	321
	81	18	15	13	23	18	23	2.7	110
97–102	5.75	2	1	--	3	1	--	2	7
	22.5	13	9	20	16	11	13	2.5	82
	81	20	22	38	26	30	64	3.1	200
103–108	5.75	--	--	--	--	--	--	--	--
	22.5	1	--	1	5	2	2	3.2	11
	81	11	7	17	16	19	67	3.6	137

TABLE 49

The Distribution of Complainants Classified by Noise Level
and Number of Aircraft Per Day. From McKennell (536).
(With permission of the Controller of Her Britanic Majesty's
Stationery Office.)

Average number per day (i)	Noise level in PNdB				
	84-90 (ii)	91-96 (iii)	97-102 (iv)	103-108 (v)	Totals (vi)
5.75	20	7	1	0	28
22.5	9	37	9	0	55
81	7	35	37	16	95
Totals ...	36	79	47	16	178

recommended in this book for use with EPNL and the Composite Noise Rating Scheme. In this latter method, in essence, $10 \log_{10} N$ is used rather than $15 \log_{10} N$.

Figure 220(a) shows the percentages of people who expressed various attitudes and reactions to aircraft noise of different CNR values in the Netherlands and in France. Van Os (815) reported that in the Netherlands there appeared to be good agreement betweeen the observed attitudes and behavior of people and those which one would predict on the basis of their noise environment as measured by NNI or CNR.

Figure 221 shows how the results of the survey compare with the results obtained in research studies conducted at Farnborough and in Los Angeles. Of significance is the fact that people were about as tolerant of aircraft noise in their own environments as subjects judged themselves to be when placed in the relatively artificial experimental situations at Farnborough and at Los Angeles. This would suggest that valid judgments of the unacceptability or perceived noisiness of noises for real-life conditions can on occasions be obtained under laboratory and semilaboratory conditions.

United States and Swedish Surveys of Attitudes Toward Military Aircraft Noise

Sociological interview techniques have been applied by Borsky (80) in a study of the noise problems in 22 communities near military air fields in the United

States. In addition to Borsky's report, the reader is referred to a report by Clark and Pietrasanta (137). The findings show the importance of socio-psychological variables upon complaint action. For example, three factors which had a statistically significant bearing on the responses of citizens around military bases were (*a*) fear of aircraft crashes, (*b*) feelings regarding the considerateness of air base officials and pilots for the comfort and safety of the citizens, and (*c*) feelings of the importance of the air base. These sociological studies showed that individuals who deemed the air base to be important and considerate and also expressed little fear about aircraft crashes would show the same complaint potential for about four times the noise exposure per day as those individuals who were fearful and had negative feelings about the air base and its importance.

As the result of his studies, Borsky proposed a method for calculating the complaint potential of a community from prior knowledge of attitudes relating to fear, aircraft considerateness, and importance plus knowledge of the noise exposure. This interview study also provided some interesting information on the question of people getting used to the noise as the result of continued exposure. In this regard it was found that, following an initial period of adaptation, the longer a person lived in a given neighborhood, the more he was bothered by noise.

Jonsson and Sorenson (426) and Cederlof *et al.* (131) have recently conducted laboratory and field studies showing comparable effects of attitudes on reactions of "inconvenience" to both road and air traffic noise. These latter investigators found that 54% of a group of citizens who were sent positively-worded information regarding the Royal Swedish Air Force and aircraft indicated an inconvenience due to aircraft noise, whereas 79% of a control group of citizens from the same neighborhood, not given this information, indicated inconvenience due to aircraft noise.

United States Survey of Attitudes Toward Civilian Aircraft Noise

In 1968, Hazard (358) gave a preliminary report of a rather extensive sociometric study conducted in residential areas at various distances from the principal airports of five major cities in the United States. Further analyses of the data (805a) revealed the existence of a number of factors that contributed to the prediction of annoyance ratings given to aircraft noise by the several thousands of people interviewed (see Tables 51 and 52).

In addition to the interviews, some physical measures were made of the noise present at certain locations covered in this study. An estimate based on a sample of physical measures of the average noise present at the four corners of about a city block area was correlated against the noise-attitude ratings of individuals within such an area. In view of the only approximate data about the noise from external sources in individual homes over the months of exposure required to shape complaint behavior and attitudes, it is surprising that the averages of the

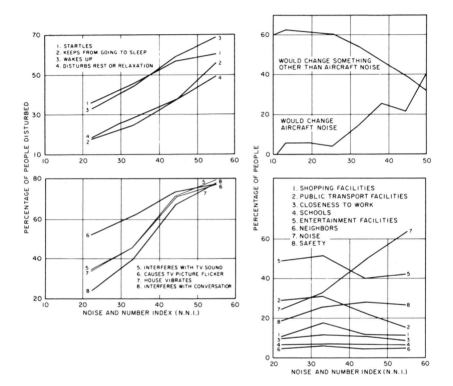

FIGURE 220.   Percentage of people disturbed by aircraft noise (left graphs) and wishing
to change their living conditions for various reasons (right graphs) as functions of NNI. From McKennell (536). (With permission of the Controller
of Her Britanic Majesty's Stationery Office.)

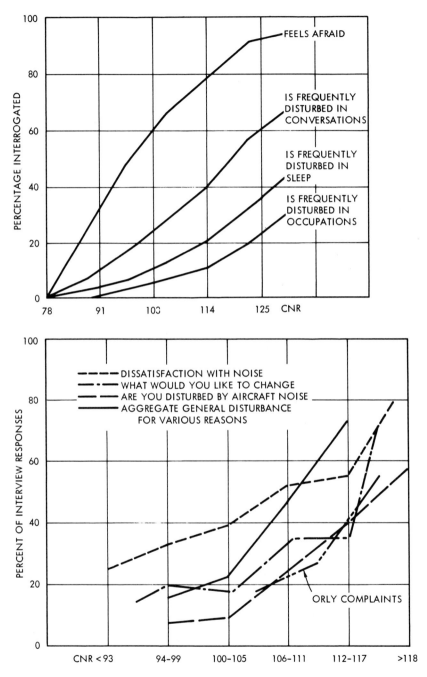

FIGURE 220a. Upper graph: Comparison of Netherlands survey results with equivalent CNR values. Lower graph: Comparison of French survey results with equivalent (CNR values. After Galloway and Bishop (272a).

TABLE 50

Data Showing the Distribution Over Noise Levels of the Total Population, and the Seriously Annoyed in the Population. From McKennell (536). (With permission of the Controller of Her Britanic Majesty's Stationery Office.)

PNdB Stratum (i)	PERCENTAGES			ABSOLUTE NUMBERS	
	% of total population in stratum (ii)	% of stratum seriously annoyed (iii)	% of total population seriously annoyed (iv)	Number of people seriously annoyed (v)	Number of people in stratum (vi)
103+	3	68	2	28,000	42,000
100-102	6	51	3	42,000	84,000
97-99	7	48	3	42,000	98,000
94-96	13	36	5	70,000	182,000
91-93	27	24	6	84,000	378,000
88-90	22	23	5	70,000	308,000
85-87	11	16	2	28,000	154,000
Up to 85	11	10	1	14,000	154,000
Totals	100	--	27	378,000	1,400,000

PNdB

103+   3%
100-102   6%
97-99   7%
94-96   13%
91-93   27%
88-90   22%
85-87   11%
Up to 85   11%

NOTES:

1. In the figure, the width of the column represents the total number in that PNdB stratum.

2. The shaded section represents the number seriously annoyed in that stratum.

3. Total in all strata is 1,400,000.

4. Total in shaded section is 27% of this, or 378,000.

NOTES:

1. For these data "seriously annoyed" refers to those having a score on the annoyance scale of 3.5 or above.

2. The population is that within a ten-mile radius of London Airport; 1,400,000 adults.

3. Entries in column (iv) are derived from those in columns (ii) and (iii); e.g. 68% of 3% gives 2% (rounded).

4. The numbers in column (vi) correspond to the percentages in column (ii), and those in column (v) to column (iv).

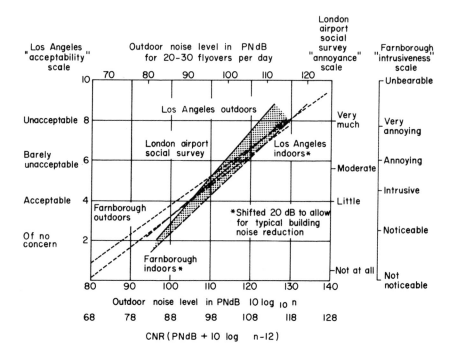

FIGURE 221.   Comparison of aircraft-noise rating scales (assuming $10 \log_{10} n$ weighting for number, n, of flyovers). After Bishop (67).

TABLE 51

Predictions of Annoyance to Aircraft Noise. From Hazard (358).

Primary predictors of noise annoyance scores for all localities as a whole, in order of importance.

(1)   Aware of aircraft between midnight and 6 A.M.
(2)   Live in high aircraft exposure areas
(3)   Have high noise susceptibility
(4)   Perceive a steady increase in the amount of air traffic
(5)   Argue that they would be unable to adapt to increased exposure
(6)   Have knowledge of how to complain effectively.

Secondary predictors of noise annoyance scores in some localities, in order of importance.

(1)   Living from 3 to 6 miles from the airport
(2)   High occupational status, high income, and expensive residence
(3)   Having fear of aircraft crashing in the neighborhood
(4)   Long-time residency in the neighborhood
(5)   Knowledge of neighbors who have moved away due to aircraft noise
(6)   Generally positive attitudes toward the aircraft industry
(7)   Belief that the airport is important to the economy of the city.

TABLE 52

Predictor Variables of Annoyance to Aircraft Noise. From TRACOR (805a).

In order of importance.

(1)   Fear of aircraft crashing in the neighborhood
(2)   Distance from the airport
(3)   Susceptibility to noise
(4)   Noise adaptability
(5)   Aircraft noise exposure (CNR)
(6)   City of residence
(7)   Belief in misfeasance by aircraft or airport operators
(8)   Extent to which the airport is considered to be important to the local economy.

outdoor measurements of the aircraft noise correlated (coefficients of 0.46) as well as they do with attitudes of individuals. The results of this study will be mentioned later under a discussion of the Composite Noise Rating procedure.

In at least two major cities, some attempt has been made to set limits on the maximum or peak level of noise that can be made in communities near airports as the result of aircraft flying overhead. The British Ministry of Aviation has set 110 PNdB during daylight and 100 PNdB at night as the maximum limit to be generated in neighborhoods near London Airport. And the Port of New York Authority has set 112 PNdB as the maximum level to be reached around New York City airports at any time of the day or night.

These specifications were made with not only the local social and economic conditions in mind but also they were based on the knowledge that the great majority of daily flight operations at each airport would create in neighboring communities noise levels considerably below the maximum levels specified. In reality, the limits were a reflection of typical levels in existence at the time the specifications were set and cannot be interpreted as necessarily tolerable levels. The question of noise codes for environmental noise will be discussed following the next section on research on sonic booms.

Booms from Supersonic Aircraft

The proposed advent of the Supersonic Transport (SST) for commercial passenger traffic would introduce, on a broad scale, a relatively new type of noise into man's environment − the sonic boom. The sonic boom problem is interesting from acoustical, psychological, sociological, and political points of view. Because the problem has so many facets and is potentially so important, the nature of the boom as an auditory stimulus and the results of laboratory, field, and real-life experiments on the effects of sonic booms will be discussed in some detail. Also, the sonic boom question deserves special attention because it represents a somewhat unique opportunity for the application of scientific knowledge to decision-making about the development and use of a potentially major vehicle of transportation that is also potentially a major source of noise pollution. Psychological research studies on the sonic boom will be presented next, and the implications of this research for the operation of the SST over populated areas is given in a later section on the Composite Noise Rating.

Energy Spectrum of the Sonic Boom

The pressure waveforms and energy spectra of typical sonic booms, called N-waves, are illustrated in Fig. 222. The sonic boom can be treated, for practical purposes, as an impulse whose spectrum can be approximately known from a knowledge of its peak overpressure, its rise time and duration. Methods for converting impulses to spectra of various bandwidth (energy per Hz, per octave and one-third octave) are given in Chapter 1.

One reason for expressing the sonic boom in terms of its energy spectrum lies in the fact that it reveals how it will be perceived by the auditory system in dimensions that are common to all sounds, be they impulsive or nonimpulsive. It is of interest to note, as shown by the work of Zepler and Harel (899) and Shepherd and Sutherland (735), that the perceived loudness or noisiness of one sonic boom vs. another boom is predictable from loudness or perceived noisiness

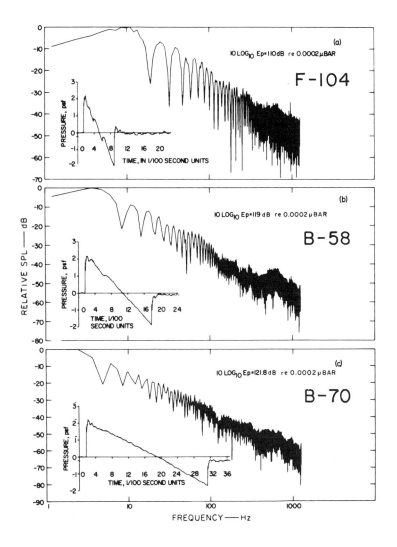

FIGURE 222. Inserts: Pressure (psf) vs. time. Full graphs: Pressure (dB) vs. frequency plots for the booms from three supersonic aircraft. Energy spectrum pressure (Ep) of most intense frequency component is indicated for booms (a), (b), and (c). From Young *et al.* (894).

levels calculated from one-third or one octave band spectra of the booms. Zepler and Harel calculated a unit proportional to Phons (Stevens) for impulses used in their tests (see Fig. 156), and we have calculated PNdB values for sonic booms of Shepherd and Sutherland, using the one-third octave band spectra obtained from computer-calculated Fourier transforms. For these stimuli, changing the rise times caused changes in calculated PNdB and Phon commensurate with changes in judged loudness or noisiness, and the differences in duration caused insignificant changes in calculated PNdB, as well as in the judgments, as shown on Fig. 223. Because of the similarity of the frequency weighting applied to the low frequency spectra involved, Phons (Stevens) and PNdB for these impulses will be highly correlated. Examples of the energy spectra of these simulated sonic booms are given in Fig. 224.

Figure 225 illustrates how judged loudness varied as a function of acoustic impulses of different waveforms. These data are particularly interesting in that they show that, for these more complex waveforms, PNdB calculated from energy spectra predict reasonably well the subjective judgments. Sample energy spectra for these waveforms are shown in Figs. 226 and 227.

A word of caution is in order regarding the validity of acoustic signals from Zepler and Harel earphones, and the loudspeaker-booths used by Pearsons and Kryter (613) and Shepherd and Sutherland for simulating sonic booms. Figure 228 shows the test chamber, which is like that used earlier by Pearsons and Kryter, employed by Shepherd and Sutherland for their simulated sonic boom tests. The pressure waveforms from these devices as present at the ear of the listeners has more or less the forms desired, that of an N-wave or sonic boom; however, the temporal variation in the air particle velocity for these pressure waves is considerably different than it is for a similar pressure waveform when generated out-of-doors, i.e., not in an enclosed, essentially airtight cavity. Zwislocki (914) suggests that the loudness of low frequency sounds is significantly affected by phase relations between particle velocity and pressure waveform. Further research on this and related questions is necessary before the results of the laboratory tests done to date of the audibility of outdoor sonic booms can be interpreted with confidence.

Sonic Boom Experiments at Selected Cities in the United States

Although research on the audibility of sonic booms provides insights into the basis of how the auditory system responds to sonic booms, of greater immediate interest is research information concerned with how people will behave when exposed to sonic booms from supersonic aircraft. Three major research studies concerned with that and related questions were conducted by the U.S.

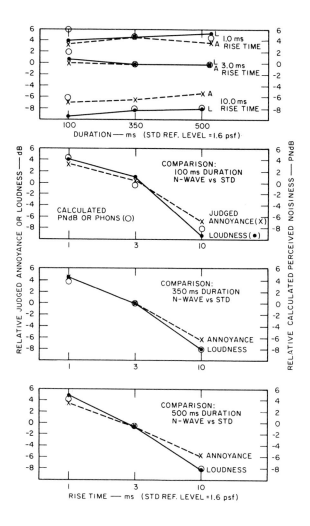

FIGURE 223.   Comparison of judged annoyance or loudness of simulated outdoor sonic booms. Judgment data from Sheperd and Sutherland (735). Calculated PNdB and Phons by Kryter and J. Young.

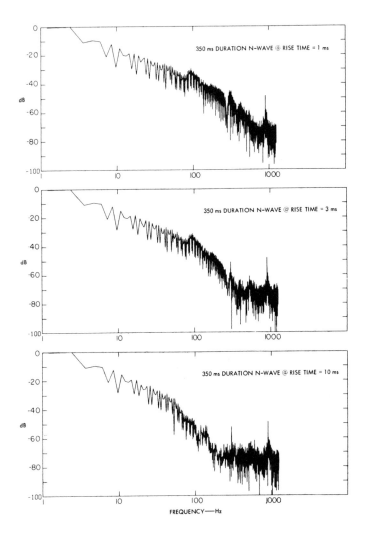

FIGURE 224.   Samples of spectral energy of idealized simulated sonic boom waveforms
used for judgment test results and calculations given in Fig. 223. From
Shepherd and Sutherland (735).

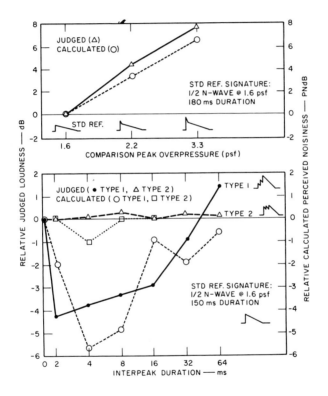

FIGURE 225. Comparison of judged loudness and calculated perceived noisiness and loudness of simulated outdoor sonic booms. Judgment data from Sheperd and Sutherland (735). Calculated PNdB and Phons by Kryter and J. Young.

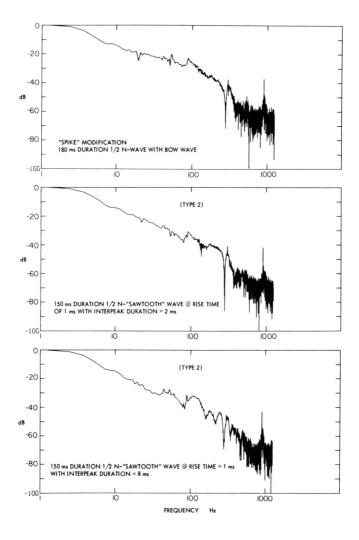

FIGURE 226.   Sample of spectral energy of various idealized acoustic impulses used for
judgment test and calculation results given in upper and lower (Type 2)
graphs of Fig. 225. From Shepherd and Sutherland (735).

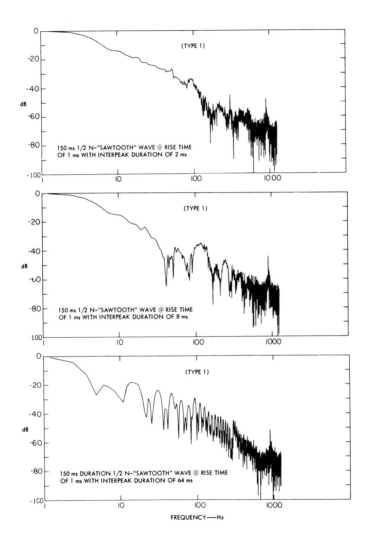

FIGURE 227.   Sample of spectral energy of some idealized acoustic impulses used for judgment test and calculation results given in lower graph (Type 1) of Fig. 225. From Shepherd and Sutherland (735).

FIGURE 228.   Sonic boom simulation chamber. Note hinged wall access. From Shepherd
                and Sutherland (735).

Government: one at Oklahoma City (81) in 1964, one at Edwards Air Force Base in California (473) in the summer and winter of 1966, and attitude surveys made in Atlanta, Dallas, Denver, and Los Angeles in the 1967-1968 period (805). Other less extensive experiments on the effects of sonic booms on people were conducted in Great Britain (853, 856), France (189), and the United States (582) during the past nine years.

Oklahoma City Study

For a six-month period in 1964, military aircraft, primarily F-104 fighter aircraft, were flown over Oklahoma City in order to create in the city about seven sonic booms per day. The average measured peak overpressure directly under the flight path of the aircraft was about 1.2 pounds per square foot (psf).

The residents of the city were advised of the nature and importance of the tests to the country and aviation, and extensive data were collected about the attitudes of the people to the sonic booms. Near the end of the six-month period, when the intensity of the booms was increased somewhat, complaints rose greatly and the City Council requested that the Federal Government cease the tests. The general findings concerning the effects of the booms on the people are summarized in Tables 53 and 54 from Borsky (81).

TABLE 53

Percent of Respondents Annoyed by
Various Sonic Boom Interferences. From Borsky (81).

Type of Interference	Oklahoma City
House shaking--rattles	54%
Startle	28
Sleep interruption	14
Rest interruption	14
Conversation interruption	10
Radio-TV interruption	6

Edwards Air Force Base Study

By means of paired-comparison tests, one should be able to determine the relative human response to sonic booms that differ with respect to their duration, rise time, or other signature variations. Paired-comparison tests between the noise from subsonic aircraft and sonic booms from supersonic

TABLE 54

Percent of Survey Respondents
Reporting Various Levels of Annoyance. From Borsky (81).

Level of Annoyance	Oklahoma City
Annoyed by booms	56%
Felt like complaining	22
Actually complained	2
"Cannot accept booms"	27
Filed damage claims	0.2

aircraft can also serve as a means of indirectly determining what people might do in the future about sonic booms from commercial supersonic aircraft. It is, of course, to be understood that the paired-comparison tests, particularly involving two sounds that differ so much in their spectral and temporal nature, require some validation before they can be accepted with confidence. Fortunately, validation data on this matter is already available to some extent for the sonic boom (81, 189, 582, 805), and particularly for the noise from commercial aircraft near busy metropolitan airports (274, 880).

With these concepts and background information in mind, the following series of experiments using military supersonic and subsonic jet aircraft were conducted at Edwards Air Force Base with subjects placed inside and outside of typical residential houses.

1. Paired-comparison tests and absolute ratings of the relative acceptability of (a) sonic booms with flyover noise from subsonic jet aircraft, (b) sonic booms from one type of supersonic aircraft with sonic booms from a second type; and (c) sonic booms from the same type of aircraft but flown under different operational conditions.

2. An attitude survey of the acceptability of the sonic booms to residents in a military community habitually exposed to sonic booms.

Boom vs. Subsonic Noise

Figure 229 shows a plot of typical results obtained from the judgment tests. The intensity level at which 50 percent of the subjects rated one of the sounds of

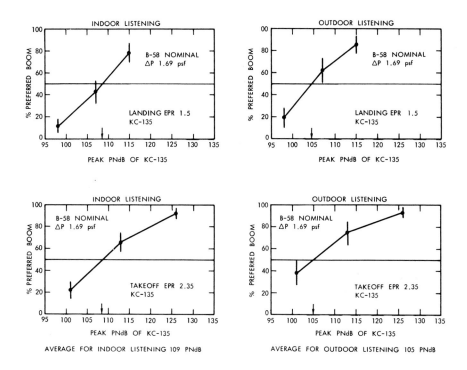

FIGURE 229.   Results of paired-comparison judgments of sonic boom vs. subsonic noise. The boom from the supersonic aircraft, B-48, was presented at nominal peak overpressure (P) of 1.69 psf. The noise of the subsonic jet aircraft, KC-135, was varied by flying that aircraft at a variety of altitudes, with the engine power (EPR) at either a landing or takeoff setting. The vertical bars mark the 90% confidence limits of plotted data points. Listeners were from Edwards Air Force Base. From Kryter *et al.* (473).

Fig. 229 (the noise from the KC-135 subsonic jet aircraft) equal in acceptability to the other sound of Fig. 229 (the sonic boom from the B-58 supersonic aircraft) at a nominal peak overpressure of 1.69 psf was taken as the point at which the sounds are equally acceptable to the subjects. Table 55 gives the intensity, in peak PNdB, required for the noise from the subsonic jet aircraft to be judged equal in acceptability to the sonic booms; the data in Table 55 are taken from graphs similar to Fig. 229.

Figure 229 and Table 55 indicate that for indoor listening the noise from a subsonic aircraft (KC-135) at a level of 109 PNdB was about equally preferred to a sonic boom of a nominal 1.69 psf from a B-58. The theory used (120, 393, 394) for the calculation of the nominal peak overpressures takes into account, relative to the generation and propagation of sonic booms, the volume and lift components of the aircraft, temperature, pressure, and density changes in the atmosphere which have some influence on boom propagation along the boom path, and effects of near-field (to the aircraft) signature characteristics. The nominal peak overpressures for supersonic aircraft, according to latest theory, agree within 1 dB, on the average, with actual measured peak overpressures (see Col. 10, Table 55). It is perhaps of interest to also note that while the median peak overpressure of five microphones spaced within 100 feet of each other in a cruciform array was, on the average, less than 1 dB from the nominal theoretical value, the variations in peak overpressure among points within that space was, on the average, about 1.5 dB (see Col. 11, Table 55). This "fuzziness" in the peak overpressure is found within distances as close as a few feet and is apparently due to normal, low-altitude atmospheric turbulence.

For indoor listening when the nominal sonic boom overpressure was increased to 2.65 psf, the PNdB level of the noise from the KC-135 had to be approximately 117 PNdB to be judged as equally acceptable as the boom. This result is not to be expected since increasing the overpressure from 1.69 to 2.65 psf represents only a 4-dB increase (see Fig. 10) in physical intensity, whereas, as judged against the noise from the KC-135, there appeared to be an effective increase in subjective noisiness of about 8 PNdB. Likewise, for indoor listening, an overall increase of about 12 dB in the physical intensity of the boom from the F-104 (from 0.75 psf to 2.8 psf) required an increase of 21 PNdB in the aircraft noise to maintain equal acceptability of the two sounds.

These results would imply that the subjective unacceptability of a sonic boom increases at a greater rate than does the sound from a subsonic jet aircraft when the intensity of the two sounds is increased by an equal amount. Broadbent and Robinson (99), using a magnetic tape recording (played back via loudspeakers) made inside a structure overflown by a supersonic aircraft, found a somewhat similar but less dramatic difference between the growth (as a function of their intensities) of the unacceptability of sonic booms and aircraft noise.

Indoor vs. Outdoor Listening – Relative Judgments

Table 55 shows the boom heard outdoors is more acceptable relative to the noise of the subsonic jet aircraft (by an amount equivalent to about 5 PNdB) than when the two sounds are heard indoors. That the results between the relative judgments indoors and outdoors should be even this similar is perhaps fortuitous in that the nature of the two sounds outdoors is so different, and because the two sounds, due to attenuation by the house, further differ from their outdoor counterparts. Also, secondary sounds or "rattles" introduced by the nonlinear response of components of the house to the boom presumably contribute substantially to the subjective unacceptability of the boom heard indoors.

It might be noted that in a previous laboratory test by Pearsons and Kryter (613) of the relative acceptability of recorded subsonic aircraft noise and a simulated "indoor" boom, a boom which measured 1.69 psf outdoors was judged to be equal to the noise of a subsonic jet at 113 PNdB measured outdoors. Broadbent and Robinson, using a sonic boom and aircraft noise recorded indoors and played back over loudspeakers to listeners, found a 1.69 psf boom, measured outdoors, to be judged as equally acceptable as an aircraft noise of about 107 to 113 PNdB, measured outdoors. These results compare well with the 109-112 PNdB for subsonic aircraft noise and nominal 1.69 psf for booms found in the study with actual aircraft.

Indoor vs. Outdoor Listening – Rating Scale

The scores on the acceptability rating scales (see Table 56) demonstrate that the booms heard indoors were, on the average, slightly more acceptable than the same booms heard by the subjects outdoors – about 34% of the indoor subjects rated the booms as unacceptable and about 47% of the outdoor subjects rated the same booms as unacceptable. The noise of the subsonic jet was also rated more acceptable indoors than it was when heard outdoors, but by a larger amount – 41% vs. 23%. Since the house structure should attenuate the aircraft noise by an average, over all frequencies, of 20 dB, and the sonic boom by about 10 dB (the major energy in the boom is at lower frequencies where the attenuation of the sound by the house is less than it is for the frequency region occupied by the aircraft noise), the trend in these results are to be expected. The relatively small improvement in the acceptability of the noise by virtue of the listeners being indoors, and therefore somewhat sheltered from the noise, has been found to be true in previous studies of road traffic and aircraft noise.

TABLE 55

Results of Paired-Comparison Judgments of Relative Acceptability of Sonic Booms vs. Subsonic Aircraft Noise. From Kryter et al. (473).

Note: All overpressure and energy values for the sonic boom and PNdB levels for subsonic aircraft noise are for outdoor measurements.

[1] Variable	[2] Subjects From	[3] A/C	[4] Nominal ΔP psf*	[4] dB**	[5] Measured ΔP for N Missions-Median of the Medians of 5 Microphones Over N Missions[4] psf	[5] dB	[6] Aircraft Noise when Judged Equal to Boom — Indoors PNdB	[6] Outdoors PNdB	[7] Number of Subjects	[8] N Missions-Number of Pairs of Booms vs. Noises
Subjects from Different Communities	Edwards AF Base	+ B-58[1]	1.69	132.14	1.94	133.34	109	105	120	25
	Fontana	B-58[2]	1.69	132.14	1.74	132.39	119	111	98	12
	Redlands	B-58[2]	1.69	132.14	1.73	132.34	118	108	148	12
Different Types of Aircraft	Edwards AF Base	+ B-58[1]	1.69	132.14	1.94	133.34	109	105	120	25
		- F-104[2]	1.40	130.50	1.40	130.50	107(108)[5]	97(100)[5]	120	13
		XB-70[3]	1.36	130.25	1.35	130.19	107(110)	98(101)	120	4
Booms of Different Intensities From Same Aircraft		F-104[2]	0.75	125.08	0.86	126.27	99(101)	87(89)	120	12
		- F-104[2]	1.40	130.50	1.40	130.50	107(108)	97(100)	120	13
		F-104[2]	2.80	136.52	2.77	136.43	121(120)	117(116)	120	12
		+ B-58[1]	1.69	132.14	1.94	133.34	109	105	120	25
		B-58[2]	2.33	134.93	2.56	135.74	114	111	120	20
		B-58	2.65	136.05	2.91	136.86	117	112	120	24

+ The data in these three lines are for the same missions.

- The data in these two lines are for the same missions.

1. Aircraft were flown on track 5 miles to one side of test facility.

2. Aircraft were flown directly over test facility.

3. Aircraft were flown on track 13 miles to one side of test facility.

4. The five microphones were arranged at the test facility in a cruciform with a spacing of 100 ft between microphones.

5. Values reported in a similar table in an Interim Report (July 1967) of the Edwards AF Base Study, if different than in the present table, are shown in parentheses. These changes are due to the availability for the present report of physical measurements of some of the aircraft noise not yet analyzed when the Interim Report was prepared.

* pounds per square foot (psf).

** $dB = 10 \log_{10} \dfrac{p_1^2}{p_0^2}$, and p_0 is 0.0002 μbar, and p_1 is peak overpressure in bars.

9 A/C	10 Difference between Median Measured ΔP and Nominal ΔP (Col. 4 minus Col. 5)		11 Average Difference between Median of 5 Microphones for a Single Mission and Nominal ΔP*		12 Average Difference between Median of 5 Microphones for a Single Mission and Median Measured ΔP for N Missions**		13 Median Measured Duration	14 Median Measured Rise Time
+ B-58[1]	0.25 psf	1.20 dB	0.38 psf	1.75 dB	0.33 psf	0.71 dB	0.171 sec	0.007 sec
B-58[2]	0.05	0.25	0.23	1.17	0.22	1.30	0.183	0.006
B-58[2]	0.04	0.20	0.37	1.60	0.37	1.60	0.197	0.008
+ B-58[1]	0.25	1.20	0.38	1.75	0.33	0.71	0.171	0.007
- F-104[2]	0		0.22	1.38	0.22	1.38	0.079	0.005
XB-70[3]	0.01	0.06	0.15	0.88	0.15	0.88	0.277	0.006
F-104[2]	0.11	1.19	0.25	2.10	0.21	1.63	0.106	0.006
- F-104[2]	0		0.22	1.38	0.22	1.38	0.079	0.005
F-104[2]	0.03	0.09	0.37	1.08	0.27	1.08	0.080	0.005
+ B-58[1]	0.25	1.20	0.38	1.75	0.33	0.71	0.171	0.007
B-58[2]	0.23	0.81	0.40	1.28	0.33	1.01	0.160	0.005
B-58[1]	0.26	0.81	0.39	1.17	0.31	1.92	0.148	0.009

+ The data in these three lines are for the same missions.

- The data in these two lines are for the same missions.

* $\dfrac{1}{N}\sum_{i=1}^{N}\left| X_i - \text{Nominal } \Delta P \right|$: where X_i is the median of 5 microphone measurements for the ith mission, and N is number of missions.

** $\dfrac{1}{N}\sum_{i=1}^{N}\left| X_i - \text{Median }(X_i) \right|$: where X_i is the median of 5 microphone measurements for the ith mission, and N is number of missions.

TABLE 56

Percentage of Persons Who Rated Sonic Booms and Subsonic Aircraft (WC-135B, KC-135) Noises
as Unacceptable (Less than Just Acceptable)
Listeners from Edwards Air Force Base. From Kryter et al. (473).

SOURCES OF BOOMS AND NOISES						LOCATIONS OF PERSONS									
A/C	Nom. Peak Overpressure (psf)	Alt.	EPR	PNdB	Number of Missions*	Out-door	Block-house**	E1&E2 In-door	E1-BR	E1-LR	E1-FK	E2-BR	E2-LR	E2-DR	E2-FK
B-58	1.69				12	33%	23%	27%	15%	25%	17%	39%	46%	28%	24%
B-58	2.06				4	51	--	37	42	68	20	11	28	73	54
B-58	2.33				11	63	--	28	34	44	6	13	51	38	39
B-58	2.52				2	64	--	49	41	67	32	18	83	92	40
B-58	2.65				8	68	55	62	32	70	52	89	73	56	59
Av.	2.25					56	--	41	33	55	25	34	56	57	43
F-104	0.70				6	2	--	2	6	0	1	0	0	3	3
F-104	1.36				2	17	--	3	7	0	4	0	0	9	0
F-104	1.40				6	30	--	16	16	12	9	11	9	51	15
F-104	1.50				4	29	--	27	10	29	23	54	43	4	22
F-104	1.69				1	75	--	29	43	38	0	11	22	67	38
F-104	2.00				2	33	--	31	0	7	17	75	57	0	39
F-104	2.80				7	74	--	63	54	50	22	62	89	100	73
F-104	3.30				2	98	--	82	63	75	79	100	79	50	100
Av.	1.83					45	--	32	25	26	19	39	36	36	36
XB-70	1.36				2	21	--	28	32	15	11	19	39	74	25
XB-70	2.06				4	53	--	25	33	32	9	6	21	68	27
XB-70	2.52				2	65	--	33	55	53	18	10	39	67	28
Av.	1.98					46	--	29	40	33	13	12	33	70	27
WC-135B		8000	1.76	85	2	1	--	1	0	0	4	0	0	9	0
KC-135		3000	1.5	95	4	2	5	2	0	0	2	3	0	0	3
WC-135B		4000	1.76	95	4	3	--	2	7	0	0	0	0	0	2
WC-135B		2000	1.76	105	9	24	--	11	17	11	5	0	4	17	14
KC-135		1000	1.5	107	4	28	33	22	6	30	21	15	16	11	38
WC-135B		1300	1.76	110	2	41	--	14	0	0	27	5	0	44	15
WC-135B		1000	1.76	113	3	70	--	35	25	50	22	33	15	65	44
WC-135B		800	1.76	115	6	77	--	43	44	56	19	47	24	55	49
KC-135		500	1.5	115	2	80	62	49	19	80	50	80	13	33	59
WC-135B		500	1.76	119	2	92	--	51	38	71	40	53	34	91	52
WC-135B		250	1.76	125	2	94	--	70	53	85	54	78	58	90	81
Av.			1.76	111		47	--	27	19	35	22	29	15	38	32
Number of Persons per Mission						40-48	9-11	51-70	6-8	5-8	8-11	8-11	6-9	5-6	13-18

* The ratings are only for the first aircraft of a pair.

Comparisons among Subjects from Different Communities

Table 55 shows that the subjects from Redlands and Fontana judged the sonic boom from the B-58 relative to the subsonic aircraft noise such that a noise of 118-119 PNdB was judged equal to the boom at 1.69 psf when heard indoors and to 108-111 PNdB when heard outdoors. Thus, to these subjects the boom was much less acceptable than it was to the subjects from Edwards Air Force Base — equivalent to a 10 PNdB change in the noise from the subsonic aircraft when heard indoors and about 5 PNdB when heard outdoors. The difference between the judgments of the subjects from Edwards Air Force Base and those from the relatively quiet communities of Fontana and Redlands is illustrated by the extrapolated curves in Fig. 230.

An aircraft noise survey showed that the median peak level of aircraft noise in typical residential neighborhoods in Redlands was about 75 PNdB (maximum peak level of about 95 PNdB), and in Fontana about 85 PNdB (maximum peak level of about 100 PNdB); also, these communities were not under or near usual flight tracks for supersonic military aircraft involved in training or test missions. An aircraft noise survey of the residential area of Edwards Air Force Base revealed that subsonic aircraft noise reached occasional peak levels of 110 PNdB; this area, however, was subjected to about 4-8 booms per day for the past three years at a median nominal peak overpressure of 1.2 psf. The subjects had lived on Edwards Air Force Base an average of two years.

It is presumed that the lesser acceptability of sonic booms to the subjects from Fontana and Redlands than to the subjects from Edwards Air Force Base may be due to adaptation to the sonic booms enjoyed by the Edwards subjects as the result of an average of two year's previous exposure to sonic booms. It was also found, as will be described more fully later, that the residents of Edwards Air Force Base, in reply to an attitude survey, in general believed that their exposure to sonic booms at Edwards made them more tolerant of the boom. Table 57 shows that age and sex were not consistently related to the acceptability rating scores given to sonic booms and the noise from subsonic aircraft.

Possible Biases

There were some factors in the tests that may have favored the boom and others that may have favored the subsonic aircraft noise. It might be pointed out here that the factors contributing to the relative unacceptability of the boom (impulsive nature of the sound, its intensity, house vibration, arousal, etc.) were presumably different than some aspects of the subsonic aircraft noise that were considered relatively unacceptable (interruption with speech, distractiveness,

TABLE 57

Comparison by Age and Sex of the Persons Who Rated Sonic Booms and as Unacceptable (Less than Just Acceptable). From Kryter *et al.* (473).

Group	Median Age	A/C	Number of Flights	Indoor Listening				Outdoor Listening				Critical Value at 10% Level of Significance	Decision
				ML vs. MG	FL vs. FG	ML vs. FL	MG vs. FG	ML vs. MG	FL vs. FG	ML vs. FL	MG vs. FG		
				(See notes for explanation of column headings and cell entries)									
Redlands	49	B-58	6	4/10 5/20 0.71	6/17 4/16 0.41	4/10 6/17 0.06	5/20 4/16 0.10	4/15 3/17 0.38*	8/28 3/14 0.25	4/15 8/28 0.02	3/17 3/14 0.07*	2.71	No Significant Difference in the Ratings
		WC-135B	6	2/10 3/20 0.12*	4/17 2/16 0.67	2/10 4/17 0.05*	3/20 2/16 0.05*	10/15 11/17 0.01	19/28 10/14 0.06	10/15 19/28 0.01	11/17 10/14 0.16	2.71	
Fontana	38	B-58	6	2/5 3/9 0.06*	14/22 11/25 1.81	2/5 14/22 0.94	3/9 11/25 0.31	1/2 2/6 0.20*	9/14 6/12 0.54	1/2 9/14 0.15*	2/6 6/12 0.45*	2.71	
		WC-135B	6	1/5 0/9 1.94*	4/22 2/25 1.09*	1/5 4/22 0.00*	0/9 2/25 0.77*	1/2 2/6 0.18*	6/14 5/12 0.00	1/2 6/14 0.04*	2/6 5/12 0.12*	2.71	
Edwards AF Base		B-58	9	2/5 3/7 0.01*	5/23 9/26 0.99	2/5 5/23 0.73*	3/7 9/26 0.16*	1/4 1/3 0.06*	8/19 5/21 1.52	1/4 8/19 0.41*	1/3 5/21 0.13*	2.71	
		KC-135	12	1/6 1/7 0.01*	5/25 4/26 0.19	1/6 5/25 0.03*	1/7 4/26 0.01*	1/4 1/3 0.06*	4/20 5/21 0.09	1/4 4/20 0.05*	1/3 5/21 0.13*	2.71	

* Inadequate sample size for statistical tests.

NOTES:

1. The comparisons are based on ratings for the first aircraft of a pair.

2. Symbols for age and sex classification: ML = males whose age is less than the median age; FL = females whose age is less than the median age; MG = males whose age is greater than or equal to the median age; FG = females whose age is greater than or equal to the median age.

3. Differences in the ratings due to age are tested in the columns headed ML vs. FL and MG vs. FG. Differences in the ratings due to sex are tested in columns headed ML vs. MG and FL vs. FG.

4. Cell entries: Upper left (or upper right) is a/a+b(or c/c+d) where a (or c) is the average number of unacceptable ratings and b (or d) is the average number of acceptable ratings for the designated class. (a+b (or c+d) is the average number of persons in the class.) The lower entry is the value of the test statistic: $\chi^2 = \dfrac{(ad - bc)^2(a+b+c+d)}{(a+b)(a+c)(b+d)(c+d)}$. Example: Third row and second column, a = 14, b = 8, c = 11, d = 14;
$$\chi^2 = \frac{(14\cdot14 - 11\cdot8)^2(47)}{(22)(25)(22)(25)} = 1.81.$$ The adequacy of the sample size depends on the values of a and c in addition to the values of a+b and c+d.

5. Significance test and decision rule: The data are used to determine whether the same percentage of unacceptable ratings occurs for two classes. The hypothesis that the ratings are the same would be rejected if the value of the test statistic equals or exceeds 2.71 at the 10% level of significance (i.e., the probability is 0.10 that the hypothesis is rejected when it is true).

etc., some of which were present with the boom). This difference in the effects of the two sounds or noises obviously means that the people from Edwards Air Force Base had probably learned to dislike, or the people from Fontana and Redlands initially disliked, the two sounds for somewhat different reasons.

People can and do, particularly in everyday life, make decisions about things that are considerably different. Choosing between "apples and oranges" may not, according to legend, be very scientific but such choices can be and are made as a matter of course in real life. One may question whether the subjects involved were intelligent and motivated enough to have made the relative judgments in a meaningful way, but it is not logical to presume (805) that because the unacceptability of the two noises were casually related to some extent in different ways to different effects, the judgments are for this reason invalid. Indeed, the Tracor report (805), presents attitude data, to be discussed later, which are not qualitatively inconsistent with the results of the tests performed at Edwards.

One condition that probably made the boom more acceptable in these tests than it would have been in real life was that the subjects invariably knew within several minutes when a boom or subsonic aircraft noise was going to occur. This factor probably reduced somewhat the amount of startle experienced because of the boom, although in real life, if there were regular SST overflights, there would likewise be a tendency for the boom to lose some of its initial effectiveness to startle. For a discussion of the pros and cons for these particular paired-comparison tests, see Kryter (460).

Sonic Booms at Different Intensities from Different Aircraft

A number of tests were conducted in which the subjects, all residents of Edwards Air Force Base, judged the relative acceptability of sonic booms from different supersonic aircraft or from the same type of supersonic aircraft flying in accordance with different or the same operational procedures. The results of these tests are given in Figs. 231 and 232. These tests did not show, for the booms tested, any consistent differences in the acceptability of one type of sonic boom vs. another type when both had the same peak overpressure.

Of particular interest is the rate at which the percent preference score changed as a function of a change in peak overpressure. Figure 231 shows that a change of 1.5 dB (about 0.25 psf at a boom intensity of 1.69 psf) for people indoors and 1.0 dB for people outdoors can cause an increase of about 12.5 percentage points in the number of people who judge the more intense boom to be less acceptable. This finding indicates that the subjective unacceptability of the sonic boom increases at a relatively rapid rate as its intensity level is increased, and at a somewhat more rapid rate for listeners outdoors compared

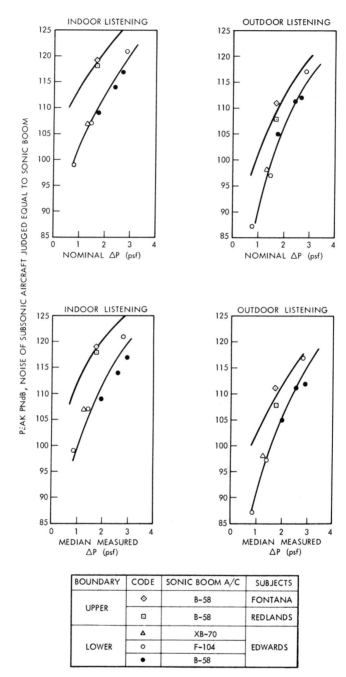

BOUNDARY	CODE	SONIC BOOM A/C	SUBJECTS
UPPER	◇	B-58	FONTANA
	▢	B-58	REDLANDS
LOWER	△	XB-70	EDWARDS
	○	F-104	
	●	B-58	

FIGURE 230.   Results of paired-comparison judgments of noise from subsonic jet aircraft and sonic booms from XB-70, B-58, and F-104 supersonic aircraft for subjects from different communities. Data obtained from Table 55. From Kryter *et al.* (473).

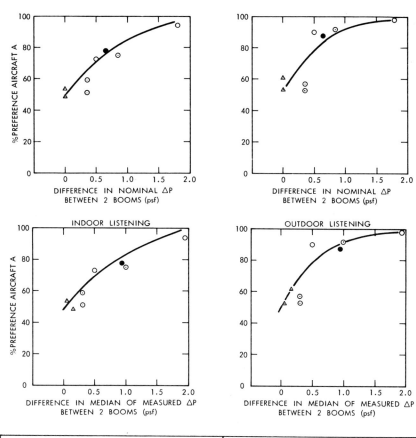

	AIRCRAFT A					AIRCRAFT B				
				%PREFERENCE					%PREFERENCE	
CODE	TYPE A/C	NOMINAL ΔP	MEDIAN OF MEASURED ΔP	INDOOR	OUTDOOR	TYPE A/C	NOMINAL ΔP	MEDIAN OF MEASURED ΔP	INDOOR	OUTDOOR
●	B-58	1.69	1.91	78%	88%	B-58	2.33	2.84	22%	12%
○	F-104	1.50	1.52	73	90	F-104	2.00	2.02	27	10
	F-104	1.50	1.63	94	98	F-104	3.30	3.56	6	2
○	F-104	2.00	2.09	51	57	B-58	2.33	2.40	49	43
	F-104	1.36	1.14	59	53	B-58	1.69	1.46	41	47
	F-104	1.50	1.20	75	92	B-58	2.33	2.18	25	8
△	XB-70	2.06	2.18	48	61	B-58	2.06	2.33	52	39
	XB-70	2.52	2.49	54	53	B-58	2.52	2.55	46	47

FIGURE 231.   Results of paired-comparison judgments of sonic booms (of the same type aircraft or two different types of aircraft) at the same and at different nominal peak overpressures in psf. Listeners were from Edwards Air Force Base. From Kryter *et al.* (473).

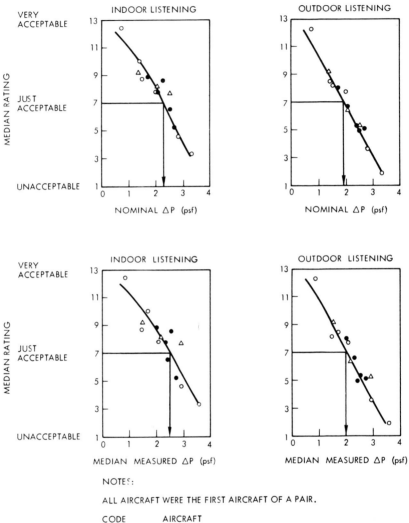

FIGURE 232.   Median ratings of XB-70, F-104, and B-58 sonic booms plotted against nominal peak overpressure and median of measured peak overpressure. Listeners were from Edwards Air Force Base. From Kryter *et al.* (473).

with listeners indoors. It was noted before that the rate of growth of unacceptability of the sonic boom appears to be greater than is the growth of unacceptability of the noise from subsonic aircraft (a 6 dB increase in the intensity of the sonic boom was found to be equivalent to a 12 PNdB increase in the level of a noise from a subsonic aircraft of equal acceptability).

Other Physical Measurements of Sonic Booms

The valid physical measurement of peak overpressure of the N wave of the sonic boom requires a microphone recording system that extends over the frequency range from about 0 to 5000 Hz or so (with most of the energy lying below 100 Hz). There is reason to believe, however, that people are not particularly sensitive or responsive to energy below 20 Hz or so. To examine this question, the energy spectra of a sample of sonic booms from the XB-70, F-104, and B-58 aircraft were obtained with the aid of an electronic computer. These energy spectra were in turn used to determine the energy in various bands as shown in Table 58.

Table 59 shows the correlations found between acceptability ratings of sonic booms and various physical measures of the booms. It is to be noted that the correlations between the ratings and the measures that exclude information regarding the energy for frequencies below 20 Hz are as high or higher than the measures, including overall peak overpressure (ΔP), which use energy below 20 Hz; compare, in particular, data columns 4 and 5, $E_{n_{0-1000}}$ vs. $E_{n_{20-1000}}$; however, the B-58 booms anlayzed in this fashion covered such a restricted range of overpressures that only a few significant correlations were obtained. It may at first seem somewhat surprising that a major portion of the energy in the boom — that below 20 Hz — and the frequencies to which the ear is most sensitive — those above 1000 Hz — can be discarded for these boom measurement and evaluation purposes. The rise time of the booms for all these aircraft were such that their nominal spectra fell off at the same rate (about 12 dB per octave) at frequencies above about 500 Hz; the frequencies in the booms that contributed the most to perceived noisiness or loudness (i.e., exceed the threshold of hearing by the greatest amount) are in the region of 200 Hz.

In our opinion (see also Chapters 8 and 9), a sound level meter set on slow and with D-weighting or a close approximation would be an appropriate instrument for measuring the perceived noisiness of sonic booms, provided an impulse correction is added (see below). Table 60 gives one-third octave band energy spectra and calculated PNdB, dB(D), and dB(A) for nominal sonic booms from XB-70 aircraft (duration about 300 msec and rise time 5 msec).

TABLE 58

Average Value and Average Deviation from Average Value for Measurements of Sonic Booms Recorded Outdoors From Kryter et al. (473).

Aircraft	Nominal ΔP dB*	Nominal ΔP psf	Number of Missions		Measurement ΔP dB*	En_{0-50} dB*	En_{0-200} dB*	En_{0-1000} dB*	$En_{20-1000}$ dB*	En_{20-200} dB*	En_{10-30} dB*	En Total dB*
XB-70	135.61	2.52	4	Avg.	136.00	123.92	124.00	124.01	111.70	111.50	113.14	124.01
				Avg. Dev.	0.37	1.16	1.14	1.15	0.19	0.17	0.58	1.15
XB-70	133.86	2.06	6	Avg.	134.22	123.23	123.27	123.28	108.73	108.53	111.47	123.28
				Avg. Dev.	1.08	0.43	0.43	0.43	0.60	0.57	0.54	0.43
XB-70	130.25	1.36	3	Avg.	130.21	117.51	117.62	117.63	104.56	104.38	107.17	117.63
				Avg. Dev.	0.42	0.38	0.31	0.31	1.32	1.26	0.82	0.31
F-104	136.52	2.80	7	Avg.	137.79	120.74	121.13	121.19	116.35	116.18	117.84	121.19
				Avg. Dev.	1.21	1.10	1.07	1.05	1.12	1.19	1.13	1.05
F-104	132.14	1.69	2	Avg.	134.83	116.98	117.47	117.53	110.98	110.83	114.08	117.53
				Avg. Dev.	1.51	0.20	0.53	0.58	2.50	2.54	0.46	0.58
F-104	130.50	1.40	10	Avg.	130.54	115.03	115.22	115.25	107.17	107.00	110.97	115.25
				Avg. Dev.	1.45	1.10	1.09	1.09	1.74	1.74	1.38	1.09
F-104	125.08	0.75	11	Avg.	126.81	111.70	111.81	111.84	101.78	101.65	104.81	111.84
				Avg. Dev.	2.08	1.59	1.59	1.60	2.50	2.46	1.99	1.60
B-58	135.61	2.52	4	Avg.	135.60	122.42	122.49	122.50	110.90	110.70	113.33	122.50
				Avg. Dev.	0.69	0.94	0.92	0.92	0.82	0.75	0.59	0.92
B-58	134.93	2.33	16	Avg.	136.04	121.71	121.81	121.82	110.70	110.51	113.43	121.83
				Avg. Dev.	1.25	0.42	0.43	0.42	1.55	1.50	0.98	0.45
B-58	133.86	2.06	5	Avg.	134.34	121.14	121.24	121.25	109.60	109.39	112.38	121.25
				Avg. Dev.	0.68	0.43	0.43	0.42	0.98	1.03	0.74	0.42
B-58	132.14	1.69	17	Avg.	132.40	119.49	119.56	119.57	106.52	106.36	109.85	119.57
				Avg. Dev.	1.18	0.89	0.91	0.91	1.79	1.75	1.28	0.91
		Total	85									
Grand Avg.	132.40				133.13	119.06	119.19	119.21	108.40	108.23	111.25	119.21
Grand Avg. Dev.					1.23	0.86	0.86	0.86	1.52	1.50	1.13	0.87

*These measures are calculated using pressures expressed in units of 0.0002 μbar. The subscripts of En (energy measures) indicate the limits of energy bands. For example $En_{20-1000}$ designates an energy measurement in the band 20 Hz to 1000 Hz and is defined as the integral of the energy spectral density function between the limits 20 Hz to 1000 Hz.

TABLE 59

Rank Correlations between Median Ratings and Various Energy Measurements of Sonic Booms Recorded Outdoors.
From Kryter *et al.* (473).

Subjects	Aircraft	Nominal ΔP (psf)	Number of Missions, N	Critical Value at 5% Level of Significance	Measure							
					ΔP	En_{0-50}	En_{0-200}	En_{0-1000}	$En_{20-1000}$	En_{20-200}	En_{10-30}	En_{Total}
Edwards Indoor →	XB-70	2.52, 2.06, and 1.36	5	0.81	0.90*	0.30	0.30	0.30	0.90*	1.00*	0.90*	0.30
	F-104	2.80, 1.40, and 0.75	18	0.40	0.93*	0.92*	0.92*	0.92*	0.94*	0.94*	0.92*	0.92*
	B-58	2.52, 2.33, and 2.06	12	0.50	0.32	0.31	0.31	0.44	0.53*	0.51*	0.27	0.44
Edwards Outdoor →	XB-70	2.52, 2.06, and 1.36	5	0.81	0.80	0.60	0.60	0.60	1.00*	0.90*	1.00*	0.60
	F-104	2.80, 1.40, and 0.75	18	0.40	0.89*	0.88*	0.87*	0.87*	0.90*	0.90*	0.88*	0.89*
	B-58	2.52, 2.33, and 2.06	12	0.50	0.32	0.42	0.42	0.39	0.17	0.17	0.14	0.39

* Rank Correlation greater than Critical Value.

Notes: 1. All judgments were made on the 1st sound of a pair of aircraft sounds.

2. The rank order correlation is defined as $1 - [6 \sum d^2 / N(N^2 - 1)]$ and d is the difference in ranks (in this case, the difference in the ranks of the median ratings and the ranks of the physical measurements). The critical value at the 5% level of significance varies with the value of N.

3. The subscripts of En (energy measures) indicate the limits of energy bands, e.g., $En_{20-1000}$ designates an energy measurement in the band 20 Hz to 1000 Hz and is defined as the integral of the energy spectral density function between the limits 20 Hz to 1000 Hz.

TABLE 60

Nominal Outdoor 1/3 Octave Band Energy Spectra and dB(A), (C), (D), and PNdB for XB-70 Sonic Boom, Rise Time 0.005 Sec From Kryter et al. (473).

Mission	ΔP		Sound Energy Level, dB--Nominal Outdoors Center Frequencies																			dB^					PNdB^
	psf	dB*	50	63	80	100	125	160	200	250	315	400	500	630	800	1000	1250	1600	2000	2500	(A)	(C)	(D_1)	(D_2)	(D_3)	
1-1	2.91	136.86	102	102	99	96	93	87	73	81	81	71	75	68	67	64	60	58	54	52	85	106	97	92	85	99
2-1	2.55	135.71	101	100	98	95	92	86	72	80	80	70	73	67	66	63	59	57	53	51	84	105	96	91	83	98
10-1	2.41	135.22	101	100	98	95	92	86	72	80	80	69	73	67	65	62	59	57	53	50	84	105	96	91	83	98
15-1	2.18	134.35	100	99	97	94	91	85	71	79	79	69	72	66	64	61	58	56	52	50	83	104	95	90	82	97
14-1	2.10	134.02	100	99	96	94	90	85	71	79	79	68	72	66	64	61	58	55	51	49	82	103	94	89	82	97
9-1	2.09	133.98	100	99	96	94	90	85	71	79	79	68	72	66	64	61	58	55	51	49	82	103	94	89	82	97
6-2	1.78	132.59	98	97	95	92	89	83	69	77	77	67	70	64	62	60	56	54	50	48	81	102	93	88	80	95
5-2	1.19	129.09	95	94	92	89	85	80	66	74	74	63	67	61	59	56	53	50	47	44	77	99	90	85	77	92

* re 0.0002 μbar

^ Calculated from nominal spectra of boom from XB-70.

Relation between Ratings of Sonic Booms and Wall Displacements in Individual Test Rooms

Measurements made by accelerometers and displacement gauges located at various points throughout the test houses indicated that the displacement of the external walls of the test rooms (so-called plate response) appear to be the largest and most sensitive measures of the effects of various sonic booms on the houses. On the bottom row of the graphs in Fig. 233 are plotted the average displacement in inches for a wall of three of the test rooms as a function of the peak overpressure, measured outdoors, of the sonic booms made by F-104, B-58, and XB-70 aircraft. It is seen for these measures that the magnitude of the wall displacement is approximately equal for the booms of equal overpressure from XB-70 and B-58 aircraft whereas there is apparently less displacement of the walls from the boom of the F-104 aircraft, but of comparable peak overpressure. As discussed elsewhere, this result is not unexpected because the boom of the B-58 and XB-70 aircraft contains more energy than the boom from the F-104 at frequencies of around 10-20 Hz, frequencies which apparently were particularly effective in moving the walls of the rooms. It is also apparent that the location and precise construction of the individual test rooms in the houses, as well no doubt as the flight direction of the aircraft, influenced the magnitude of the displacement of the individual walls of the room.

Shown in the top row of graphs on Fig. 233 are the average ratings given to the sonic booms that caused the wall displacements shown in the graphs in the bottom row of the figure. On the upper right-hand graph in Fig. 233 are plotted the ratings obtained as a function of wall displacement, independent of the test rooms in which the listeners were located. In general it appears that the greater the wall displacement, the more unacceptable is the rating received by a sonic boom. However, acoustic rather than direct vibrational stimulation was deduced to be primarily responsible for the ratings given to the booms from the fact that subjects placed on vibration-isolated chairs for some of the tests rated the booms about the same as did subjects not on these special chairs.

It might be deduced also that whereas the sonic boom from the F-104 caused somewhat less wall displacement than did the booms from the other aircraft, there were other aspects of the boom from the F-104 that contributed to its unacceptability ratings which, for a given wall displacement, appeared to be about equal to or greater than the ratings received by the other booms. It is possible that the boom from the F-104 had a slightly shorter rise time and a sharper "crack" to it than the booms from the other aircraft because the boom path to the ground from the F-104 was shorter than for the other aircraft, and the boom had less opportunity to be disturbed and distorted by atmospheric effects (the F-104 had to fly at a considerably lower altitude than the B-58 or XB-70 in order to generate on the ground a boom of comparable peak over-

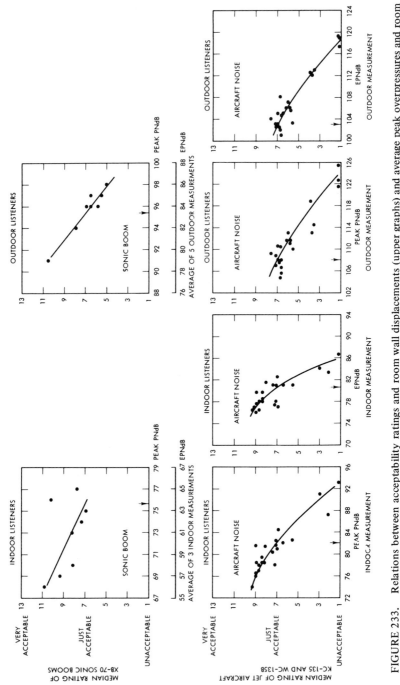

FIGURE 233. Relations between acceptability ratings and room wall displacements (upper graphs) and average peak overpressures and room wall displacements (lower graphs) to sonic booms from XB-70, F-104, and B-58 aircraft. From Kryter et al. (473).

pressure). Neither the psychological nor the physical measurements that were made were precise enough to demonstrate with any certainty some of these possible detailed effects and relations.

Comparison of Perceived Noise Levels of Sonic Booms and Aircraft Noise

One-third octave band spectra were found for recorded indoor and nominal outdoor XB-70 sonic booms. Peak (or Max) PNdBs were calculated for each of these spectra (see Table 60 for data on outdoor booms). Because of their short duration (less than 0.5 sec), by definition, EPNdB of a sonic boom equals Peak PNdB -12 (see Chapter 8). In Fig. 234, acceptability ratings are plotted against Peak PNdB and EPNdB for subsonic aircraft noise. It is clear that booms of a given EPNdB did not receive the same acceptability rating as did the noise from a subsonic aircraft having the same EPNdB — outdoor listeners: 83.5 EPNdB for the boom vs. 103 EPNdB for subsonic aircraft noise; indoor listeners: 63.5 EPNdB for the boom vs. 81 EPNdB for the noise.

Proposed Impulse Correction Factor

It is perhaps logical, on the basis of the functional model of the ear proposed earlier and the concept of effective perceived noise level, to expect agreement between judged noisiness and EPNL for the aircraft noise and sonic booms. However, as noted, there is a discrepancy between the EPNdBs of sonic booms and the aircraft noise judged to be equally noisy by the indoor and outdoor subjects from Edwards Air Force Base — subjects who were presumably equally adapted to the subsonic aircraft noise and the sonic booms. On the basis of these data, it appears that the perceived noisiness of impulses is more than a simple matter of some frequency weighting of its energy spectrum. This is probably due to a startle or suddenness aspect found in impulses that is missing (again with equally familiar sounds) from nonimpulsive sounds. The effect of this hypothecated attribute of impulsive sound could be accounted for by increasing calculated PNdB of the boom by an amount read from Fig. 174. Although Fig. 174 is little more than a plausible guess at this time, it is proposed that it be used as a correction for suddenness or startle in the calculation of the perceived noisiness of impulsive sounds, even though not novel or unusual for a given environment (see also Chapters 8 and 10).

Mail Survey Ratings of Sonic Booms, Aircraft Noise, and Street Noise by Residents of Edwards Air Force Base

Figure 235 depicts the acceptability ratings of environmental noises made by residents of Edwards Air Force Base as a function of their age and years of

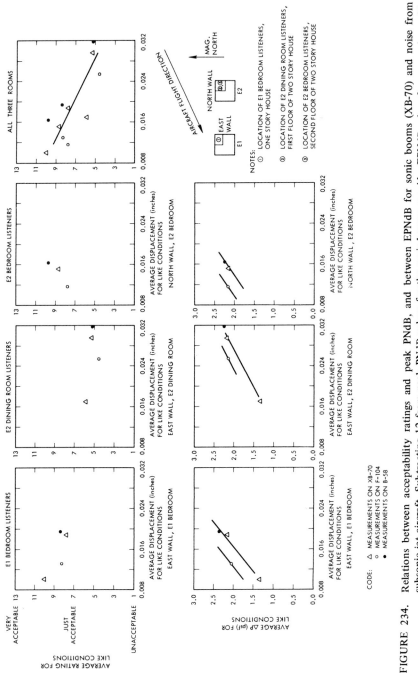

FIGURE 234. Relations between acceptability ratings and peak PNdB, and between EPNdB for sonic booms (XB-70) and noise from subsonic jet aircraft. Subtracting 12 from peak PNdB values for the sonic booms provides EPNdB values for compariosn with EPNdB of aircraft noise. After Kryter *et al.* (473).

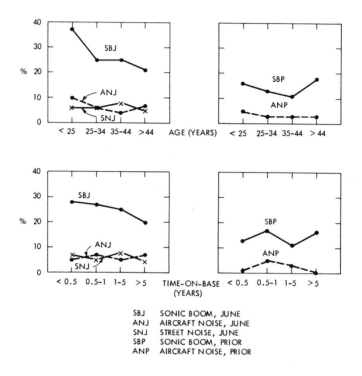

SBJ   SONIC BOOM, JUNE
ANJ   AIRCRAFT NOISE, JUNE
SNJ   STREET NOISE, JUNE
SBP   SONIC BOOM, PRIOR
ANP   AIRCRAFT NOISE, PRIOR

FIGURE 235.   Percentage of persons who rated sonic booms as unacceptable (less than just acceptable). Data obtained by mail survey of 2000 families residing at Edwards Air Force Base. From Kryter *et al.* (473).

residence at Edwards. It would appear from this figure that, particularly with respect to sonic booms, the older the person and the longer he or she had lived there, the more acceptable were the noises. Age and years of residence are obviously not independent of each other, and an analysis of the data by years of residence, keeping age constant, showed no consistent influence of age upon the ratings of sonic booms. As previously noted, no significant difference was found between the results of paired-comparison tests for different age groups of subjects.

At the same time it should be mentioned, as shown in Fig. 235, that about 14% of the people who replied to the mail questionnaire rated in retrospect the sonic boom conditions prior to the month of June as being unacceptable, compared to 26% who rated the booms heard during June as being unacceptable. Part of the explanation for this difference undoubtedly was due to the difference in boom exposures during these periods. The average nominal peak overpressure of sonic booms during a typical operational month prior to June 1966 in the residential area of Edwards is about 1.2 psf, and the average frequency is about 4 to 8 per day. During the month of June, however, about 289 booms were created, giving a daily average of about ten and a median nominal peak overpressure of about 1.69 psf.

Recent Sonic Boom Studies in Other Cities

In 1967-1968 the SR-71 military supersonic aircraft was flown repeatedly over a number of cities in the United States as part of the military training program. This aircraft creates during cruise a boom directly under its flight path that has a nominal overpressure on the ground of approximately 1.0 psf.

Under NASA sponsorship, surveys were conducted prior to, during, and following a number of the SR-71 flights. Although these training missions were curtailed, a considerable amount of sociological data related to the effects of the sonic booms was obtained (805). Some of these data are summarized in Fig. 236 and Table 60a, where it is seen that:

1. The sonic booms, an average of less than one boom per day for about three months in the cities studied, increased the annoyance level in each city. (Also, it was found in this study that the booms were considered the worst source of noise annoyance in their environment.)

2. The higher the economic level, as measured by the value of the house in which a person resided, the higher was the mean annoyance.

3. The anti- or pro-theme of articles appearing in the press about the SST and sonic boom had a relatively small overall effect on mean annoyance ratings.

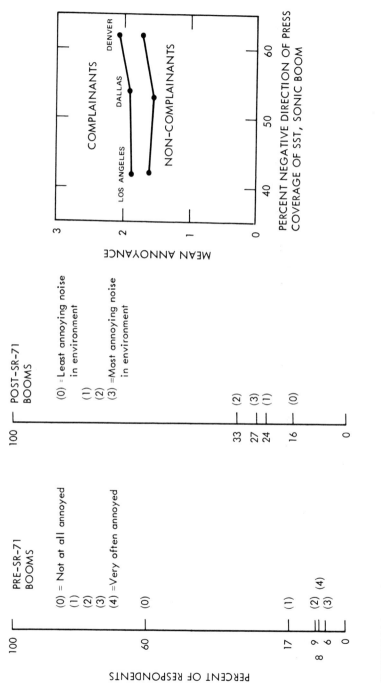

FIGURE 236. Left-hand bar graphs: Degree of annoyance from sonic booms (nominal 1.0 psf, average of 2 booms every three days). Merged attitude data for Atlanta, Dallas, Denver, and Los Angeles. After Tracor (805). Right-hand graph: Mean annoyance felt from SR-71 sonic booms as influenced by press articles. From Tracor (805).

Summary of Psychological Research on the Sonic Boom

Essentially two groups of experiments have been conducted that purport to demonstrate what the effects of sonic booms for the SST might be on people:

1. Attitude surveys and observations of behavior of residents in various cities in the United States and France when these residents were subjected to sonic booms generated by military aircraft.

2. So-called paired-comparison tests conducted in laboratories and under field conditions in the United States and in Great Britain in which subjects estimated the relative acceptability, as though heard under real-life conditions, of two sounds presented in rather rapid succession (a boom vs. flyover noise from a subsonic aircraft, and one boom vs. a second boom).

The results of all of these studies are consistent with each other and indicate that sonic booms above about 0.6 psf nominal are considered equal to or more annoying than environmental noises, such as from subsonic aircraft that are considered very obnoxious and the source of complaints by residents near airports. (More will be said later about the possible unacceptability of an SST.)

Regulatory Codes for Community Noise

Before presenting methods for specifying tolerable limits of exposure to environmental noise, a few comments on some of the political-legal aspects of the problem are perhaps appropriate. Most of the comments to follow in this regard apply to the United States, although they will often be appropriate for other countries as well.

Except within certain factory work spaces or for persons firing guns or operating or maintaining certain types of machines (aircraft, trains, tractors, etc.), noise as we now know it has not been demonstrated to be dangerous to the health of people in a residential community. Possible exceptions to this are found in people living near some airports. Included here are not only direct physiological effects but also indirect effects from loss of sleep and startle; this will be discussed in some detail in Part III. This is not to say that, in some courts of law, noise in some communities will not be considered as hazardous to health and well-being and therefore an illegal nuisance. This is certainly a possibility, and, with respect to damage or potential damage to hearing, is in some situations unquestionably justified, as was discussed in Chapter 5.

TABLE 60a

Some Results of Attitude Surveys Following Exposures to SR-71 Booms. From Tracor (805).

PERCENTAGE OF LOS ANGELES RESPONDENTS WHO SELECTED THE LISTED SOUND AS UNNECESSARY AND SHOULD BE THE FIRST ELIMINATED — Post boom period

Listed Sounds	Non-Complainants (a)	Complainants (b)
No Sound	5 %	6 %
Automobiles and/or Trucks	9	4
Motorcycles and/or Hot Rods	26	10
Aircraft Operations	20	7
Dogs or Other Pets	7	3
Sonic Booms	19	63
Neighborhood Children	0	0
Sirens	4	3
People	4	0
Lawn Mowers and/or Garbage Collection	1	0
Trains	0	1
Construction	0	0
Other Sounds	3	0
	N = 659	N = 360

PERCENTAGE OF LOS ANGELES RESPONDENTS WHO FELT BOOM AN UNNECESSARY SOUND AND WHO JUDGED THE LISTED ACTIVITIES AS BEING DISTURBED BY THE SONIC BOOM — Post boom period

Sounds	Non-Complainants (a)	Complainants (b)
Relaxing or Resting Inside	30 %	61 %
Relaxing or Resting Outside	26	49
Sleeping	19	36
Telephone	20	36
Listening to Records or Tapes	18	35
TV or Radio Reception	22	39
Reading or Concentrating	28	55
Eating	12	29
	N = 125	N = 227

This is also not to say that adverse psychological and behavioral effects, as distinct from physiological, of noise on man may not be of sufficient magnitude to warrant its control as a public nuisance. In most localities in the United States, there are no limits specified for noise but people are protected by general laws to the effect that any act that causes a public nuisance or endangers public health is an offense. An act passed by the British government in 1960 states that noise and vibration can be deemed a public health nuisance and that 3 persons occupying a property affected by the nuisance can make a complaint to public authorities.

However, public "nuisance" is a complex basis for the legal establishment of noise limits. In the first place, what bothers some people is acceptable to others; but more importantly, a nuisance can be made legal if it is in the general interest of the public to have the nuisance—and many of the significant sources of noise, for example aircraft, qualify as legalizable nuisances. This balance between different and conflicting values, such as the need for both relative quiet and economical transportation within a community, can only be settled by means of governmental processes and decisions.

Damages to property values may provide, if provable, legal grounds for limiting noise in communities. In many countries it is maintained that neither the government nor private parties can take or destroy property without adequately compensating the owner of the property. If the presence of noise at a person's house makes that house less desirable as a house, its value is reduced and property has been partly "taken" by the presence of the noise, be the noise in the public interest or not. In short, noise may damage or cause a relative reduction in the value of a property because it is not acceptable to people wishing to live on the property.

A specific example of the complexity of finding solutions to a noise pollution problem is that of aircraft noise (Golovin [308]). The airlines and aircraft manufacturers want the airport operators or the government to buy or zone the land around airports so people will not be exposed to intense aircraft noise. The operators and the government want the airplane and engine manufacturers to reduce the noise and they want the airlines to fly the aircraft in such a way that the noise will not be a major problem around airports. However, it is a major legal and social problem to force industrial or other zoning on land now used as residential areas (Goldstein [309]) and, if rezoning is accomplished, to obtain the funds required to buy or compensate people for the taking of land (even though the land may sometimes be resold for other uses at a profit). Also, the general protection afforded people relative to equitable taxation would seem to require that the costs of any such land purchase, assuming rezoning could be effected, would have to be borne by the vehicular passengers involved and not the general public (Dygert [200, 201]).

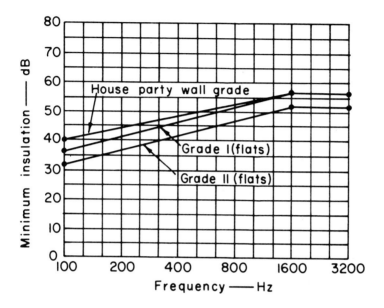

FIGURE 237.   Grade curves for airborne sound insulation. From Allen (10). (With permission of the Controller of Her Britanic Majesty's Stationery Office.)

Maximum Tolerable Levels Specified

In spite of these difficulties, tolerable levels, primarily aimed at nonpublic sources of noise, have been specified by governmental agencies in various localities, as shown in Tables 61 to 64. Even when methods of measurement and type of noise are the same, we see in Tables 61-64 considerable variation in specified tolerable levels. It might also be noted that the codes in the United States are 10 dB or so less restrictive than in some European countries. It has been suggested that this is due to the streets being narrower in Europe than in the United States.

Codes on Noise in the Home

Figure 237 and the following quotation (Allen [10] pp. 361-362) from a British standard would appear to summarize a great amount of knowledge about sound insulation requirements between flat (apartment) dwellings.

"Grade I represents the highest insulation that is practicable for flats (in United Kingdom constructional practice), and with it the noise from neighbors causes only minor distrubance to most tenants; it is no more nuisance than other minor disadvantages of living in flats.

"Grade II is a lower value of insulation and with it many of the tenants consider the noise from their neighbors to be the worst single factor about living in flats, but even so at least half the tenants will consider themselves not seriously disturbed by the noise. If the insulation is less than Grade II, then the number of tenants seriously distrubed will increase until a level of insulation as low as 8 dB worse than Grade II is reached when strong reactions, i.e., deputations, are probable."

Table 65 contains a summary of sound insulation building codes from various countries. Table 66 gives the recommendations of Waterhouse for building standards with respect to sound control for use in the United States. As Waterhouse notes, one deficiency of these codes is that they specify transmission loss for sound passing through a wall or floor that is the average loss for (usually) six test frequencies from 100 to 4000 Hz. Since some frequency regions are more important than others, and the noises in the home do not have their energy evenly distributed over this frequency region, the codes specified do not always insure adequate protection from noise. Blazier (72) suggests that the background NC noise levels due to residential heating and air-conditioning systems should be as shown in Table 67.

TABLE 61

The Maximum Permissible Standards for Motor Vehicle Noise in Different Countries. From Osipov and Korigin (599).

| Country | Unit of Measurement | Maximum Permissible Noise Values | | | | | | Conditions for the Vehicle | |
| | | Bicycle with a Motor | Motorcycles | Automobiles | Trucks | | | Standing Vehicles | Moving Vehicles |
					With Gasoline Engine	With Diesel Engine	Buses		
Finland	dB(B)*	75	82-84	85	90	90	90	Engine running at maximum acceleration with number of revolutions corresponding to the speed of 40 km/hour	
France	"	78 80	85 60	85 60	88-95 80-90	88-95 80-90	88-95 80-90		Moving on uneven road at the speed of (km/hour)
Luxembourg	"	75	80-85					Engine running at maximum number of revolutions	
German Federal Republic	"	75	80-82	82	Load capacity up to 2 tons -82; over 2t -87	87	87	Engine running at the rate corresponding to maximum speed	Moving at full acceleration. Throttled down speed - 50 km/hour
England Sweden	" "	80	80-85 85	80	88	88	88	" "	Moving at full acceleration. Throttled down speed - 40 km/hour
Sweden Czechoslovakia	" "	75-80	80-85-90	80	85 85	90 85	85 85	" "	Normal speed, full acceleration of engine

*For noise from motor vehicles, dB(A) is typically at least 5 dB less than dB(B).

TABLE 62

City Noise Limits

All levels shown here are sound levels relative to 0.0002 γbar as measured with a sound-level meter using A weighting, except where B weighting is indicated. From Loye (526).

For noise from motor vehicles, dB(A) is typically at least 5 dB less than dB(B).

City in	Noise Levels Specified by Laws	Equivalent Noise Levels at 20 ft	Equivalent Noise Levels at 50 ft
Ohio	95 dB(A) 20 ft from right rear of passing vehicle	95 dB(A)	88 dB(A)
California	95 dB(A) 5 ft from vehicle	84 dB(A)	77 dB(A)
Illinois	85 dB(A) 50 ft from vehicle	92 dB(A)	85 dB(A)
Wisconsin	95 dB(B) 20 ft from right rear of passing vehicle	89 dB(B)	82 dB(B)
Tennessee	90 dB(B) 20 ft from vehicle	84 dB(B)	77 dB(B)

NOISE TOLERATION LIMITS CITY OF SAO PAULO LAW 4805, From Levi[496]

For industrial plants

Noise allowed to vehicles	85 dB (scale B)	Strictly residential zone by day	60 dB (scale B)
Noise allowed to machines, motors, compressors and stationary generators by day	55 dB (scale B)	Strictly residential zone by night	45 dB (scale A)
		Predominantly residential by day	70 dB (scale B)
Noise allowed to machines, motors, compressors and stationary generators by night	45 dB (scale A)	Predominantly residential by night	55 dB (scale A)
		Mixed zone by day	80 dB (scale B)
Noise allowed to loud-speakers, radios, orchestras, single instruments or any such kind of device by day	55 dB (scale B)	Mixed zone by night	65 dB (scale B)
		Industrial zone by day	85 dB (scale B)
Noise allowed to loud-speakers, radios, orchestras, single instruments or any such kind of device by night	45 dB (scale A)	Industrial zone by night	65 dB (scale B)

TABLE 63

Noise Zoning Ordinance, Chicago. From Anon (32).

Performance Standards–Noise–M1 Districts

Octave band cycles per second			Along residence district boundaries maximum permitted sound level in decibels	Along business district boundaries maximum permitted sound level in decibels
0	to	75	72	79
75	to	150	67	74
150	to	300	59	66
300	to	600	52	59
600	to	1200	46	53
1200	to	2400	40	47
2400	to	4800	34	41
above		4800	32	39

Performance Standards–Noise–M2 and M3 Districts

Octave band cycles per second			Maximum permitted sound level in decibels along residence district boundaries or 125 feet from plant or operation property line		Maximum permitted sound level in decibels along business district boundaries or 125 feet from plant or operation property line	
			M2	M3	M2	M3
0	to	75	72	75	79	80
75	to	150	67	70	74	75
150	to	300	61	65	68	70
300	to	600	56	59	62	64
600	to	1200	50	53	56	58
1200	to	2400	45	48	51	53
2400	to	4800	41	44	47	49
above		4800	38	41	44	46

TABLE 64

1968 California Vehicle Code Noise Levels (113)

Vehicular Noise Limits

23130.   (a) No person shall operate either a motor vehicle or combination of vehicles of a type subject to registration at any time or under any condition of grade, load, acceleration or deceleration in such a manner as to exceed the following noise limit of the category of motor vehicle based on a distance of 50 feet from the center of the lane or travel within the speed limits specified in this section:

		Speed limit of 35 mph or less	Speed limit of more than 35 mph
(1)	Any motor vehicle with a manufacturer's gross vehicle weight rating of 6,000 pounds or more, any combination of vehicles towed by such motor vehicle, and any motorcycle other than a motor-driven cycle	88 dB(A)	92 dB(A)
(2)	Any other motor vehicle and any combination of vehicles towed by such motor vehicle	82 dB(A)	86 dB(A)

Composite Noise Rating (CNR) — a Method for Rating, Evaluating, and Predicting Effects of Environment Noise

As mentioned earlier, the overall effective value of a noise environment to which a person is exposed on the average of a daily basis has been expressed in various ways. The classical work of Rosenblith, K.N. Stevens, and Pietrasanta (702, 625, 626, 768, 769) which developed the "Composite Noise Rating" (CNR) scheme is, with some modifications, the basic method generally used or proposed for the evaluation of noise environments in the United States. The Composite Noise Rating has evolved into the U.S. Department of Defense Manual and a contractor's report to the Federal Aviation Administration on Land Use Planning Relating to Aircraft Noise (76), and is basically the same as the so-called "Noise Exposure Forecasts" (69,755). In Chapter 11, specific definitions and procedures for possible further standardization of CNR are presented.

The CNR concept consists of two ideas: (*a*) the behavioral response of people to their acoustic environment is a function of the sum, on an "energy" ($10 \log_{10}$) basis of the perceived noise levels of the noises that occur more or less regularly during each 24-hour day over a period of some months; and (*b*) there is a greater sensitivity, due to sleep interference, on the part of people during the night than the day to noise that is equivalent to a 10 dB difference in noise level.

TABLE 65

Noise-Control Requirements in Building Codes of the World. From Waterhouse (854).

Country	Date of Code	Sound-insulation requirements, dB					Other noise-control require-ments
		Dwellings			Class-rooms	Hos-pitals	
		Party walls	Party floors				
			Air-borne sound	Impact sound			
Germany.......	1938 Revised 1953	48	50		--	--	--
Canada........	1941 Revised 1953	50	50	--	40	40	--
	1954	45	45				
Britain.......	1944	55	55	15 wood 20 conc.	45	--	Site-noise figures
	Revised 1948						
	Revised 1954	I 46	46	*			
		II 41	41	*			
Sweden........	1946						
	Revised 1950	48	48	*	44	48	Yes*
Norway........	1948	50	50	12	44	50	--
Netherlands...	1952	a 52	52	--	--	--	
		ab 50	52				
		b 50	50				

The above figures are given for comparison purposes and may be read as average trans-mission-loss figures, except for the impact figures, which give the required sound-level improvement over a bare concrete floor. In the codes some of the above fig-ures are expressed differently, as explained in the Waterhouse text.

* See Waterhouse Text.

In the original version of CNR it was suggested that corrections be made to measured SPLs of a noise to take into account apparent different sensitivities people of different socioeconomic levels or neighborhoods have with respect to noise. It was later surmised that this supposed greater sensitivity was possibly an artifact due to the lesser abilities of some groups of people to complain about noise than others, and these corrections for socioeconomic factors were dropped from the calculations of CNR. The empirical fact remains that the greater numbers of complaints and legal actions with respect to noise come from the higher rather than lower economic residential areas having the same CNRs. Various objective measures of the noise have been used as the basic unit of

TABLE 66

Proposed Standards for U.S. for the Sound Insulation of Party Walls and Floors in Apartments, Semidetached Houses, and Row Houses. From Waterhouse (854).

Type of Sound	Class A (standard), dB	Class B (minimum), dB
Air-borne sound insulation		
Party walls and floors separating the living rooms and bedrooms of one dwelling from the living rooms, bathrooms, and kitchens of adjacent dwellings, shall have a transmission loss of at least......	50	40
All other party walls and floors shall have a transmission loss of at least......	45	40
Impact-sound insulation		
All party floors shall have an impact-sound insulation of at least......	15	10

Proposed Standards for the Sound Insulation of Walls and Floors in Hospitals and Schools.

Type of Sound	Class A (standard), dB	Class B (minimum), dB
Air-borne sound insulation		
Walls and floors shall have a transmission loss of at least.....	50	40
Impact-sound insulation		
Walls and floors shall have an impact-sound insulation of at least......	15	10

NOTES: a. The airborne-sound insulation figures cited in this table are to be taken as the average transmission loss values of partitions, over the frequency range 125 to 4,000 Hz.

b. The impact-sound insulation is the decibel difference between the sound levels measured under test conditions for the floor in question and a 4-in. floor of bare concrete.

TABLE 67

Relation between NC Values for Background Noise in Residential Rooms
and Complaints as Estimated by Blazier (72)

		Living Area				Bedroom		
		NC	dB(A)	dB(D)		NC	dB(A)	dB(D)
No complaints	A	--				--		
		37	40	47		30	33	40
Occasional complaints	B	37	40	47		30	33	40
		42	45	52		35	38	45
Frequent complaints	C	42	45	52		35	38	45
		52	55	62		45	48	55
Unlimited complaints	D	52	55	62		45	48	55
		--				--		

estimating the perceived noisiness of individual sounds for the computation of a CNR (*a*) Noise Rating Contour, (*b*) energy in the 300-600 Hz band, and (*c*) PNdB, dB(D'), and dB(A').

Except for certain details and terminologies, the CNR scheme is very much the same as the Noise and Number Index (NNI) procedure developed by McKennell (536) in Great Britain, and the $\overline{Q}$ method developed by Bürck *et al.* (105) in Germany. Formulae and procedures for the calculation of CNR will be given in detail in Chapter 11. The general formulae for NNI, and $\overline{Q}$, are as follows:

1. Noise and Number Index (NNI) = Average Peak PNdB-80+15 $\log_{10}N$, where N is the number of occurrences of a sound having a peak level of 80 PNdB or more. Averaging to be done on 10 $\log_{10}$ antilogs of Peak PNdB's.

2. $\overline{Q} = \frac{1}{\alpha} 10 \log_{10} [\frac{1}{T} \Sigma 10^{\alpha Q_k} X \tau_k]$

where Q_k represents a noise level, such as PNdB; $1/\alpha$ can be a function of Q_k or a constant; T is total period of time, and τ_k is the time durations of classes of k.

Traffic Noise Index (TNI)

Recently Langdon and Scholes (485) and Griffiths and Langdon (328) have proposed a method for summing, over a 24-hour period, exposures to road-traffic noise. The unit of measure, called the Traffic Noise Index (TNI), of their method is calculated according to the following formula:

TNI = 4 · (level in dB [A] exceeded by 10% of the traffic noise minus the level in dB[A] exceeded by 90% of the traffic noise) + 90% level − 30.

Note: The distribution of noise levels in percent is for the 24-hour period.

TNI provides a measure that tends to weight the extreme levels, as does CNR, but does not take explicitly into account the number or duration of the noises, as does CNR. Thus it would seem not to provide as general a description of noise environments as does CNR. Nevertheless, TNI does correlate well, (see Table 68) with attitude scores of dissatisfaction that were obtained from a number of neighborhoods (as shown in Table 69). However, the TNI procedure was derived, apparently, from a consideration of the same dissatisfaction scores also shown in Table 69 and validation of the procedure on a set of independent data is in order.

Table 68 shows that $\bar{Q}$ (which for a given period of time during the day is comparable, relative to the comparative ratings that would be given among the various noise environments, to CNR) correlates as well for certain periods of the day as TNI with the median noise dissatisfaction scores obtained from different neighborhoods. However, $\bar{Q}$, and in particular TNI, correlated best with the dissatisfaction scores when taken over a period of 24 hours.

It is interesting to speculate how well CNR, which would provide a differential weighting to the noise levels present for the hours 10 P.M. to 7 A.M., might correlate with the attitude survey scores. Unfortunately, the amount of traffic flow as a function of the time of day (as is necessary for the calculation of CNR) is not available from the published study of Griffiths and Langdon. It is also worthy to note that the absolute levels and ranges of noise levels are much less than those for aircraft noise in neighborhoods near airports.

TABLE 68

Product Moment Coefficients of Correlations of $\bar{Q}$ Values and TNI with Median Dissatisfaction Scores. From Griffiths and Langdon (328).

	Day	Evening and Night	24 Hours
Q Value	0.59*	0.57*	0.64*
TNI	0.42	0.68*	0.88*

*   p < 0.05, n = 11

** p < 0.09, n = 11

TABLE 69

Mean Sound Levels [in dB(A)]

Dissatisfaction Scores and Daily Traffic Flow by Site. From Griffiths and Langdon (328).

Site	Number Interviewed	Mean Sound Level dB(A)			Traffic Noise Index	Median Dissatisfaction Score	No. of People Described by Median	Mean Daily* Traffic Flow Range (Nov.-Dec. Figs.)
		10%	50%	90%				
Main Survey								
1	93	73	64	54.5	98.5	6.21	65	5
2	50	63	57	51	69	2.91	44	3
3	97	61.5	53.5	48	72	3.58	63	1
4	80	75.5	68	61	89	4.43	68	4
5	79	73.5	67.5	62.5	76.5	4.43	60	7
6	140	67	58	51	85	4.74	126	3
7	65	67	57.5	49	91	5.5	50	2
8	76	67	59.5	52	82	4.75	64	3
9	71	69	61.5	55.5	80.5	4.45	46	4
10	68	64	56	49	79	3.13	62	2
11	74	74.5	66.5	61	85	4.88	60	6
Pilot Survey								
12	97	69	63.5	56	77	2.63	72	4
13	101	74	66	60	84	6.08	85	7

1 * (1) 0-400   (2) 4001-8000   (3) 8001-12,000   (4) 12,001-20,000   (5) 20,001-30,000   (6) 30,001-40,000

(7) 40,001 + (vehicles/day)

Noise Pollution Level (NPL)

Robinson (686a) has made explicit the concepts involved in Griffiths and Langdon's formula for TNI that (*a*) people adapt to a steady background noise level, and (*b*) a fluctuating noise level is more annoying than a steady-state level. Robinson demonstrated, using certain assumptions regarding the duration of aircraft noises, that Estimated Effective PNLs multiplied by the standard deviation of these levels during a given period of exposure appeared to fit well attitude and annoyance ratings of road noise (Griffiths and Langdon [328]) and aircraft noise (McKennell [536]). The formula for this calculation which was designated the Noise Pollution Level (NPL) is: NPL equals the mean EPNL + k · σ where σ is the standard deviation of the EPNL values around the mean EPNL.

The use of the difference between the 10% and 90% levels in the calculation of TNI or the standard deviation value of NPL provides a measure of the noise environment not necessarily directly furnished by CNR or $\overline{Q}$, and which may be a factor in dissatisfaction with noise. In essence, the smaller the difference between the 90% level and the peak 10% level, or the smaller the standard deviation, the lower the TNI and NPL. This would imply that there is some adaptation or shift of threshold of dissatisfaction as the general noise level, keeping peak level constant, is increased in a neighborhood. This notion is consistent with the aforementioned results of Spieth (759) and Pearsons (608) and the previously discussed effects of "onset-duration" in that a high background noise would tend to reduce the onset duration of extruding noises. However, Robinson *et al.* (693) found that the absolute level of aircraft noise and not the background level controlled judgements of the noise.

In the original forms of CNR, this factor was recognized by adding "corrections" to the CNR values depending upon the level of the background noise and peak factor of the intruding noise; in the present use of the CNR no such corrections are used but rather somewhat higher tolerable limits are allowed for areas having more background noise; for example, the tolerable limits for rural vs. city residential areas are set, for this reason, at somewhat different values of CNR.

However, this concept of a generally favorable effect of background noise leads to the seemingly absurd conclusion that increasing the number of occurrences of noises that are greater than 90% or so of the peak would result in increased satisfaction with the noise environment. It is observed that people in the presence of a high level of background noise, say near a factory or on a street heavily travelled in the daytime do not notice the noise of an aircraft flying overhead during the daytime, but are bothered by the same aircraft noise at night. It is sometimes implied that this fact demonstrates that is is not the absolute but the relative level of noise that bothers people. While this contention is undoubtedly partly true, the general behavior of people over time suggests that a more likely, and previously mentioned, interpretation of the phenomenon is that the absolute level of annoyance due to noise is rather high to begin with in the high background noise level, and the aircraft noise cannot add much to the general level of annoyance.

Also, the validity and practicality of this particular aspect of TNI and NPL is probably dependent upon the long-term activities of the people involved in a given noise environment and their desire to continue their activities in a normal manner. For example, adaptation to high background noise may occur by virtue of, for example, a person not engaging in voice communications, or because the noise masks distracting or startling sounds. Such beneficial effects are probably not to be expected in those situations where the steady background noise itself masks the reception of wanted speech or other wanted signals.

Relation between CNRs and Human Behavior and Attitudes

Figures 238 and 238a summarize the general relations between CNR, and related measures of noise, for a noise environment and various human reactions to that sound environment (702, 771, 820). These relations are extrapolated from and consistent with laboratory and field research and actual behavior of people in communities, some data for which, drawn from Table 70 and Fig. 239, are plotted on Fig. 238. Somewhat different words, but conveying essentially the same meanings, have been used in the past on the vertical scale of Fig. 238 (625, 626, 702, 769, 771).

It is estimated that there is an initial adaptation or familiarization over a period of the first several months of exposure to a given noise environment that reduces reactions to the noise by an amount equivalent to a reduction of about 10 CNR. The behavioral reactions cited on Fig. 238 are those typically observed after months and years of daily or near daily exposure to the respective noise environments described by CNR.

It is presumed that the range in reactions of people to a given noise environment as illustrated on Fig. 238 is a joint function of: (*a*) individual and group differences in attitudes towards noise and the sources of the noise, and abilities in the expression of complaints and other related behavior; and (*b*) variations in CNR exposure conditions among rooms, homes, buildings, and open areas presumed to have equal CNRs. Physical measurements, in regard to this latter point, are often inadquate or insufficient. It might be again noted that the spread or range shown in Fig. 238 can be greatly reduced by applying a socioeconomic corrections to CNR such that the tolerability of high income neighborhoods is 10-15 CNR units or so less than low income neighborhoods (W.G. Galloway, personal communication, 1969).

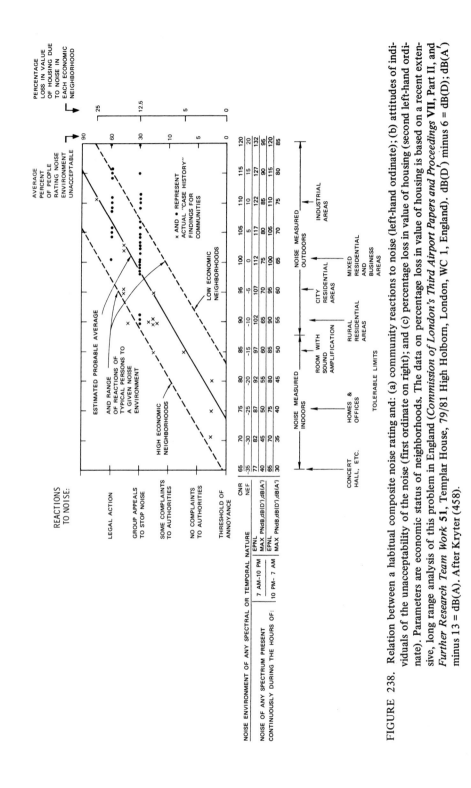

FIGURE 238. Relation between a habitual composite noise rating and: (a) community reactions to noise (left-hand ordinate); (b) attitudes of individuals of the unacceptability of the noise (first ordinate on right); and (c) percentage loss in value of housing (second left-hand ordinate). Parameters are economic status of neighborhoods. The data on percentage loss in value of housing is based on a recent extensive, long range analysis of this problem in England (*Commission of London's Third Airport Papers and Proceedings* **VII**, Part II, and *Further Research Team Work* **51**, Templar House, 79/81 High Holborn, London, WC 1, England). dB(D′) minus 6 = dB(D); dB(A′) minus 13 = dB(A). After Kryter (458).

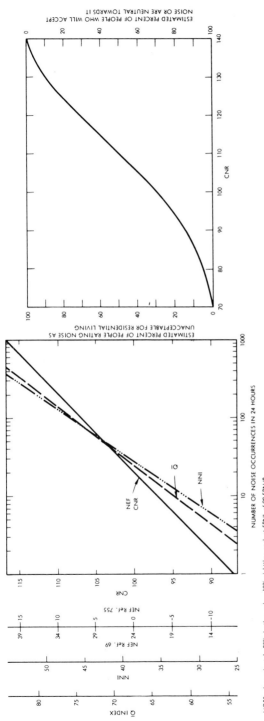

NOTE: Approximately 90% daytime noises, 10% nighttime each of EPNL of 98 EPNdB

FIGURE 238a. Left Graph: Typical relations between various composite noise rating schemes. There is some confusion in the literature concerning NEF-CNR relations caused by (a) the existence of two different constants (76 and 100) in the formulae for the calculation of NEF (refs. 69 and 755), and (b) the use in ref. 69 of Max PNL in the calculation of CNR, and EPNL in the calculation of NEF. In ref. 755, this book, and the above graph, CNR and NEF are both calculated on the basis of EPNL. Right Graph: Estimates of attitudes to be expected from nonfear provoking noise in residential living areas. Derived from data in refs. 81, 274a, 478, 536, and 805a. The reported percentages of people expressing negative attitudes about aircraft noise were reduced for this function in an attempt to reflect attitudes based on factors other than fear (speech interference, arousal, etc.—see Figs. 220 and 220a for examples).

The percentage of people on the right-hand ordinate of Fig. 238 who can be expected to have strong negative attitudes towards the noise when exposed to a given outdoor CNR is deduced and extrapolated from the attitude studies conducted at Edwards Air Force Base (473), Oklahoma City (81), London (536), Netherlands and France (see Fig. 238a). A distinction must be made between the attitudes (right-hand ordinate, Figure 238) people have towards noise in their living environment and what they do about it (left-hand ordinate, Fig. 238). The relations between CNR and attitude is usually more similar for all communities regardless of economic status (perhaps this is best represented by the solid "average" curve of Fig. 238) than is the relation between CNR and complaint behavior. See caption of Fig. 238a for further details.

Validity and Purposes of Real-life and Laboratory and Field Studies

In a recently reported study (358, 805a) it was found that certain psychological and geographic factors were about as equally important to the prediction of attitudes and complaints of aircraft noise as was CNR (see Table 52). CNR, based on Max PNdB, was slightly superior to other composite noise rating schemes, such as NNI. However, there were no statistically significant differences found between most of the noise measurement procedures. It was probably impossible to show any statistically significant differences possibly present among these units of measurement in view of (a) the lack of close correspondence between the sample of measured noises, the noise as actually heard in homes and the attitudes of individuals, and (b) constraints and attenuations of the basic correlation data due to the fact that a wide range of CNRs (or EPNLs) were not found for each value or level, taken singly and jointly, of the psychological, sociological, and demographic variables explored by attitude sampling in the community, real-life studies.

The primary purpose of these sociological studies is obviously not to determine how noise should be measured physically, i.e., dB(A) vs. (C), (D), or PNdB, etc. Rather, the major goal of such studies is to learn what factors contribute significantly to negative attitudes and complaints about noise so that some steps, if appropriate, might be taken to influence these attitudes and complaints. An examination of the list of factors in Table 52 indicates that some might lend themselves to such use by the aviation industry or the government, such as "fear of aircraft crashing into neighborhood," or the lesser important factors, "belief in misfeasance by aircraft or airport operators," and "extent to which the airport is considered to be important to the local economy" (this last factor, by the way, is to the effect that the people who were more upset about the noise also tended to consider the airport important to the local economy).

The other factors in Table 52—susceptability and adaptability to noise, distance from airport (this was not a simple matter of being closer vs. farther away), and city of residence—would appear to be not amenable to significant outside control; findings with regard to these two variables may be perhaps peculiar to the specific cities studied. The remaining significant factor listed in Table 52 influencing the attitudes and complaints, and presumably the condition necessary to the operation of all the factors, is that of noise, the CNR.

The purpose of a number of psychological studies conducted in the laboratory and the field has been to learn what physical characteristics of the noise need be measured and controlled in order to reduce the negative effects of the noise on people. It is perhaps unfortunate that real-life sociological studies are not the most efficient and effective way to develop the proper procedures for physical noise measurement in order to effect proper control of the noise. It is, at the same time, exceedingly important that there is nothing inconsistent between the results of laboratory and field studies concerned solely with the relations between the physical aspects of noise and the effects of noise on man, and the real-life studies of the effects of noise and factors other than noise on attitudes and complaint behavior. The consistency of the results between these two types of studies in this regard is perhaps surprisingly good, particularly in view of the fact that the actual noise to which individuals are exposed in the real-life sociological studies is, of necessity, only approximately known, and the importance, in real-life, of factors other than noise upon attitudes and behavior towards the noise.

Nevertheless, the question is often raised as to what is the value of using one unit of physical measurement (say dB[D]) rather than another unit (say dB[A]) that is only 1 dB more accurate in predicting the judged perceived noisiness of sounds in laboratory and field tests but not in sociological studies. As mentioned earlier, in setting tolerable limits in real-life, a difference in 1 dB does become crucial at the boundaries of the limit, and, also, a change in actual perceived noise level by 1 dB should eventually be noted so that all the curves shown on Fig. 238 would be shifted accordingly. An example of the latter situation would be the introduction of a new commercial aircraft or type of automobile whose noise measured the same as present-day comparable vehicles on dB(A), but was truly 1 dB worse according to judgment tests. It would not, in our opinion, be prudent, if that vehicle became the dominant source of noise, to take the position that in time the change in noisiness would not be reflected in the attitudes and behavior towards the noise on the part of the people exposed, even though it would be difficult to prove this by means of correlations, perhaps primarily because it is so difficult to measure the noise environment for individuals over long periods of time.

The above discussion should not be interpreted as suggesting that EPNLs based on one-third octave band spectra (PNdBs), or even overall dB(D)s, are necessary either for use in sociological survey work related to noise or for the

TABLE 70

Summary of Case Histories of Responses to Noise in Residential Areas
From Rosenblith and Stevens (702).

No.	Description of Facility and Noise	CNR*	Predicted Average Response	Actual Response
1	Large wind tunnel in mid-west	110	Vigorous legal action	Municipal authorities forced facility to shut down
2	Large wind tunnel in mid-west	100	Threats of legal action	Vigorous telephone complaints and injunction threats. Management took immediate steps to lessen noise
3	Exhaust for air pumps, factory in industrial area	95	Strong complaints	Lodging house owner entered complaints with client and with local Dept. of Health
4	Engine run ups aircraft mfg. plant	80	Less than mile annoyance	No complaints reported by management. Operations restricted to daytime only
5	Airport ground run ups	95	Strong complaints	Complaints by civic organizations, individual telephone calls and letters of complaints
6	Aircraft in flight near airport	95	Strong complaints	Vigorous complaints by letter and telephone. One town attempted to prevent passage of aircraft
7	Aircraft engine mfg. plant test cells	85	Mild annoyance	No complaints reported for daytime operation; a few for operation after 11 p.m.
8	Loading platform with trucks, men shouting, etc.	100	Threats of legal action	Vigorous complaints to management. Acoustical consultant called in by firm
9	Transformer noise in very quiet res. area	105	Between threats of legal action and vigorous legal action	Injunction threats
10	Large fan at power company; single freq. components	90	Strong complaints	Residents complained consistently, consultants called in to advise on noise control
11	Weapons range, intermittent firing, 3-sec bursts several times per day	100	Threats of legal action	Vigorous complaints from nearby residents for winter operation

*Estimated on basis of "level rank" band spectral measures as given in original reference.

purpose of monitoring or policing environmental noise. However, it does appear from both laboratory, field, and real-life tests that, for those two purposes, EPNL based on overall sound measurements with the recommended D weighting would be the simplest and, on the average, most consistently accurate unit to use.

Specific Examples of Development and Application of CNR

Table 70 from Rosenblith and K.N. Stevens (702) gives some "case history" findings that illustrate the general relation between CNR and behavior of people. It should be noted that the CNR numbers given in Table 70 were determined by translating their original "Noise Level Rank" calculations into estimated comparable CNR values. Galloway and von Gierke (274) recently prepared an excellent review of the development of procedures for evaluating aircraft noise and presented the results of recent studies of community behavior in response to aircraft noise, as shown in Fig. 239.

A more recent example of how CNR seems to predict community reaction is illustrated in Fig. 240 which shows CNR contours for small jet aircraft on takeoff and landing at a small noncommercial airport. Encircled on Fig. 240 are the approximate groupings of residences of citizens who brought legal suit (and lost) against operators of the airport for alleged reduction in the value of their property and for adverse effects upon their well-being. The complaints were against the Santa Monica Airport, with respect to the jet aircraft noise and not the noise of the propeller aircraft which constituted the greater percentage of operations but which, because of their low level of intensity, did not significantly contribute to the CNR of that community.

HL's of People Near Santa Monica Airport

Auditory tests by Gottlieb and Gottschalk (311) revealed that some of the housewives and husbands living near the Santa Monica Airport had a statistically significant hearing loss at 4000 Hz, but not at other frequencies, and that some of the husbands (those not exposed to noise in their work) showed no significant loss (see Table 71). The damage prediction procedures discussed in Chapters 5 and 6 and information pertaining to the relation between noise spectra and pattern of hearing loss (that the greatest loss occurs usually at a somewhat higher frequency than the frequency of maximum noise energy) indicate that these audiometric findings are to be expected. For example, the jet noise during run-up (see Fig. 241) measured 106 dB(A) and often lasted at least three minutes per run-up, five times per day. The CDR, calculated from the octave

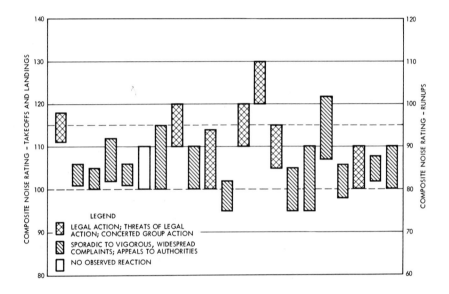

FIGURE 239.   Reactions of people in communities exposed to aircraft-noise environments
of different CNR values. The height of the bars represents the range of
CNR values taken over a given neighborhood. From Galloway and von
Gierke (274).

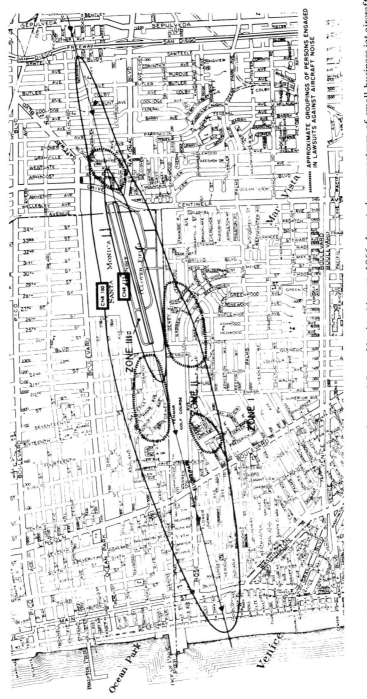

FIGURE 240. CNR contours for small jet aircraft around Santa Monica Airport, 1966. Insert shows spectrum of small business jet aircraft involved during run-up and, when over homes, during takeoff. From testimony, Santa Monica lawsuit (575).

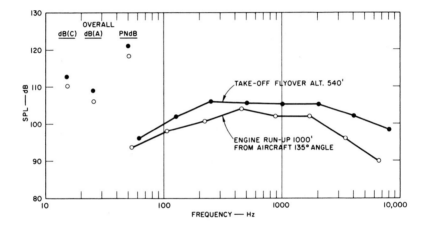

FIGURE 241.   Spectrum of run-up and flyover noise of small business jet aircraft using
               Santa Monica Airport (575).

band spectra, for the engine run-up noise, assuming 30 minutes between run-ups, would be 10, indicating that a 10 dB hearing loss beyond that of presbycusis at 4000 Hz could be expected in about 25% of the people exposed to that noise. A CDR value of 10 would be present for the flyover noise if one assumed that the peak level would be present for two seconds and that there was an average of five minutes between flights and 200 flights per day. Since there was an average of only five jet takeoff operations per day at this particular airport, and because the noise from small propeller aircraft that also used this airport was more than 20 PNdB less intense than the jet noise, it would appear that the damage risk to hearing, and the apparent actual loss, was due almost exclusively to the jet engine run-up noise.

It should be noted that the housewives, when indoors, would be protected from the noise to a considerable extent, and the number of years of exposure is not known. On the other hand, the jet engine run-up levels at 50 feet from the aircraft (about the closest distance of a home from the runways) would be about 10 dB more intense than those given in Fig. 241. Also, increases in the duration of the engine run-ups, which were claimed by some people, would likewise increase the damage risk. For reasons such as this, and the small number of people tested, it is not possible to prove that there was a causal relation between the measured hearing levels of these particular people and the aircraft noise. It would appear, nevertheless, that the hearing levels of the women and men tested are in reasonable agreement with the type and degree of hearing loss to be expected from exposure to the jet aircraft run-up noise present outdoors at some homes near the Santa Monica Airport. The defense argument in this court case was that inasmuch as a committee of the American Association of Opthomologists and Otolaryngologists (510) recommends that hearing losses at frequencies above 2000 Hz, no matter how severe, not be considered as constituting hearing impairment for speech (see Chapter 4), the possible aircraft noise-induced deafness on the part of the residents near the airport was of no consequence; in my opinion, this argument is unjustifiable, particularly in this situation.

The SST

On the basis of the aforementioned psychological research on the effects of sonic booms, it is possible to apply the CNR concept to predict the probable reaction of people to the proposed supersonic transport aircraft (Kryter [460]). To do so requires an estimation of the flight paths and number of flights of commercial SST that would be anticipated over the United States when these aircraft would be presumably placed in full operation, sometime in the 1980s. These calculations and estimates have been accomplished and Table 72 gives the number of people in the United States who would be exposed to sonic booms

TABLE 71

Mean and Median Hearing Loss and Statistical Probability of Its Significance of a Sample of Adult Persons Living Near Santa Monica Airport

Hearing loss is taken as the average hearing level of both ears minus median 1965 U.S. National Health Survey for same age of each person tested. From Gottschalk and Gottlieb (311).

	14 Housewives Exposed to Santa Monica Jet Noise		7 Men Not Exposed to Noise at Their Work		7 Men Exposed to Noise at Their Work		5 Men Exposed Primarily to Santa Monica Jet Noise	
	2000 Hz	4000 Hz	2000 Hz	4000 Hz	2000 Hz	4000 Hz	2000 Hz	4000 Hz
M = Mean Hearing Loss Beyond National Health Survey Average for Persons of Same Age	6.0	12.0	3.5	9.0	12.5	24.5	12.0	20.0
Median Hearing Loss	5.5	11.5	0	7.5	11.5	22.0		
σ = Standard Deviation of Hearing Losses	4.6	6.0	7.0	8.5	8.5	8.0	6.5	7.0
M/σ	.65	1.55	.25	.65	.75	1.65	.9	1.45
Probability of Observing Such a Large Mean Hearing Loss by Chance Fluctuation	.10	.02	.30	.14	.07	.001	.04	.002

(A probability less than .10 is considered barely significant, and a probability less than .05 is considered strongly significant.)

TABLE 72

Estimated 1975 Population under Each Sonic Boom Category for Great Circle Routing of Medium (1200-1800 Miles) and Long-Range (2000-2400 Miles) SST Routes in the U.S.A.

Because of overlapping boom paths across the country some relatively small regions of the country will receive many more booms per 24-hour period than will other regions. It is seen that about one half of the total numbers of people given in the table would receive ten or more booms per day, and the remainder would receive less than ten booms per day. From Kryter (460). (Copyright 1969 for the American Association for the Advancement of Science.)

Expected Number Booms Per 24-Hr Period	POPULATION Boom Path 50 Miles Wide	(CNR)	POPULATION Boom Path 25 Miles Wide	(CNR)
1-4	52,400,000	92-103	26,200,000	95-103
5-9	25,200,000	98-106	12,600,000	101-106
10-19	19,500,000	101-109	9,750,000	104-109
20-34	29,400,000	104-112	14,700,000	107-112
35-51	2,900,000	107-115	1,450,000	110-115
TOTAL	129,400,000		64,700,000	

from the overland operation of the SST. It was assumed for these calculations that the SST would carry 50% of the air passengers on city-pair flights in the United States longer than 1200 miles.

Also on Table 72 are given the CNR environments that would be experienced by the various numbers of people. By reference to Fig. 238 it is obvious that these CNR values indicate that there would be strong complaint and legal action involving millions of people against the sonic booms, if the presumed SST operations were carried out. Because of the anticipated negative public reaction to the sonic boom even after years of adaptation, it appears that the SST will be used primarily only over water and sparsely populated land areas.

Tolerable Limits of Noise Exposure

It is sometimes suggested that a common rating scheme for noises from different sources is not realistic because, for example, the noise from a washing machine at a given level that is acceptable in the home is not acceptable in the concert hall or the noise from an airplane outdoors is not to be compared with the noise of an air-conditioner, etc. While some large differences in tolerability to different noises are clearly attributable to the various values that the different sources have for the listeners, it is suggested that within limits the particular functions to be performed in a given space in the long run set noise tolerability limits, and not the source (barring emotional connotations) of the noise per se. The phrase "tolerable limit" is used here to indicate the maximum amount of noise, regardless of its source, that can usually be present in an area and have the area usable, but not ideal, for the typical activities performed in the area. Suggested tolerable limits for noise in various rooms or areas are indicated on the bottom of Fig. 238 and in Tables 40 and 41.

Some allowances are made in Fig. 238 and Tables 40 and 41 for background noise level, e.g., the limits for rural vs. urban areas. As discussed previously, although a direct measure of speech masking, such as AI, could possible be more appropriate for a space devoted more or less exclusively to speech communications, CNR based on units of PNdB, dB(D), and, to a lesser extent dB(A), is a good predictor of speech interference by noise because the correlations with AI are so high. Further, loudness, distractiveness, and for impulses, startle, are also contributing factors to ratings of unacceptability of the noise environment.

Value judgments rendered by people in real-life situations about (a) the detrimental effects of noise, and (b) the beneficial effects to a community of the sources producing the noise are represented by the curves and data points shown in Fig. 238 and Tables 40 and 41. However, the interpretation and application of these data for the setting of environmental noise limits that are economical and, at the time, acceptable to the public in a specific locality is a task that may often require special information and judgment about the specific locality or neighborhood.

CNR (and NEF) Tolerable Limits of Exposure to Aircraft Noise

Attention is invited to the following chart which is incorporated in the airport planning guide developed in the United States (76). This chart proposes that three zones be identified from CNRs as measured or predicted for given neighborhoods:

Zone I: CNR<100. Essentially no complaints would be expected. The noise may, however, interfere occasionally with certain activities of the residents.

Zone II: $100 \leqslant CNR \leqslant 115$. Individuals may complain perhaps vigorously. Concerted group action is possible.

Zone III: CNR>115. Individual reactions would likely include repeated, vigorous complaints. Concerted group action might be expected.

The above zones, particularly Zone I, would appear to greatly underestimate typical behavior of people exposed to noise as shown in Fig. 238 and Table 69. The rating of noise environments by these Zones can lead also to implausible conclusions. For example, a neighborhood exposed to 32 day-time aircraft flyover noises each having an EPNL of 97 EPNdB, would have, from that noise source, a CNR of 100 and be classified as falling in Zone II; however, a CNR of 114, also Zone II, would be reached from near 1000 exposures to the same aircraft noises. It is hard to believe that increasing the daily number of noises of this level, 97 EPNdB, from 32 to 1000 would be accepted with equal equanimity by a given neighborhood.

The originators of the above zone chart did not necessarily intend the zones to be used as "criteria." Nevertheless civil and legal agencies are likely to make decisions in terms of these zones, and as a result, as with previously discussed damage risk limits, the highest noise level exposure of a zone becomes, in fact, the tolerable limit to be allowed for noise control codes or regulations. We would submit that the weight of the sociological, psychological and political evidence is that in typical residential communities an appreciable percentage (approximately 10%, see Fig. 238) of the people will complain, or feel like complaining, vigorously when the CNR reaches 90 and that legal and other group actions against the noise will start with CNRs of 90 and be nearly universal with CNRs above 100, unless suppressed because of strong economic or political forces, or sparseness of people exposed. Further, when some percentage of the people, such as 10%, feel the noise is unacceptable, the remaining 90% will dislike the noise or be neutral towards it (see Fig. 238a). It would appear that the zoning given in the subject "planning guide" is based primarily on what are descriptors of the behavior in low economic or depressed areas and not that in the average or higher economic residential areas.

A new land use compatibility chart was recently prepared, under the auspices of the Federal Aviation Administration, for use with aircraft noise, as shown in Table 72a. The Relation Between NEF and CNRs, as given in Table 72a, are explained in Fig. 238a. Table 72a appears to be more realistic than the earlier airport planning guide, but the suggestion of a real compatibility of a CNR of 90-100 with residential dwellings is not consistent with the experience of average and better than average residential areas.

Rational Criteria for Estimating Tolerable Limits of Noise Exposure

It is perhaps constructive to speculate what noise levels the average person might be expected to select as acceptable if no unusual conflicting or competing

TABLE 72a

Land Use Compatibility Chart for Aircraft Noise (Takeoff and Landing). After Ref. 69,755.

NOISE SENSITIVITY ZONE	CNR	NEF 755	NEF 69	LAND USE COMPATIBILITY								
				Residential	Commercial	Hotel, Motel	Offices, Public Buildings	Schools, Hospitals Churches	Theaters, Auditoriums	Outdoor Amphi-Theaters, Theaters	Outdoor Recreational (Non-spectator)	Industrial
I	Less Than 90	Less Than -10	(Approx.) Less Than 24	yes	yes	yes	yes	yes	Note (A)	Note (A)	yes	yes
II	90-100	-10-0	24-34	yes	yes	yes	yes	Note (C)	Note (C)	no	yes	yes
III	100-115	0-15	34-49	Note (B)	yes	Note (C)	yes	no	no	no	yes	yes
IV	Greater Than 115	Greater Than 15	49	no	Note (C)	no	Note (C)	no	no	no	yes	Note (C)

NOTE (A) - A detailed noise analysis by qualified personnel should be undertaken for all indoor or outdoor music auditoriums and all outdoor theaters.

(B) - Case history experience indicates that individuals in private residences may complain, perhaps vigorously. Concerted group action is possible. New single dwelling construction should generally be avoided. For high density dwellings (apartments) construction, Note (C) will apply.

(C) - Avoid construction unless a detailed analysis of noise reduction requirements is made and needed noise control features are included in building design.

values are involved, and he had no special antipathies against the source or cause of the noise. It is suspected from a consideration of the general pattern of the various data on the subject that the following might be the result:

1. *Speech Interference Noise Indoors, Daytime.* Objections will start when the noise reaches levels that interfere somewhat with the reception of conversational level speech: a noise level of about 50 PNdB, 37 dB(A), or 44 dB(D). Webster (860) notes that men aboard U.S. Naval vessels "get used to" and are eventually not bothered by noise; however, Beranek (comments at a conference on noise, University of Washington, March 27, 1969) has observed that under such conditions the Naval personnel learn how to perform their tasks without talking or by using a minimum of signals.

2. *Sleep Interference Noise Indoors, Nighttime.* It is surmised that noise reaching levels of 40 PNdB, 27 dB(A), or 34 dB(D) will be resented as somewhat interfering with normal sleep or the process of going to sleep.

3. *Damage Risk to Hearing Noise Outdoors or Indoors Where Speech Communication Requirements Are Modest.* Noise reaching levels 80 PNdB, 67 dB(A), or 74 dB(D) at the ear will be resented. Noise present continuing at this level can eventually cause some 10 dB or so of hearing loss at frequencies above 2000 Hz. It is hypothesized that somehow people sense the potential auditory fatigue effect and are naturally, whether with or without conscious knowledge of its long-term damage risk, concerned about noise above these levels regardless of any masking of speech.

4. *Startle to Noise Indoors or Outdoors.* Noise that increases at rate of 40 or more dB per 0.5 sec will be heard as impulsive and will cause some startle reflex or reaction. It is hypothesized that complete psychological and possibly physiological (see Part III) adaptation to this startle does not occur and that such impulses will be resented.

Whether one can gainfully work and live in higher or lower noise levels than these suggested natural tolerable levels is, of course, influenced by special conditions, needs, and factors of a given situation. However, these purely rationalized tolerable levels are not inconsistent with the tolerable levels of indoor and outdoor noise that would seem to be dictated by the statistical sociological data and behavior of people in real life, as shown in Fig. 238 and Tables 40 and 41.

Finally, a comment might by made about the fact that persons who are found by personality tests to be unusually anxious and perhaps not well adjusted consider noise to be a strong and consistent source of irritation (52). It does not appear, however, that the majority of the complaints and actions taken by individuals and groups against noise are anything but normal reactions. Indeed, the greatest force against noise comes from the higher socioeconomic groups who presumably enjoy generally better mental and physical health and higher standards of living.

Chapter 10

Summary of Methods of Predicting Certain Responses to Noise

Introduction

Part I and the preceding chapters of Part II have attempted to bring together the experimental basis and rationale of procedures for estimating from physical measures of noise certain of its effects on man. Perhaps a summarization of the state-of-the-art is in order. It appears that the state of psychoacoustic research justifies the following conclusions and recommendations.

1. The response of man's auditory system to a sound can probably be adequately described from a knowledge of the rms pressures, taken over 0.5 sec, and measured every 0.5 sec during the sound's occurrence in the one-third octave bands covering the frequency range from 45 to 11,020 Hz (band center frequencies of 50 to 10,000 Hz).

2. It is possible from these band spectral measures to predict with practical accuracy the effects of sound with regard to damage to hearing, masking of speech, perceived noisiness, and community behavior.

3. It is recommended that a set of standard procedures utilizing band spectral measure be used for estimating the effects listed in (2) above.

4. It is recommended that the same set of standard procedures, but utilizing weighted, overall frequency sound level measures, be used as a secondary means for estimating the effects listed in (2). The relation between the secondary and primary units should be determined for each type or class of sound to be evaluated by means of a secondary unit.

A large amount of psychoacoustic data, as is reviewed above, was involved in the development and validation of the primary and the secondary standard procedures to be cited below. It is unfortunate, of course, that there are not available more data that would show beyond doubt whether or not these procedures are valid in all respects. There is no question but that further research will lead to improvements in the understanding of the effects of noise on man and in methods of its measurement. However, it will probably require 5 to 10 years of active research at the rate that such research is presently accomplished, to provide results that would be of sufficient amount to successfully challenge the mass of results obtained to date and to lead to significant changes, if needed, in the concepts and procedures involved.

Summary of Methods of Calculating Units of Noise Related to Quantities of Human Response

Table 73 lists those sound measurement and evaluation procedures that appear (*a*) to provide the physical measures of sound or noise that best relate to the response quantities or attributes listed, and (*b*) to meet recommendations (3) or (4) of the introduction to this chapter. The basic response quantities are the damage to hearing, masking of speech, and annoyance (Preceived Noisiness). The so-called primary physical quantities of predicting these response quantities are: Effective Damage Risk Level (EDRL) and Composite Damage Risk (CDR); Articulation Index (AI); Effective Preceived Noise Level (PNL) and Composite Noise Rating (CNR).

Various units, some labeled, some not, have been developed for expressing degrees or amounts of these quantities. The rank ordering of the units on Table 73, with respect to their accuracy for estimating the various quantities specified, is based on a consensus of experimental data, where available. Where no validation data are available, the rank order is based on an estimate of how well a given unit of measurement would seem to describe the appropriate response characteristics of the auditory system as known from basic research data. The consistency with which a unit operates over all pertinent laboratory and field studies is an important consideration and is taken into account in the rank ordering of the efficacy of the units listed in Table 73. Data supporting these rankings are to be found earlier in the text and in some paragraphs following.

Often the difference between the relative values measured for two noises by two given units may often be no more than 1 dB or so; this raises again the question of whether a difference of no more than 1 dB is significant. As previously discussed, from a statistical point of view, differences of about 0.5 dB for tests of preceived noisiness, probably about 0.25 dB for speech masking and 1.0 dB for shifts in auditory thresholds, can be statistically significant provided proper measurement procedures are followed. From a practical point of view, it can be noted that a 1 dB increase in sound pressure level of noise represents about a 25% difference in acoustic power, and, over typical levels, an increase of about 5-10% in the subjective level of preceived noisiness, a 5% decrease in the intelligibility of speech (PB Word Tests), and a 1 dB increase in the threshold level of hearing.

Comparison of Secondary and Primary Physical Units

The greater specificity of the measurements involved in the units of physical measurement listed first in Table 73, as well as empirical test results, would suggest that these units provide the best basis for predicting the respective human response attributes. If this conclusion is accepted, it is possible to evaluate somewhat the merit of the various proposed secondary units of physical measurement using the primary unit as a reference standard.

In Table 7 are listed the octave band levels of the noise from the following sources:

1. Thermal Noise Generator (spectrum approximately −6 dB/oct above 100 Hz)
2. Thermal Noise Generator (spectrum approximatley uniform)
3. Motor Generator
4. Commercial Jet Aircraft
5. Wood Planer
6. Trolley Buses
7. Automobiles

The magnitudes of each of the primary and secondary physical units for the evaluation of noise that are listed in Table 7 were determined for these noises when their overall sound pressure was set at various levels.

From these results, tables were made showing the magnitudes of the secondary units for each of the noises when the EDRL, CDR, AI, and EPNL values, as estimated by the respective primary units, were the same for each of the noises. A secondary unit that was the same for each of the seven noises when the primary unit was the same would obviously be a perfect substitute for the primary unit. The greater the differences among the values of a secondary unit under the circumstances described, the less effectively could that unit be used in the place of the primary unit.

Tables 74, 75, and 76 show the average deviations, the standard deviations, and the rank order for each of the secondary units from perfect agreement with each of the primary units, EDRL, AI, and EPNdB respectively. It is seen that the rank ordering of the various units is comparable to the rank ordering of merit of these units according to tests, as cited in previous chapters of the text, of auditory threshold shift, masking of speech, and perceived noisiness. This, of

TABLE 73

Section I. HUMAN RESPONSE QUANTITIES OF DAMAGE TO HEARING, MASKING OF SPEECH AND ANNOYANCE (Perceived Noisiness) AND UNITS OF SOUND MEASUREMENT TO BE USED FOR ESTIMATING THESE QUANTITIES FROM AN EXPOSURE TO A GIVEN NOISE

HUMAN RESPONSE QUANTITY:	Damage to Hearing	Masking of Speech	Annoyance (Perceived Noisiness)
PRIMARY UNIT OF MEASUREMENT OF RESPONSE:	Complaints and Behavior of People	Complaints and Behavior of People	Complaints and Behavior of People and Communities
SECONDARY UNITS OF MEASUREMENT OF RESPONSE:	Pure-tone Thresholds, Speech Test Scores	Speech Intelligibility Test Scores	Relative Judgments, Ratings and Attitude Surveys
TYPE OF SOUND:	Broadband Sound / Narrow Band or Broadband Sound with tones	Broadband Sound / Narrow Band or Broadband Sound with tones	Broadband Sound / Narrow Band or Broadband Sound with tones
UNITS OF SOUND MEASUREMENT: Listed in Measured or Estimated Rank Order of Accuracy. Units within a Numbered Box of about Equal Accuracy	Effective Damage Risk Level (EDRL) 1. EDRL from Band Spectra applied to DR Contours 2. EDRL based on PNdB, dB(D), dB(A), or Phon	Articulation Index (AI) 1. AI from 1/3 or Octave Band Spectra 2. AI based on: SIL 600-4800 Hz or 700-5600 Hz 3. AI based on: PNdB, dB(D), dB(A), or Phon 4. AI based on: SIL 355-2800 Hz or 300-2400 Hz	Effective Perceived Noise Level (EPNL) 1. EPNdB from 1/3 or Full Octave Band Spectra, EdB(D') 2. EPhon', EdB(A') 3. EEPNdB, EEdB(D'), EEdB(A'), EEPhon' 4. Max PNdB, Max Phon, Max dB(D'), Max dB(A')
	1. EDRL from 1/3 Octave or Narrower Band Spectra applied to DR Contours 2. EDRL, based on PNdB, dB(D) + 5 dB, dB(A) + 5 dB, or Phon +5 dB	1. AI from 1/3 or Narrower Band Spectra	1. EPNdB from 1/3 Octave or Narrower Band Spectra 2. EEPNdB 3. Max PNdB

Section II. HUMAN RESPONSE AND SOUND MEASUREMENT UNITS TO BE USED FOR THE EVALUATION OF TOTAL DAILY SOUND ENVIRONMENTS

HUMAN RESPONSE QUANTITY:	Damage to Hearing		Annoyance (perceived noisiness) and General Masking of Speech	
TYPE OF SOUND:	Broadband Sound	Narrow Band or Broadband Sound with tones	Broadband Sound	Narrow Band or Broadband Sound with tones
UNITS OF PHYSICAL MEASUREMENT: Listed in Measured or Estimated Rank Order of Accuracy. Units within a numbered box are of about equal accuracy	1. Composite Damage Risk (CDR) based on DR Contours 2. CDR based on: PNdB, dB(D), dB(A), or Phon	1. Composite Damage Risk (CDR) based on DR Contours 2. CDR based on: PNdB, dB(D) +5 dB(A) +5, or Phon +5	1. Composite Noise Rating (CNR) based on EPNdB or EdB(D') 2. CNR based on EPhon', EdB(A')	1. CNR based on: EPNdB

Note 1: A "tone" is said to be present in a sound when the sound pressure level in any one-third octave band exceeds the level in immediately adjacent bands by 3 or more dB.

Note 2: Only those measurements listed for a given purpose are deemed appropriate and sufficiently accurate for the use specified. dB(B), dB(C), NC, NCA, NR, and related "tangent band" measurements are not recommended for any of the above purposes.

TABLE 74
Speech Masking

Values of secondary units of physical measurement when for each of the seven noises the value of the primary unit of physical measurement of speech masking, AI, equals 0.5. The speech level for the calculation of AI kept constant at 80dB. The spectra of the seven noises are given in Table 7.

Primary unit of physical measurement AI = 0.5, speech 80dB.

Noise		dB(A)	dB(C)	dB(D₁)	dB(D₂)	dB(D₃)	PNdB	Phon(S)	Phon(Z)	SIL 600-4800 or 700-5600 Hz	SIL 355-2800 or 300-2400 Hz	NC
Thermal -6 dB/Oct	1	76	89	83	81	74	89	89	92	63	69	77
Thermal "Flat"	2	72	71	80	80	75	85	82	87	64	66	69
Motor	3	71	74	78	78	73	84	83	88	64	63	67
Jet Aircraft	4	68	68	75	75	71	80	79	81	62	61	64
Planes	5	70	71	77	77	72	83	80	84	64	64	65
Trolley Buses	6	72	78	77	76	71	83	84	89	63	68	70
Automobile	7	74	83	80	79	73	85	86	92	63	67	71
Mean		71.9	76.3	78.6	78.0	72.7	84.1	83.3	87.6	63.3	65.4	69.0
A.D.		1.9	6.0	2.1	1.7	1.2	1.9	2.6	3.1	0.6	2.4	3.1
S.D.		2.6	7.5	2.6	2.2	1.5	2.7	3.5	4.0	0.7	2.9	4.4
Rank Order		5	11	4	3	2	6	8	9	1	7	10

TABLE 75

Values of secondary units of physical measurement when for each of the seven
noises the value of the primary unit of physical measurement of damage risk to
hearing, DRL, equals 25 (HL at 500, 1000, and 2000 Hz will equal 25 dB). The
daily exposure duration is taken to be 480 minutes continuous. The spectra of the
seven noises are given in Table 7.

DRL = 25, duration 480 minutes continuous exposure.

Noise	dB(A)	dB(C)	dB(D_1)	dB(D_2)	dB(D_3)	PNdB	Phon(S)	Phon(Z)	NC
1	99	112	106	104	97	112	112	115	100
2	97	96	105	105	100	110	107	112	96
3	100	103	107	107	102	113	112	117	97
4	96	96	102	103	99	108	107	109	92
5	96	97	103	103	98	109	106	110	91
6	94	100	99	98	93	106	106	111	92
7	96	105	102	101	95	107	108	114	93
Mean	96.9	101.3	103.6	103.0	97.7	109.3	108.3	112.6	94.4
A.D.	1.6	4.6	2.1	2.0	2.3	2.0	2.1	2.4	2.8
S.D.	2.0	5.9	2.7	2.9	3.0	2.5	2.6	2.9	3.3
Rank Order	1	9	4	5	7	2	3	6	8

course, is not to be unexpected since some of the same test data served as the
basis for setting the character of the secondary units. The analyses shown in
Tables 74-76 do not lend themselves to meaningful statistical analysis because,
among other things, the seven noises involved are not necessarily truly
representative of the noises in general. However, if one wishes to consider these
noises as representative of the range of typical spectra to be found in industry
and from transportation vehicles, it is tempting to deduce that the differences
between the means of the secondary units of measurement for each of the
quantities evaluated represent a constant that can be used to convert one
secondary unit into the proper value of another secondary unit, and, in the case
of PNL, into the value of the primary unit. To some extent the limits of this
assumption are illustrated in the paragraphs following.

Constants for Converting Units of PNL to a Common Base

Table 77 is a summarization fo the relations between Max PNdB, dB(A), (C),
(D_1), (D_2), and (D_3) for a variety of noises. These noises have been set to be
equal to each other in PNdB; Table 77 shows how the other units differ in their

TABLE 76

Perceived Noisiness (Annoyance)

Values of secondary units of physical measurement when for each of the seven noises the value of the primary unit of physical measurement of perceived noisiness or annoyance is 100 PNdB peak. The spectra of the seven noises are given in Table 7. PNL = 100 PNdB.

Noise	dB(A)	dB(C)	dB(D_1)	dB(D_2)	dB(D_3)	Phon(S)	Phon(Z)	NC
1	87	100	94	92	85	98	101	86
2	87	86	95	95	90	97	102	84
3	87	90	94	94	89	98	102	82
4	88	88	95	95	91	99	101	84
5	87	88	94	94	89	97	101	82
6	89	95	94	93	88	100	105	86
7	89	98	95	94	88	100	106	85
Mean	87.7	92.1	94.4	93.9	88.6	98.4	102.7	84.1
A.D.	0.8	4.7	0.5	0.8	1.3	1.1	1.7	1.3
S.D.	0.9	5.5	0.5	1.1	1.9	1.3	2.0	1.7
Rank Order	2	8	1	3	6	4	7	5

average values under this condition. Averaging over all types of noises, it is seen that adding 2 to (C), 11 to (A), 4 to (D_1), 6 to (D_2), and 12 to (D_3) would make the numerical values about equal to each other. Earlier in the text, a constant of 13, rather than 11.2 shown in Table 77, was recommended for converting dB(A) to PNdB for purposes of estimating PNL and Damage Risk on a common numerical base. This was done because the constant 13 is appropriate for the aircraft and industrial noise more typically involved in problems of noisiness and damage risk. However, it is to be emphasized that different types of noises would require somewhat different valued constants to convert, with as little error as possible, the various units to a common base.

It should also be noted that the variability of the magnitude of these conversion constants is greater for some units than for others, indicating that less error on the average would be experienced with those units having the least variability in estimating (presuming this is desirable) the PNdB of any particular noise regardless of source. Of the overall frequency-weighted units, dB(D_2) shows the least variability in this regard, with a largest difference of 8 dB. Data for "old" PNdB and proposed PNdB' are given as a matter of academic interest.

TABLE 77

Differences Between Max PNdB and Max "Old" PNdB, PNdB', dB(A), (C), (D_1), (D_2), and (D_3) Units for a Variety of Noises

The differences are those between the latter units and PNdB when the noises were at the levels present in the specified figures and tables. PNdB and PNdB' calculated in accordance with Chapter 11, "old" PNdB in acordance with refs. 466, 467, and 468.

		PNdB	"Old" PNdB	PNdB'	dB(A)	dB(C)	dB(D_1)	dB(D_2)	dB(D_3)
Office Noises	Aver.	0	- 1.0	- 2.8	- 7.7	+ 5.7	+ 0.2	- 2.1	- 8.7
(16, Fig. 188)	Diff.								
	Largest Diff.	0	- 5.3	- 6.1	-12.9	+15.5	+ 8.0	- 7.3	-14.5
Noises in the Home	Aver.	0	- 0.2	- 4.9	-10.6	- 4.1	- 3.6	- 4.6	-10.0
(25, Figs. 190 and 198)	Diff.								
	Largest Diff.	0	- 6.3	- 7.0	-13.9	-13.1	+ 6.6	- 8.0	-14.9
Ground Transportation	Aver.	0	+ 0.8	- 5.3	-11.6	- 1.8	- 5.0	- 6.7	-12.2
(10, Figs. 194 and 197)	Diff.								
	Largest Diff.	0	+ 2.5	- 7.0	-13.9	- 9.4	- 6.9	- 8.9	-15.4
Seven Representative Noises	Aver.	0	+ 0.8	- 5.7	-12.2	- 7.9	- 5.5	- 6.1	-11.2
(Table 7)	Diff.								
	Largest Diff.	0	+ 1.1	- 6.5	-13.2	-13.7	- 5.9	- 8.4	-14.7
Aircraft									
Fixed-Wing Jet	Aver.	0	+ 1.2	- 5.4	-10.3	- 6.2	- 6.1	- 6.5	-10.9
Takeoff (9, Fig. 209)	Diff.								
	Largest Diff.	0	+ 1.7	- 6.8	-13.5	-10.4	- 8.3	- 8.4	-13.1
Fixed-Wing Jet	Aver.	0	+ 0.7	- 5.4	-14.9	-13.4	- 8.1	- 8.2	-13.1
Landing (6, Fig. 209)	Diff.								
	Largest Diff.	0	+ 0.9	- 6.3	-19.5	-19.6	-11.4	-11.4	-16.6
Fixed-Wing Prop	Aver.	0	+ 0.8	- 6.2	-14.6	- 3.0	- 7.2	- 9.1	-16.4
(2, Fig. 209)	Diff.								
	Largest Diff.	0	+ 0.9	- 6.7	-15.5	- 4.9	- 7.7	- 9.3	-16.5
Helicopter - Takeoff	Aver.	0	+ 1.4	- 5.7	-11.9	- 6.7	- 5.7	- 6.2	-11.0
(6, Fig. 215)	Diff.								
	Largest Diff.	0	+ 2.2	- 6.1	-13.2	-11.9	- 6.5	- 7.6	-13.2
Helicopter - Cruise	Aver.	0	+ 1.3	- 5.4	-12.2	- 5.8	- 5.8	- 6.5	-12.0
(6, Fig. 215)	Diff.								
	Largest Diff.	0	+ 1.8	- 5.7	-13.5	-12.6	- 6.6	- 8.2	-13.9
All Aircraft - (Unweighted for	Aver.	0	+ 1.1	- 5.7	-12.8	- 7.0	- 6.6	- 7.3	-12.7
Number)	Diff.								
All Aircraft and the Other	Aver.	0	+ 0.3	- 4.9	-11.8	- 4.8	- 5.2	- 6.2	-11.7
Noises - (Unweighted for Number)	Diff.								
Range of Averages		0	2.4	3.4	7.2	19.1	8.3	7.0	7.7
Range of Largest Differences		0	7.8	1.3	7.3	35.1	19.4	4.1	3.5

Recommended Units of Sound Measurement

On the basis of the information developed in Tables 74-76 and the results of the various research findings reported earlier in the text, it is recommended that:

1. For estimating the masking of speech by noise use AI based on band spectral, signal-to-noise ratio procedures or, less accurately, equivalent AI based on SIL, or PNdB or dB(D_2).

2. For estimating damage risk to hearing of more or less continuous or intermittent workday noises over a specified number of years, use CDR based on DRs or, less accurately, PNdB or dB(D_2).

3. For estimating the perceived noisiness of and human reactions to community noise environments or the noises in various types of living areas and work rooms, use the unit CNR based on EPNdB or, less accurately, EdB(D_2).

4. For estimating, with practical accuracy and on the basis of the same unit of noise measurement, either speech masking, damage risk to hearing, or the perceived noisiness of a noise or noise environment, use (E)PNdB or (E)dB(D_2).

5. For showing which parts of a particular noise spectrum contribute the most to the perceived noisiness or damage risk to hearing, plot the band spectra of the noise on graphs that show equal noisiness or damage risk contours as a function of frequency.

6. Use (E)dB(A) as a secondary, less accurate, substitute for (E)dB(D_2).

Chapter 11

Proposed Procedures for the Evaluation of Environmental Noises

Introduction

It is the purpose of this chapter to describe precisely the procedures to be followed to obtain accurate physical units as appear to be now available for evaluating the perceived noisiness of nonimpulsive and impulsive sounds and sound environments. In addition to the more accurate units, the unit dB(A) is included because of its general wide use and the existence of this A-weighting on standard sound level meters. It is possible that D_2 or some similar weighting will be standardized and incorporated into sound level meters. If one wishes to use, for the evaluation of the perceived noisiness of a noise or noise environment, a sound level meter with D- or A-weighting and/or to measure only maximum noise levels, the following material includes appropriate steps and definitions. If, on the other hand, one wishes to take into account additional acoustical and temporal factors concerning the noise, the necessary steps are also given. It may appear while reading the following that a wide variety of units are to be calculated. There are, however, but three basic quantities: (a) Perceived Noise Level (PNL) based on one-third octave, full octave, or overall frequency-weighted sound levels measured in successive 0.5-sec intervals of time during the occurrence of a sound, and the Max PNL in any single 0.5-sec interval of time during the occurrence of a sound; (b) EPNL, consisting of the integrated, on a 10 $\log_{10}$ antilog basis, PNLs of each 0.5-sec interval of sound divided by a reference time; and (c) Composite Noise Rating, based on EPNL's integrated over 24 hours.

There is some indication that the noy band summation method first used for loudness level (Stevens) and adopted for PNdB may not be as good a band summation procedure for perceived noisiness of broadband noises as a power summation of band SPL adjusted first in accordance with the noy contours. To encourage the evaluation and possible use of this modified band summation procedure, we have included it as a possible step in the calculation of PNdB. It is

471

suggested that the units from this possible alternative method be designated as PNdB', Max PNdB', or EPNdB'.

These calculation procedures are derived from psychological judgment tests; verification of and changes to these procedures will rest upon additional judgment test data. For this reason, definitions of terms for judged perceived noisiness are included.

Finally, for noise measurements to be useful, limits of noise exposure must be set with respect to some criteria of human behavior and tolerance; graphs and tables are referred to that can be used for this purpose.

Definitions of Terms

Sound. For present purposes, sound is defined as airborne acoustic energy in the frequency region from 45 to 11,020 (see Table 2).

Impulse Intervals of Sound. When the overall sound pressure level changes, during any 0.5-sec interval of time, 40 or more dB, the sound during that interval is called implusive.

Nonimpulsive Intervals of Sound. All 0.5-sec intervals of sound that are not impulsive.

Sound Pressure Level (SPL) in Decibels (dB). The sound pressure level re 0.0002 μbar as measured by means of a meter or recording device that meets the specifications of a sound level meter (SLM) set on "slow" is called dB when the flat-frequency weighting is used.

One-third Octave and Octave Band Level. The SPL re 0.0002 μbar as measured on a SLM set on "slow" and flat-frequency weighting in conjunction with one-third octave or octave band filters.

Sound in 0.5-sec Intervals of Time. The sound pressure level, band, or overall bands, as read, or would be read, on a SLM set on slow and flat weighting is taken for purposes of this document as the sound present during a 0.5-sec interval.

Judged Preceived Noisiness. The attribute of a familiar expected sound that is judged as "unwanted" or "unacceptable" for everyday living conditions as a standard reference sound independently of any cognitive meaning conveyed by the sound, it called "Judged Perceived Noisiness," The term judged perceived noisiness is synonymous, for purposes of this document, with the term annoyance.

Standard Reference Sound. The standard reference sound against which other sounds may be judged with respect to perceived noisiness is as follows: a band of random pink noise centered at 1000 Hz and with frequency skirts sloping at the rate of 48 dB or more per octave below 710 Hz, and 51 dB or more per octave above 1400 Hz that has a steady maximum sound pressure level for 2 sec with the increase in pressure to the steady maximum pressure level and the decline in pressure following the steady maximum pressure level at a rate of 2.5dB per 0.5-sec. (Note: A secondary reference sound may be substituted for the standard reference sound for certain relative comparisons in order to provide a reference sound that is more similar in character than the standard reference is to the noises with which it is to be compared.)

Standard Reference Background Noise. The standard reference sound and a comparison sound to be judged shall both be presented in the presence of a random band of pink noise extending from about 50 to 8000 Hz at a sound pressure level such that it is at least 15 dB below, at all frequencies, the level of the standard and reference comparison sound.

Noy. The subjective unit of perceived noisiness is called the Noy. One noy is the value assigned to the standard reference sound during an interval of 0.5 sec when the sound is at a level of 40 dB. Noy values, as the result of judgment tests conducted in the laboratory, have been assigned to the SPL of bands of frequencies present during an interval of 0.5 sec as shown in Fig. 242 and Table 78.

Calculated Perceived Noise Level (PNL) in PNdB and Maximum PNL in Max PNdB. A means of estimating the Judged Perceived Noisiness for a 0.5-sec interval of a given sound from the noy value for that 0.5 sec of the given sound. The sum, as calculated according to prescribed procedures, of the noy values of a frequency band or frequency bands of sound is designated as Perceived Noise Level in PNdB. The highest value of the PNdBs calculated for each 0.5-sec interval during the occurrence of a sound is called the Max PNdB of the sound. (*Note*: Two alternative methods of calculating the unit PNdB will be given below.)

PNL in dB(D) and dB(A), and Maximum PNL in Max dB(D) and Max dB(A). The level, plus a constant, as read on an SLM with a D- or A-frequency weighting characteristic and set on "slow" meter action is designated as the PNL in dB(D) or dB(A) respectively for each 0.5-sec interval during the occurrence of a sound. The highest valued dB(D) on dB(A) in any 0.5-sec interval is called Max dB(D) or Max dB(A) respectively of a given sound. D- and A-weighting characteristics are specified in Table 2. Weighting D_2 is recommended above D_1 or D_3 for this purpose. *Note*: In order to make the units PNdB, dB(D), and dB(A) numerically equal, on the average, to each other, a constant of 6 is added to $dB(D_2)$ and 13 to dB(A). The results are designated as dB(D') and dB(A') respectively.

Threshold of Perceived Noisiness. A level measured during the day (the hours of 7 A.M. to 10 P.M.) and indoors of 40 PNdB, dB(D'), or dB(A'), or a level measured outdoors of 60 PNdB, dB(D'), or dB(A'), is specified as the threshold of perceived noisiness. This threshold during the night (the hours 10 P.M. to 7 A.M.) is 10 PNdB, dB(D), or dB(A) lower than during the day.

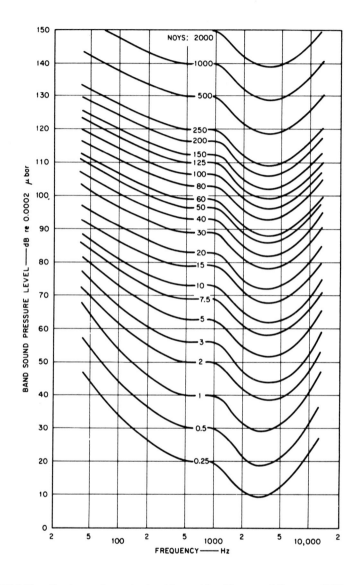

FIGURE 242.   Contours of perceived noisiness. After Kryter and Pearsons (466).

Practical Threshold of Perceived Noisiness. For the purpose of the measurment or calculation of perceived noisiness of occurrences of individual sounds, it is found sufficiently accurate to define, as the threshold of perceived noisiness, the level that is 15 PNdB, dB(D), or dB(A) below the highest level when the highest level is greater during the day than 55 (45 at night) PNdB, dB(D'), or dB(A') when measured indoors, and greater during the day than 75 (65 at night) PNdB, dB(D'), or dB(A') when measured outdoors. *Note*: 10 PNdB, dB(D), or dB(A) below the highest (Max) level has generally been used in the past as the practical threshold of perceived noisiness, partly because of the limited range of physical noise measurements, and partly because typical noises from passing aircraft and highway vehicles tests show this to be a reasonably satifactory threshold. However, with the advent of helicopter noises or other noises having a more erratic or slowly changing level in time, it is believed that the 15 dB range is a more realistic range to use if physical measurements permit.

Duration of the Occurrence of a Sound. The time in seconds between the moment a sound starts to rise above the threshold or practical threshold of perceived noisiness and the next succeeding moment in time it recedes to the threshold or practical threshold of noisiness.

Onset Duration. The time between the first 0.5-sec interval a nonimpulisve sound is at Max PNL and the last preceding 0.5-sec interval the sound was at the PNL of the background noise, or the threshold of noisiness, or the practical threshold of noisiness, whichever is higher, is taken as the onset duration of a nonimpulsive sound.

Onset Correction. The onset duration in seconds is used to determine a correction (called *oc*) to be applied in the calculation of EPNdB, EdB(D), or EdB(A) (see Fig. 173).

Impulse Level. The difference in PNL in PNdB, dB(D), or dB(A), of an impulse from the PNL of the background noise, or the threshold of perceived noisiness, whichever is the higher, is called the impluse level.

Impulse Level Correction. The impulse level in PNdB, dB(D'), or dB(A') is used to determine a correction (called *ic*) to be applied in the calculation of EPNdB, EdB(D'), or EdB(A') as appropriate (see Fig. 174).

Calculated Effective Perceived Noise Level (EPNL) in EPNdB, EdB(D), and EdB(A). The sum as calculated by formulae of PNdBs, dB(D)s, or dB(A)s in successive 0.5-sec intervals during the occurrence of a sound −12 plus a correction for onset duration or impulse level, as appropriate. The sum of these calculations is called EPNdB, EdB(D), or EdB(A) respectively. The value −12 comes from the choice of sixteen 0.5-sec intervals, a total of 8 seconds, as a standard duration to which all effective levels are referred.

Composite Noise Rating (CNR) from EPNdB, EdB(D'), or EdB(A'). The sum as measured to calculated according to the prescribed formulae, of the EPNLs during a 24-hour time cycle at a given location is designated as the Composite Noise Rating for that location.

TABLE 78

Antilog (Base 10) of SPL/10 and NOYS as a Function of SPL

This formulation of Table 73 for noys represents an approximation to the contours of Fig. 242 and is used as a practical convience for computer calculation of preceived noisiness. R. A. Pinker, Note 684, Feb. 1968, NGTE, Great Britain.

BAND CENTER FREQUENCY

SPL-dB	LOG_{10}^{-1} (SPL/10) (EXPON. NUMERIC FORM. EXAMPLE: 7.94E 02 ≡ 7.94 x 10^2.)	50HZ	63HZ	80HZ	100HZ	125HZ	160HZ	200HZ	250HZ	315HZ	400HZ	500HZ	630HZ	800HZ	1KHZ	1.2KHZ	1.6KHZ	2KHZ	2.5KHZ	3.1KHZ	4KHZ	5KHZ	6.3KHZ	8KHZ	10KHZ
29	7.94E 02																			1.0	1.0				
30	1.00E 03																		1.0	1.1	1.1				
31	1.26E 03																		1.1	1.1	1.1	1.0			
32	1.58E 03																	1.0	1.1	1.2	1.2	1.1			
33	2.00E 03																	1.1	1.2	1.3	1.3	1.2	1.0		
34	2.51E 03																1.0	1.2	1.3	1.4	1.4	1.3	1.1		
35	3.16E 03																1.1	1.3	1.4	1.5	1.5	1.4	1.2		
36	3.98E 03																1.2	1.3	1.5	1.6	1.6	1.5	1.4		
37	5.01E 03																1.3	1.4	1.6	1.8	1.8	1.6	1.5	1.0	
38	6.31E 03															1.0	1.3	1.5	1.8	1.9	1.9	1.7	1.6	1.1	
39	7.94E 03															1.1	1.4	1.6	1.9	2.0	2.0	1.9	1.8	1.2	
40	1.00E 04										1.0	1.0	1.0	1.0	1.0	1.2	1.5	1.7	2.0	2.2	2.2	2.0	1.9	1.4	
41	1.26E 04										1.1	1.1	1.1	1.1	1.1	1.3	1.6	1.8	2.2	2.4	2.4	2.2	2.0	1.5	1.0
42	1.58E 04									1.0	1.1	1.1	1.1	1.1	1.1	1.3	1.7	2.0	2.4	2.6	2.6	2.4	2.2	1.7	1.1
43	2.00E 04									1.1	1.2	1.2	1.2	1.2	1.2	1.4	1.8	2.2	2.6	2.8	2.8	2.6	2.4	1.8	1.2
44	2.51E 04								1.0	1.1	1.3	1.3	1.3	1.3	1.3	1.5	2.0	2.4	2.8	3.0	3.0	2.8	2.6	2.0	1.4
45	3.16E 04								1.1	1.2	1.4	1.4	1.4	1.4	1.4	1.6	2.1	2.6	3.0	3.2	3.2	3.0	2.8	2.2	1.5
46	3.98E 04							1.0	1.2	1.3	1.5	1.5	1.5	1.5	1.5	1.7	2.3	2.8	3.2	3.4	3.4	3.2	3.0	2.4	1.7
47	5.01E 04							1.1	1.3	1.4	1.6	1.6	1.6	1.6	1.6	1.8	2.4	3.0	3.4	3.6	3.6	3.4	3.2	2.6	1.8
48	6.31E 04						1.0	1.2	1.4	1.5	1.7	1.7	1.7	1.7	1.7	2.0	2.6	3.2	3.6	3.9	3.9	3.6	3.4	2.8	2.0
49	7.94E 04						1.1	1.3	1.5	1.6	1.9	1.9	1.9	1.9	1.9	2.1	2.8	3.4	3.9	4.1	4.1	3.9	3.6	3.0	2.2
50	1.00E 05						1.2	1.4	1.6	1.7	2.0	2.0	2.0	2.0	2.0	2.3	3.0	3.6	4.1	4.4	4.4	4.1	3.9	3.2	2.4
51	1.26E 05					1.0	1.3	1.5	1.7	1.9	2.1	2.1	2.1	2.1	2.1	2.4	3.2	3.9	4.4	4.7	4.7	4.4	4.1	3.4	2.6
52	1.58E 05					1.1	1.4	1.6	1.9	2.0	2.3	2.3	2.3	2.3	2.3	2.6	3.5	4.1	4.7	5.0	5.0	4.7	4.4	3.6	2.8
53	2.00E 05				1.0	1.2	1.5	1.7	2.0	2.1	2.5	2.5	2.5	2.5	2.5	2.8	3.7	4.4	5.0	5.3	5.3	5.0	4.7	3.9	3.0
54	2.51E 05				1.1	1.3	1.6	1.9	2.1	2.3	2.6	2.6	2.6	2.6	2.6	3.0	4.0	4.7	5.3	5.7	5.7	5.3	5.0	4.1	3.2
55	3.16E 05				1.2	1.4	1.7	2.0	2.3	2.4	2.8	2.8	2.8	2.8	2.8	3.2	4.3	5.0	5.7	6.1	6.1	5.7	5.3	4.4	3.5
56	3.98E 05			1.0	1.3	1.5	1.9	2.2	2.4	2.6	3.0	3.0	3.0	3.0	3.0	3.5	4.6	5.3	6.1	6.5	6.5	6.1	5.7	4.7	3.7
57	5.01E 05			1.1	1.4	1.7	2.0	2.4	2.6	2.8	3.2	3.2	3.2	3.2	3.2	3.7	5.0	5.7	6.5	7.0	7.0	6.5	6.1	5.0	4.0
58	6.31E 05			1.2	1.5	1.8	2.2	2.6	2.8	3.0	3.5	3.5	3.5	3.5	3.5	4.0	5.3	6.1	7.0	7.5	7.5	7.0	6.5	5.3	4.3
59	7.94E 05			1.3	1.7	2.0	2.4	2.8	3.0	3.2	3.7	3.7	3.7	3.7	3.7	4.3	5.7	6.5	7.5	8.0	8.0	7.5	7.0	5.7	4.6
60	1.00E 06		1.0	1.4	1.8	2.2	2.6	3.0	3.2	3.5	4.0	4.0	4.0	4.0	4.0	4.6	6.1	7.0	8.0	8.7	8.7	8.0	7.5	6.1	5.0
61	1.26E 06		1.1	1.5	2.0	2.4	2.8	3.2	3.5	3.7	4.3	4.3	4.3	4.3	4.3	5.0	6.5	7.5	8.7	9.3	9.3	8.7	8.0	6.5	5.3
62	1.58E 06		1.2	1.7	2.2	2.6	3.0	3.5	3.7	4.0	4.6	4.6	4.6	4.6	4.6	5.3	7.0	8.0	9.3	10.0	10.0	9.3	8.7	7.0	5.7
63	2.00E 06		1.3	1.8	2.4	2.8	3.2	3.7	4.0	4.3	4.9	4.9	4.9	4.9	4.9	5.7	7.5	8.7	10.0	11.0	11.0	10.0	9.3	7.5	6.1
64	2.51E 06	1.0	1.5	2.0	2.6	3.0	3.5	4.0	4.3	4.6	5.3	5.3	5.3	5.3	5.3	6.1	8.0	9.3	11.0	11.0	11.0	11.0	10.0	8.0	6.5
65	3.16E 06	1.1	1.6	2.2	2.8	3.2	3.7	4.3	4.6	5.0	5.7	5.7	5.7	5.7	5.7	6.5	8.7	10.0	11.0	12.0	12.0	11.0	11.0	8.7	7.0
66	3.98E 06	1.2	1.8	2.4	3.0	3.5	4.0	4.6	5.0	5.4	6.1	6.1	6.1	6.1	6.1	7.0	9.3	11.0	12.0	13.0	13.0	12.0	11.0	9.3	7.5
67	5.01E 06	1.4	2.0	2.6	3.3	3.7	4.3	5.0	5.4	5.9	6.5	6.5	6.5	6.5	6.5	7.5	10.0	11.0	13.0	14.0	14.0	13.0	12.0	10.0	8.0
68	6.31E 06	1.6	2.2	2.8	3.6	4.0	4.6	5.4	5.9	6.4	7.0	7.0	7.0	7.0	7.0	8.0	11.0	12.0	14.0	15.0	15.0	14.0	13.0	11.0	8.7
69	7.94E 06	1.8	2.3	3.0	3.9	4.3	5.0	5.9	6.4	6.9	7.5	7.5	7.5	7.5	7.5	8.7	11.0	13.0	15.0	16.0	16.0	15.0	14.0	11.0	9.3
70	1.00E 07	2.0	2.5	3.3	4.2	4.6	5.4	6.4	6.9	7.5	8.0	8.0	8.0	8.0	8.0	9.3	12.0	14.0	16.0	17.0	17.0	16.0	15.0	12.0	10.0
71	1.26E 07	2.2	2.8	3.6	4.6	5.0	5.9	6.9	7.5	8.0	8.6	8.6	8.6	8.6	8.6	10.0	13.0	15.0	17.0	19.0	19.0	17.0	16.0	13.0	11.0
72	1.58E 07	2.3	3.0	3.9	5.0	5.4	6.4	7.5	8.0	8.7	9.2	9.2	9.2	9.2	9.2	11.0	14.0	16.0	19.0	20.0	20.0	19.0	17.0	14.0	11.0
73	2.00E 07	2.5	3.3	4.2	5.4	5.9	6.9	8.0	8.7	9.3	9.8	9.8	9.8	9.8	9.8	11.0	15.0	17.0	20.0	21.0	21.0	20.0	19.0	15.0	12.0
74	2.51E 07	2.8	3.7	4.6	5.9	6.4	7.5	8.7	9.3	10.0	10.6	10.6	10.6	10.6	10.6	12.0	16.0	19.0	21.0	23.0	23.0	21.0	20.0	16.0	13.0
75	3.16E 07	3.0	4.1	5.0	6.4	6.9	8.0	9.3	10.0	11.0	11.3	11.3	11.3	11.3	11.3	13.0	17.0	20.0	23.0	24.0	24.0	23.0	21.0	17.0	14.0
76	3.98E 07	3.3	4.5	5.4	6.9	7.5	8.7	10.0	11.0	11.0	12.0	12.0	12.0	12.0	12.0	14.0	19.0	21.0	24.0	26.0	26.0	24.0	23.0	19.0	15.0
77	5.01E 07	3.7	5.0	5.9	7.5	8.3	9.3	11.0	11.0	12.0	13.0	13.0	13.0	13.0	13.0	15.0	20.0	23.0	26.0	28.0	28.0	26.0	24.0	20.0	16.0
78	6.31E 07	4.1	5.4	6.4	8.3	9.1	10.0	11.0	12.0	13.0	14.0	14.0	14.0	14.0	14.0	16.0	21.0	24.0	28.0	30.0	30.0	28.0	26.0	21.0	17.0
79	7.94E 07	4.5	5.9	6.9	9.1	10.0	11.0	12.0	13.0	14.0	15.0	15.0	15.0	15.0	15.0	17.0	23.0	26.0	30.0	32.0	32.0	30.0	28.0	23.0	19.0
80	1.00E 08	5.0	6.4	7.5	10.0	11.0	11.0	13.0	14.0	15.0	16.0	16.0	16.0	16.0	16.0	19.0	24.0	28.0	32.0	35.0	35.0	32.0	30.0	24.0	20.0
81	1.26E 08	5.5	6.9	8.3	11.0	11.0	12.0	14.0	15.0	16.0	18.0	18.0	18.0	18.0	18.0	20.0	26.0	30.0	35.0	37.0	37.0	35.0	32.0	26.0	21.0
82	1.58E 08	6.1	7.5	9.1	11.0	12.0	13.0	15.0	16.0	17.0	19.0	19.0	19.0	19.0	19.0	21.0	28.0	32.0	37.0	40.0	40.0	37.0	35.0	28.0	23.0
83	2.00E 08	6.8	8.3	10.0	12.0	13.0	14.0	16.0	17.0	19.0	20.0	20.0	20.0	20.0	20.0	23.0	30.0	35.0	40.0	42.0	42.0	40.0	37.0	30.0	24.0
84	2.51E 08	7.5	9.1	12.0	13.0	14.0	15.0	17.0	19.0	20.0	21.0	21.0	21.0	21.0	21.0	24.0	32.0	37.0	42.0	45.0	45.0	42.0	40.0	32.0	26.0
85	3.16E 08	8.3	10.0	13.0	14.0	15.0	16.0	19.0	20.0	21.0	23.0	23.0	23.0	23.0	23.0	26.0	35.0	40.0	45.0	47.0	47.0	45.0	42.0	35.0	28.0
[illegible]	[illegible]	[illegible]	[illegible]	13.0	15.0	16.0	17.0	20.0	21.0	23.0	24.0	24.0	24.0	24.0	24.0	28.0	37.0	42.0	47.0	50.0	50.0	47.0	45.0	37.0	30.0

BAND CENTER FREQUENCY

LOG⁻¹ (SPL/10) (EXPON NUMERIC) FORM. EXAMPLE: 7.94 x 10²) SPL-dB	50HZ	63HZ	80HZ	100HZ	125HZ	160HZ	200HZ	250HZ	315HZ	400HZ	500HZ	630HZ	800HZ	1KHZ	1.2KHZ	1.6KHZ	2KHZ	2.5KHZ	3.1KHZ	4KHZ	5KHZ	6.3KHZ	8KHZ	10KHZ
89 7.94E 08																							45	37
90 1.00E 09																						55	47	40
91 1.26E 09															37	47	55	63	67	60	60	63	47	42
92 1.58E 09																							50	45
93 2.00E 09																							55	47
94 2.51E 09																							63	50
95 3.16E 09																							67	55

Tolerable Limit. The maximum amount of noise that will permit effective utilization of a space for its normal use by the average person who is adapted to the noise as the result of repeated near daily exposure to the noise. The use to which a space is put and, within limits, the socioeconomic status of the users of of the space and other social and economic factors, determine the amounts of noise that are tolerated.

Calculation Procedures for Perceived Noise Level (PNL) and Effective Perceived Noise Level (EPNL)

PNL in PNdB

Formula 1: $PNdB_i = 40 + \dfrac{10}{0.30103} \left(\log_{10} PN_i \right)$

where i is a 0.5-sec interval of time.

Formula 2: $PN_i = N_{i_m} + f \left(\Sigma \, N_i - N_{i_m} \right)$

where i is a 0.5-sec interval of time, $\Sigma \, N_i$ is noys in all bands, N_{i_m} is maximum number of noys in any band, and f is 0.15 for one-third octave bands and 0.3 for octave bands.

Formula 3: $PNdB' = 10 \log_{10} \sum\limits_{i}^{x} 10^{(SPL_i'/10)}$

where x is the number of band filters, adjusted for critical bandwidth of the ear below 355 Hz, and SPL' is the SPL of 1000 Hz band having the same noy value as that for SPL in the i band.

Step 1. Determine the sound pressure level that occurs in each one-third or full ocatve band in each successive 0.5-sec interval of time.

Step 2—One-third Octave Bands. Add on a 10 $\log_{10}$ antilog basis the band levels of the one-third octave bands having the center frequencies of:

(*a*) 50, 63, 80, and 100 Hz, and assign the result to the band center frequency having the greatest intensity.

Note: The procedures herein described for PNL are, except for the summation in Step 2 of the SPLs of frequency bands below 315 Hz and the tone-corrections of Step 3, essentially identical with procedures described in documents of ISO (29), SAE (756), and FAA (758) for the evaluation of aircraft noise. In the FAA document a different tone-correction procedure is used (see Chapter 8); in the ISO and SAE documents no tone correction is used.

(*b*) 125, 160, and 200 Hz and assign the result to a band center frequency having the greatest intensity.

(*c*) 250 and 315 Hz assign the result to the band center frequency having the greatest intensity.

Note 2: If the greatest intensity in Step 2*a, b,* and *c* is present in more than one band within a step, assign the sum to the band with the highest frequency and a highest SPL in *a, b,* and *c.*

Step 2. Full Octave Bands. Add on a 10 $\log_{10}$ antilog basis the band levels of the octave bands having the center frequencies of 63 and 125 Hz, and assign the result to the band center frequency having the greatest intensity. (If the intensity is the same in the two bands, assign the sum to the center frequency 125 Hz. Steps 2 and 5 are given for both one-third octave and full octave bands and are to be used according to which bands are used for the band spectrum analysis of a given sound.)

Step 3. If any band above 400 Hz of a nonimpulsive sounds is abutted above and below by bands that are both less intense than the in-between band, a correction is determined from the appropriate abscissa on Fig. 243 and added to the SPL of the respective bands or summed bands.

Note 1: In Fig. 243, the abscissa is $L_B - \left[L_{B-1} + L_{B+1} / 2 \right]$ where L_B is the SPL in dB of band B. B-1 is the abutted lower frequency band, B+1 is the abutted higher frequency band. The addition of L_{B-1} to L_{B+1} is arithmetic.

Note 2: When the highest frequency band of a sound is 3 dB more intense than the band immediately below it, L_{B+1} is taken as 3. When the lowest frequency band of a sound is 3 dB more intense than the band immediately above it, L_{B-1} is taken as 3.

Note 3: Care must be taken to insure that the presence of a pure tone or very narrow band (less than one-third octave wide) of concentrated energy is not overlooked because the center frequency of the tone or narrow band of sound is at or near the crossover frequencies between two adjacent filter bands. When there are pure tone or very narrow band spectral components at or near filter crossover points between two adjacent filter bands, add the appropriate amount found in Fig. 243 to the band of higher intensity, or to the band of higher frequency when the two adjacent bands are of equal intensity.

Step 4. Find the noy values from Table 78 for (*a*) the summed band levels at the assigned center frequencies at and below 355 Hz as obtained in Step 2, and (*b*) the band levels present in each band having center frequencies at and above 355 Hz, as corrected in Step 3.

Step 5. One-third Octave Bands. Add to the largest noy value obtained for any single band in Step 4, the sum of the noy values for all the other bands as

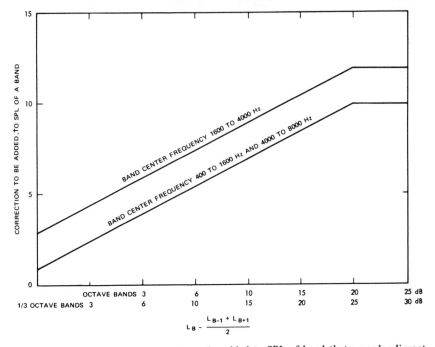

FIGURE 243. Showing dB correction to be added to SPL of band that exceeds adjacent bands by amount on abscissa. The parameter is band-center frequency.

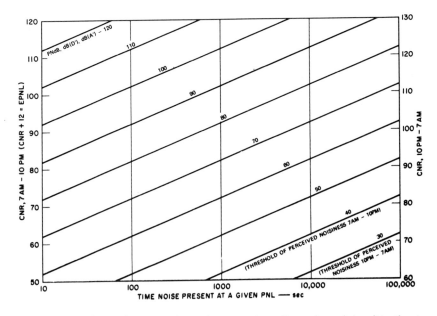

FIGURE 244. Graph for converting noises present continuously or intermittently at a given PNL during the hours 7 A.M. to 10 P.M. (left ordinate) or 10 P.M. to 7 A.M. (right ordinate) to an equivalent CNR. EPNLs for noises occurring any time during a 24-hour period can be found for a given PNL and duration by reading corresponding value on left ordinate and adding 12.

found in Step 4 multiplied by 0.15. The result is called PN for what 0.5-sec interval of a given sound.

Step 5—Octave Bands. Add to the largest noy value obtained for any single band in Step 4 the sum of the noy values for all the other bands as found in Step 4 multiplied by 0.3. The result is called PN (Oct.) for that 0.5-sec interval of a given sound.

Alternative Step 5. Find from Table 78 the antilog$_{10}$ value for the SPL of the band centered at 1000 Hz that has the same, or closest, noy value as each of the bands, or summed bands below 355 Hz, as corrected in Step 3. Sum these 10 antilog$_{10}$ values. (It is anticipated that alternative Step 5 will replace present Step 5 if further tests demonstrate its superiority.)

Step 6. Convert the PN for each 0.5-sec interval of sound into PNdB by reference to Table 79. The result is called PNL in PNdB for each 0.5-sec interval of sound. (All units of PNL and EPNL calculated from octave band spectra are to be designated as Oct., i.e., PNdB Oct. Units without qualification are those calculated from one-third octave band spectra, i.e., PNdB.)

Alternative Step 6. Convert the sum found in Alternative Step 5 into "dB" by reference to the left-hand columns of Table 78. Add to this value the constant number 5.5. The result is called PNL in PNdB' for each 0.5-sec interval of time. (It is anticipated that Alternative Step 6 will replace present Step 6 if further tests demonstrate its superiority; at that time PNdB' can be written as PNdB, and present PNdB can be designated as obsolete.)

PNL in dB(D) or dB(A)

Formula 4—For overall SLM: $dB(D) = 10 \log_{10} \left\{ \left| \int_{45}^{11,020 \text{ Hz}} W(f) \cdot S(f) \, df \right| \right\}$

where W is a complex frequency power weighting for perceived noisiness (40 noy contour), and S is a complex power spectrum of a given sound, and the variable of integration is frequency in Hz.

Formula 5—For band spectra: $dB(D) = 10 \log_{10} \left(\sum_{i}^{x} 10^{(SPL_i + W_{40_i})/10} \right)$

where W_{40_i} is the band weight, for the i band, adjusted for critical bandwidth for the ear below 355 Hz (see D_2 of Table 2 and Fig. 8) of the 40 noy contour.

Step 1. Read the highest value reached in each 0.5-sec interval of sound on a SLM with D- or A-frequency weighting and set on slow meter action. The result is called PNL in dB(D) or dB(A) for each interval of sound.

Step 2. Add a constant to these meter readings in accordance with Table 30, as appropriate. The result is called PNL in dB(D') on dB(A') for each interval of sound.

TABLE 79

Perceived Noise Level in Steps of 1 PNdB as Function of Total Perceived Noisiness of a Sound

PN			PNL in PNdB	PN			PNL in PNdB
Lower	Mid	Upper		Lower	Mid	Upper	
1.0	1.0	1.0	40	43.8	45.2	46.8	95
1.1	1.1	1.1	41	46.9	48.5	50.2	96
1.1	1.1	1.2	42	50.3	52.0	53.8	97
1.2	1.2	1.3	43	53.9	55.7	57.7	98
1.3	1.3	1.4	44	57.8	59.7	61.8	99
1.4	1.4	1.5	45	61.9	64.0	66.3	100
1.5	1.5	1.6	46	66.4	68.6	71.0	101
1.6	1.6	1.7	47	71.1	73.5	76.1	102
1.7	1.7	1.8	48	76.2	78.8	81.6	103
1.9	1.9	1.9	49	81.7	84.4	87.4	104
2.0	2.0	2.1	50	87.5	90.5	93.7	105
2.1	2.1	2.2	51	93.8	97.0	100.4	106
2.3	2.3	2.4	52	100.5	104.0	107.6	107
2.5	2.5	2.5	53	107.7	111.4	115.3	108
2.6	2.6	2.7	54	115.4	119.4	123.6	109
2.8	2.8	2.9	55	123.7	128.0	132.5	110
3.0	3.0	3.1	56	132.6	137.2	142.0	111
3.2	3.2	3.4	57	142.1	147.0	152.2	112
3.5	3.5	3.6	58	152.3	157.6	163.1	113
3.7	3.7	3.9	59	163.2	168.9	174.8	114
4.0	4.0	4.1	60	174.9	181.0	187.4	115
4.2	4.3	4.4	61	187.5	194.0	200.8	116
4.5	4.6	4.7	62	200.9	207.9	215.3	117
4.8	4.9	5.1	63	215.4	222.8	230.7	118
5.2	5.3	5.5	64	230.8	238.8	247.3	119
5.6	5.6	5.8	65	247.4	256.0	265.0	120
5.9	6.1	6.3	66	265.4	274.4	284.0	121
6.4	6.5	6.7	67	284.1	294.0	304.4	122
6.8	7.0	7.2	68	304.5	315.2	326.3	123
7.3	7.5	7.7	69	326.4	337.8	349.7	124
7.8	8.0	8.3	70	349.8	362.0	374.8	125
8.4	8.6	8.9	71	374.9	388.0	401.7	126
9.0	9.2	9.5	72	401.8	415.8	430.5	127
9.6	9.8	10.2	73	430.6	445.7	461.4	128
10.3	10.6	10.9	74	461.5	477.7	494.5	129
11.0	11.3	11.7	75	494.6	512.0	530.0	130
11.8	12.1	12.5	76	530.1	548.7	568.1	131
12.6	13.0	13.5	77	568.2	588.1	608.9	132
13.6	13.9	14.4	78	609.0	630.3	652.6	133
14.5	14.9	15.4	79	652.7	675.5	699.4	134
15.5	16.0	16.6	80	699.5	724.1	749.6	135
16.7	17.1	17.7	81	749.7	776.0	803.3	136
17.8	18.4	19.0	82	803.4	831.7	861.1	137
19.1	19.7	20.4	83	861.2	891.4	922.9	138
20.5	21.1	21.8	84	923.0	955.4	989.1	139
21.9	22.6	23.4	85	989.2	1024.0	1060.1	140
23.5	24.2	25.1	86	1060.2	1097.5	1136.1	141
25.2	26.0	26.9	87	1136.2	1176.2	1217.7	142
27.0	27.8	28.8	88	1217.8	1260.6	1305.1	143
28.9	29.8	30.9	89	1305.2	1351.1	1398.8	144
31.0	32.0	33.1	90	1393.9	1448.2	1499.1	145
33.2	34.3	35.5	91	1499.2	1552.1	1606.7	146
35.6	36.8	38.1	92	1606.8	1663.4	1722.1	147
38.2	39.4	40.8	93	1722.2	1782.8	1845.7	148
40.9	42.2	43.7	94	1845.8	1910.7	1978.2	149

Max PNL

Step 1. Find the highest valued PNL for any 0.5-sec interval during the occurrence of a given sound. This value is called the Max PNL. (By definition PNL and Max PNL are the same for inpulsive sounds)

EPNL for Impulsive and Nonimpulsive Sounds

Formula 6: $\quad EPNL = 10 \log_{10} \left[\Sigma_i \log_{10}^{-1} [PNL_i/10] \right] - 12 + oc, + ic$

where i are successive 0.5-sec intervals of time, oc is an onset-duration correction, and ic is an impulse level correction.

Step 1. Sum on a $10 \log_{10}$ antilog basis the PNLs found occurring in 0.5-sec intervals between points in time when the level is above the threshold or the practical threshold of perceived noisiness.

Note 1: The practical threshold of perceived noisiness should be used as a starting point only when it exceeds the threshold of perceived noisiness.

Note 2: The practical threshold of perceived noisiness should be used only when considerations related to sound measurement procedures and indeterminate knowledge about background noise conditions makes the use of the threshold of perceived noisiness impractical.

Step 2. Subtract 12 from the number found in Step 1.

Step 3. Find the onset duration of the sound in seconds above the PNL of the background noise. (The practical threshold of noisiness shall be used in place of the PNL of the background noise when the latter is not known or has not been measured.)

Step 4. Enter Fig. 173 with this duration and read the correction, oc. Add the correction to the number found in Step 2.

Step 5. Find the difference in PNL between the level reached during the impulsive interval of sound and the PNL of the background noise.

Step 6. Find from Fig. 174 the impulse level correction, ic, for the difference found in Step 5. Add ic to the result of Step 4 above. The result is called EPNL in EPNdB, EPNdB', EdB(D), EdB(A), EdB(D'), or EdB(A') depending upon the basic unit of measurement used.

Recommended Units for Estimating Judged Noisiness

It is recommended that EPNL in EPNdB (or EPNdB') be used as the basic unit for estimating the judged effective perceived noisiness of sounds. For general noise survey and monitoring purposes, EPNL in EdB(D) as measured on a frequency-weighted sound level meter is suitable for estimating the perceived noisiness of a sound or sound environment. EdB(A) is often adequate but not generally as accurate in this regard as EPNdB or EdB(D). PNL values based on one-third octave band spectra are to be preferred to those based on full-octave band spectra.

Calculation Procedures for Composite Noise Rating (CNR)

Calculation of CNR from EPNL Values

Formula 7:

$$CNR = \left[\left[\overbrace{[EPNL_1 + 10 \log_{10} 0_1]}^{\text{7 A.M.–10 P.M.}} \mathbin{|+|} [EPNL_2 + 10 \log_{10} 0_2] \right. \right.$$
$$\left. \mathbin{|+|} \cdots \mathbin{|+|} [EPNL_n + 10 \log_{10} 0_n] \right] - 12 \mathbin{|+|} \left[\overbrace{[EPNL_{1p} + 10 \log_{10} 0_{1p}]}^{\text{10 P.M.–7 A.M.}} \right.$$
$$\left. \left. \mathbin{|+|} [EPNL_{2p} + 10 \log_{10} 0_{2p}] \mathbin{|+|} \cdots \mathbin{|+|} [EPNL_{np} + 10 \log_{10} 0_{np}] \right] \right] - 2$$

where $0_1 \ldots 0_n$ are numbers of occurrences of sounds of EPNLs 1 through n during the hours of 7:00 A.M. to 10:00 P.M., and $0_{1p} \ldots 0_{np}$ are occurrences of sounds of EPNLs 1 through np during the hours of 10:00 P.M. to 7:00 a.m. See Fig. 17 for nomograph of $10 \log_{10} N$; $|+|$ is addition of $10 \log_{10}$ anti-log basis.

Step 1. Add arithmetically to the EPNL of each given value $10 \log_{10}$ of number of occurrences of sounds for each given EPNL value.

Step 2. Sum on a $10 \log_{10}$ antilog basis the results of Step 1 for the time period of 7:00 A.M. to 10:00 P.M. and subtract 12 from the sum.

Step 3. Sum on a $10 \log_{10}$ antilog basis the results of Step 1 for the time period of 10:00 P.M. to 7:00 A.M. and subtract 2 from the sum.

Step 4. Sum on a $10 \log_{10}$ antilog basis the results of Steps 2 and 3. The result is called the Composite Noise Rating in EPNdB, EPNdB', EdB(D'), or EdB(A') depending on the units of sound measurement used.

Calculation of CNR from PNL Values Taken Every 0.5 Sec

$$\text{Formula 8: } CNR = \left[10 \log_{10} \left[\overbrace{[\Sigma_i \log_{10}^{-1} PNL_i/10]}^{\text{7 A.M.–10 P.M.}} \right] - 24 \right.$$
$$\left. \mathbin{|+|} 10 \log_{10} \left[\overbrace{[\Sigma_i \log_{10}^{-1} PNL_i/10 + 10]}^{\text{10 P.M.–7 A.M.}} \right] \right] - 24$$

where $|+|$ is addition on $10 \log_{10}$ antilog basis and i is successive 0.5-sec intervals of time.

Step 1. Sum on a 10 $\log_{10}$ antilog basis, the PNLs of all sounds that exceed 60 at a given location outdoors, or 40 indoors between the hours of 7:00 A.M. and 10:00 P.M.

Step 2. Sum on a 10 $\log_{10}$ antilog basis, the PNLs of all sounds that exceed 50 at a given location outdoors, or 30 indoors between the hours of 10:00 P.M. and 7:00 A.M., and then add 10 to the sum.

Note: The addition of 10 to the sum of PNLs for the hours of 10:00 P.M. to 7:00 A.M. is based on the finding that people tend to complain more about environmental noise in those hours than for the hours 7:00 A.M. to 10:00 P.M.

Step 3. Sum on a 10 $\log_{10}$ antilog basis, the results of Steps 1 and 2 and subtract 24. The result is called the Composite Noise Rating (CNR) from EPNdB, EPNdB', EdB(D'), or EdB(A') depending on the units of sound measurement used.

Note: The number 24 is a constant equivalent in the present formulation to an arbitrary constant of 12 that has traditionally been included in the calculation of CNR.

CNR Obtained from a Graph

Figure 244 provides a graphic means of converting noises of a given PNL present continuously or intermittently during a 24-hour period into their equivalent, approximate CNR and EPNL value.

The CNR and EPNL values thus obtained from only the Max PNLs of noises will be closely equivalent, within one unit, to those calculated from procedures given with formulas 7 and 8 whenever the rise and decay time of the noises is shorter than the duration of the noise at its Max PNL (for example, a duration at Max PNL of 5 seconds with a rise and decay time of less than 5 seconds). When the rise and decay times of the noise to and from their maximum levels is appreciably long compared to the duration at maximum level, it is advisable to use the procedures given with formulas 7 and 8 above or to enter Fig. 244 with each PNL level present during a 24-hour period for given durations.

CNRs from Fig. 244 for noises present during the 24-hour period at different PNLs are combined into the total CNR for the 24-hour period by summing the individual CNRs on a 10 $\log_{10}$ antilog basis. Figure 15 for CNRs of different values, and Fig. 17 for a number of CNRs of the same value, can be used as aids for this calculation.

For example, in a given neighborhood, if the CNR from background industrial noise is 100 and that from traffic noise is 104, the total CNR (see Fig. 15) from these two sources is 105. Another example: in a given neighborhood, if the CNR from the flyover of an aircraft is 90, and there are 100 such flyovers per day, the total CNR from this source would be 110 (see 10 $\log_{10}$ function on Fig. 17).

The procedure given with formula 7 is the one normally to be used for measuring the CNR of an environment containing a variety of sounds from possibly unspecified sources—for example, the noise environment near a highway or airport that is used by unspecified numbers and types of vehicles operating according to a variety of more or less unspecified procedures.

The procedures given with formulas 7 and 8 are the ones normally to be used for calculating the CNR of an environment from knowledge of the PNLs or EPNLs of specified sources—for example, the noise environment to be expected near a highway or airport that will be used by specified numbers and types of vehicles operating according to specified procedures.

PART III

NONAUDITORY SYSTEM RESPONSES TO NOISE

Introduction

Casual introspection reveals that stimulation of the ear has effects on parts of the body and nervous system other than those concerned with what can be called the "hearing" phenomena described earlier. These other effects, called herein nonauditory system responses are, for the most part, the result of the stimulation by the auditory system of three neural systems that are not devoted exclusively to audition:

1. The so-called autonomic nervous system which controls general somatic responses and the state of arousal of the body—the glands, viscera, heart, blood vessels, etc.

2. The so-called reticular nervous system which appears to be involved in the state of arousal of the higher brain centers of the central nervous system and with sensory inputs related to pain and pleasure.

3. The cortical and subcortical brain centers concerned with cognition, consciousness, task performance, "thinking," etc.

Figure 245 is a schematic diagram of this somewhat arbitrarily simplified and not completely understood functional neuroanatomy of the auditory and related nonauditory systems. There are, of course, many other interconnections among the parts of man's neural structure that are not suggested in Fig. 245.

Biological Considerations

It is probably not necessary for present purposes to discuss in more detail the physiology of the nervous system or other body structures and organs. However,

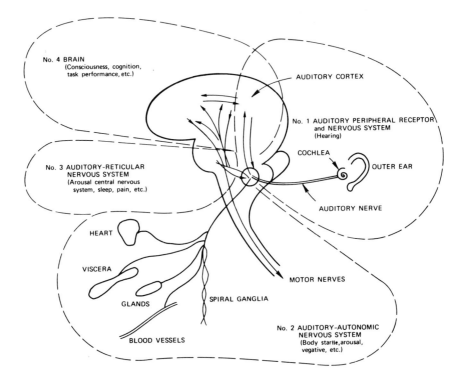

FIGURE 245.   Schematic diagram of primary auditory (hearing) and secondary auditory (nonauditory) systems.

it will be helpful to the interpretation of some of the results of research studies to be described below to have in mind the general schema of Fig. 245 and the following principles that seem to underlie the functioning of successful biological organisms:

1. A sensory receptor, such as the ear, will be much more sensitive to the form of energy that normally activates it than will be any other part of the organism.

2. A sensory receptor will not transmit the energy it receives to other parts of the organism, but will transmit only signals (usually nerve impulses) that the receptor system generates, i.e., the physical energy of the stimulus, as such, is not transmitted.

3. A sensory receptor will be damaged by excessive amounts of the physical energy to which it is especially sensitive before any other component of the organism adversely affected by that energy.

4. A sensory receptor will not generate signals (either electrical impulses or chemicals) that can harm any part or component of the normal organism.

5. The integrity of parts and mechanisms of the organism will tend to be maintained or strengthened as the result of responding to normal stimulation.

6. Responses of parts and mechanisms of the organism that serve no useful purpose will tend to be inhibited by the organism (i.e., adaptation will take place).

Health

These postulates, which are drawn from biological and physiological theory, have at least one important implication for the evaluation of the effects of noise on man, namely, sound or noise, as defined, will not have any harmful physiological effects on man other than to the auditory receptor mechanism. The hypothesis that it is not possible through normal stimulation of the ear to activate directly neural, muscular or glandular mechanisms in man, to his longtime detriment, appears agreeable, though not unequivocably, with present research data.

Adaptation

The auditory system appears to have developed phylogenetically to serve as a warning system for approaching objects, as attested by the nearly direct connections between the ear and the autonomic nervous system which help prepare the organism for bodily action. Sound or noise, being omnidirectional, capable of bending around obstacles, and present both day and night, is highly suitable as a warning signal. Justifiably, sound, as a warning signal, has a special role and significance that would seemingly mitigate against adaptation if it is to be reliable in this regard. However, adaptation to nondangerous stimuli is probably also a necessary condition for the maintenance of both general body health and the functional usefulness of the receptor involved. Without adaptation to noises signifying nondangerous sources, the organism or its receptors would perhaps indeed be too fatigued to respond to auditory signals from dangerous sources.

Performance

It is sometimes asserted that sound or noise may have adverse effects upon man's ability to perform nonauditory mental and motor tasks, even though the noise is not physiologically harmful. This possibility, which at first thought may seem reasonable, is not necessarily true. Indeed, as we shall see, it is a commonplace observation that, with respect to most nonauditory system responses, noise is sometimes reported to have a good effect, a bad effect, or no effect. Further, the correct meaning and interpretation of the results of some of these experiments are open to debate and the magnitude of the effects, when found, is often so small their measurement is difficult.

Habitual Noise

Finally, by way of introduction to Part III, it should be mentioned that, as with auditory system responses, it is man's nonauditory system responses and behavior to regular, habitual environmental noises as heard over days or months, that is of the most practical interest to society and to those wishing to control noise and set tolerable limits. Much of the research, as will be discussed later, on the nonauditory system responses to noise, although of great importance, has unfortunately been concerned only with the effects of relatively brief exposures to noise. More appropriate has been the practice in studies of nonauditory system responses to avoid using sounds or noises that have meanings and associations peculiar to a few given individuals. This has not ruled out, of course, concern with the obviously important questions of basic individual or group difference.

The research studies to be discussed have been divided into two, not independent groups: Chapter 12—those concerned with nonauditory physiological responses, and Chapter 13—those concerned with performance on nonauditory mental and motor tasks.

Chapter 12

General Physiological Responses to Noise

Somatic Responses

Studies of R.C. Davis and colleagues (182-186), and others in the United States, and Tamm (790), Meyer-Delius (504), Lehmann (491-494), Jansen (401-406), Oppliger and Grandjean (596-597), and Rossi *et al.* (705), to name a few in Europe, have been concerned with measuring certain nonauditory physiological reactions in man when stimulated with sound. The works of Davis and Jansen have been particularly extensive and important. In general, underlying these studies has been the hypothesis that these physiological reactions, particularly those classified as somatic responses, could possibly be related to feelings and emotion, to bodily health, and to the ability of the person to perform mental or motor tasks.

Davis *et al.* (186) questioned whether it is constructive to develop any biological philosophy as to whether or not somatic responses to noise represent attempts on the part of the organism to achieve honeostasis, or to overcome "counterrelevant forces" within the body. More important is that these responses are widespread in the body, vary considerably among individuals, are greater for intense than weak sound, and cease or adapt out with continued stimulation. However, the exact course and degree of adaptation of all these responses has not been very thoroughly studied.

The N-Response

Davis *et al.* called the following complex of responses to sound the N-response:

1. A blood circulatory response dominated by vasoconstriction of the peripheral blood vessels with other adjustments of blood pressure throughout the

body and minor changes in heart rate. One general result of this response would be to increase blood flow to the brain whose blood vessels show no constriction with such stimulation.

2. Slow, deep breathing.

3. Galvanic skin response (GSR, a change in the resistance of the skin to electricity).

4. A brief change in skeletal-muscle tension.

To this list might be added, among others, the changes that occur in gastrointestinal motility (Davis and Berry [184] and Stern [764]), and chemical change in the blood and urine from glandular stimulation (Hale [335], Levi [498], and others). Davis *et al.* note that these responses cannot necessarily be called "fear," "startle," or "anxiety" because some of them can be associated with emotion-arousing, and some with emotion-suppressing, activities of the autonomic nervous system.

These various responses appear to be highly interrelated. As a result, some measure of blood circulation (usually blood volume or pulse in the skin of a finger) and skeletal muscle response (such as from the muscles of the forearm) have much to recommend them—they can be readily sensed and recorded by available electronic means without greatly inconveniencing the subject, they represent somewhat different motor and nervous systems of the organism, and they appear to be more reliably related to sound variables than some of the other somatic responses.

Figures 246 and 247 show, respectively, the effect of variation in the intensity of a tone on skeletal muscle action potential (measured by an electromyograph, EMG, from the forearm) and vasoconstriction of peripheral blood vessels (measured as the difference between maximum and minimum volume pulse from a plethysomgraph on the tip of the index finger). Also shown on Fig. 246 is the adaptive effect of stimulus repetition as occurred during successive quarters of a test session. Figure 248 shows the general adaptive behavior of various somatic responses to repetitons of a tonal stimulus.

These responses are not peculiar to auditory stimulation. Although the pattern of the N-response may be different for different types of stimuli, comparison of Fig. 246 with Fig. 249 shows that at least for skeletal arm muscle potentials, the response is somewhat similar for various cutaneous stimuli and for the tone.

It is perhaps interesting to note that at a sound pressure level of somewhere near 70 dB or so, the 1000 Hz tone starts to become effective in the elicitation of the N-responses; the tone at this level is at a level that, if continued for a long enough time, will cause TTS and NIPTS, and is near the level (or comparable level in terms of perceived noisiness) that broadband noise may become somewhat significantly aversive to people.

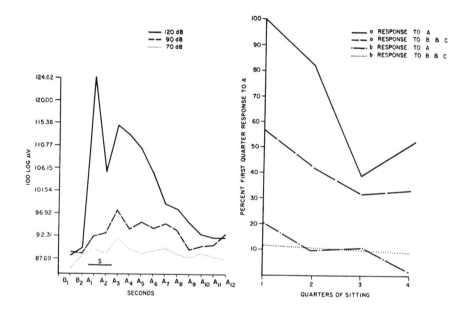

FIGURE 246.   Left graph: Mean muscle-action potential (left forearm) response to 1000-cycle tone of varying intensities. Horizontal line indicates stimulus duration. On the abscissa, B indicates periods before and A indicates periods during and after stimulation. Right graph: Adaptation of muscle-action potential responses, (a) brief latency and (b) long latency, to various intensities of a 1000-cycle tone presented in mixed order during a 15 min sitting. A = 120 dB, B = 90 dB, C = 70 dB. From Davis *et al.* (186).

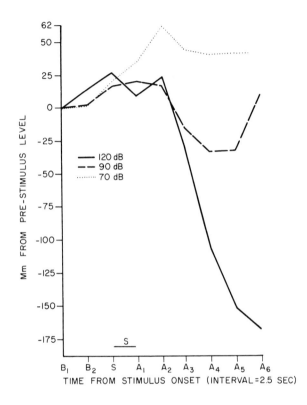

FIGURE 247.   Effect of 1000-cycle tone on finger volume. Downward change represents
constriction. Two-sec stimulus begins at S on base. Davis *et al.* (186).

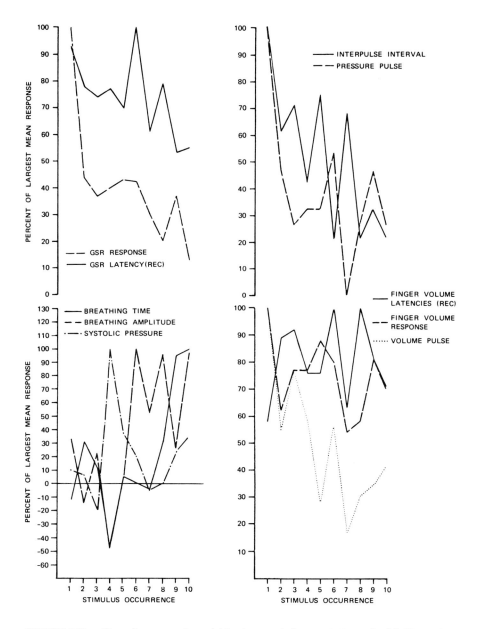

FIGURE 248.   Size of response in variables to repeated presentations of a 98 dB rough tone of approximately 800 cycles. From Davis *et al.* (186).

Breathing Amplitude

Davis *et al.* found that adaptation of the magnitude of chest movement during breathing, called breathing amplitude, followed a somewhat different course than did the adaptation of the other N-responses. Figure 250 shows the relation between breathing amplitude and tone intensity and, on the insert, changes in the magnitude of the response with continued repetition of the tone. It is seen in the insert to Fig. 250 that, in the presence of the 120 dB tone, the amplitude increased rather than decreased with continued repetition. Whether this represents some significant phenomenon or is due to some experimental error cannot be adequately assessed on the basis of present data.

G.I. Motility

Numerous investigators suggest that it is the change, either an increase or decrease, in man's environment that elicits the somatic, and perhaps other, responses rather than merely an increase in sound level. This is perhaps a fact, as will be shown to some extent in the discussion later of task performance in noise, but it has not been well demonstrated experimentally. Figure 251 shows that a change from low to moderate level of stimulation caused an increase in gastrointestinal (G.I.) motility, whereas a decrease from a high level caused a decrease. Although the study purported to show the importance of a "contrast" effect, it can also be interpreted as revealing that (*a*) higher intensities of sound (plus in this study, light) cause more G.I. motility than do lower levels, and (*b*) continued exposure in any level, at least those studied, which were rather modest, results in adaptation to stimulation.

Response-Contingent Effects of Noise

Davis and Berry (184) and Stern (764) found that humans who could avoid an 80 dB, 10-sec 800 Hz tone by pushing a switch at the correct time (every 5 sec) exhibited greater G.I. motility during the tone (i.e., when they had failed to push the switch) than did subjects who had no means of avoiding the same tone burst given at random times. The tone caused some G.I. motility that adapted out for the no-task group, but the motility to the tone tended to be large or to increase for those times that the tone avoidance task group failed to push the switch at the correct time. The noise thus became an aversive stimulus primarily because it signified incorrect behavior on the part of the subject; its aversive effects without the task contingencies were very small. Noise is not, as surmised by Brady *et al.* (86) in commenting on the Davis and Berry study, a significantly aversive, unconditoned stimulus.

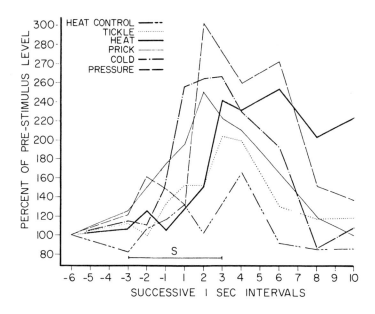

FIGURE 249. Muscle-action potentials in the stimulated arm. Mean percentages of pre-stimulus level are plotted for each stimulus. From Davis *et al.* (186).

As mentioned earlier, responses to noise that are the result of differential conditioning of people by some means of reward or punishment (in real life or in the laboratory) are considered outside the purview of this document. The particular study just discussed was included because of possible misinterpretation of the meaning of the results obtained.

Vasoconstriction to Noise During Exercise

The studies of Davis *et al.* and others cited above were for the most part conducted with the subject in a state of rest or partial rest in the laboratory. Figure 252 from a study by Jansen (403) shows finger pulse amplitude (constriction of blood vessels causing a decrease in amplitude) of subjects during periods of rest and work (a bicycle ergrometer) when in quiet and in the presence of noise. These data show several important phenomena:

1. The flow of blood in the finger is more significantly changed by the work task than by the noise (the work caused a dilation of the blood vessels, the noise a constriction).

2. There was apparently little or no adaptation of the dilation response from the work, even some continued increase as work continued, as perhaps is to be expected due to heating of body and metabolism.

3. There was at least some adaptation to the constriction response from the noise, as indicated by the reversal of the constriction after the first minute or less of the noise.

With respect to the latter point, it should be noted that the duration of the noise exposures were for the most part too brief (1 to 2 minutes) to permit complete adaptation. Jansen states that in another experiment the subjects showed constriction of the blood vessels when exposed to noise for the successive times they returned to the laboratory for testing.

Startle

The average somatic arousal responses to sound discussed above are not what one would call a startle reaction to an unexpected frightening stimulus. While many of the somatic responses cited above are involved in the fright-startle reaction, the pattern is different. In particular, there is usually a change in rate of heart beat and blood pressure from startle that is not found in the arousal noted from sounds or noises that are more or less expected or normal to an experimental or real-life situation.

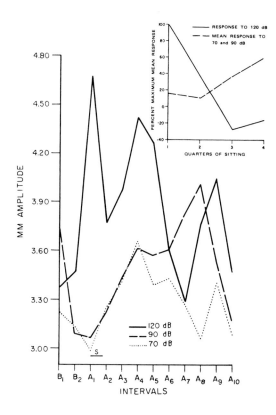

FIGURE 250. The effect of a 1000-cycle tone of varying intensities on breathing amplitude. Horizontal line indicates stimulus duration. Intervals are 2.5 sec each. Each point is the mean amplitude of 30 Ss to four presentations of a stimulus. Insert: Adaptation of breathing-amplitude responses to a 1000 Hz tone of several intensities presented in mixed order. From Davis *et al.* (186).

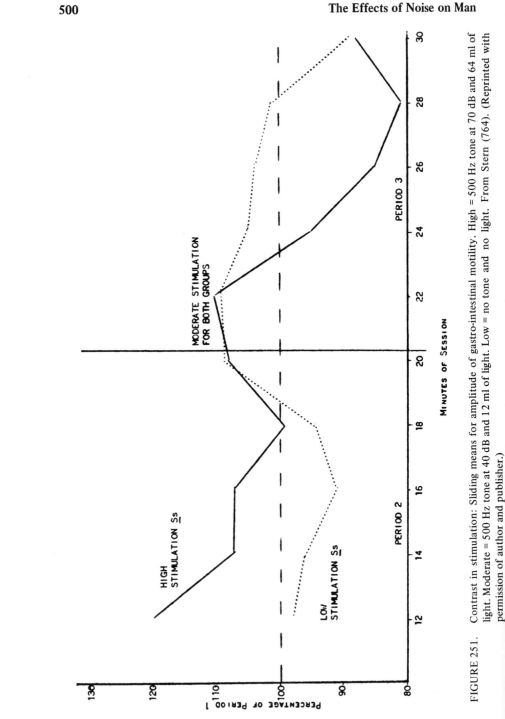

FIGURE 251.   Contrast in stimulation: Sliding means for amplitude of gastro-intestinal motility.   High = 500 Hz tone at 70 dB and 64 ml of light. Moderate = 500 Hz tone at 40 dB and 12 ml of light. Low = no tone and no light. From Stern (764). (Reprinted with permission of author and publisher.)

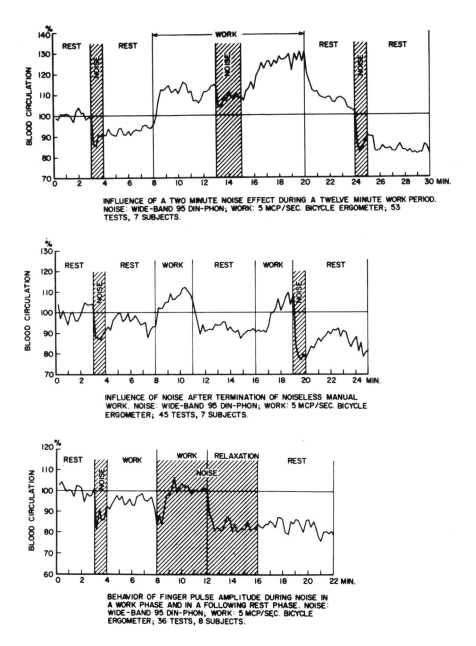

FIGURE 252. Effects of combinations of work, rest, and noise on blood circulation. A reduced percentage indicates vasoconstriction of the peripherical blood vessels. From Jansen (403).

The possible difference and interplay between these two response patterns—auditory arousal and startle—was perhaps shown in a study by Hoffman and Fleshler (386). This investigator found that animals (rats) who were in a background of either steady or pulsed noise reacted to a gunshot-like burst of acoustical energy much more violently (movement measured in an automatically recording activity cage) than did animals who were in silence or pulsed noise, as shown in Fig. 253. These results are somewhat inexplicable in terms of the effect being merely an addition of arousal with startle because the pulsed noise would seem to also have caused arousal; perhaps the explanation is to be found in the fact that some rats respond to sounds of sufficient intensity and duration with so-called audiogenic seizures and that these various patterns of acoustic stimulation also involved to some extent this phenomenon. Hoffman and Fleshler found adaptation to the startle response as shown in the right-hand graph of Fig. 253. Davis (182) and Davis and Van Liere (185) found similar adaptation of human subjects to repeated loud noises or shots from a blank pistol, and Pearsons and Kryter (613) found similar adaptation to a startle response (heart rate) to simulated sonic booms. Illustrative of the pervasive principle of adaptation of organisms to sound is the finding of Bartoshuk (44) that the acceleration of the heart beat rate in unborn babies to bursts of acoustic clicks (85 dB level) adapt out by the end of 40 trials. Ando and Hattori report that babies of mothers from neighborhoods subjected to aircraft noise were much less aroused from sleep by aircraft noise than were babies from mothers who lived in quiet neighborhoods during pregnancy.

It is perhaps worthwhile noting that while man (and no doubt most other hearing animals) adapts to background or "regular" noise, he readily responds when the auditory stimulus is changed in level or character. Thus this adaptation, in accordance with the usual definition given the word, must be recognized as some neural inhibition process and not a "fatiguing" of the organism. For example, Rossi *et al* (705) found that the adaptation of vasoconstriction in subjects exposed to a background noise (500 Hz tone at 70 dB) did not reduce vasoconstriction to superimposed 2000 Hz tones at levels of 80 to 105 dB.

Although physiological adaptation to impulsive noise stimuli appears to occur, this does not mean that people psychologically find impulsive sounds pleasant or acceptable, particularly when very intense. Data relevant to this point was presented in the earlier discussion of the effects of sonic booms on people.

Stress and Health

The word stress, implying an actual or eventual debilitating effect, is often loosely used to signify a state of arousal in an organism. Whether the arousal is stressfull in a debilitating sense is often a moot question. The physiological facts

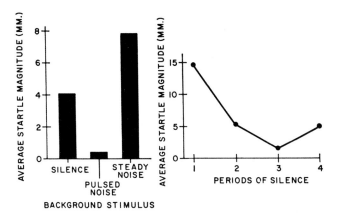

FIGURE 253.   Effects of background stimulation on the startle reaction to an intense
sound. Overall effects of background stimulation are summarized for all
subjects in the bar graph. The curve to the right shows the effects of
repeated presentation of the intense sound during the four periods of
silence. From Hoffman and Fleshler (386). (Copyright 1963 by the American Association for the Advancement of Science.)

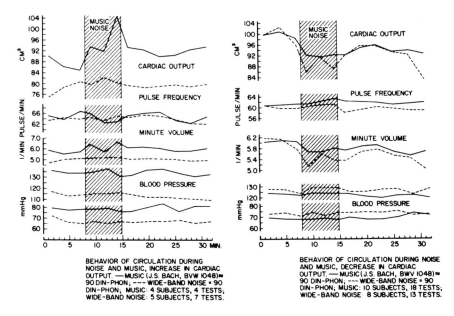

FIGURE 254.   Differential response of subjects to music and noise. Cardiac increase (left
graph), cardiac decrease (right graph). From Jansen and Klensch (405).

of arousal, as suggested earlier in the biological postulates, probably do not mean that when a person is in a state of arousal he is less able to perform mental and motor tasks or that the arousal "wears him out" so that in the long run he is less productive.

To distinguish physiologically between conditions of stress and arousal is difficult, if not impossible. For example, Jansen and Klensch (405) found that either random noise or music of somewhat equal intensity (a loudness or perceived noise level of about 90 Phon or PNdB) caused very similar blood circulatory responses in subjects, as shown in Fig. 254. The circulation response reported in Fig. 254 was measured by whole body movement from heart pulsations, the so-called ballistocardiogram.

Figure 254 also shows that while the majority of the subjects (18 vs. 9) showed decreased caridac output and minute volume flow (presumably to the periphery of the body) during the noise or music, some behaved otherwise. Unquestionably, individual differences with respect to some of these somatic responses to sound can be significant. None of the subjects showed significant change in pressure or pulse frequency, as is consistent with other studies of somatic responses to sound.

The importance of the similarity between the effects of music and the noise lies, or course, in the fact that it was apparently the intensity of the sound and not its aversive (presumably noise) or its pleasurable (presumably music) aspects that controlled the somatic responses. It is possible, however, that the music, although not abnormally intense for concerts, was stressful because of its level.

Related and similar results on "stress" were obtained by Levi (497-498). Levi used as a measure of stress the presence in the urine of his subjects of certain chemicals that result from excretions of the endocrine glands. These glands are regulated to some extent (and to some extent vice versa) by the autonomic nervous system. The N-complex of responses are presumed by most investigators to be related in certain ways to these glandular responses.

In any event, Levi found the following:

1. Pleasant stimuli (motion pictures evoking amusement) were nearly as potent as unpleasant stimuli (motion pictures evoking anger) in causing increased excretion of catecholamines.

2. Work in industrial noise and office work caused increased excretion of catecholamines.

3. Noise, light, or task have less influence on the catecholamine excretion levels than does the subject's attitude.

4. Emotionally vulnerable people as a group do not excrete more catecholamines than do normal people under experimental stress.

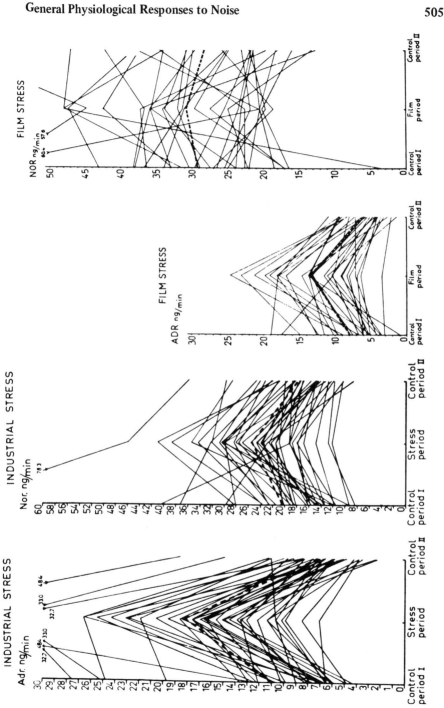

FIGURE 255. Urinary excretion of adrenaline (ADR) and noradrenaline (NOR) before, during, and after two hours of simulated industrial work (sorting steel balls in presence of realistic industrial noise) by healthy soldiers (left graphs), and two-hour film program depicting violence (right graphs). Short-dash line indicates mean values. From Levi (498).

Figure 255 shows an example of the endocrine output data collected by Levi. It is seen in Fig. 255 that the pattern and magnitude of change in some adrenal gland excretions was similar for the subjects in the noisy work situation and the subjects seated in the movie theater. The higher base output of the men who worked, relative to those viewing the movie, is no doubt related to the general attitudes of the subjects to the two situations.

Adaptation to Stress

By and large the physiological experiments that have been done on stress have been rather short-term or have not included all the same physiological measures so that conclusive statements are not justified, particularly with respect to individual differences. Data obtained by Levi seem to show some adaptation, or no large fluctuation caused by continued work, during a day, and examination of the data from the previously reported laboratory studies seems to show that a given response mechanism does tend to adapt to the acoustic environment.

Anthony and Ackerman (34) found that rats, mice, and guinea pigs all showed physiological arousal or stress to intense broadband noise (110 dB to 140 dB), but that behavioral and histological data showed no harmful effects, and that satisfactory adaptation had occurred. However, while the noise by itself was not harmful, it appeared to reduce survival time when coupled with food restriction for those animals showing audiogenic seizures (muscular spasms and "fits", sometimes coupled with unconsciousness that are peculiar to certain strains of mice and rats, particularly when suffering from middle ear infections). Because of the audiogenic seizure syndrome, it is not justified to generalize to man the results of auditory-stress experiments performed with rats. It appears that rats or mice that experience these seizures suffer abnormal auditory systems and/or infections (429a) and must be specially bred (111a).

We know of no laboratory tests (those of Harmon [338] and Finkle and Poppen [228] done in the 1940s on noise still appear to remain valid) that show that in man complete physiological adaptation does not occur to noise. It must be emphasized that not a great deal of research data is available on this question, but what data are available seem to support that notion; as Levi (498) pointed out, the attitude of the person is probably more important to stress or arousal than is the task, and the task is more important in this regard than the noise environment. The graphs on Fig. 256 from Helper (367) show the relatively greater effect on pulse interval, skin conductance (GSR), and skeletal muscle tension of confronting the subjects with a mental performance task, such as arithmetic, than with 110 dB of random noise alone.

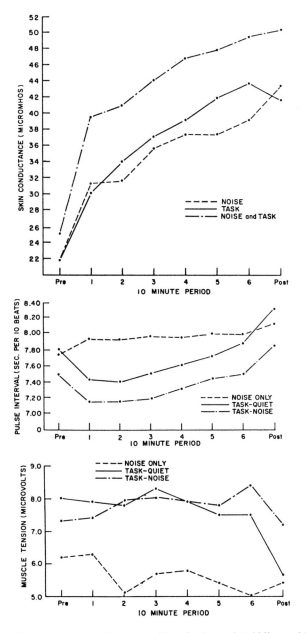

FIGURE 256.   Skin conductance (upper graph), pulse interval (middle graph), and mean muscle tension (lower graph), in successive 10-min periods, of three types of experimental hours. Pre and Post represent 10-min control periods. From Helper (367).

Data from Industry

It might be expected that in some industries one would find possible harmful effects to man's nonauditory systems as the result of long-term exposures to intense noise. Perhaps the adaptation to the noise arousal as measured in the laboratory is not complete and there is an accumulated effect over the years that becomes significant. Some studies have been published that purport to show such effects or the possibility of such effects.

Figure 257 from Jansen (402) shows some greater percentage of blood circulation, heart, and equilibrium problems in workers from the more intense noise than from the less intense noise environments. Table 80 from Shatalov *et al* (734) compared various cardiovascular characteristics of persons who worked in a moderate noise (spinning mill, 85-95 dB) and in a ball bearing plant (114-120 dB). Figure 258 (Andriukin [15]) presents data which show that the incidence of hypertension (arterial blood pressure) tends to be greater in workers exposed to high frequency shrill lathe noise and to the very intense broadband noise found in ball bearing producing shops than in men working in less intense noise. Additional data collected, for the most part, in Russia (16, 734, 784) indicates that workers in heavy, noisy industries (foundries, etc.) suffer unusually high percentage of circulatory, digestive, metabolic, neurological, and psychiatric difficulties.

All of the authors responsible for the industrial data just discussed conclude, or imply, that intense noise is probably responsible to a significant extent for the physiological problems exhibited by the workers. However, the presence of other possibly important factors in these industrial work situations must not be overlooked: (*a*) poor ventilation, heat and light; (*b*) danger from accidents; (*c*) anxiety over job security; and (*d*) personnel selection (on the average those men in the noisier jobs were perhaps less well off in terms of general health and economic and social status prior to and during employment, than were the men in the less noisy occupations). While it is not possible to rule out noise as a major or contributing factor to the reported adverse physiological conditions of some industrial workers, other nonnoise factors are also reasonable explanations for the differences found among the workers. The study summarized in the next paragraph may throw some light on this question.

Tables 81, 82, 83, 84, 86, and 87 are taken from a study made of some personnel involved in aircraft launch operations aboard a U.S. Navy aircraft carrier (172, 173). Figure 259 illustrates the level and duration of the noise present at the most exposed operational positions. While Tables 81-83 show no completely clearcut differences between the general abilities of the men least and most exposed to the aircraft noise, there appears to be a consistent tendancy for the most exposed to perform the least well. Also, psychiatric examination of the men revealed that the most exposed men had somewhat greater feelings of

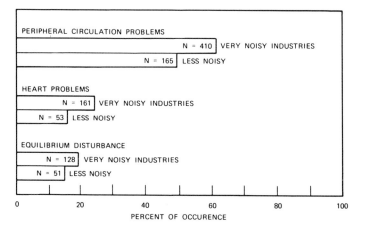

FIGURE 257. Differences in percentage of occurrence of physiological problems in 1005 German industrial workers. The differences in peripheral circulation and heart problems in the two classes of industry were statistically significant. After Jansen (402).

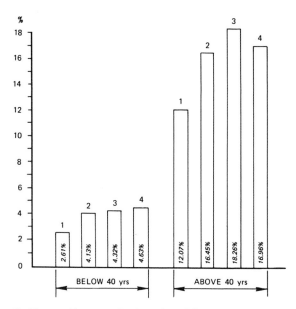

FIGURE 258. Incidence of hypertension in male and female workers (in age groups under and above 40 years) in noisy workshops: (1) tool making workshop; (2) sorting workshop; (3) workshop with automatic lathes; (4) workshop producing ball bearings. From Andrukin (15).

TABLE 80

Comparative Data on Certain Indices of Cardiovascula State in Spinning Mill and Ballbearing Plant Workers
From Shatalov et al. (734).

Cardiovascular indices	Spinning mill			Ballbearing plant		
	Number of subjects	Changes detected	Percentage of subjects	Number of subjects	Changes detected	Percentage of subjects
Uncomfortable heart sensations	156	62	30.9	144	68	47.2
Bradycardia	156	28	17.9	86	39	27.0
Heightened maximal blood-pressure	156	14	8.9	144	9	6.9
Lowered maximal blood-pressure	156	11	7.0	144	26	18.0
Retardation of intravasicular conductivity in the electrocardiogram	144	55	38.1	86	8	9.3
Depression of T-wave in the electrocardiogram	144	16	11.1	86	11	12.7
Depression of T-wave after stress	144	48	33.3	85	23	27.0
Depression of T-wave after work	33	7	21.2	9	5	55.5

TABLE 81

For the "Most-Exposed" and "Least-Exposed" Groups on the Large-Sample Tests, Analyzed According to Each of the Criteria for Hazardous Noise Exposure

The score of the group exhibiting the poorest performance for each comparison has been underlined. From Davis (172).

TEST	TYPE OF TEST*	CRITERION OF HAZARDOUS NOISE EXPOSURE					
		AUDITORY THRESHOLD SHIFT		ESTIMATE OF NOISE EXPOSURE		SQUADRON ASSIGNMENT	
		MOST EXPOSED	LEAST EXPOSED	MOST EXPOSED	LEAST EXPOSED	MOST EXPOSED	LEAST EXPOSED
Critical Flicker Frequency	H Score = / N	38.9** / 7	40.8 / 30	40.6 / 15	41.3 / 12	40.5 / 24	39.6 / 55
Tapping Speed	H Score = / N	20.8 / 7	20.8 / 30	19.5 / 14	20.1 / 12	20.7 / 24	20.1 / 54
Reaction Time	L Score = / N	204.9 / 7	191.6 / 30	218.2 / 14	201.1 / 12	199.6 / 24	199.5 / 54
Fine Hand Steadiness	L Score = / N	32.5 / 7	26.7 / 29	25.2 / 15	31.9 / 10	28.8 / 21	26.0 / 48
Gross Hand Steadiness	L Score = / N	17.9 / 7	15.7 / 30	15.2 / 15	14.7 / 10	18.7*** / 22	14.8 / 49
Steadiness of Standing	L Score = / N	35.4 / 7	28.9 / 30	30.9 / 15	27.3 / 11	31.4 / 21	31.0 / 50

* For a test designated as "H", a high score indicates good performance and a low score indicates poor performance; for a test designated "L", a low score indicates good performance and a high score indicates poor performance.

512

TABLE 82

For the "Most-Exposed" and "Least-Exposed" Groups on the Small-Sample Tests, Analyzed According to Each of the Criteria for Hazardous Noise Exposure

The score of the group exhibiting the poorest performance for each comparison has been underlined. From Davis (172).

TEST	TYPE OF TEST*	CRITERION OF HAZARDOUS NOISE EXPOSURE							
		ESTIMATE OF EXPOSURE				SQUADRON ASSIGNMENT			
		MOST-EXPOSED	N	LEAST-EXPOSED	N	MOST-EXPOSED	N	LEAST-EXPOSED	N
Tactual Threshold	L	_10.4_	11	5.8	13	8.2	14	_8.5_	14
Knox Cube Test	L	_2.7_	11	2.4	13	2.1	14	2.1	14
Dexterity Test	H	_15.9_	11	17.4	13	_16.4_	14	16.8	14
Digit Symbol Test	H	_52.6_	11	52.8	13	52.3	14	_50.7_	14
Taylor Anxiety Scale	L	_10.7_	11	6.3	12	_10.1_	14	7.5	12
Saslow Screening Inventory	L	1.6	11	_2.5_	13	_2.6_	14	2.0	13
Cornell Index	L	_4.7_	11	3.1	13	_5.1_	14	3.5	13
Visual Acuity	H	_21.1_	11	21.8	13	21.0	14	_20.7_	14
Visual Phoria	H	4.8	11	_5.1_	12	_5.1_	14	_6.3_	12
Depth Perception	H	6.1	10	_5.1_	12	_5.2_	14	5.5	13

* For a test designated as "H", a high score indicates good performance and a low score indicates poor performance; for a test designated as "L", a low score indicates good performance and a high score indicates poor performance.

TABLE 83

Adjusted Mean Scores on Factored Aptitude Tests, with the "Most-Exposed" and "Least-Exposed" Groups Statistically Matched in Terms of Intelligence Level

A high score indicates good performance for all tests. The poorest performance for each comparison has been underlined. From Davis (172).

TEST*	ESTIMATE OF NOISE EXPOSURE			
	MOST-EXPOSED	N	LEAST EXPOSED	N
Aiming	63.7	8	70.4	8
Number Facility	21.5	9	26.2	8
Speed of Closure	13.0	9	17.5	8
Speed of Symbol Discrimination**	17.7	9	26.7	8
Deduction	15.3	8	15.4	8
Associative Memory	5.6	9	8.6	8
Induction	9.5	9	10.0	8
Visualization	10.2	9	8.7	8
Motor Speed	10.0	9	11.4	8

* Tests are denoted by the aptitude factors which they purportedly measure.

** For this test the difference between means was significant at the 5 per cent level of confidence.

TABLE 84

Percentage of Subjects in Each Noise-Exposure Group who Expressed Negative Reactions to Their Job, to Jet Noise, and to Shipyard Duty. From Davis (172).

NOISE EXPOSURE	EXPRESS ANXIETY ABOUT JOB	STATE JET NOISE DISTURBING	DISLIKE SHIPBOARD DUTY
Very High	92.3	53.8	69.2
High	61.1	55.5	55.5
Moderate	71.4	61.9	76.1
Low	21.4	35.7	42.9

anxiety than the others. It is tempting to blame the generally poorer showing of the most exposed men on the fact that they suffered more anxiety than did the men less exposed to the noise and aircraft because their jobs were inherently more dangerous or difficult. This seems borne out by the data in Table 84 which show that the men most exposed to the noise did not rate the jet aircraft noise as more disturbing than did some of the other groups, but did express the most anxiety about their jobs.

It must not be forgotten that the presence of intense noise usually signifies the nearby operation of powerful machinery that may be dangerous or requires skillful control to avoid danger. This anxiety by itself, rather than the arousal effects of the noise, may be the source of the nonauditory physiological problems found in some industries. To the worker this distinction is, of course, academic. However, as shown in the Navy study cited above, men involved in nonauditory work in noise often do not rate noise as a significant problem or bother once they "are used to it" even though tests may show that permanent damage to their acuity of hearing is being accrued.

Miscellaneous Health Problems Related to Noise

Brewer and Briess (88) report that one nonauditory health problem created by noise is that people working in noise develop coughs, hoarseness, lesions, and pains in their throats from the strain of talking in the noise. Also, Buyniski (112) found that deaf (not defined) employees in a large company (some or perhaps most of whom presumably suffered noise-induced deafness) made four to five times as many trips to the company dispensary per year and suffered greater medical pathologies than did the employees with normal hearing.

Bredenberg (87), Denzel (192), and Minckley (559) discuss the question of the effect of noise in a hospital upon the patients. These authors deplore the high noise levels (telephones, talking, public address systems, machinery, etc.) and suggest that recovery of health would be quicker with less noise. They present no data to support their concern although Minckley reports that the ratio of persons receiving medication to those not receiving medication was somewhat larger in the section of a ten-bed ward room where the noise level was the highest (60-70 dB) than for the beds in the average, quieter locations (40-60 dB). However, the lack of controls regarding possible relations between illnesses of the particular patients and bed assignments, relations between noise levels and the act of administration of medication and the relatively small size of the room (30' X 18'), make the results of this study suggestive rather than definitive.

Goshen (310) comments that Denzel's contention that hospital noise is deleterious to health is based on the erroneous conception, in his opinion, that because ill health produces discomfort, discomfort can produce ill health.

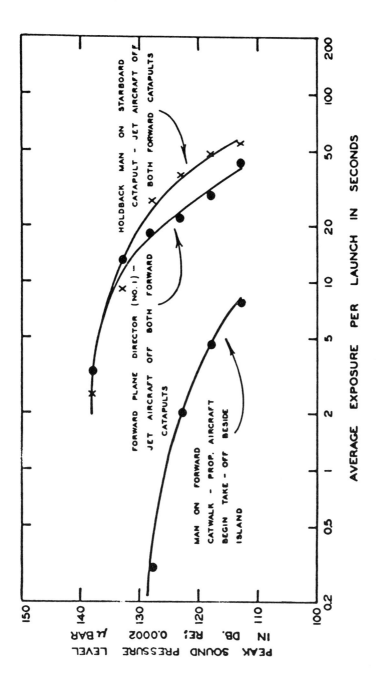

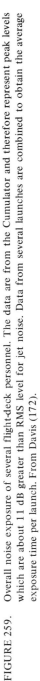

FIGURE 259.   Overall noise exposure of several flight-deck personnel. The data are from the Cumulator and therefore represent peak levels which are about 11 dB greater than RMS level for jet noise. Data from several launches are combined to obtain the average exposure time per launch. From Davis (172).

Goshen argues further that sensory stimulation, rather than being harmful, probably contributes to health. We think it could also be contended that adaptation to noise would occur very rapidly in an organism which, for some physiological or psychological reason of health, should not be aroused.

In short, it is conceivable and may be fortunately true, that the more isolation a person needs from noise or other sensory stimulation because of a physiological or psychological health reason, the less sensitivity will that person have to his noise environment. This is not to say, of course, that this suggested ability should be exercised to its limits or that the persons involved psychologically appreciate the fact. And there is the not implausible possibility that sometimes the discomfort felt by persons when ill or when trying to sleep is generalized and blamed on the noise because the noise attracts their attention rather than the noise being the basic source of their discomfort. Also, it would seem improbable, as noted earlier, that a biological system would not have internal mechanisms to protect itself from overstimulation by its own receptors. In spite of the comfort of such rationalizations, and the support of some emperical data and lack of uncontestable contrary data, the notion of possible harmful effects to some nonauditory systems as the result of auditory stimulation cannot be completely dismissed at this time.

Sleep

Much remains to be learned about sleep in man, but the following general relations appear to be established:

1. As revealed by the electroencephalogram (EEG), there are four stages of sleep, one of which looks in general pattern like the EEG of an awake person but is accompanied by rapid eye movements (REM) as well as other muscle responses.

2. Man typically spends various portions of a night of sleep in these different stages in a cyclic pattern as is shown in Fig. 260 from Williams *et al*. Figure 260 shows the percentage of time spent in different stages by a subject during: (*a*) nights called B_1 and B_2 (or base nights) of sleep when he had not been earlier deprived of sleep; and (*b*) after he had been deprived of sleep for the 64 previous hours (called R_1 and R_2, recovery nights). The greatest change between the B and R nights is with respect to the Awake (A) and REM stages.

Because the REM stage and awake EEG patterns are similar, it may indicate a state of normal cortical activity. However, cats (Jouvet [428]), man (Williams *et al*. [879]), and chimpanzees (Adey *et al*. [7]) are usually insensitive to auditory or other stimulation during REM stage; stage REM is sometimes called paradoxial sleep for this reason. Adey *et al.* suggest that in paradoxial sleep the

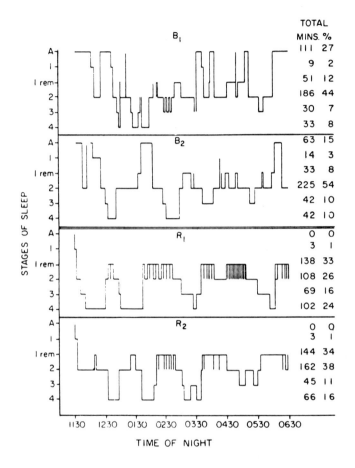

FIGURE 260.   Effect of sleep loss on the distribution of EEG stages of sleep for one
subject (Be). Sleep loss produced a decrease in the time taken to go to
sleep as well as in the amounts of stage 1 and stage 2. The amounts of
stages 3, 4, and 1_{rem} increased following sleep loss. From Williams *et al.*
(879).

organism has "internalized his attention," tends to dream, and is not in a stage of deep unconsciousness as suggested by the elevated threshold to sound.

Figures 261, 262, and 263, very clearly reveal a number of important features of the effects of noise during sleep. To obtain these data Williams *et al.* periodically exposed the subjects to 5-sec bursts of recorded random noise at various levels above the awake threshold of the listeners. These investigators recorded three types of responses that occurred in each sleep stage to the various levels of noise: (*a*) EER (sum of EEG events evoked by each stimulus) (Fig. 261); (*b*) BR (Behavioral Response) where the subject awakened and pushed a signal switch (Fig. 262); and (*c*) VCR (vasoconstriction response) (Fig. 263).

It is seen that:

1. With respect to the brainwaves (EER) and behavioral awake responses (BR), the subjects are more responsive in certain stages of sleep than in others.

2. As the intensity of the stimulus is increased, the number or magnitude of the EER and BR responses increases.

3. Following 64 hours of sleep deprivation, the number or magnitude of the EER and BR responses are less during all stages and all levels of stimulation than during the base nights.

4. The vasoconstriction of the peripheral blood vessels (VCR) behaves somewhat differently than the brainwaves and behavioral awakening. In particular, the VCR was only slightly less during the recovery nights than during the base nights, and did not differ during the different stages as much as did EER and BR. The VCR even showed greater response in stage 1 REM than in the lighter stages of sleep. It might be noted that Jansen and Schulze (406) report a similar finding for the vasoconstrictive response to noise during sleep.

One interesting finding in this important study of Williams *et al.* (879) is that the threshold level for behavioral awakening (Fig. 262) during the base of normal nights is 20 dB or so above that threshold when the subject is awake, and that after sleep deprivation the noise even at 35 dB (the highest tested by Williams *et al.*) above normal threshold seldom elicited an awakening. Kryter and C.E. Williams (unpublished data), in tracing the auditory threshold, as measured by EEG changes, to bands of random noise, pure tones, and aircraft noise (each of about 5-sec duration), found in a somewhat sleep-deprived college student an elevation relative to awake thresholds of about 30 dB in stage 2, about 50 dB in stage 3, and about 80 dB in stage 4. In the Kryter and C.E. Williams study, behavioral awakening occured to levels about 5 to 10 dB higher than those which caused a change in EEG pattern. The change in EEG and behavioral awakening was a function of the sensation level (dB above its threshold in the awake subject) of the different tones, propeller and jet aircraft noises.

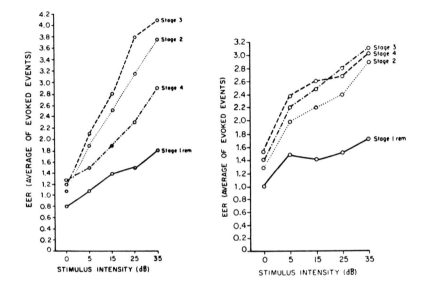

FIGURE 261.   Effect of stimulus intensity on the EER during four stages of sleep during
baseline nights B_1 and B_2 (left graph) and recovery nights R_1 and R_2
(right graph) after 64 hours of sleep deprivation. EER is the sum of EEG
events evoked by each stimulus. From Williams *et al.* (879).

Adaptation During Sleep

It is not unreasonable to suppose that some adaptation to noise takes place during sleep. It is a common observation that one sleeps better in a familiar environment than an unfamiliar environment containing unfamiliar sounds. Also, awake people can be instructed to respond or not to awaken to certain sounds when they are asleep and to ignore others. Oswald *et al.* (600) found that persons would awaken more readily to the sound of their own name than to other names.

On the other hand, the large differences in sensitivity to noise during different sleep stages, the increased insensitivity with sleep deprivation, the adaptation to arousal to specific sounds and to the environment *prior* to sleep suggest that possibly this adaptation is not simply the type of inhibitory action, conscious or unconscious, presumed to operate in the adaptation to arousal and startle responses when awake. It is proposed, as a working hypotheses, that people: (*a*) cannot learn to adapt, or not to adapt, to a noise when asleep unless they can engage in cognition relative to the input stimulus; (*b*) cannot cognate unless that stimulus exceeds their threshold of cognitive arousal, i.e., unless they are awake; (*c*) conditioning, when awake, to be particularly responsive to some sounds and not responsive to others will influence their threshold of awakening arousal when asleep in stages 1 and 2, but not in 3, 4, or REM.

Intuitively, it would seem that a reasonable compromise between the requirements to give the brain or "thinking" mechanism a rest and yet maintain some reasonable contact between the organism and the world outside would be for the organism to somehow "turn-off" (increase the auditory cognitive arousal threshold) by varying amounts during a night of sleep. This hypothesis is consistent with the studies of Simon and Emmons (221 and 744) who found that so-called sleep-learning (information recorded on magnetic tape and played via an earphone under a person's pillow) only occurred when the listeners exhibited brainwave activity associated with being behaviorally awake. To be sure, sometimes the listeners were not consciously aware (particularly the next morning) of the particular times they were awake and capable of cognitive behavior and hearing.

Also these notions for the most part are supported by studies conducted by Williams *et al.* (879) and by Rechsthafen *et al.* (662). Williams *et al.* instructed subjects prior to going to sleep that if they did not awaken to a tone to which the subjects were particularly responsive when awake (called critical stimulus), they would be aroused by a fire alarm and electric shocks to their legs. Further, the subjects were told that they would not be thus awakened if they failed to respond to another, equally loud but different pitched tone (neutral tone). They were instructed not to respond to the neutral tone even when awake. The results are shown in Fig. 264 where it is seen that the relative arousal or awakening effects of the two tones remain relatively unchanged whether or not the failures

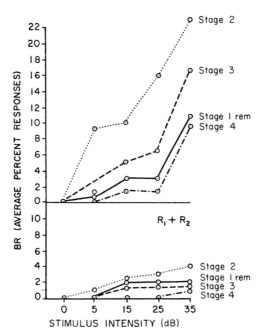

FIGURE 262. Effect of stimulus intensity on behavioral response (BR) during four stages of sleep for baseline nights B_1 and B_2 and recovery nights R_1 and R_2 after 64 hours of sleep deprivation. From Williams *et al.* (879).

to respond were punished, except in stage REM, i.e., when asleep the subjects did not learn with punishment in stages 1, 2, 3, and 4 to discriminate more effectively between the two tones but did so in stage REM. In the deep sleep stages of 3 and 4 the subjects did not discriminate, with or without punishment, between the critical and neutral tones.

Rechesthafen *et al.* likewise found that punishing the subject by shaking him awake when he failed to awaken to a sound did not tend to increase arousal to subsequent exposures to the sound except in stage REM. As commented above, there is some similarity between REM and awake stages with respect to (*a*) the state of activity of the cortical brainwaves (they show fast, low voltage activity in both cases), and (*b*) the prevalence of conscious cognition when awake and dreaming when in REM.

Rowland (706) found that dysynchronization of the EEG (presumed to mean awakening) in sleeping cats occurred to an aversive tone (conditioned in the awake animal prior to sleep through electric shocks given via electrodes applied deep in the brain), but not to an unconditioned tone. The results would mean, of course, that, as with people, discriminations between sounds learned when the cats were awake were also made when the cats were in a stage of sleep.

Consistent with the findings of Williams *et al.* and Rowland are the results of an experiment conducted with cats by Buendia *et al.* (101). These latter investigators found that cats conditioned, similarly to the procedures used by Rowland, to a tone showed discriminative awakening responses between conditioned and unconditioned tones when asleep and in "high voltage" stage sleep, but not when in the "paradoxial" stage of sleep thought to be the counterpart of stage REM in man. As noted in stage REM, as in stage 3 and 4, Williams *et al.* also found no discrimination (without punishment during sleep) between the critical and neutral tones in sleeping human subjects. Apparently, then: (*a*) people and cats can be more responsive when asleep in the high voltage, slow brainwave stages 1, 2, and perhaps 3 to sounds that have special meanings than to meaningless sounds; and (*b*) people (this has not been tested in cats) can learn this differential responsiveness only when awake or in stage REM.

Research data showing that a person, in some stages of sleep, can discriminate among auditory stimuli in terms of their meaning is consistent with anecdotes that one can "listen" for certain sounds when asleep and ignore others. This apparently is a form or recognition that is readily learned through previous awake exposure to a noise or change in the acoustic environment. Also, it is perhaps possible that one can, to some extent, control for periods his general state of arousal so that he spends more time in "light" stages of sleep than in "deep," thereby increasing the probability of hearing sounds because of his lower threshold of auditory arousal.

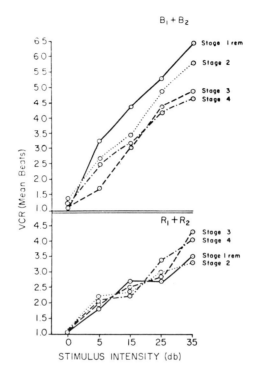

FIGURE 263.   Effect of stimulus intensity on vasoconstriction response (VCR) during four stages of sleep for baseline nights B_1 and B_2 and recovery nights R_1 and R_2 after 64 hours of sleep deprivation. From Williams *et al.* (879).

Sensitivity of Different Age Groups to Noise When Asleep

Lukas and Kryter (528) have found that older persons are much more sensitive, particularly with respect to behavioral awakening, to simulated sonic booms and recorded subsonic aircraft noise than are younger persons, as shown in Fig. 265. Indeed, the youngest subjects, age 7 to 8, were not aroused by sonic booms more intense than booms that awakened the 67- to 72-year-old men nearly 70% of the time. It is possible, but of course debatable, that older people need less "deep" sleep and are therefore more sensitive to arousal (even though their hearing acuity is less acute) than the younger people.

Sleep and Health

Grandjean (312), Lehmann (492), Richter (676), Jansen and Schulze (406), and others surmise that perhaps one of the greatest hazards of noise to man's health is that of stimulating the sleeping person. It is often hypothesized that noise damages health through sleep disturbance because of the presence of small transient changes that occur in the EEG (sometimes called the K-complex, although it is somewhat different in different stages) and the vasoconstrictive responses (VCR) and perhaps other somatic responses which can be evoked in the sleeping person by sounds at certain levels of intensity. Richter, for example, observed that a sleeping subject exhibited EEG and VCR reactions about every 30 seconds due to cars, trains, and motorcycles passing the test room, even though the person slept quietly and upon awakening had no recollection of any disturbances. Richter views these responses as indicating that the vegetive system is withdrawn from the recovery and strength-gathering process of sleep and further believes this withdrawal is detrimental to normal health.

There is no question that, when deprived of sleep, people become irritable, and may show some irrational behavior and a desire to sleep (870a). Also, when suffering most illnesses, people need sleep and rest. However, quantitative evidence that regular (or, for that matter, irregular) environmental noise causes any physiological or mental ill health appears to be completely lacking. Further, it is possible that the so-called vegetive system is not the system requiring sleep so much as the higher nervous systems and skeletal muscles, and that unless behavioral awakening occurs or even if it does, within limits, it may be unreasonable to surmise that the beneficial effects of sleep are not realized. Indeed it might be presumed that activity of the vegetive system is required more or less continuously 24 hours per day. The fact that the VCR arousal (data obtained by Williams *et al.*) is more or less similar for sleep-deprived and nondeprived subjects during all stages of sleep would support this idea (see Fig. 263).

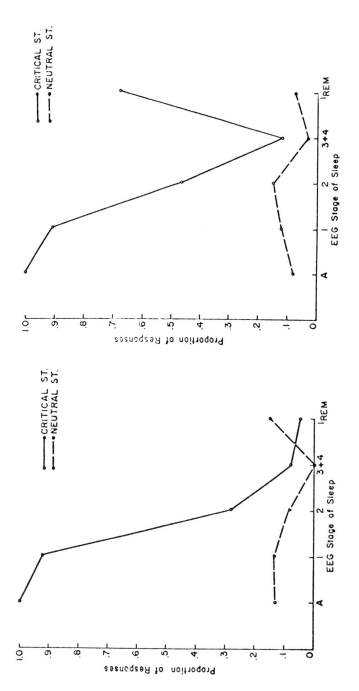

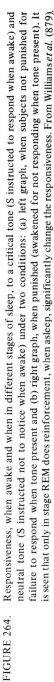

FIGURE 264.   Responsiveness, when awake and when in different stages of sleep, to a critical tone (S instructed to respond when awake) and neutral tone (S instructed not to notice when awake) under two conditions: (a) left graph, when subjects not punished for failure to respond when tone present and (b) right graph, when punished (awakened for not responding when tone present). It is seen that only in stage REM does reinforcement, when asleep, significantly change the responsiveness. From Williams *et al.* (879).

The data in Fig. 260 indicate that, when previously deprived of sleep, people seldom "wake up" in their sleep. Further, it is seen in Fig. 260 that normal sleep patterns are cyclic, so that some awakenings during sleep are not abnormal. The aforementioned conjecture that the more an organism needs sleep, the less sensitive it will be to auditory arousal is appropriate to this question. As noted earlier, tests show that this arousal threshold can be elevated during sleep by at least 80 to 90 dB for fairly prolonged periods, and that during sleep, following a period of sleep deprivation, the person spends an abnormally greater time in stage REM where he is very insensitive to all auditory stimulation. Whether this represents some protective arrangement whereby the organism can (a) maintain periodic monitoring of his environment, (b) insure a physiological deep sleep condition, and (c) avoid somatic and behavioral arousal except for certain sounds, is, of course, speculation. In brief, parallel to the previous comments regarding health and feelings of discomfort, the reasoning that (a) sleep is needed for health, (b) noise can arouse people from sleep, and (c) therefore noise can cause ill health, is perhaps not valid.

On the other hand, it seems clear that, when persons are going to sleep, a noise will serve to arouse and to prevent going to sleep. Indeed, as noted in the earlier discussion on community reaction to aircraft noise, the strongest complaints arose from noise occurring in the late evening hours. Also, as noted earlier under "stress and health", even though adaptation or accommodation to sleep arousal effects of noise may be expected, there are not sufficient data available to prove that this occurs as a matter of course in all people.

Noise-Induced Sleep

As mentioned earlier, a more or less steady level of a broadband background sound can mask the distraction of intermittent sounds. It is thought by some that sound can induce sleep, either because of its prevention of other distractions or by serving as a monotonous, hypnotic focus for one's attention. Little experimental research seems to have been done on this phenomenon, although Olsen and Nelson (593) claim a tone of 320 to 350 Hz calms crying babies and puts them to sleep, and some devices for making soothing sounds for inducing sleep are on the commercial market.

Audioanalgesia

Gardener and Licklider (280) developed a device that permitted a dental patient to listen, via earphones, to stereophonic music or filtered random noise. The typical procedure was for the patient to relax by listening to the music and to switch to the noise when he felt any dental pain, increasing the intensity of the noise as necessary to "kill" the pain. Specifications of tolerable durations and intensities of the noise to prevent undue auditory fatigue from the device were established. The experience of a few thousand dentists using the device was found (personal communication from the inventors, based on an independent users survey) to be about as follows: a third or so of the dentists found complete analgesia; a third found partial, but significant, analgesia; and a third found essentially no analgesic effects in their patients. Those who found analgesic effects claimed that the operative and postoperative conditions of the patients were better than with local or general anesthetics, which apparently is a general medical finding relative to the use of anesthetics. It was clear from the surveys made that when the device was a successful analgesic in dentistry, the confidence of the patient and the dentists in the efficacy of the procedure was an important, if not a necessary requirement.

Laboratory experiments conducted by the inventors of the device and by others (Gardner *et al.* [281], Melzak [538], Carlin *et al.* [119], Camp *et al.* [117], and Robson and Davenport [696]) revealed that pain from such things as heat and cold applied to the hand, electrical stimulation ("tingle" threshold) applied to the teeth, pressure applied to the arm, etc., were not suppressed by the presence of intense random noise, although a reduction in the sensation level of the deep muscle pressure was found. Some observations and tests have reportedly been made of possible pain suppression by noise in persons during child-birth, surgical operations other than dental, diseased conditions, etc., with mixed success.

It is well established (see Fig. 266) that hypnotic suggestion can be effective in suppressing pain in some people. However, the psychological and neurological mechanisms are not well understood. For example, the effectiveness of morphine in the relief of postoperative pain is, statistically, apparently not much different than the apparent effectivenss of audioanalgesia in clinical dentistry: a third of the patients gain relief through morphine that is greater than relief from a placebo, about a third get as much relief from a placebo as from the morphine, and the final third get no relief from either the morphine or a placebo (Beecher [48]). The essentially negative results from laboratory experiments on audio-analgesia notwithstanding, it is not justified to conclude that, in the clinical situation, the audioanalgesia does not involve physiological as well as psychological mechanisms in the suppression of pain during dentistry. Also, Beecher has pointed out that morphine and other analgesic agents that do not give consistent suppression of pain at its threshold do provide consistent relief at suprathreshold levels.

It can perhaps be concluded that, when audioanalgesia is effective, one or more or perhaps all of the following explanations are valid:

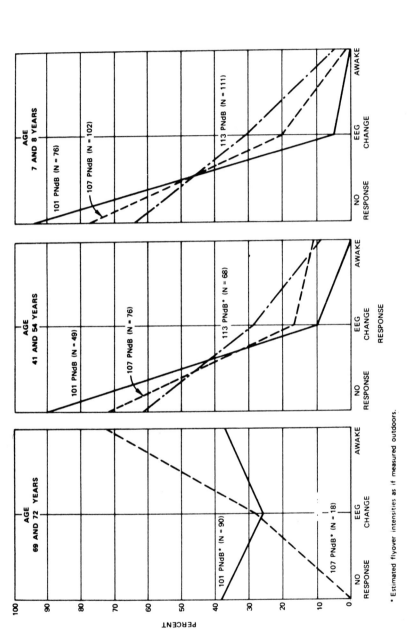

FIGURE 265.   Results of exposure of three different age groups, while asleep, to recorded subsonic aircraft noise (above graphs) and simulated sonic booms (facing-page graphs). From Lukas and Kryter (528).

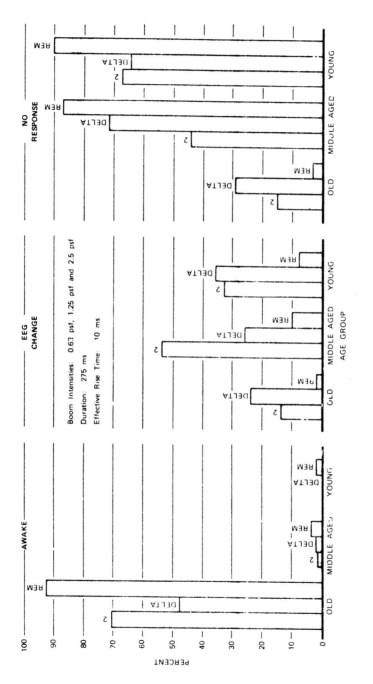

FIGURE 265. Continued.

1. Suggestion, enhanced by attention to the music or noise (distraction from any pain).

2. Perhaps in given cases no pain would actually be felt except that due to enhancement from the anxiety of the patient, i.e., the music and the noise relax the patient and reduce false anxiety.

3. Neural impulses from the auditory system serve to preempt to some extent the activity of centers in the reticular formation of the brain stem that are involved in the processing of pain impulses to the higher nerve centers. The neurological evidence and theory for this type of activity is meager, but not without some substance (ref. 281, 538).

Effects on Other Senses

Cutaneous Sensations from the Ear

Ades *et al.* (6) and Plutchik (632) have evoked sensations, other than audition, from ears exposed to intense noise; they obtained results in essential agreement with the curves for "touch," "tickle," and "discomfort" shown in Fig. 267. Also, Plutchik reported that pulses at a rate of 3 per second were rated as more unpleasant than slower or faster rates.

Figures 267 is from the interesting experiment conducted by Ades *et al.* in which they exposed persons who were totally deaf to very intense tones and noise. Because of their deafness, the subjects could be exposed to levels that would be harmful to the normal ear. It seems likely that the discomfort or pain thresholds reported for the deaf and normal ear (see Fig. 72) are attributable to stimulation of the eardrum or some middle ear receptors. In fact, Ades *et al.* (6) found that persons without eardrums reported no pain sensations with levels up to 170 dB.

Vestibular

Connected to the cochlea of the inner ear are the so-called sacculus, utricle, and semicircular canals. These structures, called the vestibular organs, share certain fluids with the cochlea and their innervations are closely connected (see Fig. 2). These vestibular organs are involved in maintaining body balance and orientation in space. When stimulated in certain ways, a person may loose his sense of balance, become dizzy, his eyes may show nystagmus movements (a fast movement back and forth of the eyeballs) and under extreme conditions he may become nauseated.

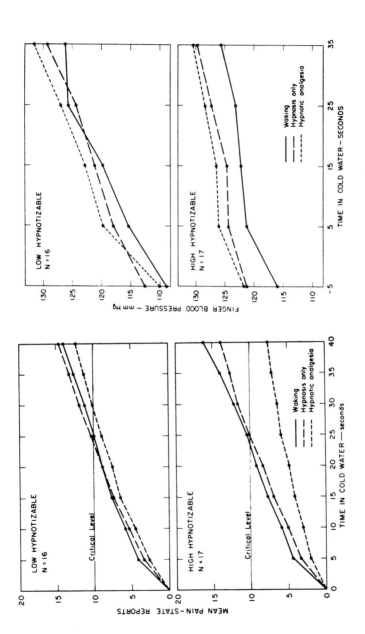

FIGURE 266. Pain (left graph) and blood pressure (right graph) as a function of time in water of $0°C$ in waking state, and following attempted hypnotic induction without analgesia instructions and with analgesia instructions. From Hilgard (369).

Because of their close proximity and fluid connections, it is not surprising to find that intense sounds affect the cochlea and vestibular system. Figure 267 shows the intensity required for various tones to reach the threshold of nystagmus in one subject. Dickson and Chadwick (193) report that, in jet air-craft noise over 140 dB or so, a person may feel a sense of disturbance in his equilibrium. Roggeveen and van Dishoeck (697) note that in persons who exper-ience vestibular reactions to relatively weak intensity sounds (considerably less than those shown in Fig. 267) there are usually present lesions in the bony walls of the vestibular system.

Vision

Noise has been thought to influence visual acuity and field, color vision, and the so-called critical flicker frequency (CFF). The latter phenomenon refers to the fact that alternating dark and light visual fields will become blurred (cease to flicker) at some frequency of alternation.

Visual contrast thresholds (bright target on less bright field), and minimum visual acuity for lines and discs (see Fig. 268) are generally apparently not affected by noise levels up to 140 dB or so (Broussard *et al.* [100], Krauskopf and Coleman [438]), although Dorfmann and Zajone (196) found some effect of sound level on background brightness, but no effect on size estimation of objects such as coins in children of different economic status. Loeb (515) found that broadband noise at a level of 115 dB had no effect on visual acuity; however, Rubenstein (708) reported adverse effects from noise 75-100 dB, and Chandler (132) reported a shift of verticality of a visual line away from the ear stimulated with noise. Benko (51a,b) reports a narrowing of the visual field in noise.

McCrosky (534) reports that random noise at levels from 85 to 115 dB reduced the CFF from 25 to 22 per sec. Ogilive (590) found no change in CFF with steady-state random noise of 80 to 90 dB, an increase in CFF with noise "fluttered" out-of-phase with the visual flicker, and a decrease in CFF with noise "fluttered" in-phase. Walker and Sawyer (821), however, were not able to dupli-cate Ogilive's findings and got negative results except for a small difference in CFF between steady and in-phase noise (see Table 85).

The effect of steady noise on CFF when the color of the light was varied has also been studied, but the results are very inconsistent. For example, Maier *et al.* (530) found that, when the light was orange-red, CFF decreased with increased loudness but no change occurred with green light. All in all, it would appear that noise can sometimes effect a 10% or so change, usually a decrease, in CFF from the CFF found in quiet, but the exact effects as a function of various noise and light conditions are highly variable and perhaps a matter of experimental chance and error.

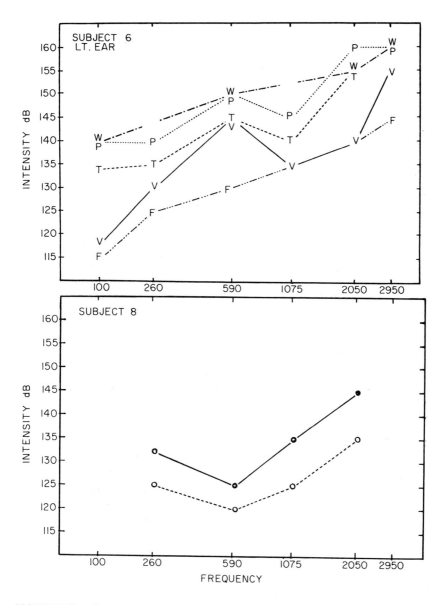

FIGURE 267. Upper graph: Threshold curves for several sense modalities for left ear of subject 6. V = vibration; T = Tickle; P = Pain; W = Warmth; F = Feeling. Lower graph: Small eye movements (broken line) and marked nystagmus (solid line). From Ades *et al.* (6).

TABLE 85

CFFs Under Four Noise Conditions. From Walker and Sawyer (821).

A. WITH ARTIFICIAL PUPIL

Condition	N	Mean	S.D.
No Noise	13	32.62	6.20
Steady Noise	13	31.94	4.75
In-phase Noise	13	32.34	4.78
Out-of-phase Noise	13	32.78	5.44

B. WITHOUT ARTIFICIAL PUPIL

Condition	N	Mean	S.D.
No Noise	13	38.12	3.43
Steady Noise	13	38.01	3.48
In-phase Noise	13	38.42	3.28
Out-of-phase Noise	13	38.72	3.73

The converse effects, that of light on auditory threshold, are small and perhaps fortuitous. For example, O'Hare (591) claimed some colors of light caused a 1-2 dB increase, some a 1-2 dB decrease in auditory threshold.

Low Frequency and Infrasound

It is conceivable that intense low frequency sound and acoustic energy at frequencies below about 20 Hz (infrasonic) could have particularly adverse effects on man. In addition to possible stimulation of the vestibular system and pain in the ear, sound in the region of 10-75 Hz or so could cause resonant vibration in the chest, throat, and nose cavities of the body, and the resonant frequency of the eyeball is near 5 Hz.

A considerable amount is known about the effects from vibrations at these frequencies when imposed directly on the body through mechanical contact. Because of the impedance mismatch between airborne acoustic energy and the body, acoustic energy has little or no effects upon the body other than the ear until the levels become quite intense. Much of what is known about the effects of very intense airborne sound on the body comes from a series of somewhat qualitative tests conducted by U.S. Air Force and NASA research personnel, and reported by Mohr et al. (561).

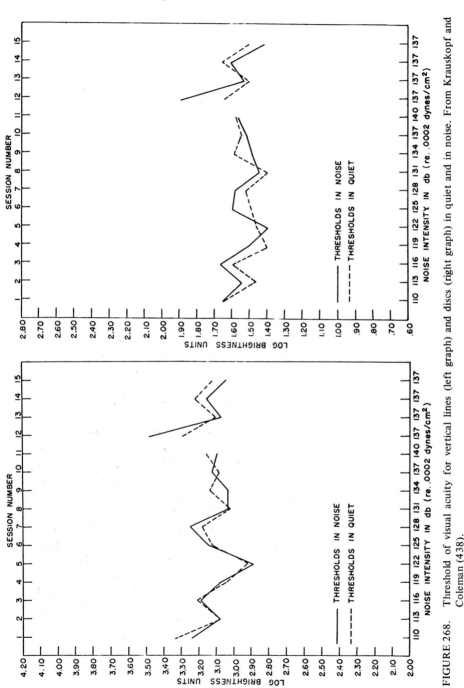

FIGURE 268. Threshold of visual acuity for vertical lines (left graph) and discs (right graph) in quiet and in noise. From Krauskopf and Coleman (438).

The stimuli used in these tests, which were discussed earlier with respect to damage risk to hearing, are shown in Fig. 134. No nonauditory effects were noted until the spectrum levels exceeded approximately 125 dB. At various higher levels, decrement in visual acuity, some vestibular reactions, and chest, nose, and throat responses occurred, and if no ear protective devices were worn, ear pain and middle-ear fullness was felt. The observations of the subjects for tests 5 through 16 are given in Table 86. The results of tests 1 through 4 conducted in levels lower than 125 dB revealed no significant effects of the noise on the subjects.

It is clear from these tests that the nonauditory and, to some extent, auditory effects of airborne sound and infrasound become significant only at very high intensities, at spectrum levels in excess of 130 dB or so. Except in the vicinity of unusual sources of noise, such as near heavy rocket engines or special test sirens, one seldom finds steady-state low frequency acoustic energy at these intensities.

Ultrasonics

Acoustic energy in the frequency region above 20,000 Hz is called ultrasonic because it is inaudible to man. Actually, for most adults, acoustic energy above 10,000 Hz is ultrasonic. As noted in a review of the effects of ultrasound by Parrack (604), the advent of the jet aircraft engine, high-speed dental drills, and so-called ultrasonic cleaners provided relatively common sources of high intensity ultrasonics. Tables 87, 88, and 89 show the spectra of the noise from representative samples of these devices.

It is to be noted that for each of the spectra, while there is considerable energy in the bands above 20,000 Hz, there is energy in the audible frequency region that often exceeds the damage risk values specified as tolerable, for long exposures, in Fig. 133. For this reason the tinnitus, dizziness, headache, nausea, and fullness of the ears often reported by some persons exposed to these noises probably are due not to ultrasonic but to audible components in the noise.

In Parrack's opinion, the reactions of dizziness, nausea, and headache listed above, are presumably psychosomatic and engendered by unwarranted apprehension. Although, as we shall see, the energy above 20,000 Hz (or possibly at any frequency that is inaudible to a given individual) is probably not the source of the reactions of dizziness, nausea, and headache experienced by persons exposed to the noise sources given in Tables 87, 88, and 89, the reactions in question are perhaps not "psychosomatic" in origin either. At the same time, the belief of Davis (168), Davis et al. (181), Parrack (604), and others, that any acoustic energy at high frequencies that significantly affects man does so only through his inner ear, appears to be very true.

The physical arguments against ultrasound, except from some specially

TABLE 86

Summary of Effects of Acoustic Stimuli Shown in Fig. 136 upon a Group of Five Persons

Exposure durations 1-2 minutes. Ear protective devices usually worn. From Mohr *et al.* (561).

Test 5 – *Broadband noise. Peak spectrum level 128 dB at 50 Hz.*   The speech signals recorded were completely masked despite the noise reduction provided by microphone and shield. Pulse rates were increased 10 to 40 percent over resting levels. Two subjects reported mild chest wall vibration; two others noted mild nasal cavity vibration; and one of these perceptible throat fullness.

Tests 6, 7, and 8 – *Broadband noise at about same levels as Test 5 with relatively less energy in the higher frequencies.*   All subjects considered the exposures tolerable for the short durations involved. Speech signals were completely masked, nevertheless, except those of one subject who was stationed inside a vehicle which afforded appreciable attenuation of the high frequencies. His speech was definitely modulated but the poor intelligibility achieved was attributed to the masking. All subjects reported mild to moderate chest wall vibration; two subjects noted throat pressure; three subjects experienced perceptible though tolerable interference with the normal respiratory rhythm. Pulse rates measured during Test 7 exhibited no significant changes during the exposure.

   Throughout these tests visual acuity, hand coordination and spatial orientation were subjectively normal.

Tests 9-11 – *Narrow bands, center frequencies, 2 to 10 Hz, spectrum level 142-153 dB.* The most prominent effects attributable to the infrasonic noise spectra (Tests 9-11) occurred during exposure without ear protection. An uncomfortable sensation reflecting pressure build-up in the middle ear was elicited which required frequent Valsalva swallowing to relieve. This effect was almost entirely absent when insert earplugs were used. Earmuffs alone helped prevent the middle ear pressure changes. Three subjects described an occasional tympanic membrane tickle sensation during these exposures without protection and one subject observed marked nostril vibration. Another noted mild abdominal wall vibration during exposure to the test 10 spectrum (5-10 Hz). No shifts in hearing threshold were detectable one hour following these exposures. When ear protectors were worn to lessen the middle ear pressure changes, exposures to infrasound of these levels were judged well within tolerance.

Tests 12-14 – *Narrow bands, center frequencies 15 to 50 Hz, spectrum level at 140 dB.* The maximum intensity low sonic exposures produced moderate chest wall vibration, a sensation of hypopharyngeal fullness (gagging) and perceptible visual field vibration in all subjects. Two subjects experienced mild middle ear pain during brief periods without ear protection but a third had no sensation of tickle or pain. Recorded speech sounds exhibited audible modulation. Post-exposure fatigue was generally present after a day of repeated testing. The exposures as a group were not considered pleasant; however, all subjects concurred that the environments experienced were within the tolerance range.

Test 15 – *Pure tones 3 to 40 Hz, spectrum level 145-153 dB.*   Exposures to 24 discrete frequency noise fields showed both objective and subjective responses qualitatively similar to those elicited by the corresponding narrow band spectra. Pressure build-up in the middle ear was not a factor at 30 Hz and above, but the gag sensation was magnified for at least one subject. Although all exposures were judged tolerable, it was noted that the subjective sensations rose to intensity very rapidly as sound pressure levels were increased above 145 db.

TABLE 86 (Continued)

Test 16 – *Pure tones 40 to 100 Hz, spectrum level 150-155 dB.*   Voluntary tolerance of
the subjects was reached at 50 Hz (153 dB), 60 Hz (154 dB), 73 Hz (150 dB), and
100 Hz (153 dB). The decision to stop exposures at these levels was based on the
following subjectively alarming responses: mild nausea, giddiness, subcostal discomfort,
cutaneous flushing and tingling occurred at 100 Hz; coughing, severe substernal pressure,
choking respiration, salivation, pain on swallowing, hypopharyngeal discomfort and
giddiness were observed at 60 Hz and 73 Hz. One subject developed a transient headache
at 50 Hz; another developed both headache and testicular aching during the 73 Hz
exposure.

A significant visual acuity decrement (both subjective and objective) occurred for all subjects during the 43, 50, and 73 Hz exposures. Speech sounds were perceptible modulated during all exposures. All subjects complained of marked post-exposure fatigue. No
shifts in hearing threshold were measurable two minutes post exposure; the earplug and
muff combinations worn are known to provide sufficient protection against the higher
harmonics of the noise fields and were apparently effective to an appreciable degree
in attenuating the fundamental tones. Recovery from most of the symptoms was complete upon cessation of the noise. One subject continued to cough for 20 minutes, and
one retained some cutaneous flushing for approximately four hours post exposure.
Fatigue was resolved by a night's sleep.

---

designed device, entering or stimulating man except possibly through normal
stimulation of the inner ear, are:

1. The absorption coefficient of the skin for sound above 20,000 Hz is less
than 0.1 to 1%, and levels at these frequencies that would cause any even slightly
noticeable local heating effects would have to be in excess of 100 dB or so. (The
absorption coefficient of acoustic energy at 20,000 Hz in small furred animals
[rats, guinea pigs] is of the order of 21% so that lethal heating can occur in these
animals at levels of ultrasonics that go unnoticed by or are harmless to man.)

2. Ultrasonic frequencies generated by crystals (20-108 kHz, Haeff and
Knox [333]; 25-62.5 kHz, Belluci and Schneider [51]; 50 kHz, Deatherage *et
al.* [188]) applied to bones and tissues of the head, if sufficiently intense,
resulted in the person perceiving an audible high-pitched tone usually around
8000 to 10,000 Hz, depending upon his upper limit of hearing. Deaf subjects in
these experiments heard nothing. It is to be noted that crystals have many
resonant modes so that an audible subharmonic could have been detected by the
rubbing of the crystal against the surface of the skin and then radiating as an
acoustic signal (analogous to the mode of detection of the presumed "electrical"
stimulation of the ear mentioned in Chapter 2). Also, von Gierke (personal
communication) suggests that subharmonics falling in the normal frequency
range of audibility are generated in the middle ear when the ear is exposed to
intense ultrasonics.

TABLE 87

Sound Pressure Level Around a Jet Aircraft in One-Third Octave Bands. From Parrack (604).

Positions and Operating Conditions	Band Center Frequencies in KiloHertz														
	2	2.5	3.15	4	5	6.3	8	10	12.5	16	20	25	31.5	40	50
500 ft Idle	57	67	63	54	53	51	44	38	37	37	35	--	--	--	--
500 ft Military	98	95	92	90	88	84	82	77	76	75	74	74	--	--	--
500 ft A B	105	103	103	100	99	96	95	91	89	87	83	80	78	--	--
100 ft Idle	75	83	82	74	77	77	74	73	70	64	60	57	56	55	59
100 ft Military	111	114	112	111	109	108	106	104	102	100	97	95	92	88	--
100 ft A/B	123	122	123	121	119	118	118	115	115	113	112	111	110	107	106
25 ft forward Idle	90	96	94	96	94	92	92	90	90	88	86	84	80	76	73
25 ft forward Military	102	107	104	101	102	100	100	97	96	93	91	89	86	83	80
25 ft forward A/B	109	110	107	107	105	104	103	101	100	97	95	93	90	87	87
Maintenance Idle	91	95	94	94	93	90	93	85	85	83	79	77	73	70	--
Maintenance Military	117	115	114	116	112	112	111	108	108	105	103	101	99	96	--
Maintenance After Burner	121	120	120	123	118	117	117	114	113	111	108	106	103	100	--

Aircraft is F-102: Position 500 ft is on radius located 125° from nose. Idle, Military and After Burner Power.

Position at 100 feet is on radius located 120° from nose. Power settings as before.

Position at 25 feet intake about 30° from nose of aircraft.

Maintenance Position is just off of main landing gear and is under fuselage.

TABLE 88

Sound Pressure Levels Measured in Air Around the Weber High-Speed Dental Drill. From Parrack (604).

Measurement position*	OASPL	31.5	63	125	250	500	1000	2000	4000	8000	16000	31500
				Center frequencies of octave bands measured in Hz								
1.	97 dB	46 dB	44 dB	50 dB	48 dB	52 dB	55 dB	64 dB	87 dB	84 dB	93 dB	95 dB
2.	89	44	43	47	43	46	47	54	77	82	83	83
3.	82	47	51	43	45	44	45	52	73	76	78	76
4.	81	48	54	41	42	44	45	51	74	76	77	74

* Position 1. Location of patient's ear – 6 in. from source

Position 2. Location of patient's ear – 20 in. from source

Position 3. Farfield on radius – 65 in. from source

Position 4. Farfield-on radius 90% to position 3,–65 in. from source

Drill was suspended in fixed position. Drill and microphone were 54 in. from floor for these measurements.

TABLE 89
Sound Pressure Levels in One-Third Octave Bands

Center frequencies as shown. Six positions near the Bendix model sec 1825A sonic energy cleaning system. From Parrack (604).

Positions	Band Center Frequencies in KiloHertz																			
	1.0	1.25	1.6	2.0	2.5	3.15	4	5	6.3	8	10	12.5	16	20	25	31.5	40	50	63	80
Operator Position Cover Closed	56	55	55	57	62	68	70	72	73	83	96	83	83	101	85	91	89	86	85	--
Operator Position Cover Open	--	62	63	63	66	70	74	77	81	91	104	91	93	109	93	102	99	95	95	--
At Desk Cover Closed	45	43	42	42	47	51	54	56	57	64	77	64	64	80	63	71	69	64	62	--
At Desk Cover Open	49	48	49	49	52	55	58	61	64	73	86	73	73	89	72	79	78	72	71	--
Office Cover Closed	45	45	44	43	41	42	42	41	40	44	57	45	45	64	46	45	42	38	--	--
Office Cover Open	46	45	44	43	42	42	42	43	43	51	63	50	50	57	49	52	48	42	--	--

Operator's Position is immediately adjacent to one end of cleaner system.

Desk Position is about 15 feet from edge of tank where operator stands in laboratory work area.

Office Position is adjacent room (door open) about 12 feet from cleaner system - also about 13 feet from operator position.

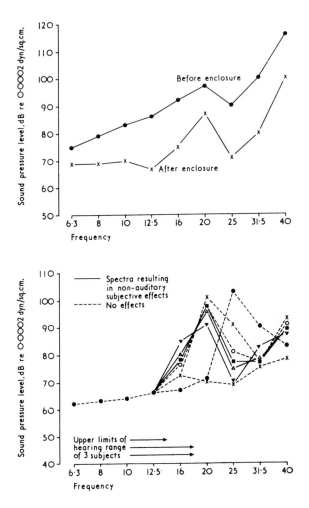

FIGURE 269.   Upper graph: Band spectra of noise from an ultrasonic washer, freq kHz. Lower graph: Spectra of noise and tones that were used in laboratory tests, freq kHz. From Acton and Carson (3). (By permission of the authors and editors of the *British Journal of Industrial Medicine*.)

Acton and Carson (3) (see also Skillern [740]) found convincing evidence that unless a person's range of hearing extended to about 17,000 Hz, and unless the energy in that frequency region exceeded 70 dB, no subjective effects (tinnitus, headaches, fatigue, etc.) were experienced. Figure 269 (right-hand graph) shows the results of laboratory tests with ultrasonic signals. The upper curve, left-hand graph of Fig. 269, shows the spectra of noise which caused complaints of adversive subjective effects in female attendants of an industrial ultrasonic washer, and the lower curve, left-hand graph of Fig. 269 shows the spectra of noise which did not cause in the same workers significant complaints. Acton and Carson noted that women had adverse symptoms more often than men, and young men more often than older men. This was presumed to be due to the auditory acuity of the people involved and not to their sex or age per se.

It also appears that perhaps even the hearing in the higher frequency regions of most industrial workers sooner or later is reduced by the upward spread of a noise-induced permanent threshold shift from lower frequency noise as much, if not more, than it is by the acoustic energy above 10,000 Hz or so. It might even be conjectured from an examination of the curves of Fig. 133 that the upturn in so-called normal threshold of audibility from 4000 Hz to say 20,000 Hz is severely influenced by everyday noise in the frequency region 2000 to 8000 Hz.

In any event, it is concluded that the subjective effects of "ultrasonics" are not due to apprehension on the part of the listener, but are due to sound that exceeds 78 dB or so in the frequency region of about 16,000 Hz and that is audible to the listener. Continued exposure to sufficiently intense sound below or at 16,000 to 20,000 Hz results in an elimination of the earlier mentioned subjective and audible effects because, apparently, noise-induced threshold shifts increase in those frequency regions.

The question remains unanswered as to why these adverse subjective effects are more often noticed from ear damaging exposures to these higher frequencies than from damaging exposures to noise at lower frequencies, say below 2000 Hz. Of course there may be no fundamental difference since some comments regarding headache, unusual fatigue, and certainly tinnitus are also sometimes reported from initial exposures to lower frequency noise when sufficiently intense and for sufficiently long exposures.

Chapter 13

Effects of Noise on Mental
and Motor Performance

Introduction

It is obvious from a consideration of the masking effects of noise that any task, be it mental (involving primarily thinking and secondarily muscular activity) or motor (involving primarily muscular activity and secondarily thinking) that requires the perception of auditory signals for correct performance, will be adversely affected by noise. It is not so obvious that the effects of noise on man's nonauditory systems, as defined above, are sufficient to have significant effects on mental or motor work that does not require audition for its performance. Nevertheless, the practical importance to overall work efficiency and production of even a small fraction in degradation of performance has led to a considerable amount of research on the effect of noise on work performance, both in the laboratory and in the factory.

The physiological effects discussed in the preceding chapter could probably fairly be summarized by saying that, within broad limits, noise below roughly 40 PNdB has no real physiological somatic arousal effects, between 40 to 80 PNdB it has slight effects, and from 80 PNdB to 130 PNdB it has some definite degree of somatic arousal; at higher levels it also causes pain or cutaneous stimulation of the ear. However, it must be noted that, except possibly for exposures at the highest levels, these somatic arousal effects adapt with continued steady-state or interrupted exposures to the noise, and the arousal from noise to begin with is relatively small and transitory compared to that evoked by task motivation. As noted previously, there is often some adaptation in the threshold of ear pain or feeling from intense noise; however, this adaptation is possibly the result of some direct physiological effect on the receptors involved. In addition, the general objectionableness and somatic arousal from noise may be a secondary reaction due to masking of unwanted auditory signals or arousal from sleep or rest.

Because of adaptation, one could anticipate that regular, expected noise may in general have no adverse effects on nonauditory mental or motor work performance or output. Indeed, in our opinion, the experimental data to be presented show this to be the general fact of the matter. This general conclusion is not shared by D.E. Broadbent who has been a major contributor to the fund of research data and theory on the effects of noise on the mental and motor performance of man. Special attention is invited to his work and writings (89-99).

Problems in the Interpretation of Research

All who have reviewed the research literature in this problem area have been impressed by some inconsistencies in the results obtained by different experimenters. It may be constructive, before reviewing studies considered particularly relevant to this matter, to discuss a number of specific findings which serve to illustrate some of the pitfalls facing the research workers and some of the confounding factors available to mislead the theortician.

Specification of Noises

Although the relations have not been worked out for each type of response involved, there is every indication that it is the loudness level or the perceived noise level of a sound that, to the first order of approximation, determines its arousal effects on man's nonauditory system and work performance. Unfortunately, many early studies report only an overall sound level meter reading, with the meter usually set on flat- or C-weighting. The band spectra or the spectrum levels of the noises reaching the listener's ears were often unspecified.

This creates no particular problem when pure tones were the sounds involved, but random or background noises could often, in spectrum, be far from white noise containing equal energy at all frequencies. In some studies "quiet" if specified as being as high as 85 dB, and in others the noise is specified as low as 64 dB; it is suspected that usually the "quiet" condition, unless otherwise specified, consisted of low frequency ambient background noise which tended to give high, but not loud, sound level meter readings (on flat or C scale), and that the unspecified "noises" were relatively higher frequency random sound and much more audible to the average listener. Be that as it may, it is not possible to interrelate many of these studies for this reason.

As examples: Broadbent (96) concludes that the noise must reach a level of about "90 dB" before it can affect nonauditory work performance, although he notes that high frequency sounds, above 2000 Hz, have more effect on people than low frequency sounds below 2000 Hz. Kirk and Hecht (431) find that noise, spectra unspecified, of a variable but average level of 64.5 dB is detrimental to performance in contrast to a steady noise of 64.5 dB, or a "quiet" condition, spectra unspecified, of 60 dB. Even though the relative data within the study may be meaningful, it is not much help for purposes of generalizing among studies to report the noise levels in terms of voltages present across an undesignated type of earphone, as was done by Reiter (667a).

Comparisons with Control Group Responses

In many studies on the effects of noise on work output or errors, the performance of a group of subjects, called the experimental group, exposed to noise has been compared with the performance of a group of subjects, called the control group, not exposed to the noise or to a less intense noise. The difference in the performance of the two groups on the task at hand is attributed to the presence of the noise. Efforts were usually made in these studies either (a) to match the groups with respect to some independent criterions such as age, intelligence, motor ability, etc., or (b) to select the experimental and control groups randomly from a larger group, thereby, hopefully, obtaining subgroups of equal capabilities on the average. This does not always happen, of course; for example, Broadbent (89) found that individual differences were such that one subject in an experimental group made five times as many errors on a test as all the other subjects in the group.

Studies of Broadbent (96), Shambaugh (731), Barrett (43), and Cohen et al. (146) revealed that, in general, subjects who were found on personality tests or by other means of personality rating to be "anxious," "introverted," and "somatic responsive" were more adversely affected by noise in the performance of mental (I.Q. tests and arithmetic) and motor tasks (reaction time and tracking) than were the better adjusted subjects. Table 90 from Cohen et al. shows examples of this finding. On the other hand, Blau (71) found no difference in the effects of 103 dB random noise on performance on tests of mental ability between groups of well-adjusted and less well-adjusted (according to a personality test) college students, and Angelino and Mech (17) (a tone at 85 dB) and Auble and Britton (39) (recorded speech at 80 dB) found that less well-adjusted students generally did better than well-adjusted students in the "noise." However, in the Angelino and Mech study, the less adjusted were also apparently more capable in both the quiet and the noise, and in the Auble and Britton investigation the scores in the quiet were also significantly different for the two groups. Tables 91 and 92, from these two latter studies, show that the high anxiety groups improved their performance on the mental tests in going from the quiet to the "noise" conditions; this is opposite of the findings of Barrett

TABLE 90

Comparison of Mean Vigilance Scores (Percent Correct Detections)

for 12 "Most Normal" and 12 "Most Deviant" Subjects on the Personality Measures for Each Background Noise Condition

Control (quiet) was random noise at 75 dB, high level at 95 dB, and variable from 70-95 dB. From Cohen, *et al.* (146).

Background Noise Condition	Personality Grouping	Anxiety (Welsh)	Neuroticism (Winne)	Manifest Anxiety (Taylor)	Psychoasthenia	Introversion
Control	Normal	94.00	92.71	93.09	93.75	94.11
	Deviant	93.00	94.70	92.78	95.58	92.27
	Difference	1.00	-1.99	0.31	-1.83	1.84
High Level	Normal	96.00	95.00	94.90	95.65	93.46
	Deviant	93.00	92.41	91.51	95.01	87.93
	Difference	3.00	2.59	3.39	0.64	5.53
Variable	Normal	94.00	94.70	95.51	95.63	94.66
	Deviant	93.00	92.72	89.19	93.50	88.39
	Difference	1.00	1.98	6.32	2.13	5.27

TABLE 91

Scores Obtained on Arithmetic Tests by Persons Having Highest and Lowest Adjustment Scores on Personality Test When in Quiet and in Noise (Tone at 85 dB)
From Angelino and Mech (17).

Adjustment Group	Quiet	Minutes in Noise			
	1-5	6-10	11-15	16-20	21-25
	No. of Correct Additions Per 5 Minutes				
Highest	48.07	37.64	42.14	43.64	45.85
Lowest	55.74	44.57	64.35	67.92	69.35

TABLE 92

Scores Obtained on Number and Name Checking
Test by Persons with Lowest and Highest Anxiety
Scores on Personality Test in Quiet and Noise-Speech
Recording at 80 dB. From Auble and Britton (39).

Anxiety Group	Quiet	Noise
Highest	141.9	154.2
Lowest	152.4	145.9

and Shambaugh, and Cohen *et al.* However, in the latter three studies, the noise
was random noise at levels from 95 to 105 dB.

It would appear from the studies just reviewed that the performance (as
measured from a few hours of testing in the laboratory) of so-called "somatic or
anxious" types of people tends to be affected by rather intense noise and the
performance of the more stable is not. Because of the brief duration and artificial laboratory environment situation, these results may not be generalizable to
the relative performance of these two personality types in real-life work situations.

It has been found that the performance of simple motor or mental tasks by
persons rather severely handicapped in personality or intelligence is not apparently or necessarily adversely affected by the presence of noise. Barnett *et al.*

(42) and Pascal (607), working with mental defectives with very low intelligence, and Guertin (331a) with psychiatric patients in a hospital, found no difference in the performance on reaction time or simple mental tasks of their subjects in the quiet compared with their performance of these tasks in the presence of random noise. The general level of performance in either the quiet or the noise was, of course, below that found with normals.

In conclusion, comparing the average performance of a group of subjects exposed to noise with a group of subjects exposed to quiet or relative quiet may lead to invalid conclusions regarding possible effects of noise unless the two groups are carefully matched in terms of personality and mental and motor abilities and are relatively large in numbers. Unfortunately, demonstration of adequate matching of control and experimental groups was not always possible in many of the studies reported in the literature on this problem.

Results from Small Groups

Figure 270 from Broadbent (94) purports to show that being in the noise on the first day affects subsequent performance on a difficult arithmetic task that required short-term memory, i.e., that there is an after effect of the noise (N) from the first day that adversely affects the second day's performance in the quiet (Q), (NQ), group. However, it would have been as reasonable to say that the quiet from the first day carried over to the second day, since the group that had quiet on the first day and noise on the second day (the QN group) performed as well as the QQ group having quiet on both days.

Actually, in our opinion, it would appear that on the second day all groups showed some improvement and that the NQ group by its somewhat similar deterioration in performance during the N and during the Q days, interacted differently to the experimental procedures and the continuation of working on the tasks during each day than did the QQ or QN groups, independently of whether or not there was noise present. Since there were only six subjects in each group, and taking into consideration the aforementioned influence of personality differences upon the reaction of people to noise, such a possible artifact cannot be ignored.

Some conclusions between data obtained from different groups of subjects appear to be based on rather tenuous interpretation of experimental data. For example, Table 93 is from a study of Broadbent's (93) in which was measured (a) the reaction times of different groups of subjects to noise of different intensities and frequency content; and (b) the error made in touching a contact in front of five flashing lights in the presence of the noises at different intensities. From these data, Broadbent concludes (93, p.21) that: "It thus appears that sounds more likely to interfere with work also produce a faster reaction when them-

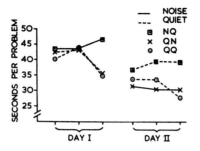

FIGURE 270.   Performance of groups in noise and in quiet. Note that noise causes deterioration on Day I, while on Day II there is an after-effect from the noise of
Day I. NQ = Noise on first day, quiet on second day; QN = Quiet on first
day, noise on second day; QQ = Quiet on both days. Three groups, each
with ten arithmetic problems requiring memory. From Broadbent (94).

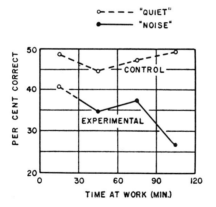

FIGURE 271.   Performance of the nine subjects during the experimental and control sessions of the test. The experimental session was for 0.05 hr in quiet (83 dB
masking noise) followed by 1.5 hr in noise (114 dB). The control session
was in quiet throughout. Note unusually high performance by control
group in quiet on last sessions. From Jerison and Wing (422).

TABLE 93

Mean Reaction Time in MSEC to Different Stimuli at Different Times After Change of
Stimulus. From Broadbent (93).

Stimulus		Position of response in group of 5					
Intensity	Frequency	1	2	3	4	5	Average
75	Low	209	199	187	196	189	196
75	High	196	198	193	187	187	192
100	Low	203	180	187	190	190	192
100	High	201	187	197	192	189	193

Mean errors with noise of various spectra and intensities. The level
of noise given is the intensity of the high frequency noise: the
other noise was at equal loudness (= +3 dB physical intensity)

Stimulus		
Intensity	Frequency	Errors
80	Low	22
	High	24
100	Low	32
	High	49

selves acting as signals, confirming a view already advanced about noise effects:
that the effect is due to competition between various stimuli to control
response." Although it would seem likely that the noise had some effect on the
performance of the light-discrimination task, there do not appear in Table 93 to
be sufficient systematic relations, in our opinion, between noise condition and
reaction time to justify the conclusion reached.

In general it has been found that on tasks of vigilance (wherein the subject
must remain alert for relatively long periods during which he must note small
changes on one of a complex of dials, lights, or clocks), as the period of vigilance
is prolonged, the performance of the subject becomes progressively worse.
Broadbent (96) and Jerison and Wing (422) proposed that in noise this degradation with time is accelerated relative to the degradation found in the quiet.
Jersion (418) concluded that an anomaly existed in the data collected earlier by
Jersion and Wing (see Fig. 271). Rather than conclude that the high frequency
random noise (114 dB) to which the experimental group was exposed caused a
different effect over time of exposure than did the "quiet" (83 dB low frequency masking noise) for the control group, Jerison later chose to deduce that

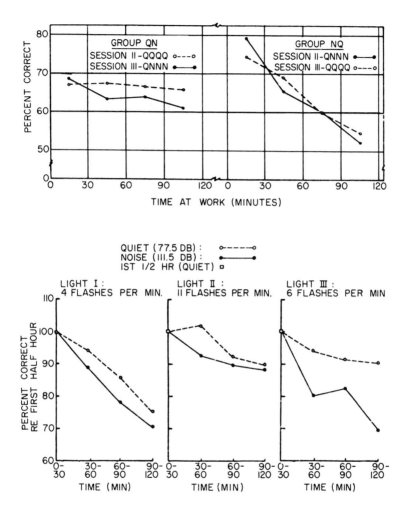

FIGURE 272. Upper graphs: Performance as a function of time compared for the experimental (QNNN) and control (QQQQ) sessions for the two groups of subjects. The QN and NQ labels for the groups refer to the order of the experimental (N) and control (Q) sessions; i.e., four successive 0.05-hr periods in QNNN, and four in QQQQ for the NQ group, and the reverse for the QN group. Lower graphs: Performance as a function of time during the experimental and control sessions presented separately for each light. All performance is considered relative to that during the first half-hour which was in quiet. From Jerison (416).

the last data point plotted on the right-hand side of the figure for the control group was anomalous, and was due to some experimental factor unknown to the experimenter. That is, the *improvement* at the end, for this type of experiment by the "quiet" group, is too contrary to the usual trend of results to be used as a basis for showing that noise, over time, accelerates errors more than quiet. So, taking into account the difference between groups implied by the left-hand (first) points, noise had no effect on the subjects' performance.

Scoring Problems

Occasionally experimental findings are reported that permit one to reach opposing conclusions. A possible example is to be found by comparing the top curves of Fig. 272 with the lower curves of Fig. 272. Both sets of curves are from Jerison (416) and are based on the same data, the lower curves being somewhat more detailed, and derived as explained in the caption. (Each letter of the series indicated 30 minutes of testing in either in Q [quiet, 77.5 dB ambinet noise] of N [111.5 dB random noise] ; the NQ group received series QNNN first, and then on the second day, series QQQQ; the QN group received the series in the reverse order.)

In our opinion, the upper curves show that the noise has, at the most, a slight depressing effect on the performance (a complex counting task), but that for some reason of chance both the NQ and QN groups performed better during the initial Q portion of the test for the QNNN series than the QQQQ series. However, the lower curves would seem to indicate that the effect of the noise is to depress the performance scores rather severely and uniformly.

The analysis represented by the lower curves of Fig. 272 was done on the assumption that individual differences in the basic abilities of the two groups of subjects were the cause of the lack of an apparent effect of noise upon performance, and that this difference was removed by using the data in the first session in the quiet as a reference performance. It is, of course, a matter of judgment whether one such data point for each group is sufficient for this purpose, but the rather striking importance of this assumption to the conclusions to be drawn from the data is clear.

Grimaldi (329) also notes that, by averaging certain performance scores, he was able to show that reaction time tasks were performed somewhat better in noise than in the quiet, but that a greater percentage of people showed better performance in the quiet than in the noise. This inconsistency follows, of course, from the fact that one or two of the subjects contributed very heavily to the error scores in the noise.

Test Differences

As indicated above, it is sometimes reasonable to question the comparability of different groups of subjects used to evaluate the differential effects of noise and quiet or degrees of noise on people. Occasionally, tests are performed in which different forms or versions of a given intelligence or mental test are used as being equal to each other. Form A of a subtest of logical reasoning, for example, in a given intelligence test is supposedly equal in difficulty to form B of the same subtest. Because the effects of noise compared to quiet are generally so small, in terms of individual test scores, this assumption can be risky.

We believe that such an assumption was unwarranted in a study reported by Lienert and Jansen (509). In this experiment, subjects were administered one form of a battery of intelligence tests in the quiet and a second form of the same test in the presence of a broadband random noise at a level of 75 dB. The test results are shown in Fig. 273.

The authors conclude:

1. There is a significant effect of noise upon mental task performance; one task being improved (by 2 scoring units), three tasks adversely affected (by 1 or 2 scoring units), and five tasks unaffected by the noise.

2. The interaction was due to relative overarousal of the organism by the noise for the best performance of some tasks, and relative underarousal in the quiet for the best performance of some of the tasks. This under-overarousal effect on some of the intelligence subtest scores is shown in the hypothetical functions on the right-hand graph of Fig. 273. Lienert and Jansen credit Hormann and Todt (390) with this concept.

It would seem that an alternative explanation would be that noise essentially has no more effect on the test scores than quiet, the differences noted in the left graph of Fig. 309 being due to unequal difficulties between some of the alternate test forms used in the quiet (control) and in the noise. The analysis of variance given in Fig. 273 would support the notion that unequal difficulty among the test forms contributed to the results, as shown by the fact that by far the most significant F-test was for the various subtests.

Response Contingency Effects

Finally, in discussing problems of interpreting the effects of noise on task performance, attention is invited to an important paper and set of definitive experiments prepared and conducted by Azrin (41). Azrin makes clear the fact that sound can be:

1. *Response contingent;* that is, the noise can be made an aversive stimulus if it becomes associated with incorrect behavior (see also previous study of Stern on G.I. motility), thereby improving arousal and work output, but perhaps increasing errors. Or the noise can become a kind of reward by being associated with correct behavior, thereby tending to increase productivity in the noise.

2. *Stimulus (task) contingent*; that is, the task is made easier because of the noise, i.e., the noise contains information relative to the task. In this case, the sound in question is not in all aspects "noise" in the sense of being unwanted sound.

3. *Not task (stimulus) or response contingent*; this state is, to be sure, the condition thought by most experimenters to prevail in studies on the effects of noise on mental and motor work performance. In this state, the presence of the noise is presumably not a function of the task or the proficiency of performance.

Some of the research that is concerned with the effects of noise on so-called vigilance tasks (see McCann [533]) is a mixture of (a) the arousal or awakening effects of noise during a monotonous vigilance or monitoring task (to be discussed to some extent later); and (b) noise response and stimulus contingency effects which involve learning as to the meanings of the noise and, as such, are largely outside the scope of this document, as mentioned earlier.

However, some of the inconsistencies in the results of laboratory and field tests of the effects of noise on work, particularly in small groups, could conceivably be attributable to unintended, by the experimenter, stimulus and response contingency interpretations of the meaning of the noise by the subjects (such as the subjects viewing the noise as signifying the "more important part of the experiment" and that he should work harder, or that the noise is punishment and the "quiet" reward, etc.). The occasional possible role and importance of these factors in the explanation of the effects of noise on work performance will be discussed later.

Theories of Effects of Noise on Work Performance

In view of some of the above discussion, it might be suggested that there are no real effects of noise per se on nonauditory mental or motor work. Whether there are or not, attempts have been made to organize theories that would explain some of the research data that have been obtained, as will be discussed next along with additional research findings.

Because of the procedural problems mentioned and the wide differences in experimental conditions involved, it is difficult to organize, for purposes of discussion, the research studies in this problem area. To aid in this endeavor we

CAUSES OF VARIANCE	SUM OF THE SQUARED DEVIATIONS	DEGREES OF FREEDOM	SQUARED	F
SUBJECTS	11415	39	292.69	3.35[1]
SUBTESTS	8141	8	1017.63	11.64
EXPERIMENTAL CONDITIONS	2	1	2.00	0.02
SUBJECTS·SUBTESTS	9007	312	28.87	0.33
SUBTESTS·EXPERIMENTAL CONDITIONS	4598	8	574.75	6.57[1]
SUBJECTS·EXPERIMENTAL CONDITIONS	1324	39	33.95	0.39
REMAINDER	27272	312	87.41	
ALL TOGETHER	61759	719		

[1]SIGNIFICANT AT THE 1% LEVEL

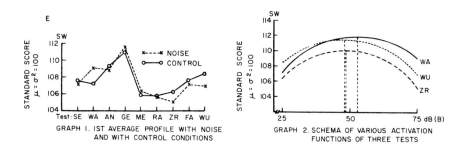

GRAPH 1. 1ST AVERAGE PROFILE WITH NOISE AND WITH CONTROL CONDITIONS

GRAPH 2. SCHEMA OF VARIOUS ACTIVATION FUNCTIONS OF THREE TESTS

FIGURE 273.   The table is the results of analysis of variance of intelligence tests given in noise [75 dB(B)]. Graph 1 shows profile of various subtest scores and Graph 2 is hypothetical relation between test scores for noise conditions between 25 and 75 dB(B). From Lienert and Jansen (509).

will group together research that is more or less directed towards certain hypotheses or theories, even though the word theory may suggest a more formal construct than was intended by the original investigator.

Theory of Blinks

In 1954 Broadbent (90) conducted a study that, in conjuction with other data, led him to propose a theory that the stimulation of the auditory system with noise set up a condition in the perceptual nervous system that attracted attention from the other sensory perceptions periodically for a very brief period (about one second). This was deemed somewhat analogous to the blinking of eyes. It is not entirely clear whether this "blinking" is presumed to occur regularly but is accentuated in the presence of steady-state noise, or whether the noise induces a condition of blinking. It is made clear, however, that the phenomenon is presumed to occur in the internal nervous system.

The evidence for the theory is somewhat inconclusive. Table 94 shows the results of Broadbent's study of 1954. Two series of tests were conducted, one in which 10 subjects noted whenever a pointer on one of 20 dials moved, and one in which the subjects pressed a key whenever one of 20 dim lights came on. The dials and lights were distributed on three walls of a 12 foot-wide room. A dial change, or light during the light tests, was given on an average of once every 6 minutes (1 to 12 minutes between stimuli) over a period of 1½ hours (a total of 15 stimuli per test run). Only one experimental run was given in one day.

It is seen in Table 94 that, for the third and fourth runs (the noise runs), there were approximately an average of 22% "quick founds" compared to an average of about 36% "quick founds" in runs two and five (the quiet runs). A "quick found" was a response in nine seconds or less. How many responses occurred simultaneously with the dial movement or light flashing is not specified, but it was reported that that response was about equal in both the noise and quiet runs, i.e., the noise had no adverse effect on this response. Also, the number of "never seen" is not given but was presumably zero in both the quiet and the noise. In any event, the proportions in Table 94 are based on a total of something less than 15, perhaps 10 or so, responses.

Broadbent concludes that the noise caused a decrement of about 40% in "quick founds" relative to performance in the quiet. But, with respect to 20-Dial Test and blink theory, it should be noted that the data in Table 94 indicated that only somewhere, on the average, between one and possibly three more "quick founds" (22% vs. 36% of 10 to 15 "quick founds") were achieved in quiet than in noise; this could mean that possibly for only about 27 seconds out of 4500 seconds (1½ hours) were the subjects less quick in the noise than in the quiet.

TABLE 94

The Proportion of "Quick Founds" in Noise and Quiet for the 20-Dial and 20-Light
Vigilance Tests of 1-1/2-Hr Duration. From Broadbent (90).

Task and Group	Run				
	1	2	3	4	5
Task and Group	Quiet (Practice)	Quiet	Noise	Noise	Quiet
Dials	0.215	0.339	0.193	0.244	0.381
Lights: the markedly improving subjects	0.424	0.574	0.328	0.374	0.454
Lights: the least improving subjects	0.432	0.226	0.482	0.300	0.414

Note: On the Lights Test the subjects shown are those treated similarly to the Dials
subjects, so as to facilitate comparison. See text for the control of order
effects by other groups. The "markedly improving" subjects are the five showing the greatest value of (Run 2—Run 1) and the "least improving" the five
remaining.

While the difference between the averages of the noise and quiet runs is
statistically significant, the differences between the runs on the lights in the
presence of quiet and noise are not significant.

It should be noted that:

1. Greater confidence could be placed in the results had the control of
Noise-Quiet-Quiet-Noise runs also been conducted to obviate any possible
sequence effects.

2. Since immediate founds (called seens) were equal for the noise and quiet
conditions, the use of founds within 9 seconds or less, as a measure of merit,
seems somewhat arbitrary. The average time required for founds, for example,
would also be of interest.

3. Of more importance is the fact that the data for the lights show no
"blink" effect as postulated by the theory.

Broadbent (94) suggests that blinks only work for more difficult tasks and cites
Jerison's (417) findings that, for a "one clock" test, noise did not affect
performance but, with a "three clock" test (Jerison and Wing [422]), the noise
at a level of 114 dB caused a decrement relative to the quiet.

It would seem, however, that if the noise served as an ever-present and waiting
source of distraction, there would be more blinks for the easier light tests, i.e.,

less competition for perceptual attention. Boggs and Simon also report that more complex tasks (actually two tasks) were more adversely affected by noise than simpler tasks. However, one of Boggs' and Simon's tasks involved hearing words, the presentation of which, while not occurring synchronously with the interrupted broadband noise at 92 dB, could have been influenced ty TTS from the noise. On the other hand, Park and Payne found that the harder tasks (mental arithmetic) were less affected by noise than were the easier.

Broadbent (89) also suggests that the results he obtained on a more continuous task, in which the subject taps a contact in front of one of live lights, at which time another light comes on, supports the blink theory. The support comes from the finding that the noise did not paralyze or block in any way the organism on the motor side. In this study it was found that, in the noise, the subjects made a few more errors than in the quiet, but they had the same number of corrects, i.e., the subjects responded more often and quickly. (However, it would appear that the error score was the result of but one subject in the experimental group making five times as many errors in the quiet as in the noise.) Although the blinks from the noise do not affect adversely, relative to performance in the quiet, the correct perception of the lights, Broadbent surmises that the relative increase in errors made to the light with continued exposure to noise is due to an increase in blinks. As noted above (see Fig. 271) Jerison (417) questions whether there is a greater increase in errors in the noise than in the quiet during the relatively long vigilance test conducted by Jerison and Wing. Also, McBain found fewer errors on a repetitive task in the noise than in the quiet.

Other Experimental Findings Related to Blink Theory

Cohen *et al.* (146), using relatively large groups of subjects, conducted a study of the effects of noise on a vigilance task, a psychomotor coordination tracing task, and a mental anagram task. Figure 274 shows the results on the tracing tasks. The principal results of the mental tests are shown in Tables 95 and 96. It is seen that the only consistent effect of an increase in broadband background noise level from 75 dB to 95 dB appeared to be on the anagram mental test (Table 96) were about one-half word more was completed in the low level control (called Quiet) than in the high level noise condition. An analysis of variance of the data from the anagram test revealed that the differences between high and low level noise were not statistically significant, and that (somewhat as with the intelligence test forms used by Lienert and Jansen [509]) supposedly equally difficult test forms were actually somewhat different. As noted earlier (see Table 90), Cohen *et al.* found that while the differences in performance in two noise conditions were, on the average, negligible, there was a tendency for

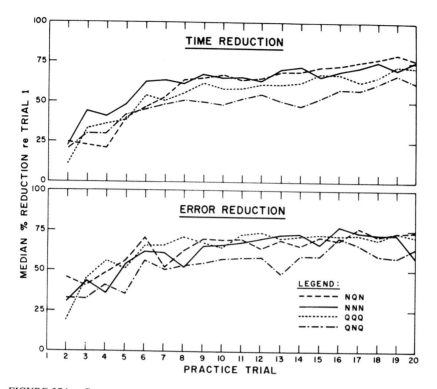

FIGURE 274. Percentage reductions in time and error scores on star-tracing test between first practice trial and each succeeding trial in the 20-trial practice period for each experimental group. Group QNQ had better performance on trial 1 than did the other groups so that the error and time reduction for that group tended to be lower than for the other groups. Differences among groups were not significant. N is random noise at 110 dB (earphones) and Q is random noise at 75 dB (earphones). From Cohen *et al.* (146).

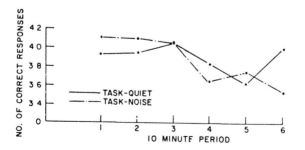

FIGURE 275. Performance in successive 10-min periods during quiet and 110 dB random noise of complex counting task. See also Fig. 256. From Helper (367).

TABLE 95

Results of Detection Tests on 10 Dials

Random noise at 75 dB (quiet, control) and 95 dB high level, and variable, 90-95 dB.
From Cohen *et al.* (146).

Mean Percent Detection Scores ($\bar{X}$) and Standard Deviations (SD)
by Background Noise Conditions and Time Period

		Time Period							
		1	2	3	4	5	6	7	8
Elapsed Time (minutes)		0-15	16-30	31-45	46-60	61-75	76-90	91-105	106-120
Control	$\bar{X}$	98.79	97.25	95.49	97.59	99.14	97.60	98.76	97.29
	SD	2.71	3.25	7.60	6.26	1.35	.61	2.95	3.97
High Level Noise	$\bar{X}$	99.55	98.69	95.48	98.12	97.73	96.58	98.40	97.39
	SD	1.82	3.35	6.89	5.71	5.72	5.45	3.77	7.40
Variable Noise	$\bar{X}$	99.58	96.55	97.27	95.70	95.90	95.57	95.58	97.19
	SD	2.31	5.19	6.15	5.69	4.59	5.43	4.54	5.39

subjects, classified by personality tests as being what might be called "anxious," to perform the vigilance 10-Dial task slightly less well in the control, low level, noise than in the higher or variable noise, whereas those subjects classified by the personality tests as being normal, performed slightly better in the high than the low noise levels.

Except for the aforementioned results of the 20-Dial (Broadbent [96]) and Three-Clock Test (Jerison and Wing [422]),), the findings of Cohen *et al.* are in general agreement with most studies on this problem done before 1950 (442) as well as with more recent studies. For example, Broadbent (94), Brewer and Briess (88), Helper (367), Loeb *et al.* (525), Miller (550), Smith (753), Plutchik (631), Saul and Jaffe (722), Sanders (715), and Park and Payne (602) found that steady-state or interrupted noise up to levels of 120 dB or so had no average discernible effects on the performance of a wide variety of mental and motor tasks, although some of these investigators did find somewhat greater variability in noise than in quiet. Ohwaki (589) found no significant effects of interrupted noises of 60 and 80 Phons on psychomotor tests (pursuit rotor and finger dexterity) but some degrading effect of the 80 Phon noise on a group doing mental word formation. However, the selection and matching of the subjects assigned to the different control and experimental groups may have contributed some unknown error to the tests.

Humel *et al.* (396) note that the results of a study by Loeb and Jeantheau (523), that is sometimes cited as in support of the blink theory of noise interference with a motor task, is confounded by the fact that the noise in this study was accompanied by vibration which could have affected task performance.

Theory of Distraction-Arousal

As noted in the preceding chapter, one effect of a change in the acoustic environment, particularly if it consists of an increase in noise level above about 70 to 80 dB above the threshold level of the noise, is an initial arousal of the somatic responses followed by adaptation. Further, as shown in Fig. 256, this arousal to expected 110 dB random noise is rather smaller or no greater than the arousal from the task itself. Arousal is usually viewed as necessary for good performance and therefore the noise could be beneficial to performance. However, comparison of Fig. 256 with Fig. 275 does not show any obvious systematic relation between somatic arousal as measured and task performance (continuous counting of independent flashes of three lights) or between performance in noise and quiet as found by Helper (367).

The experimental findings of Corcoran (160) and Wilkinson (877) shown in Fig. 276 are pertinent to this question. Here it is seen that sleep-deprived subjects (57, 32, and 24 hours) generally performed better in 100 dB and 90 dB random noise than in the quiet, particularly in the later parts of the test sessions. In general, although not consistently, the subjects within a group (sleep deprived or nonsleep deprived) tended to perform better in the quiet than in the noise.

Although the blink may be a questionable distraction phenomenon in steady-state noise, the notion of some short (about 20 seconds) distraction to the noise or to somatic reactions to the noise appears to fit rather well with some performance data taken shortly after the onset of intermittent noise. Woodhead (883-888) has studied the effect of intermittent bursts (70 dB and 110 dB, 0.97-sec duration) of a recorded "rocket" noise upon making rapid decisions with respect to various kinds of visual tasks (counting symbols on a number of simultaneously presented cards, marking off different symbols on a paper, etc.). Typical results are shown in Fig. 277 where we see that there was an apparent brief temporal degrading effect of the noise on performance. However, as shown in Table 97, there was absolutely no degradation in percent correct, wrong, or omitted in the test items taken over at 15-minute test period in a similar study.

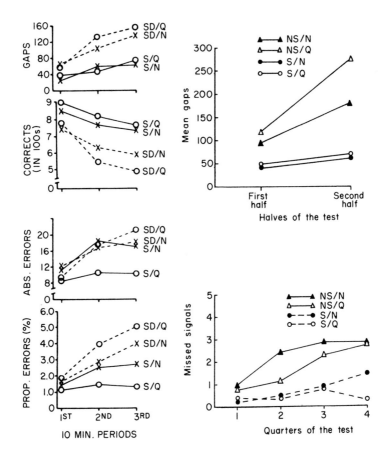

FIGURE 276.   Left graphs: Performance on 30-min serial reaction test (5 lights) of non-sleep deprived (S) and 32-hr sleep-deprived (NS) subjects in the quiet and 10 dB random noise. From Wilkinson (877). Upper right graph: Signals missed on an auditory vigilance task of nonsleep deprived (S) and 57-hr sleep deprived (NS) subjects in quiet and in 90 dB random noise. Lower right graph: Same task as left graphs. 24-hr sleep-deprived and nonsleep-deprived subjects in quiet and 90 dB random noise. Right graphs from Corcoran (160).

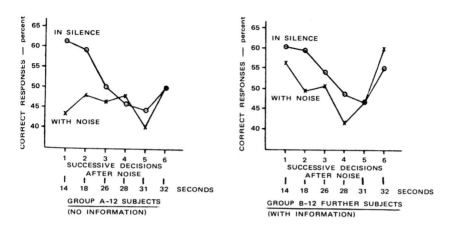

FIGURE 277. Effects of burst of recorded rocket noise (110 dB) and silence upon performance on rapid symbol comparisons and decisions during 32 sec after noise, and comparable time periods for subjects not receiving the burst of noise. Group A was given no information regarding accuracy of performance; Group B was given information. From Woodhead (883).

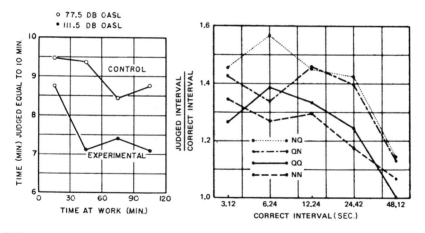

FIGURE 278. Left graph: Change in time judged equal to 10 min during successive half-hours of the experimental control sessions. The first half-hour of the experimental session was, as indicated by the open circle, in quiet. From Jerison and Smith (420). Right graph: Relative error for various rates of movement of the target pip. Rates are converted into correct interval measures which reflect the duration of the invisible portion of the target's course from the disappearance point to the vertical crosshair. Correct responses would yield a value of 1.0 on the ordinate. Random noise at 108.5 dB, quiet at 69 dB. From Jerison and Arginteanu (419).

TABLE 96

Results of Anagram Test in High-Level (105 dB) and Low-Level (75 dB) Random Noise. From Cohen *et al.* (146).

Means ($\bar{X}$) and Standard Deviations (SD) of Anagram Solving Performance by Noise Condition and Test Session

Noise Condition	Session 1		Session 2		Session 3		Session 4		Session 5	
	$\bar{X}$	SD	$\bar{X}$	SD	$\bar{X}$	SD	$\bar{X}$	SD	$\bar{X}$	SD
Exp. - (High Level (N = 40))	18.5	2.99	20.0	3.18	17.6	3.91	19.8	2.74	19.9	2.73
Cont. - (Low Level (N = 50))	19.1	3.52	20.6	3.13	18.3	4.61	21.4	3 15	20.6	3.18

TABLE 97

Performance on Rapid Decision Task over a 4-Minute Test Period
in Quiet and in Intermittent Bursts of Noise. From Woodhead (896).

	% Correct	% Wrong	% Omitted
Noise (110 dB)	57	6	37
Quiet (20 dB)	56	6	38

Although not so done by Woodhead, it appears necessary to interpret these data to mean that the noise arouses the organism so that it is somehow able to perform somewhat better starting 30 or so seconds after the noise comes on, but that this improved performance is, on the average, reduced or negated by the momentary distraction that occurs with the onset of the noise. Presumably the distraction from the burst of noise is followed by increase-beyond-normal attention to the signal or some increase in ability to respond for some period of time.

When persons are exposed to brief bursts of noise such as sonic booms, it is possible that there will be startle reactions that would interfere with performance of either mental or motor tasks. Woodhead (888) found (see Table 98) that a recorded sonic boom presented to listeners at various levels caused an increase in the number of omitted responses (rapid decisions regarding symbols on visually displayed cards) that occurred within 30 seconds after the sonic boom. The effects on overall performance for periods beyond 30 seconds are not presented.

Sanders (715) found no difference in errors made on a cancellation test (subjects crossed out certain specified numbers and groups of dots on sheets with many numbers and dots) between a steady noise condition 90 dB (18 tones 85 to 1360 Hz) and the same noise varying in level from 65 to 95 dB with an average level of 75 dB. There was some indication of greater variability during the varying noise which would suggest agreement with the distraction-arousal compensation hypothesis.

As mentioned earlier, McBain (532) found that noise (recorded speech played in reverse) reduced the number of errors relative to the number obtained in the quiet, when performing a monotonous task—handprinting of pairs of letters. These findings are attributed, by McBain, to arousal.

TABLE 98

Results of Fast Visual Symbol Discrimination Test in the Presence of Recorded Sonic Booms Presented at Different Intensities. From Woodhead (888).

		2·53	1·42	0·80
Pressure level in lb/ft²:		2·53	1·42	0·80
Number of subjects:		36	36	36
		a	*b*	*c*
Number of subjects	with decrement	23	16	15
	with improved performance	6	12	15
	showing no change	7	8	6
Statistical levels of probability	for each group	0·01	0·14	0·64
	two groups, $a + b$ (72 subjects)	——0·005——		
	two groups, $b + c$ (72 subjects)		——0·18—— (insignificant)	
	three groups, $a + b + c$ (108 subjects)	——0·032——		

Theory of Time Judgment

A theory of the effects of the subjective judgment of time on mental and motor work performance in noise has not been specifically formulated. However, Jerison and Smith (420) and Jerison and Arginteanu (419) have studied the effects of noise on the estimation of time and discussed the general meaning of these effects. In one experimental procedure, Jerison and Smith had subjects estimate when three lights blinking at different rates would come on, and Jerison and Arginteanu had subjects watch a visual dot which moved across a visual plane in front of the subject, and then disappeared behind an opaque screen. When the subject estimated the dot had reached a position exactly behind a crosshair on the front of the screen, he pushed a switch. The task was performed in noise (108.5 dB) and in quiet (69 dB). Some of the results are shown in Fig. 278 where it is seen that, in general, the noise caused a greater degree of overestimation of passing time in noise than in quiet. However, the results of condition NN compared with QQ, the right-hand graph, are somewhat inconsistent with that conclusion.

Jerison (414, 415, 416) found that counting and marking every-so-often how many times each of three lights had flashed was more adversely affected over a two-hour period in 110 dB random noise than in the quiet. Loeb (516) obtained results that partially confirmed Jerison's findings. Loeb had subjects estimate 3- and 10-minute intervals while completing a jigsaw puzzle in 80 dB and 110 dB noise. The results of these studies could be interpreted to mean either that noise effects short-term memory, or the ability to estimate time.

Jerison and Arginteanu suggest a two-factor description of the effect of noise on time estimation:

1. As a neutral acoustic stimulus, noise fills and expands the time scale, i.e., as was found by Hirsh *et al.* (380), in the presence of a neutral background sound, less time is judged to have passed than actually has.

2. As an adverse stimulus, noise contracts the time scale, i.e., more time is judged to have passed than actually has.

It is clear that noise, at least for relatively short exposures, does affect subjective estimates of the passing of time. How this effect of noise influences mental or motor work performance, if indeed it does, has not been fully studied.

Theory of Vestibular Involvement

C.S. Harris (340) has conducted research aimed at assessing the possible influencing role of the vestibular system upon mental and motor performance in the presence of noise. The hypothesis of the studies was that the first sensory system, after the auditory, to be assaulted by intense noise is the vestibular.

The noise via loudspeakers, was presented equal at both eardrums (the listeners wore earplugs in both ears) or unequal (the listeners wore earplugs in both ears and a muff over one ear); these conditions were called, respectively, symmetrical and asymmetrical. The results are shown in Fig. 279 (left-hand graph) where it is seen that the higher noise levels resulted in more visual discriminations (boxes completed), but more errors being made, when the noise was asymmetrical. As noted in other studies, noise sometimes increases output and also errors.

Figure 279 also shows the results for the hand-tool dexterity test (involving sorting and manipulation of nuts and bolts). It is seen that exposure at 120 dB noise level, the asymmetrical and symmetrical exposures, cause about the same degradation (time to complete) in the task. However, Harris reports that at the higher noise levels, particularly 140 dB, some of the nuts and bolts in the hand-tool dexterity tests were shaken on the table due to mechanical vibration, causing the subjects some difficulty with the test; this shaking, rather than the noise, may have been the basis of the poorer performance at the highest noise level. The noise spectra for these tests as present in the room, and the listeners' ear canals are shown in Fig. 280.

One conclusion in this study was that a hand-tool dexterity test involving manipulation of nuts and bolts revealed greater sensitivity to the noise than did a test that involved little proprioceptive activity, a discrimination test involving a comparison of printed visual symbols. It is not clear from an examination of Fig. 280 that this conclusion is justified. Psychologically, the subjects did not rate

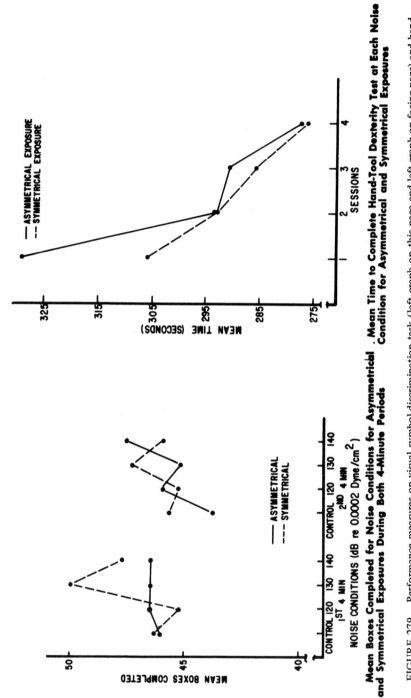

FIGURE 279. Performance measures on visual symbol-discrimination task (left graph on this page and left graph on facing page) and hand-dexterity test. From Harris (340).

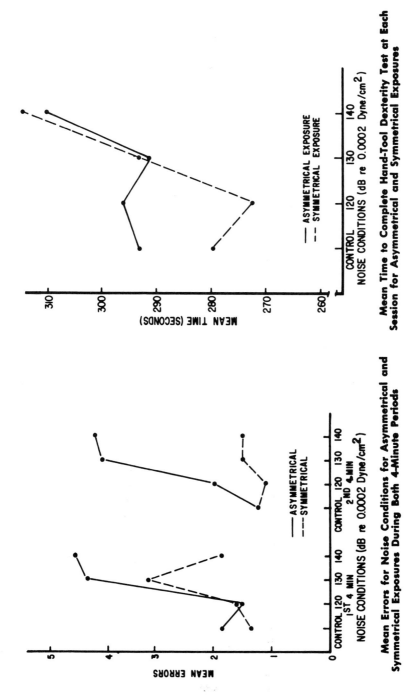

FIGURE 279. Continued.

the symmetrical noise as being more bothersome to the hand-tool dexterity than the visual task, and they rated the asymmetrical noise as being more bothersome to the discrimination than to the hand-tool dexterity test, as shown in Fig. 280.

All in all, it would appear that no appeal to any vestibular involvement need be made to explain the data obtained in these experiments. Rather, the asymmetry of aural stimulation would seem to have affected the mental test more than the task requiring some proprioceptive involvement. The general increase, which for some unknown reason is greater with asymmetrical noise, in errors on the mental tests was offset by an increase in the number of tests completed (in agreement with the kind of results obtained by Woodhead) so that the number of correct items for the noise conditions remained about the same. The increase in time required to complete the hand-dexterity test appears to be due to the vibration shaking of the test items to be manipulated at the higher levels of noise.

The importance of the asymmetry of aural stimulation found in these tests is reminiscent of the results of an earlier study (Nixon *et al.* [583]) where it was found that the act of balancing on a rail was adversely affected by asymmetrical aural stimulation but not symmetrical. Whether this finding is attributable to some involuntary turning of the head toward a sound source as an act of auditory localization, or to some effect of the noise on the vestibular system, is an open question.

Recent Experiments on Distraction-Arousal

Explicit in much of the thinking underlying the planning and interpretation of many of the research studies discussed above has been the thought that noise is a distractive stimulus that competes for a person's attention, and arouses the organism because it is an intense, aversive stimulus. Consciousness of the arousal reactions in the body may be another (besides the direct auditory) sensation that somehow, either at a neural or muscular level, or both, inhibits or enhances the organized activity required to perform a task. Accordingly, much of the previous research material presented could be subsumed under this very general theory.

Without attempting to specify the underlying physiological mechanism, Teichner *et al.* (795) stated a distraction-arousal hypothesis and designed two experiments to test it. Visual search tasks (finding certain letter combinations appearing on displays before the subject) were used for these experiments. The subject responded by throwing a switch which measured how quickly he had reached a decision. The time required for the decision-response was taken as the basic measure of performance.

In one experiment, the subjects responded to 150 displays over about a 45-minute period in the presence of a broadband random noise at 81 dB. At that

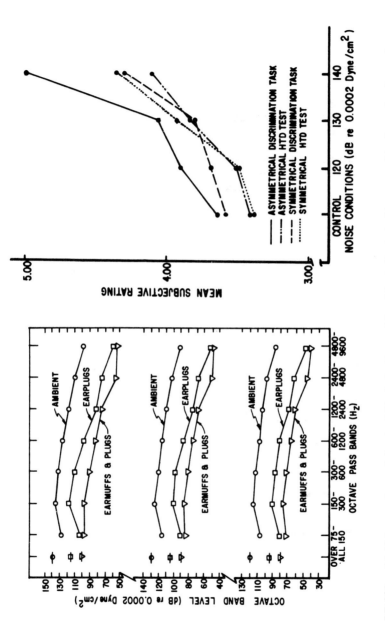

FIGURE 280. (a) Noise spectra used in visual-discrimination and hand-dexterity test; (b) mean subjective rating for each noise condition. From Harris (340).

time, without warning, the noise level changed to 57, 69, 93, or 105 dB for 50 or more displays over a 15-minute period. The results are shown in Fig. 281 where it is seen that the change in noise level increased the time required for decisions (expressed in terms of a percent decrease in bits per second in decision information relative to that found in the 81 dB noise). Actually, Teichner *et al.* found that efficiency of task performance increased during the last 50 displays, when the noise level either stayed at 81 dB or was shifted, but that the shift reduced the rate at which the subjects improved relative to the rate maintained when the noise level stayed constant at 81 dB. The loss was slightly less for the upward shift in noise level than the downward, which would be indicative, according to the theory, that arousal of the upward shift was more significant than that due to the downward shift, or else the distraction for the upward shift was less severe.

Schoenberger and Harris (737) also performed an experiment designed to test the effect of a sudden change in background noise level upon the performance of a psychomotor test. In this test, the subject, as rapidly as possible, serially connected by a line numbers that were scattered at random on a sheet of paper. Unlike the Teichner *et al.* study, the change, when one occurred, was always from a lower to a higher noise level. The results are shown in Fig. 282.

In Fig. 282 it is seen that there appeared to be a positive improvement in performance during the trials 26 through 75 except when the acoustic background changed after trial 50 from quiet to 110 dB of broadband random noise. For each test session, 1 to 75 trials lasted 15 minutes; trials 1 through 25 were considered as practice trials and not included in the data analysis. Of considerable importance is the fact that the amount of learning or improvement in performance increased as the amount of change in background noise level decreased.

To perhaps show this point more clearly, on Fig. 283 we have plotted the gain in mean performance of trials 51 to 75 over 26 to 50 as a function of change in noise level that occurred between trial 50 and 51. Also on Fig. 283 are comparable data for the study of Teichner *et al*. It is seen that the effect of change in background noise level is to slow down, at least for the periods involved, the rate of learning or task improvement.

The data of Teichner *et al.* and Schoenberger and Harris do not speak to the perhaps more practical or common question of whether habituation, with respect to interference effects of noise on learning or the performance of a well-learned task, will take place in the presence of intermittent noise regularly present over extended periods of time. Intuitively, it seems likely that habituation to an irregular acoustic environment would be less complete during the learning of a new task than for maintenance of performance on a previously well-learned task. In any case, these two studies do show that the degree of sudden change in noise level has some affect upon the rate of learning.

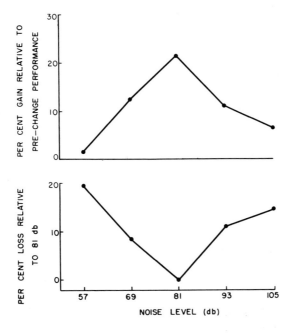

FIGURE 281.   Effects on information transmission related to changing noise level from 81 dB to higher and lower noise levels. From Teichner *et al.* (795).

Teichner *et al.* report data (see Fig. 284) showing that search and identification time of alphabet letters to be identified from a slide projection is a complex function of (*a*) the percent time that random noise at 100 dB is on, and (*b*) the duration time of the test. The authors deduce that a distraction-arousal hypothesis explains the results. However, these data may reflect complex arousal-habituation phenomena limited to the several hours of testing received by the subjects on a task possibly undergoing some improvement due to learning, the task being not too dissimilar to that used in the study of Teichner *et al.* that was discussed in previous paragraph.

Summary of Distraction-Arousal Theory

Except for possibly a momentary distraction of some seconds when the noise is turned on, with the level being higher probably than 80 PNdB or so, followed by lessened distraction or improved ability for some seconds, the data reviewed suggest that regular, steady or intermittent noise may have no effect on the performance of a well-learned nonauditory mental of psychomotor task. The arousal, if viewed as a general somatic response to the noise, could be viewed as causing first, distraction, either perceptually to the "new" stimulus (the noise) or to the somatic responses themselves, and second, a slight improvement in sensory-motor proficiency due to the increased blood supply to the periphery of the body. All effects are probably transient according to direct physiological measures and also by measures of task performance. The apparent adverse effects of changes in noise levels upon learning may be due to a lack of habituation to, or "learning" of, the irrelevance to the task of irregular noise patterns in specific studies.

It is conceivable that even during the period of somatic responses, the organism is physically and mentally fully capable, perhaps more so, of performing tasks; perhaps the organism is thrown into a state of perceptual and bodily awareness in order to best recognize and respond to the new stimuli. In short, his attention is fully directed to the new situation. The time required to recognize the relevance or irrelevance of the new stimulus (say, sudden noise) probably represents the distractive effect of noise, the arousal per se being of some benefit to the organism.

Related to this thought is the finding by Thackray (797) that, with subjects highly practiced on a visual reaction time test, an unexpected stimulation with a previously experienced 440 Hz tone at 120 dB caused sharp somatic startle arousal, but did not influence the average reaction time of subjects who had practiced the reaction time task. However, in the practiced subjects, the unexpected tone resulted in some subjects having reduced, and some subjects lengthened, reactions times, showing again the importance of individual differences.

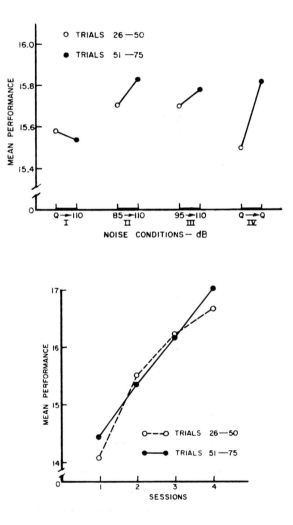

FIGURE 282. Upper graph: Mean performance for each noise condition for trials 26-50 (initial noise level) and trials 51-75 (final noise level). Lower graph: Mean performance for each session for trials 26-50 and trials 51-75. From Schoenberger and Harris (737).

Productivity in Industry

As noted in my previous review (442) noise per se does not appear to reduce nonauditory work productivity in the factory and office, it even improves some performance by apparently isolating the person from being interrupted by certain distracting auditory signals or speech. Felton and Spencer (225) comment that ego involvement in a high-status occupation offsets concern about noise (94-119 dB). Ganguli and Rao (275) believe, but present no data, that productivity in most workers is not affected by noise of 100 dB or lower. However, de Almida (187) found absenteeism from the work room dropped when noise level was reduced.

Two rather recent studies purport to show deleterious effects of noise on work production. Broadbent and Little (98) reported that the reduction in the noise level from 99 to 89 db in a factory work space (bay) resulted in fewer numbers of broken rolls of film and equipment shutdowns than were experienced by the same workers when they worked in an untreated bay (the workers moved from one bay to another during the work day). The work performance improved in both the sound-treated and non-treated bays after some of the bays were treated, apparently due to general improved morale. The data are shown in Table 99.

TABLE 99

Comparison of Acoustically Treated and Untreated Work Bays Before and After
Treatment Was Carried Out

The treatment was applied to the "Treated Bays" at end of 1957. Therefore, the 1956-1957 data for those bays are for prior to treatment. The "Untreated Bays" remained unchanged from 1956-1958. The workers moved from one bay to the other during normal work procedures. From Broadbent and Little (98).

	TREATED BAYS		UNTREATED BAYS	
	1956/7	1957/8	1956/7	1957/8
Broken rolls (attributed to operator)	75	5	25	22
Other shutdowns (attributed to operator)	158	31	75	56
Calls for maintenance (excluding first six week period in each year)	746	597	516	468
Point hour	84.5	89.6	91.2	95.25
Absenteeism (time as % of possible hours worked)	5.18	4.43	2.72	1.556
Labour turnover (Mean per six weeks)	1956/7 = 6.2%		1957/8 = 0	

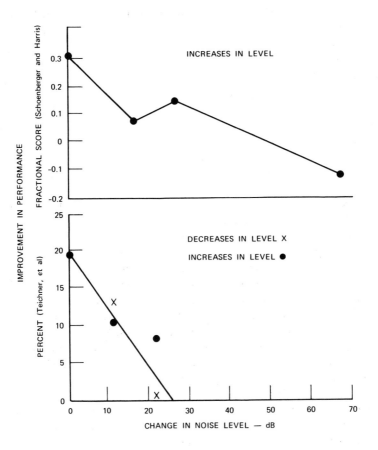

FIGURE 283.   Change in task improvement as function of sudden change in noise level
near end of learning session. Data from Teichner *et al.* (795).

Broadbent and Little propose that these findings support the laboratory tests, discussed earlier, of "blinks" on task performance. This may indeed be involved, but there is the possibility that there was an auditory component to the work (threading film on spools) that aided the workers in threading the film and in detecting films slipping from sprockets or malfunctions in the machinery. The reduction of the noise, if this were the case, should, of course, lead to improved work performance.

Kovrigin and Mikheyev (437) found that increasing the noise level, via loudspeakers, in the room used for postal letter sorters increased the number of sorting errors (see Fig. 285). The increase in errors was systematic with increase in noise level. These results cannot be taken to necessarily mean that the noise per se caused the increase in errors because of some basic physiological or psychological distractive effect, but could be due to personnel viewing the noise as aversive because it bothered their hearing and/or represented a degradation in the concern of management with their comfort and well-being. Also, the measured effects quite possibly could have disappeared with continued stimulation.

It is perhaps unfortunate that industrial work situations do not lend themselves to nicely controlled experimental programs. But, other than the study of Broadbent and Little, which we question on other grounds, it is not possible in our opinion to demonstrate that habitual noise, as such, reduces or interferes with nonauditory work productivity in industry. This is not to say, of course, that noise reduction does not improve workers' morale and thereby increase production, or that it does not lessen temporary and permanent damage to hearing, or that it does not improve work output because helpful auditory information that is unknowingly present and masked by the noise now becomes available to the worker.

As mentioned earlier, a sound or noise may on occasion mask other sounds or noises that can disturb or distract a worker thereby reducing productivity. For some purposes, in generally quiet surroundings, a low-level broadband random noise may be introduced to increase a sense of privacy with some possible beneficial effects. Music has also been used in work situations, not so much perhaps to mask other sounds, but to provide some pleasant stimuli to persons doing nonauditory work. The presumptions have been that work output will be increased because of improved morale, or that people are kept more aroused and alert than they otherwise become in monotonous jobs. Figure 286 shows some data obtained in one study (891) on this matter. There appears to be some, but no consistent, relation between the presence of music and work output. The clearly cyclic characteristics of the work output makes firm interpretation of the data difficult.

It is not our purpose in this document to review research on the effects of music on work performance. It might be noted, however, that in general it has

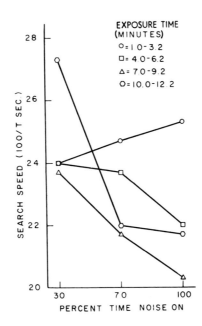

FIGURE 284. Effects of on-off ratio of random noise (100 dB) on response speed after exposure times indicated by legend. From Teichner *et al.* (795).

been difficult to quantify the beneficial effects of music in industry, partly for some of the same reasons that it is hard to show any detrimental effects of noise on work output:

1. In many cases the effects are transitory and related to temporary changes in worker morale.

2. There are no beneficial effects, or perhaps the opposite.

3. There are beneficial effects that are relatively small compared to other task and motivational factors in the situation.

Physiological and Psychological Factors Theory

Most of the theorizing about the effects of noise that is not meaningfully related to the stimulus or response of a given mental or motor task is naturally oriented towards innate physiological mechanisms that do not involve meaningful cognitions. It is concluded that most of the data, where familiar noise environments are involved, show no differences in work performance between noise and quiet conditions because of habituation of possibly disruptive physiological responses. However, even after disregarding studies that have possibly run afoul of one or more experimental errors, there appear to be data from both the laboratory and factory that seem to disagree with this negative conclusion.

It seems unlikely that the distraction-arousal-adaptation-theory can or need be elaborated to explain on a direct physiological level these other usually degrading effects of noise on mental and motor work performance. Rather, it is suggested that these results are obtained because the basic assumption underlying the usual interpretation of laboratory and factory studies of noise effects is not, in these particular cases, valid. The assumption in question is that the task and its completion are not dependent upon the presence of the noise.

In line with the studies and formulation of Azrin (41), it is hypothesized that certain of the following psychological factors are at play in these "discrepant" performance data:

1. *Stimulus Contingency.* In order to explain individual differences in the reactions to noise, it is presumed that the noise is considered more harmful, disagreeable, or aversive to some persons than to others. The general finding that the performance of the more anxious personality types is more affected by noise than that of nonanxious types would attest to the existence of a stimulus-contingency factor. In terms of learning or conditioning, the task becomes disliked and is performed relatively poorly because it is related to or contingent upon the aversive noise. Habituation can be expected to take place

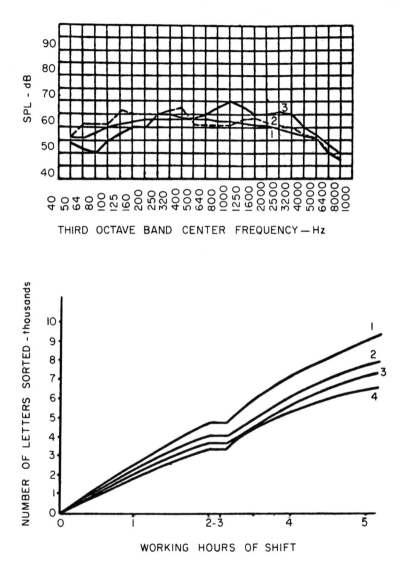

FIGURE 285.   Upper graph: Normal noise spectra at postal sorting machines. Lower graph: Effect of change in noise level on working efficiency. 1. 78-80 dB; 2. 85 dB; 3 90 dB; 4. 95 dB. From Kovrigin and Mikheyev (437).

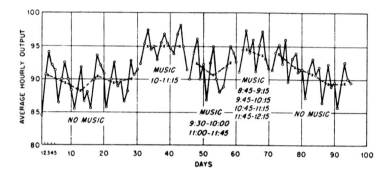

FIGURE 286.   Output in a light manual task (rolling paper novelties) under various conditions of music presentation. From Wyatt and Langdon (891), after Broadbent (92).

(or counter motivation factors, such as necessity to work for compensation) but at different rates for different people.

2. *Response-Related Factors.* This rationalization would say that some subjects or workers may view the noise as a punishment or the lack of noise as a reward for work performance. Here, regardless of what the intentions of the experimenter or the factory manager might be, the person will be more highly motivated and perform better if he thinks his responses will tend to result in or maintain reduced noise, or conversely, he will be less motivated if he thinks his work will increase or not reduce the noise. As with stimulus-contingency, habituation can be expected to take place if the rewards and punishments are truly not response-contingent. This is, of course, in agreement with the so-called "Hawthorne" effect. This effect refers to the fact that improving a worker's environment, such as reducing the background noise, results in improved morale and work performance. The improvement to nonauditory work may disappear in time and may actually be unrelated to interference effects of the noise per se.

The distillation of physiological and psychological factors given above is perhaps still too complicated to provide a framework for understanding the fascinating mass of data and observations that are available about the effects of noise on nonauditory mental and motor work. At the same time, it appears that anything less complex provides too limited a view to explain all of the effects that have been reported.

PART IV

A SUMMARY

A possible teaching of much of the data presented in this book is that, other than as a damaging agent to the ear and as a masker of auditory information, noise will not directly harm people or interfere with psycho-motor performance. Man should be able, according to this concept, to adapt physiologically to his noise environment, with only transitory interference effects of physiological and mental and motor behavioral activities during this period of adaptation. This concept, or its converse, is difficult to substantiate by scientific research and must be recognized as being hypothical at this time.

All other noticed effects of noise, including physiological stress reactions, are taken to be due to psychological factors related to stimulus and response contingencies as associated with the noise by individuals. Further, it can be expected that inappropriately interpreted stimulus and response contingencies leading to individual differences in behavior to noise would tend to be eliminated with learning and experience; indirectly aroused physiological stress reactions (fear, frustration, etc.) to many repeated exposures to a noise, if not eliminated through learning would undoubtedly be harmful to mental and physical health.

The striking, considering its dynamic range and complex functioning, similarity of the organ of hearing among people makes possible the prediction of the masking and, at least on a statistical basis, the auditory damaging effects of noise with considerable accuracy. The general similarity in work and play activities, and social values within large groups of people makes feasible the implementation of such concepts as perceived noisiness for the measurement and prediction of the effects of noise on the general reactions and behavior of people and society.

Because nonauditory physiological stress responses in an organism to environmental noise are apparently primarily the result of interactions between specific behavioral activities and the noise, rather than the noise per se, the results of research on the effects of noise on lower animals, except for damage to

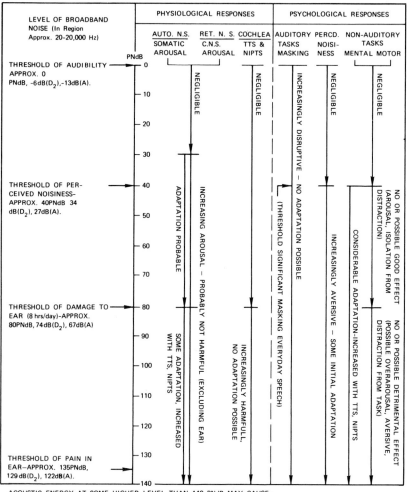

FIGURE 287.   Basic physiological and psychological responses of man to habitual environmental noise. Auto. N.S., Ret. N.S., and C.N.S. stand for autonomic,
reticular, and central nervous systems, respectively.

the ear, cannot usually be generalized to man. Research in this area, with rodents and rabbits, is particularly specious because of the presence in some of these animals of so-called audiogenic behavior and responses.

Quantitative methods for the physical measurement, setting of tolerable limits, and management of environmental noises are reported and developed in this book, review in particular Fig. 61 on masking of speech, Fig. 133, and Table 25 on damage risk to hearing, and Fig. 238 and Table 40 on tolerability and community behavior. An attempt has been made in Fig. 287 to summarize general guidelines to follow in this regard; these are consistent with the theory and facts outlined above. In this figure significant aversion to noise is presumed to be primarily a joint function of (*a*) some learned criteria of what levels of noise will significantly interfere with the reception of speech and other auditory signals, (*b*) an unwantedness due to arousal, distraction, loudness, and startle, and (*c*) some sensed, through somatic and auditory responses, perhaps both consciously and unconsciously, levels that are, with continued exposure, damaging to the ear.

References

1. Abey-Wickrama, I., A'brook, M.F., Gattoni, F.E.G., and Herridge, C.F., Mental hospital admissions and aircraft noise. *Lancet* 1275-1277 (December 13, 1969).
1a. Acton, W.I., Effects of ear protection on communication. *Ann. Occup. Hyg.* **10**, 123-429 (1967).
2. Acton, W.I., A review of hearing damage and risk criteria. *Ann. Occup. Hyg.* **1**, 143-153 (1967).
2a. Acton, W.I., A criterion for the prediction of auditory and subjective effects due to air-borne noise from ultrasonic sources. *Ann. Occup. Hyg.* **11**, 227-234 (1968).
3. Acton, W.I. and Carson, M.B., Auditory and subjective effects of airborne noise from industrial ultrasonic sources. *Brit. J. Ind. Med.* **24**, 297-304 (1967).
4. Acton, W.I., Coles, R.R.A., and Forrest, M.R., Hearing hazard from small-bore rifles. *Rifleman* **74**, 9-12 (1966).
5. Ades, H.W., Davis, H., Eldredge, D.H., von Gierke, H.E., Halstead, W.C., Hardy, J.D., Miles, W.R., Neff, W.D., Rudnick, I., Ward, A.A., Jr., and Warren, D.R., "Benox Report—an Exploratory Study of the Biological Effects of Noise." University of Chicago Press, Chicago, 1953.
6. Ades, H.W., Graybiel, A., Morrill, S., Tolhurst, G., and Niven, J. "Non-auditory Effects of High Intensity Sound Stimulation on Deaf Human Subjects." Joint Rept. 5. University of Texas Southwestern Medical School, Dallas, Texas and U.S. Naval School of Aviation Medicine, Pensacola, Florida, 1958.
7. Adey, W.R., Kado, R.T., and Rhodes, J.M., Sleep: cortical and subcortical recordings in the chimpanzee. *Science* **141**, 932-933 (1963).
8. Aikens, A.J. and Lewinski, D.A., Evaluation of message circuit noise. *Bell Syst. Tech. J.* **39**, 879-909 (1960).
9. Albrite, J.P., Shutts, R., Whitlock, M., Jr., Cook, R., Corliss, E., and Burkhard, M., Research in normal threshold of hearing. *Arch. Otolaryngol.* **68**, 194-198 (1958).
10. Allen, W.A., Criteria for dwellings and public buildings. Paper F-4 *in* "The Control of Noise: National Physical Laboratory Symposium Number 12," pp. 359-371. Her Majesty's Stationary Office, London, 1962.
11. Allen, C.H., Jackson, F.J., Kryter, K.D., and Weaver, H.R., Temporary hearing threshold shift produced by high level tone bursts. Paper B-31. "Proceedings of the Fifth International Congress on Acoustics, Liége, Belgium, September, 1965."
12. Anderson, F.C., Subjective noise meters. *Noise Control* **6**, 7-10 (1960).
12a. Ando, Y. and Hattori, H., Effects of intense noise during fetal life upon postnatal adaptability (statistical study of the reactions of babies to aircraft noise). *J. Acoust. Soc. Am.* **47**, 1128-1130 (1970).
13. Andrews, B. and Finch, D.M., "Truck-Noise Measurement." Reprint 17. University of California Institute of Transportation and Traffic Engineering, Berkeley, 1951.
14. Andrews, B. and Finch, D.M., Truck-noise measurement. *Highway Res. Board Proc.* **31**, 456-465 (1957).
15. Andriukin, A.A., Influence of sound stimulation on the development of hypertension. Clinical and experimental results. *Cor. Vassa.* **3**, 285-293 (1961).
16. Andrukovich, A.I., Effect of industrial noise in winding and weaving factories on the arterial pressure in the operators of the machines. *Gig. Tr. Prof. Zabol.* **9**, 39-42 (1965).

17. Angelino, H. and Mech, E.V., Factors influencing routine performance under noise. II. An exploratory analysis of the influence of adjustment. *J. Psychol.* **40**, 397-402 (1955).

18. Anon., "Forschung und Versuchsanstalt." Rept. 22, 637 and Appendix. Generaldirektion PTT, Switzerland, 1959.

19. Anon., "American Standard for Audiometer for General Diagnostic Purposes." ASA Z24.5-1951. American National Standards Institute, New York.

20. Anon., "Real-Ear Attenuation of Ear Protectors at Threshold, Method for Measurement." ASA Z24.22-1957. American National Standards Institute, New York.

21. Anon., "American Standard Criteria for Background Noise in Audiometer Rooms." ASA S3.1-1960. American National Standards Institute, New York.

22. Anon., "American Standard Measurement for Monosyllabic Word Intelligibility." ASA S3.2-1960. American National Standards Institute, New York.

23. Anon., "American Standard for Preferred Frequencies for Acoustical Measurements." ASA S1.6-1967. American National Standards Institute, New York.

24. Anon., "American Standard Acoustical Terminology." ASA S1.1-1960. American National Standards Institute, New York.

25. Anon., "American Standard Specification for General-Purpose Sound Level Meters." ASA S1.4-1961. American National Standards Institute, New York.

26. Anon., "American Standard Method for the Physical Measurement of Sound." ASA S1.2-1962. American National Standards Institute, New York.

27. Anon., "Measurement of Noise Emitted by Vehicles." ISO Recommend. R362-1964. American National Standards Institute, New York.

28. Anon., "A Standard Reference Zero for the Calibration of Pure-Tone Audiometers." ISO Recommend. R389-1964. American National Standards Institute, New York.

29. Anon., "Procedure for Describing Aircraft Noise Around an Airport." ISO Recommend. R507-1966. American National Standards Institute, New York.

30. Anon., "Method for Calculating Loudness" and "Method for Calculating Loudness Level." ASA S3-1965. ISO Recommend. R532-1967. American National Standards Institute, New York.

31. Anon., "Methods for the Calculation of the Articulation Index." ANSI S3.5-1969. American National Standards Institute, New York.

32. Anon., The noise performance standards of the Chicago zoning ordinance. *Noise Control* **3**, No. 6, 51-52, 1957.

33. Anon., "Noise: Its Effects on Man and Machine." Hearings before the Special Investigating Subcommittee on Science and Astronautics, U.S. House of Representatives, 86th Congress, August 23-25, 1960. Rept. 13. U.S. Government Printing Office, Washington, D.C., 1960.

34. Anthony, A. and Ackerman, E. "Stress Effects of Noise in Vertebrate Animals." Rept. WADC-TR-58-622. Aerospace Medical Laboratory, Wright-Patterson Air Force Base, Ohio, 1959.

35. Apps, D.C., The AMA 125-sone new-vehicle noise specification. *Noise Control* **2**, 13-17 (1956).

36. Arkad'evskii, A.A., Effect of strong white noise on hearing function. *Biofizika* **4**, 166-169 (1959).

37. Arkad'evskii, A.A. "Hygienic Standards for Constant Noise In Industry." Rept. JPRS-27559, TT-64-51807, N65-11933. Joint Publications Research Service, Washington, D.C., 1964.

38. Atherley, G.R.C., Monday morning auditory threshold in weavers. *Brit. J. Indust. Med.* **21**, 150-153 (1964).

39. Auble, D. and Britton, N., Anxiety as a factor influencing routine performance under auditory stimuli. *J. Gen. Psychol.* **58**, 111-114 (1958).

40. Aubry, M., Grognot, P., and Brugeat, M., Proposition de niveaux d'intensité sonore maxima non traumatiques pour l'audition pendent huit heures d'exposition. (Proposal for maximum sound intensity levels for eight hours of exposure without threshold shift.) *Rev. Corps Santé Armées* **2**, 653-657 (1961).

41. Azrin, N.H., Some effects of noise on human behavior. *J. Exp. Anal. Behav.* **1**, 183-200 (1958).

42. Barnett, C.D., Ellis, N.R., and Pryer, M.W., Absence of noise effects on the simple operant behavior of defectives. *Percep. Motor Skills* **10**, 167-170 (1960).

43. Barrett, A.M., "Personality Characteristics under the Stress of High-Intensity Sound." Ph.D. thesis. Pennsylvania State University, University Park, 1950.

44. Bartoshuk, A.K., Response decrement with repeated elicitation of human neonatal cardiac acceleration to sound. *J. Comp. Physiol. Psychol.* **55**, 9-13 (1962).

45. Bauer, B.B. and Torick, E.L., Researches in loudness measurement. *IEEE Trans. Audio Electroacoustics* **AU-14**, 141-151 (1966).

46. Bauer, R.A., "Human Response to the Sonic Boom." Subcommittee on Human Response for Committee on SST-Sonic Boom, National Academy of Sciences, 1968; and "Some Thoughts on Human Response to Sonic Boom." Address to the American Institute of Aeronautics and Astronautics, October 22, 1968.

47. Baughn, W.L., Noise control – percent of population protected. *Intl. Audio.* **5**, 331-338 (1966).

48. Beecher, H.K., Pain: one mystery solved. *Science* **151**, 840-841 (1966).

49. Békésy, G. von, Über akustische Reizung des Vestibularapparates. *Pflüg. Arch. Ges. Physiol.* **236**, 59-76 (1935).

49a. Békésy, G. von, Über die Resonanzkurve und die Abklingzeit der verschiedenen Stellen der Schneckentrennwand. *Akust. Z.* **8**, 66-76 (1943). Translation: On the resonance curve and the decay period at various points on the cochlear partition. *J. Acoust. Soc. Am.* **21**, 245-254 (1949).

50. Békésy, G. von and Rosenblith, W.A., The mechanical properties of the ear. *In* "Handbook of Experimental Psychology," Chap. 27 (S.S. Stevens, ed.). Wiley, New York, 1964.

51. Bellucci, R.J. and Schneider, D.E., Some observations on ultrasonic perception in man. *Ann. Otol. Rhinol. Laryngol.* **71**, 719-726 (1962).

51a. Benko, E., Objekt-und Farbengeisichtsfeldeinengung bei Chronischen Larmschaden. *Ophthalmologica* **138**, 449-456 (1959).

51b. Benko, E. Further information about the narrowing of the visual fields caused by noise damage. *Opthalmologica* **140**, 76-80 (1962).

52. Bennett, E., Some tests for the discrimination of neurotic from normal subjects. *Brit. J. Med. Psychol.* **20**, 271-282 (1944-1946).

53. Benson, R.W. and Hirsh, I.J., Some variables in audio spectrometry. *J. Acoust. Soc. Am.* **25**, 499-505 (1953).

54. Beranek, L.L., The design of speech communication systems. Reprint from *Proc. Inst. of Radio Engrs.* **35** (1947).

55. Beranek, L.L., Noise control in office and factory spaces. *15th Annual Mtg. Chem, Engr. Conf., Trans. Bull.* **18**, 26-33 (1950).

56. Beranek, L.L., "Acoustics," McGraw-Hill, New York, 1954.

57. Beranek, L.L., Criteria for office quieting based on questionnaire rating studies. *J. Acoust. Soc. Am.* **28**, 833-851 (1956).

58. Beranek, L.L., Revised criteria for noise in buildings. *Noise Control* **3**, No. 1, 19-27 (1957).

59. Beranek, L.L. (ed.), "Noise Reduction," McGraw-Hill, New York, 1960.

60. Beranek, L.L., Kryter, K.D. and Miller, L.N., Reaction of people to exterior aircraft noise. *Noise Control* **5**, 23-31 (1959).

61. Beranek, L.L., Marshall, J.L., Cudworth, A.L., and Peterson, A.P.G., The calculation and measurement of the loudness of sounds. *J. Acoust. Soc. Am.* **23**, 261-269 (1951).

62. Bergman, M., Hearing in the Mabaans. A critical review of related literature. *Arch. Otolaryngol.* **84**, 411-415 (1966).

63. Betz, E., Readjustment reactions of cerebrovascular circulation to chronic noise. *Arch. Phys. Ther. (Leipzig)* **17**, 61-65 (1965).

64. Bilger, R.C., Remote masking in the absence of intra-aural muscles. *J. Acoust. Soc. Am.* **39**, 103-108 (1966).

65. Bilger, R.C., Intensive determinants of remote masking. *J. Acoust. Soc. Am.* **30**, 817-824 (1958).

66. Bilger, R.C. and Hirsh, I.J., Masking of tones by bands of noise. *J. Acoust. Soc. Am.* **28**, 623-630 (1956).

67. Bishop, D.E., Judgments of the relative and absolute acceptability of aircraft noise. *J. Acoust. Soc. Am.* **40**, 108-122 (1966).

68. Bishop, D.E., "Frequency Spectrum and Time Duration Descriptions of Aircraft Flyover Noise Signals." Rept. DS-67-6, Contract FA-65-WA-1260, N67-31325. Bolt Beranek and Newman, Inc., Van Nuys, California, 1967.

69. Bishop, D.E. and Horonjeff, R.D., "Procedures for Developing Noise Exposure Forecast Areas for Aircraft Flight Operations." Rept. DS-67-10, Contract FA67WA-1705. Bolt Beranek and Newman Inc., Van Nuys, California, 1967.

69a. Bishop, D.E. and Horonjeff, R.D., "Noise Exposure Forecast Areas for John F. Kennedy Airport, New York, for Chicago O'Hare International Airport, and for Los Angeles International Airport." Rept. FAA-DS-67, 11, 12 and 13, Contract FA67-WA-1750. Bolt Beranek and Newman, Inc. Van Nuys, California, 1969.

70. Blaesser, H.A., A loudness analyser for computation of the subjective loudness. *In* "Acoustic Noise and Its Control." Publ. 26. Institute of Electrical Engineers, London, 1967.

71. Blau, T.H., "Effects of High Intensity Noise on Certain Psychological Variables." Ph.D. thesis, Pennsylvania State University, University Park, 1951.

72. Blazier, W.E., Jr., Criteria for residential heating and air-conditioning systems. *Noise Control* **5**, No. 1, 48-53 (1959).

73. Blazier, W.E. Jr., and Wells, R.J., "A Procedure for Noise Evaluation of Sound Power Data." Paper L23 presented at the Fourth International Congress on Acoustics, Copenhagen, 1962.

74. Blodgett, H.C., Jeffress, L.A., and Whitworth, R.H., Effect of noise at one ear on the masked threshold for tone at the other. *J. Acoust. Soc. Am.* **34**, 979-981 (1962).

75. Boggs, D.H. and Simon, J.R., Differential effect of noise on tasks of varying complexity. *J. Appl. Psychol.* **52**, 148-153 (1968).

76. Bolt Beranek and Newman Inc. "Land Use Planning with Respect to Aircraft Noise." Rept. AFM 86-5, TM 5-365, Navdocks P-98. Department of Defense, Washington, D.C., 1966. Also issued as Tech. Rept. by Federal Aviation Administration, Washington, D.C., 1966.

77. Bolt, R.H. and MacDonald, A.D., Theory of speech making by reverberation. *J. Acoust. Soc. Am.* **21**, 577-580 (1949).

78. Bonvallet, G.L., Levels and spectra of transportation vehicle noises. *J. Acoust. Soc. Am.* **22**, 201-205 (1950).

79. Bonvallet, G.L., Levels and spectra of traffic, industrial, and residential area noise. *J. Acoust. Soc. Am.* **23**, 435-439 (1951).

80. Borsky, P.N., "Community Reactions to Air Force Noise. I. Basic Concepts and Preliminary Methodology. II. Data on Community Studies and Their Interpretation." Rept. TR 60-689 (II), Contract AF 41(657)-79. National Opinion Research Center, University of Chicago, Chicago, 1961.

81. Borsky, P.N., "Community reactions to sonic booms in the Oklahoma City area." I and II. Rept. AMRL-TR-65-37, AD 613620. Wright-Patterson Air Force Base, Ohio, 1965.

82. Botsford, J.H., Simple method for identifying acceptable noise exposures, *J. Acoust. Soc. Am.* **42**, 810-819 (1967).

83. Botsford, J.H., Prevalence of impaired hearing and sound levels at work, *J. Acoust. Soc. Am.* **45**, 79-82 (1969).

83a. Botsford, J. and Laks,B., Noise hazard meter. *J. Acoust. Soc. Am.,* **47**, 90 (1970).

84. Bowsher, J.M. and Robinson, D.W., On scaling the unpleasantness of sounds. *Brit. J. Appl. Phys.* **13**, 179-181 (1962).

85. Bowsher, J.M., Johnson, D.R., and Robinson, D.W., A further experiment on judging the noisiness of aircraft in flight. *Acustica* **17**, (1966).

86. Brady, J.V., Porter, R.W., Conrad, D., and Mason, J.W., Avoidance behavior and the development of gastroduodenal ulcers. *J. Exp. Anal. Behavior* **1**, 69-72 (1958).

87. Bredenberg, V.C., Quiet please. *Hosp. Prog.* **42**, 104-108 (1961).

88. Brewer, D.W. and Briess, F.B., Industrial noise: laryngeal considerations. *N.Y. State J. Med.* **60**, 1737-1740 (1960).

89. Broadbent, D.E., Noise, paced performance, and vigilance tasks. *Brit. J. Psychol. Gen. Sect.* **44**, 295-303 (1953).

90. Broadbent, D.E., Some effects of noise on visual performance. *Quart. J. Exp. Psychol.* **6**, 1-5 (1954).

91. Broadbent, D.E., Growing points in multichannel communication. *J. Acoust. Soc. Am.* **28**, 533-535 (1956).

92. Broadbent, D.E., Effects of noise on behavior. *In* "Handbook of Noise Control," Chap. 10 (C.M. Harris, ed.). McGraw-Hill, New York, 1957.

93. Broadbent, D.E., Effects of noises of high and low frequency on behavior. *Ergonomics* **1**, 21-29 (1957).

94. Broadbent, D.E., Effect of noise on an intellectual task. *J. Acoust. Soc. Am.* **30**, 824-827 (1958).

95. Broadbent, D.E., "Perception and Communication." Pergamon Press, London, 1958.

96. Broadbent, D.E., The twenty dials and twenty lights test under noise conditions. Rept. 160/51. Medical Research Council, Cambridge, England, 1961.

97. Broadbent, D.E., Differences and interactions between stresses. *Quart. J. Exp. Psychol.* **15**, 205-211 (1963).

98. Broadbent, D.E. and Little, E.A.J., Effects of noise reduction in a work situation. *Occup. Psychol.* **34**, 133-140 (1960).

99. Broadbent, D.E. and Robinson, D.W., Subjective measurements of the relative annoyance of simulated sonic bangs and aircraft noise. *J. Sound Vib.* **1**, 162-174 (1964).

100. Broussard, I.G., Walker, R.Y., and Roberts, E.E., Jr., "The Influence of Noise on the Visual Contrast Threshold." Rept. 101. U.S. Army Medical Research Laboratories, Fort Knox, Kentucky, 1952.

101. Buendia, N., Sierra, G., Goode, M., and Segundo, J.P., Conditioned and discriminatory responses in wakeful and in sleeping cats. *Electroencephalog. Clin. Neurophysiol.* Suppl. **24**, 199-218 (1963).

102. Bugard, P., The effect of high-intensity noise and ultrasonics on the neuroendocrine system. Conclusions for industrial noise. *Arch. Mal. Prof.* **19**, 21-28 (1958).

103. Building Research Station, Department of Scientific and Industrial Research PNdB, "Stevens Phons and Zwicker Phons: and "A", "B", and "N" Weighted Sound Pressure Levels." Note B237 Building Research Station, Garston, Watford, Herts. England, 1961.

104. Bürck, Von W., Kotowski, P., and Lichte, H., Die Lautstärke von Knacken, Geräuschen und Tonen. *Elek. Nachr.–Techn.* **12**, 278-288 (1935).

105. Bürck, W., Grützmacher, M., Meister, F.J., and Müller, E.-A., Fluglàrm. "Seine Messung und Bewertung, seine Berucksichtigung bei der Siedlungsplanung, massnahmen zu seiner Minderung." Göttingen, 1965.

106. Burgeat, M. and Hirsh, I.J., Changes in masking with time. *J. Acoust. Soc. Am.* **33**, 963-965 (1961).

107. Burns, W., Noise as an environmental factor. *Roy. Soc. Health J.* **81**, 86-87 (1961).

108. Burns. W., "Noise and Man." Lippincott, Philadelphia, 1969.

109. Burns. W., Hearing and the ear. *In* "The Control of Noise: National Physical Laboratory Symposium Number 12," pp. 19-40. Her Majesty's Stationery Office, London, 1962.

110. Burns, W. and Littler, T.S., Noise. *In* "Modern Trends in Occupational Health" (R.S.F. Schilling, ed.). Butterworth, London, 1960.

111. Burns, W., Hinchcliffe, R., and Littler, T.S., An exploratory study of hearing and noise exposure in textile workers. *Ann. Occup. Hyg.* **7**, 323-333 (1964).

111a. Busnel, R.G. (ed.), "Psychophysiologie Neuropharmacologie et Biochimie de la Crise Audiogene." Editions du Centre National de la Recherche Scientifique, Paris (1963).

112. Buyniski, E.F., Noise and employee health. *Noise Control* **4**, No. 6, 45-46 (1958).

113. California Department of Industrial Relations, "Noise Control Safety Orders." California Department of Industrial Relations, Division of Industrial Safety, San Francisco (undated).

114. Callaway, D.B., Spectra and loudness of modern automobile horns. *J. Acoust. Soc. Am.* **23**, 55-58 (1951).

115. Callaway, D.B., Measurement and evaluation of exhaust noise of over-the-road trucks. *Soc. of Auto. Engrs. Trans.* **63**, 151-162 (1954).

116. Callaway, D.B. and Hall, H.H., Laboratory evaluation of field measurements of the loudness of truck exhaust noise. *J. Acoust. Soc. Am.* **26**, 216-220 (1954).

117. Camp, W., Martin, R., and Chapman, L.F., Pain threshold and discrimination of pain intensity during brief exposure to intense noise. *Science* **135**, 788-789 (1962).

118. Carhart, R., The problem of isolating the noise susceptible individual. *Am. Ind. Hyg. Assoc. Quart.* **18**, 335-340 (1957).

119. Carlin, S., Ward, W.D., Gershon, A., and Ingraham, R., Sound stimulation and its effect on dental sensation threshold. *Science* **138**, 1258 (1962).

120. Carlson, H.W., Sonic boom. *Int. Sci. Technol.* **55**, 70-72, 74, 78-80 (1966).

121. Carlsson, G. and Ronge, H., "Attitude and Opinion Studies on Human Reactions to Aircraft Noise." University of Lund, Sweden, 1962.

122. Carpenter, A., The effects of noise on work. *Ann. Occup. Hyg.* **1**, 42-54 (1959).

123. Carpenter, A., Effects of noise on performance and productivity. *In:* "The Control of Noise: National Physical Laboratory Symposium Number 12." Her Majesty's Stationery Office, London, 1962.

124. Carter, N.L., Loudness of triangular transients. *J. Auditory Res.* **3**, 255-289 (1963).

125. Carter, N.L., Effect of repetition rate on the loudness of triangular transients. *J. Acoust. Soc. Am.* **37**, 308-312 (1965).

126. Carter, N.L. and Kryter, K.D., "Equinoxious Contours for Pure Tones and Some Data on the "Critical Band" for TTS." Rept. 948. Bolt Beranek and Newman Inc., Cambridge, Massachusetts, 1962.

127. Carter, N.L. and Kryter, K.D., Masking of pure tones and speech. *J. Auditory Res.* **2**, 68-98 (1962).

128. Carter, N.L. and Kryter, K.D., "Studies of Temporary Threshold Shift Caused by High Intensity Noise." Rept. 949. Bolt Beranek and Newman Inc., Cambridge, Massachusetts, 1962.

129. Cavanaugh, W.J., Farrell, W.R., Hirtle, P., and Watters, B.G., Speech privacy in buildings. *J. Acoust. Soc. Am.* **34**, 475-492 (1962).

130. Cederlöff, R., Johnsson, E., and Kajland, A., Annoyance reactions to noise from motor vehicles. An experimental study. *Acustica* **13**, 270-279 (1963).

131. Cederlöf, R., Jonsson, E., and Sörensen, S., On the influence of attitudes to the source on annoyance reactions to noise. *Nord. Hyg. Tidsk.* **48**, 16-59 (1967).

132. Chandler, K.A., The effect of monaural and binaural tones of different intensities on the visual perception of verticality. *Am. J. Psychol.* **74**, 260-265 (1961).

133. Chavasse, P. and Lehmann, R., Contribution à l'étude zero absolu des audiometers. *Acustica* **7**, 132-136 (1957).

134. Chistovich, L.A. and Ivanova, V.A., Backward masking by short sound pulses. *Biofizika* **4**, 170-180 (1959).

135. Cochelle, R., On the acceptable limits of noise. *In* "Proceedings of the Third International Congress on Acoustics" (L. Cremer, ed.). Elsevier, New York, 1961.

136. Churcher, B.G. and King, A.J., The performance of noise meters in terms of the primary standard. *J. Inst. Elec. Engr.* **81**, 59-90 (1937).

137. Clark, W.E. and Pietrasanta, A.C., Community and industrial noise. *Am. J. Public Health* **51**, 1329-1337 (1961).

138. Clarke, F.R. and Bilger, R.C., The theory of signal detectability and the measurement of hearing. *In* "Modern Developments in Audiology," Chap. 10 (J. Jerger, ed.). Academic Press, New York, 1963.

139. Cohen, A., "Temporary Hearing Losses for Protected and Unprotected Ears as a Function of Exposure Time to Continuous and Impulse Noise." Rept. EP-151. U.S. Army Quartermaster Research Engineering Center, Natick, Massachusetts, 1961.

140. Cohen, A., Damage risk criteria for noise exposure: aspects of acceptability and validity. *Am. Ind. Hyg. Assoc. J.* **24**, 227-238 (1963).

141. Cohen, A. and Ayer, H.E., Some observations of noise at airports and in the surrounding community. *Am. Ind. Hyg. Assoc. J.* **25**, 139-150 (1964).

142. Cohen, A. and Baumann, K.C., Temporary hearing losses following exposure to pronounced single-frequency components in broadband noise. *J. Acoust. Soc. Am.* **36**, 1167-1175 (1964).

143. Cohen, A. and Scherger, R.F., "Correlation of Objectionability Ratings of Noise with Proposed Noise-Annoyance Measures." Rept. RR-3. U.S. Department of Health, Education, and Welfare, Public Health Service, Division of Occupational Health, Cincinnati, Ohio, 1964.

144. Cohen, A., Anticaglia, J.R., and Jones, H.H., "Noise Induced Hearing Loss— Exposures to Steady-State Noise." Presented at the American Medical Association Sixth Congress on Environmental Health, Chicago, Illinois, 1969.

145. Cohen, A., Klyin, B., and LaBenz, P.J., Temporary threshold shifts in hearing from exposure to combined impact/steady-state-noise conditions. *J. Acoust. Soc. Am.* **40**, 1371-1380 (1966).

146. Cohen, A., Hummel, W.F., Turner, J.W., and Duker-Dobos, F.N., "Effects of Noise on Task Performance." Rept. RR-4. U.S. Department of Health, Education, and Welfare, Occupational Health Research and Training Facility, Cincinnati, Ohio, 1966.

147. Cole, J.N. and Powell, R.G., "Estimated Noise Produced by Large Space Vehicles as Relates to Establishing Tentative Safe Distances to Adjacent Launch Pads and the Community." U.S. Air Force Memo. M-2, AD276204. Wright-Patterson Air Force Base, Ohio, 1962.

148. Cole, J.N. and von Gierke, H.E., "Noise from Missile Static Firing and Launch Sites and the Resultant Community Response." Proj. 7210-71705, AD131025. Aerospace Medical Research Laboratory, Wright-Patterson Air Force Base, Ohio, 1957.

149. Coleman, P.D. and Krauskopf, J., "The Influence of High Intensity Noise on Visual Thresholds." Rept. 222. U.S. Army Medical Research Laboratories, Fort Knox, Kentucky, 1955.

150. Coles, R.R.A., "Some Considerations Concerning the Effects on Hearing of the Noise of Small-Arms." Paper H21 presented at the Fourth International Congress on Acoustics, Copenhagen, 1962; and *J. Roy. Nav. Med. Serv.* **69**, 1-5 (1963).

151. Coles, R.R.A. and Knight, J.J., Auditory damage in young men after short exposure to industrial noise. *Ann. Occup. Hyg.* **1**, 98-103 (1958).

152. Coles, R.R.A. and Knight, J.J., Unpublished studies on PTS from 7.62-mm rifle noise. Royal Naval Personnel Research Commission, London, 1962, 1965.

153. Coles, R.R.A. and Rice, C.G., High-intensity noise problems in the Royal Navy and Royal Marines. *J. Roy. Nav. Med. Serv.* **51**, 184-192 (1965).

154. Coles, R.R.A. and Rice, C.G., Speech communications effects and temporary threshold shift reduction provided by V51R and selectone-K earplugs under conditions of high intensity impulsive noise. *J. Sound Vib.* **4**, 156-171 (1966).

155. Coles, R.R.A. and Rice, C.G., Auditory hazards of sports guns. *Laryngoscope* **76**, 1728-1731 (1966).

156. Coles, R.R.A., Garinther, G.R., Hodge, D.C., and Rice, C.G., Hazardous exposure to impulse noise. *J. Acoust. Soc. Am.* **43**, 336-343 (1968).

157. Collins, E.G., Aural trauma caused by gunfire. *J. Laryngol. Otol.* **63**, 358-390 (1948).

158. Collins, G., "The Design of Frequency Selective Earplugs." Rept. ARL/D/R12. Admiralty Research Laboratory, Teddington, Middlesex, England, 1964.

159. Copeland, W.C.T., Davidson, I.M., Hargest, T.J., and Robinson, D.W., A controlled experiment on the subjective effects of jet engine noise. *J. Roy. Aeronaut. Soc.* **64**, 33-36 (1960).

160. Corcoran, D.W.J., Noise and loss of sleep. *Quart. J. Exp. Psychol.* **14**, 178-182 (1962).

161. Corso, J.F., Historical note on thermal masking noise and pure tone pitch changes. *J. Acoust. Soc. Am.* **26**, 1078 (1954).

162. Corso, J.F., "Survey of Hearing in a Low Industrial Noise-Level Population." Rept. WADC TR 58-442, AD 204799. Aerospace Medical Laboratory, Wright-Patterson Air Force Base, Ohio, 1958.

163. Corso, J.F., "The Effects of Noise on Human Behavior." Air Research and Development Command, Wright-Patterson Air Force Base, Ohio (undated).

164. Corso, J.F., Age and sex differences in pure-tone thresholds. *Arch. Otolaryngol.* **77**, 385-405 (1963).

165. Cremer, L., Plenge, G., and Schwarzl, D., Kurven gleicher Lautstärke mit Oktave-gefiltertem Rauschen im diffusen Schallfeld. (Equal loudness contours for octave band noise in a diffuse sound field.) *Acustica* **9**, 65-75 (1959).

166. Cross, G.W., Davison, I.M., Hargest, T.J., and Porter, M.J., A controlled experiment on the perception of helicopter rotor noise. *J. Roy. Aeronaut. Soc.* **64**, 629-632 (1960).

167. Dadson, R.S. and King, J.H., A determination of the normal threshold of hearing and its relation to the standardization of audiometers. *J. Laryngol. Otol.* **66**, 366-378 (1952).

168. Davis, H., Biological and psychological effects of ultrasonics. *J. Acoust. Soc. Am.* **20**, 605-607 (1948).

169. Davis, H., The articulation area and the social adequacy index for hearing. *Laryngoscope* **58**, 761-768 (1948).

170. Davis, H., Psychophysiology of hearing and deafness. *In* "Handbook of Experimental Psychology," Chap. 28 (S.S. Stevens, ed.). Wiley, New York, 1951.

171. Davis, H., The hearing mechanism. *In* "Handbook on Noise Control," Chap. 4 (C.M. Harris, ed.). McGraw-Hill, New York, 1957.

172. Davis, H., "Project ANEHIN: Auditory and Non-auditory Effects of High Intensity Noise." Joint Rept. 7. Central Institute for the Deaf and U.S. Naval School of Aviation Medicine, Pensacola, Florida, 1958.

173. Davis, H., Effects of high intensity noise on naval personnel. *U.S. Armed Forces Med. J.* 9, 1027-1048 (1958).

174. Davis, H., The ISO zero-reference level for audiometers. *Arch. Otolaryngol. (Chicago)* 81, 145-149 (1965).

175. Davis, H. and Glorig, A., Audio analgesia: a new problem for otologists. *Arch. Otolaryngol.* 75, 498-501 (1962).

176. Davis, H. and Hoople, G.D., Hearing level, hearing loss, and threshold shift. *J. Acoust. Soc. Am.* 30, 478 (1958).

177. Davis, H. and Silverman, S.R., "Hearing and Deafness," rev. ed. Holt, New York, 1960.

178. Davis, H. and Silverman, S.R., "Auditory and Non-auditory Effects of High Intensity Noise." Contract Nonr-1151(02), AD 204651. USN School of Aviation Medicine, Florida, 1958.

179. Davis, H. and Usher, J.R., What is zero hearing loss? *J. Speech Hearing Dis.* 22, 662-690 (1957).

180. Davis, H., Morgan, C.T., Hawkins, J.E., Galambos, R., and Smith, F., "Temporary Deafness Following Exposure to Loud Tones and Noise." Contract OEMcmr-194. Committee on Medical Research, OSRD, Harvard Medical School, Boston, 1943.

181. Davis, H., Parrack, H.O., and Eldredge, D.H., Hazards of intense sound and ultrasound. *Ann. Otol. Rhinol. Laryngol.* 58, 732-738 (1949).

182. Davis, R.C., Motor effects of strong auditory stimuli. *J. Exp. Psychol.* 38, 257-275 (1948).

183. Davis, R.C., Response to "meaningful" and "meaningless" sounds. *J. Exp. Psychol.* 38, 744-756 (1948).

184. Davis, R.C. and Berry, T., Gastrointestinal reactions to response-contingent stimulation. *Psychol. Rep.* 15, 95-113 (1964).

185. Davis, R.C. and Van Liere, D.W., Adaptation of the muscular tension response to gunfire. *J. Exp. Psychol.* 39, 114-117 (1949).

186. Davis, R.C., Buchwald, A.M., and Frankman, R.W., Autonomic and muscular responses and their relation to simple stimuli. *Psychol. Monographs* 69, No. 405 (1955).

187. de Almeida, H.R., Influence of electric punch card machines on the human ear. *Arch. Otolaryngol.* 51, 215-222 (1950).

188. Deatherage, B., Jeffress, L.A., and Blodgett, H.C., A note on the audibility of intense ultrasonic sound. *J. Acoust. Soc. Am.* 26, 582 (1954).

189. de Brisson, A., "Opinion Study on the Sonic Bang." Study 22, Royal Air Est. Lib. Trans. 1159, AD483066. Centre d'Études et d'Instruction Psychologiques de l'Armée de l'Air, Versailles, France, 1966.

190. de Boer, E., Note on the critical bandwidth. *J. Acoust. Soc. Am.* 34, 985-986 (1962).

191. Delaney, M.E. and Robinson, D.W. (eds.), "The Control of Noise: National Physical Laboratory Symposium Number 12." Her Majesty's Stationery Office, London, 1962.

192. Denzel, H.A., Noise and health. (L) *Science* 43, 992 (1963).

193. Dickson, E.D.D. and Chadwick, D.L., Observations on disturbances of equilibrium and other symptoms induced by jet engine noise. *J. Laryngol. Otol.* 65, 154-165 (1951).

194. Diercks, K.J. and Jeffress, L.A., Interaural phase and the absolute threshold for tone. *J. Acoust. Soc. Am.* **34**, 981-984 (1962).

195. Dieroff, H.G., Problems of deafness due to noise. *Z. Ges. Hyg.* **11**, 352-361 (1965).

196. Dorfman, D.D. and Zajone, R.B., Some effects of sound, background brightness, and economic status of the perceived size of coins and discs. *J. Abnormal Social Psychol.* **66**, 87-90 (1963).

197. Dreher, J.J. and Evans, W.E., Speech interference level and aircraft acoustical environment. *J. Human Factors Soc.* **2**, 18-27 (1960).

198. Dunn, H.K. and White, S.D., Statistical measurements on conversational speech. *J. Acoust. Soc. Am.* **11**, 278 (1940).

199. Dunnell, G., Owens, K., and Miller A., Effect of patterned auditory stimulation on autokinetic motion. *Percep. Motor Skills* **18**, 311-312 (1964).

200. Dygert, P.K., A public enterprise approach to jet aircraft noise around airports. *In* "Alleviation of Jet Aircraft Noise Near Airports," pp. 107-116. Rept. of the Jet Aircraft Noise Panel Office of Science and Technology, Executive Office of the President, Washington, D.C., 1966.

201. Dygert, P.K., "Allocating the Cost of Alleviating Subsonic Jet Aircraft Noise." Rept. AD 618748. University of California Institute of Transportation and Traffic Engineering, Berkeley, 1967.

202. Egan, J.P., The effect of noise in one ear upon the loudness of speech in the other ear. *J. Acoust. Soc. Am.* **20**, 58-62 (1948).

203. Egan, J.P., Articulation testing methods. *Laryngol.* 955-991 (1948).

204. Egan, J.P., Independence of the masking audiogram from the perstimulatory fatigue of an auditory stimulus. *J. Acoust. Soc. Am.* **27**, 737-740 (1955).

205. Egan, J.P., Perstimulatory fatigue as measured by heterophonic loudness balances. *J. Acoust. Soc. Am.* **27**, 111-120 (1955).

206. Egan, J.P. and Hake, H.W., On the masking pattern of a simple auditory stimulus. *J. Acoust. Soc. Am.* **22**, 622-630 (1950).

207. Egan, J.P. and Meyer, D.R., Changes in pitch of tones of low frequency as a function of the pattern of excitation produced by a band of noise. *J. Acoust. Soc. Am.* **22**, 827-833 (1950).

208. Egan, J.P. and Wiener, F.M., On the intelligibility of bands of speech in noise. *J. Acoust. Soc. Am.* **18**, 435-441 (1946).

209. Egan, J.P., Clarke, F.R., Carterette, E.C., On the transmission and confirmation of messages in noise. *J. Acoust. Soc. Am.* **28**, 536-550 (1956).

210. Ehmer, R.H., Masking patterns of tones. *J. Acoust. Soc. Am.* **31**, 1115-1120 (1959).

211. Ehmer, R.H., Masking by tones vs. noise bands. *J. Acoust. Soc. Am.* **31**, 1253 (1959).

212. Eldred, K.M., Gannon, W.J., and von Gierke, H., "Criteria for Short Time Exposure of Personnel to High Intensity Jet Aircraft Noise." Rept. WADC-TN-355. Aerospace Medical Laboratory, Wright-Patterson Air Force Base, Ohio, 1955.

213. Elliott, E., Attenuating properties of ARL-tuned earplugs. Joint Rept. ARL/D/22. Admiralty Research Laboratory, Teddington, Middlesex, England, 1965.

214. Elliott, L.L., Backward masking; monotic and dichotic conditions. *J. Acoust. Soc. Am.* **34**, 1108-1115 (1962).

215. Elliott, L.L., Backward and forward masking of probe tones of different frequencies. *J. Acoust. Soc. Am.* **34**, 1116-1117 (1962).

216. Elliott, L.L., Prediction of speech discrimination scores from other test information. *J. Auditory Res.* **3**, 35-45 (1963).

217. Elliott, L.L., Note on predicting speech-discrimination scores. *J. Acoust. Soc. Am.* **36**, 1961-1962 (1964).

218. Elwell, F.S., "Experiments to Determine Neighborhood Reactions to Light Airplanes with and without External Noise Reduction." Rept. 1156. National Advisory Committee for Aeronautics (now National Aeronautics and Space Administration), Washington, D.C., 1953.

219. Elwood, M.A., Brasher, P.F., and Croton, L.M., "A Preliminary Study of Sensitivity to Impulsive Noise in Terms of Temporary Threshold Shifts." British Acoustical Society Meeting on impulse noise, Southampton, 1966.

220. Embleton, T.F.W., Dagg, I.R., and Thiessen, G.J., Effect of environment on noise criteria. *Noise Control* **5**, No. 6, 37-40 (1959).

221. Emmons, W.H. and Simon, C.W., "The Non-recall of Material Presented During Sleep." Rev. Rept. P-619. The Rand Corporation, Santa Monica, California, 1955.

222. Fairbanks, G., Test of phonemic differentiation: the rhyme test. *J. Acoust. Soc. Am.* **30**, 596-601 (1958).

223. "Federal Register, Safety and Health Standards," Vol. 34, No. 12, Part II. U.S. Department of Labor, Washington, D.C., 1969.

224. Feldtkeller, R., Zwicker, E., and Port, E., Lautstärke, Verhältnislautheit und Summenlautheit. (Loudness, relative loudness and additive loudness.) *Frequenz* **13**, 108-117 (1959).

225. Felton, J.S. and Spencer, C., "Morale of workers exposed to high levels of occupational noise." University of Oklahoma School of Medicine, Norman, Oklahoma, 1957.

226. Ferris, K., The temporary effects of 125 cps octave-band noise on stapedectomized ears. *J. Laryngol.* **80**, 579-582 (1966).

227. Finck, A., Low-frequency pure tone masking. *J. Acoust. Soc. Am.* **33**, 1140-1141 (1961).

228. Finkle, A.L. and Poppen, J.R., Clinical effects of noise and mechanical vibrations of a turbo-jet engine on man. *J. Appl. Physiol.* **1**, 183-204 (1948).

229. Flanagan, J.L. and Guttman, N., Estimating noise hazard with the sound-level meter. *J. Acoust. Soc. Am.* **36**, 1654-1658 (1964).

230. Fleagle, R.G., The audibility of thunder. *J. Acoust. Soc. Am.* **21**, 411-412 (1949).

231. Fleer, R., "Protection Afforded against Impulsive Noise by Voluntary Contraction of the Middle Ear Muscles." Rept. 576. U.S. Army Medical Research Laboratories, Fort Knox, Kentucky, 1963.

232. Fletcher, H., Auditory patterns. *Rev. Mod. Phys.* **12**, 47-65 (1940).

233. Fletcher, H., A method of calculating hearing loss for speech from an audiogram. *J. Acoust. Soc. Am.* **22**, 1-5 (1950).

234. Fletcher, H., The perception of speech sounds by deafened persons. *J. Acoust. Soc. Am.* **24**, 490-497 (1952).

235. Fletcher, H., "Speech and Hearing in Communication." Van Nostrand, New York, 1953.

236. Fletcher, H. and Galt, R.H., The perception of speech and its relation to telephony. *J. Acoust. Soc. Am.* **22**, 89-151 (1950).

237. Fletcher, H. and Munson, W.A., Loudness, its definition, measurement and calculation. *J. Acoust. Soc. Am.* **5**, (1933).

238. Fletcher, H. and Munson, W.A., Loudness, its definition, measurement and calculation. I. II. *Audio* **41**, 40, 53 (1957).

239. Fletcher, H. and Steinberg, J.C., Loudness of complex sounds, *Phys. Rev.* **24**, 306 (1924).

240. Fletcher, J.L., "Acoustic Reflex Response to High Intensities of Impulse Noise and to Noise and Click Stimuli." Rept. 527, AD271071. U.S. Army Medical Research Laboratory, Fort Knox, Kentucky, 1961.

241. Fletcher, J.L., TTS following prolonged exposure to acoustic reflex eliciting stimuli. *J. Auditory Res.* **1**, 242-246 (1961).

242. Fletcher, J.L., Comparison of attenuation characteristics of the acoustic reflex and the V-51R earplug. *J. Auditory Res.* **1**, 111-116 (1961).

243. Fletcher, J.L., Reflex responses of middle-ear muscles: protection of the ear from noise. *Sound* **1**, No. 2, 17-23 (1962).

244. Fletcher, J.L. (ed.), "Middle Ear Function Seminar." Rept. 576. U.S. Army Medical Research Laboratory, Fort Knox, Kentucky, 1963.

245. Fletcher, J.L., Hearing losses of personnel exposed to impulse and steady state noise. *J. Auditory Res.* **3**, 83-90 (1963).

246. Fletcher, J.L., Protection from high intensities of impulse noise by way of preceding noise and click stimuli. *J. Auditory Res.* **5**, 145-150 (1965).

247. Fletcher, J.L., Criteria for assessing risk of hearing damage. *J. Occup. Med.* **7**, 281-283 (1965).

248. Fletcher, J.L. and King, W.P., Susceptibility of stapedectomized patients to noise induced temporary threshold shifts. *Ann. Otol. Rhinol. Laryngol.* **72**, 900-907 (1963).

249. Fletcher, J.L. and Loeb, M., "Contralateral Threshold Shift and Reduction in Temporary Threshold Shift as Indices of Acoustic Reflex Action." Rept. 490. U.S. Army Medical Research Laboratory, Fort Knox, Kentucky, 1961.

250. Fletcher, J.L. and Loeb, M., The influence of different acoustical stimuli on the threshold of the contralateral ear: a possible index of attenuation by the intratympanic reflex. *Acta Oto-Laryngol.* **54**, 33-47 (1962).

251. Fletcher, J.L. and Loeb, M., "Changes in the Hearing of Personnel Exposed to High Intensity Continuous Noise." Rept. AMRL-566. U.S. Army Medical Research Laboratory, Fort Knox, Kentucky, 1963; and *Milit. Med.* **128**, 1137-1141 (1963).

252. Fletcher, J.L. and Loeb, M., Relationship for temporary threshold shifts produced by three different sources. *J. Auditory Res.* **5**, 41-45 (1965).

253. Fletcher, J.L. and Loeb, M., The effect of pulse duration on TTS produced by impulse noise. *J. Auditory Res.* **7**, 163-167 (1967).

254. Fletcher, J.L. and Loeb, M., "Exploratory Study of the Effect of Pulse Duration on Temporary Threshold Shift Produced by Impulse Noise." Rept. 680. U.S. Army Medical Research Laboratories, Fort Knox, Kentucky, 1967.

255. Fletcher, J.L. and Riopelle, A.J., Protective effect of the acoustic reflex for impulsive noises. *J. Acoust. Soc. Am.* **32**, 401-404 (1960).

256. Flottrop, G. and Quist-Hanssen, S., The effect of ear protectors against sound waves from explosions. *Acta Oto-Laryngol.* Suppl. **158**, 286-294 (1960).

257. Flynn, W.E. and Elliott, D.N., Role of the pinna in hearing. *J. Acoust. Soc. Am.* **38**, 104-105 (1965).

258. Forester, R.S., Legislation regarding noise. *In* "The Control of Noise, National Physical Laboratory, Symposium Number 12," pp. 325-332. Her Majesty's Stationery Office, London, 1962.

259. Fowler, E.P., The percentage of capacity to hear speech and related disabilities. *Laryngol.* **57**, 103-113 (1947).

260. Fox, M.S., They are doing something about noise. Details of the new Wisconsin compensation legislation. *Noise Control* **2**, No. 1, 51-52, 61 (1956).

261. Fozard, J.L., Bacon, W.E., and Small, A.M., Jr., Masked thresholds for octave band noise. *J. Acoust. Soc. Am.* **31**, 1681-1682 (1959).

262. Frazier, F.E., The unsolved factors in loss of hearing claims. *Insurance Law J.* **388**, 335-345 (1955).

263. Frazier, F.E., Effects of research upon legislation and claims. *Noise Control* **3**, No. 5, 43-49, 68 (1957).
264. Frazier, F.E., Current trends in criteria for noise levels in industry. *A.M.A. Arch. Ind. Health* **19**, 288-297 (1959).
265. Frazier, F.E., Compensation claims for loss of hearing: Impact of standards. *Arch. Environ. Health* **10**, 572-575 (1965).
266. Freedman, S.J. and Pfaff, D.W., Trading relations between dichotic time and intensity differences in auditory localization. *J. Auditory Res.* **2**, 311-317 (1962).
267. French, N.R. and Steinberg, J.C., Factors governing the intelligibility of speech sounds. *J. Acoust. Soc. Am.* **19**, 90-119 (1947).
268. Frick, F.C. and Sumby, W.H., Control tower language. *J. Acoust. Soc. Am.* **24**, 595-596 (1952).
269. Fricke, J.E., Auditory fatigue and mental activity. *J. Auditory Res.* **6**, 283-287 (1966).
270. Gallo, R. and Glorig, A., Permanent threshold shift changes produced by noise exposure and aging. *Am. Ind. Hyg. Assoc. J.* **25**, 237-245 (1964).
271. Galloway, W.J., "Selection of a Unit for Specification of Motor Vehicle Noise." Tech. Appendix 1 of Interim Rept. Highway Research Board, Washington, D.C., 1965.
272. Galloway, W.J., Frequency analyses of short-duration random noise. *Sound* **1**, 31-34 (1962).
272a. Galloway, W.J. and Bishop, D., "Noise Exposure Forecast: Evolution, Evaluation, Extension, and Land Use Interpretation." FAA Contractor's Rept. Bolt Beranek and Newman, Inc., Van Nuys, California, 1970.
273. Galloway, W.J. and Clark, W.E., "Prediction of Noise from Motor Vehicles in Freely Flowing Traffic." Paper No. L28 presented at Fourth International Congress on Acoustics, Copenhagen, 1962.
274. Galloway, W. and von Gierke, H.E., "Individual and Community Reaction to Aircraft Noise: Present Status and Standardization Efforts." Paper prepared for International Conference on the Reduction of Noise and Disturbance Caused by Civil Aircraft, London, November 1966.
275. Ganguli, H.C. and Rao, M.N., Noise and industrial efficiency: a study of Indian jute weavers. *Arbeitsphysiol.* **15**, 344-354 (1954).
276. Gardner, M.B., Binaural detection of single-frequency signals in the presence of noise. *J. Acoust. Soc. Am.* **34**, 1824-1830 (1962).
277. Gardner, M.B., Effect of noise on listening levels in conference telephony. *J. Acoust. Soc. Am.* **36**, 2354-2362 (1964).
278. Gardner, M.B., Effect of noise, system gain, and assigned task on talking levels in loudspeaker communication. *J. Acoust. Soc. Am.* **40**, 955-965 (1966).
279. Gardner, M.B., Historical background of the Haas and/or precedence effect. *J. Acoust. Soc. Am.* **43**, 1243-1248 (1968).
280. Gardner, W. and Licklider, J.C.R., Auditory analgesia in dental operations. *J. Am. Dental Assoc.* **59**, 1144-1149 (1959).
281. Gardner, W., Licklider, J.C.R., and Weisz, A.Z., Suppression of pain by sound. *Science* **132**, 32-33 (1960).
282. Garner, W.R., The effect of frequency spectrum on temporal integration on energy in the ear. *J. Acoust. Soc. Am.* **19**, 808-815 (1947).
283. Garner, W.R., Auditory thresholds of short tones as a function of repetition rates. *J. Acoust. Soc. Am.* **19**, 600-608 (1947).
284. Garner, W.R., The loudness of repeated short tones. *J. Acoust. Soc. Am.* **20**, 513-527 (1948).

285. Garner, W.R., The loudness and loudness matching of short tones. *J. Acoust. Soc. Am.* **21**, 398-403 (1949).

286. Garner, W.R., Some statistical aspects of half-loudness judgments. *J. Acoust. Soc. Am.* **24**, 153-157 (1952).

287. Garner, W.R., A technique and a scale for loudness measurement. *J. Acoust. Soc. Am.* **26**, 73-88 (1954).

288. Garner, W.R., Half-loudness judgments without prior stimulus context. *J. Exp. Psychol.* **55**, 482-485 (1958).

289. Garner, W.R., On the lambda loudness function, masking, and the loudness of multicomponent tones. *J. Acoust. Soc. Am.* **31**, 602-607 (1959).

290. Garner, W.R., The development of context effects in half-loudness judgments. *J. Exp. Psychol.* **58**, 212-219 (1959).

291. Garner, W.R. and Miller, G.A., The masked threshold of pure tones as a function of duration. *J. Exp. Psychol.* **37**, 293-303 (1947).

292. Gassler, G., On the threshold of hearing for sounds with different spectrum width. *Acustica* **4**, 408-414 (1958).

293. Gjaevenes, K., Measurements on the impulsive noise from crackers and toy firearms. *J. Acoust. Soc. Am.* **39**, 403-404 (1966).

294. Glorig, A., Audiometric testing in industry. *In* "Handbook of Noise Control," Chap. 6 (C.M. Harris, ed.). McGraw-Hill, New York, 1957.

295. Glorig, A., "Noise and Your Ear." Grune and Stratton, New York, 1958.

296. Glorig, A., A report of two normal hearing studies. *Ann. Otol. Rhinol. Laryngol.* **67**, 93-102 (1958).

297. Glorig, A., The effects of noise on hearing. *J. Laryngol. Otol.* **75**, 447-478 (1961).

297a. Glorig, A. and Nixon, J. Hearing loss as a function of old age. *Laryngoscope* **72**, 1596 (1962).

298. Glorig, A. and Roberts J., "Hearing Levels of Adults by Age and Sex, United States 1960-1962." National Center for Health Statistics, Series 11, Number 11. U.S. Department of Health, Education and Welfare, Public Health Service, Washington, D.C., 1965.

299. Glorig, A., Grings, W., and Summerfield, A., Hearing loss in industry. *Laryngoscope* **68**, 447-465 (1958).

300. Glorig, A., Summerfield, A., and Nixon, J., Distribution of hearing levels in non-noise-exposed populations. "Proceedings Third International Congress on Acoustics," Vol. 1, pp. 150-154 (L. Cremer, ed.). Elsevier, New York, 1961.

301. Glorig, A., Summerfield, A., and Ward, W.D., Observations on temporary auditory threshold shift resulting from noise-exposure. *Ann. Otol. Rhinol. Laryngol.* **67**, 824-847 (1958).

302. Glorig, A., Ward, W.D., and Nixon, J., Damage risk criteria and noise-induced hearing loss. *Arch. Otolaryngol.* **74**, 413-425 (1961).

303. Glorig, A., Ward, W.D., and Nixon, J., Damage-risk criteria and noise-induced hearing loss. *In* "The Control of Noise: National Physical Laboratory Symposium Number 12," pp. 263-283. Her Majesty's Stationery Office, London, 1962.

304. Glorig, A., Wheeler, D., Quiggle, R., Grings, W., and Summerfield, A., "1954 Wisconsin State Fair Hearing Survey: Statistical Treatment of Clinical and Audiometric Data." American Academy Ophthalmology and Otolaryngology and Research Center Subcommittee on Noise in Industry, Los Angeles, California, 1957.

305. Goldberg, J.M., The voice interference analysis set, an instrument for determining the degradation of signal quality of a voice communication channel. *J. Audio Engr. Soc.* **11**, 115-120 (1963).

306. Golden, P. and Clare, R., "The Hazards to the Human Ear from Shock Waves Produced by High Energy Electrical Discharge." United Kingdom Atomic Energy Authority, Berkshire, England, 1965.

307. Golikov, E.E., Calculating the articulation in noisy rooms. *Soviet Physics-Acoustics* 6, 407-408 (1961).
308. Golovin, N.E., "Alleviation of Aircraft Noise." Office of Science and Technology, Executive Office of the President, Washington, D.C., 1966. (Based on March 1966 report of the Office of Science and Technology entitled "Alleviation of Jet Aircraft Noise Near Airports.")
309. Goldstein, S., A problem in federalism, property rights in air space and technology. "Alleviation of Jet Aircraft Noise Near Airports," pp. 132-142. Jet Aircraft Noise Panel, Office of Science and Technology, Executive Office of the President, 1966.
310. Goshen, C.E., Noise, annoyance and progress. *Science* 144, 487 (1964).
311. Gottlieb, P. and Gottschalk, A.H. Personal communication, Los Angeles, 1968.
312. Grandjean, E., "Biological Effects of Noise." Paper presented at Fourth International Congress on Acoustics, Copenhagen, 1962.
313. Grandjean, E. and Kryter, K.D., "Les Effets du Bruit sur l'Homme." Mensch und Umwelt, Basle, J.R. Geigy S.A., 1961.
314. Grandjean, E., Perret, E., and Lauber, A., Experimentelle Untersuchungen über die Störwirkung von Flugzeuglärm. *Int. Z. Ang. Physiol. Einschl. Arbeitsphysiol.* 23, 191-202 (1966).
315. Gravendeel, D.W. and Plomp, R., The relation between permanent and temporary noise dips. *Arch. Otolaryngol.* 69, 714-719 (1959).
316. Gravendeel, D.W and Plomp, R., Permanent and temporary diesel engine noise dips. *Arch. Otolaryngol.* 74, 405-407 (1961).
317. Greatrex, F.B., Take-off and landing of the supersonic transport. *Aircraft Engr.* 1-5 (August 1963).
318. Greatrex, F.B., "The Economics of Aircraft Noise Suppression." Paper ICAS No. 66-5 presented at Royal Aeronautical Society Centenary Congress in conjunction with Fifth Congress of International Council of the Aeronautical Sciences, London, September 1966.
319. Green, D.M., Detection of multiple component signals in noise. *J. Acoust. Soc. Am.* 30, 904-911 (1958).
320. Green, D.M., Masking with two tones. *J. Acoust. Soc. Am.* 37, 802-813 (1965).
321. Green, D.M., Additivity of masking. *J. Acoust. Soc. Am.* 41, 1517-1525 (1967).
322. Green, D.M., Sine and cosine masking. *J. Acoust. Soc. Am.* 44, 168-175 (1968).
323. Green, D.M., Bishop, D., and Parnell, J.E., Exponent in the power law and loudness or noisiness calculation. (L) *J. Acoust. Soc. Am.* 44, 633 (1968).
324. Greenwood, D.D., Auditory masking and the critical band. *J. Acoust. Soc. Am.* 33, 484-502 (1961).
325. Greenwood, D.D., Critical bandwidth and the frequency coordinates of the basilar membrane. *J. Acoust. Soc. Am.* 33, 1344-1356 (1961).
326. Gregg, L.W. and Brogden, W.J., The effect of simultaneous visual stimulation on absolute sensitivity. *J. Exp. Psychol.* 43, 179-186 (1952).
327. Greissen, L., Comparative investigations of different auditory fatigue tests, *Acta Oto-Laryngol.* 39, 132-135 (1951).
328. Griffiths, I.D. and Langdon, F.J., "Subjective Response to Road Traffic Noise." Building Research Station, Current paper 37/68, Ministry of Public Building and Works, Garson, Watford, Herts, England, 1968; *J. Sound Vibration* 8, 1 (1968).
329. Grimaldi, J.V., Sensorimotor performance under varying noise conditions. *Ergonomics* 2, 34-43 (1958).
330. Grognot, P. and Perdriel, G., Effect of noise on color vision and night vision. *Compt. Rend. Soc. Biol.* 153, 142-143 (1959).

331. Grubb, C.A., Van Zandt, J.E., and Bockholt, J.L., "Report on Data Retrieval and Analysis of USAF Sonic Boom Claim Files." Rept. 4, Contract AF 49 (638)-1696. National Sonic Boom Evaluation Office, Arlington, Virginia, 1966.

331a. Guertin, W.H., Auditory interference with digit span performance. *J. Clin. Psychol.* 15, 349 (1959).

332. Gzhesik, Ya., Lempkovski, A., Turchinski, B., Fazonovich, Ya., and Shimchik, K., Comparison of methods for evaluating loudness from data published during the period 1930-1957. *Soviet Physics-Acoustics* 6, 421-441 (1961).

333. Haeff, A.W. and Knox, C., Perception of ultrasound. *Science* 139, 590-592 (1963).

334. Hamilton, P.M., Noise masked thresholds as a function of tonal duration and masking noise bandwidth. *J. Acoust. Soc. Am.* 29, 506-511 (1957).

335. Hale, H.B., Adrenalcortical activity associated with exposure to low frequency sounds. *Am. J. Physiol.* 171, 732 (1952).

336. Hardy, H.C., Engineering and zoning regulation of outdoor industrial noise. *Noise Control* 3, No. 3, 32-38 (1957).

337. Hargest, T.J., Knowler, A.E., and Robinson, D.W., "Aircraft Noise Units." Ministry of Aviation and Ministry of Technology, Great Britain, 1967.

338. Harmon, F.L., The effects of noise upon certain psychological-physiological processes. *Arch. Psychol.* 147, 1-81 (1933).

339. Harris, C.M. (ed.), "Noise Control." McGraw-Hill, New York, 1957.

340. Harris, C.S., "The Effects of High Intensity Noise on Human Performance." Rept. AMRL-TR-67-119. Air Force Systems Command, Wright-Patterson Air Force Base, Ohio, 1968.

341. Harris, C.S. and Schoenberger, R.W., "Human Performance During Vibration." Rept. AMRL-TR-65-204. U.S. Air Force Aerospace Medical Research Laboratories, Wright-Patterson Air Force Base, Ohio, 1965.

342. Harris, J.D., On latent damage to the ear. *J. Acoust. Soc. Am.* 27, 177-179 (1955).

343. Harris, J.D., "An Evaluation of Ear Defender Devices; Two Earplugs, Four Cushions, and Three Combinations." Rept. 271. U.S. Naval Medical Research Laboratory, New London, Connecticut, 1955.

344. Harris, J.D., "Auditory Fatigue Following High Frequency Pulse Trains." Rept. 306. U.S. Naval Medical Research Laboratory, New London, Connecticut, 1959.

345. Harris, J.D., Combinations of distortion in speech. *Arch. Otolaryngol.* 72, 227-232 (1960).

346. Harris, J.D., A factor analytic study of three signal detection abilities. *J. Speech Hear. Res.* 7, 71-78 (1964).

347. Harris, J.D., Hearing-loss trend curves and the damage-risk criterion in diesel-engine room personnel. *J. Acoust. Soc. Am.* 37, 444-452 (1965).

348. Harris, J.D., Pure-tone acuity and the intelligibility of everyday speech. *J. Acoust. Soc. Am.* 37, 824-830 (1965).

349. Harris, J.D., Relations among after effects of acoustic stimulation. *J. Acoust. Soc. Am.* 42, 1306-1324 (1967).

350. Harris, J.D., Haines, H.L., and Myers, C.K., A new formula for using the audiogram to predict speech hearing loss. *Arch. Otolaryngol.* 63, 158-176 (1956).

351. Harris, J.D., Haines, H.L., and Myers, C.K., The importance of hearing at 3 kc for understanding speeded speech. *Laryngoscope* 70, 131-146 (1960).

352. Haugen, S., "Measurement of Perceived Noise Level in PNdB Made Possible." Central Institute for Industrial Research, Oslo, Norway, 1965.

353. Hawel, W., A model of a network for weighting the annoyance of noise with respect to its parameters on sound-level-meters. "Acoustic Noise and Its Control." Publication 26. Institute of Electrical Engineers, Savoy Place, London, 1967.

354. Hawel, W. and Starlinger, H., Effect of repeated 4-hour intermittent (so-called) pink noise on catecholaminar separation (in urine) and pulse frequency. *Int. Z. Angew. Physiol. Einschol. Arbeitsphysiol.* **24**, 351-362 (1967).

355. Hawkins, J.E., Jr. and Stevens, S.S., The masking of pure tones and of speech by white noise. *J. Acoust. Soc. Am.* **22**, 6-13 (1950).

356. Hawley, M.E., Noise shield for microphones used in noisy locations. *J. Acoust. Soc. Am.* **30**, 188-190 (1958).

357. Hawley, M.E. and Kryter, K.D., Effects of noise on speech. *In* "Handbook of Noise Control," Chap. 9 (C.M. Harris, ed.). McGraw-Hill, New York, 1957.

358. Hazard, W.R., Community reactions to aircraft noise—public reactions. "Conference on NASA Research Relating to Noise Alleviation of Large Subsonic Jet Aircraft." NASA Langley Research Center, Virginia, 1968.

359. Hecker, M.H.L. and Kryter, K.D., "A Study of Auditory Fatigue Caused by High-Intensity Acoustic Transients." Rept. 1158, Contract DA-49-007-MD-985. U.S. Army Medical Research and Development Command, Office of Surgeon General, Washington, D.C., 1964.

360. Hecker, M.H.L. and Kryter, K.D., A study of the acoustic reflex in infantrymen. *Acta Oto-Laryngol. (Stockholm)*, Suppl. **207**, 1-16 (1965).

361. Hecker, M.H.L. and Kryter, K.D., "Comparisons between Subjective Ratings of Aircraft Noise and Various Objective Measures." Tech. Rept. FAA NO-68-33, Contract FA67WA-1696. Stanford Research Institute, Menlo Park, California, 1968.

362. Hecker, M.H.L., Von Bismarck, G., and Williams, C.E., Automatic evaluation of time-varying communication systems. *IEEE Trans. Audio Electroacoustics, AU-16* **1**, 100-106 (1968).

363. Hellman, R. and Zwislocki, J., Some factors affecting the estimation of loudness. *J. Acoust. Soc. Am.* **33**, 687-694 (1961).

364. Hellman, R. and Zwislocki, J., Monaural loudness function at 1000 cps and interaural summation. *J. Acoust. Soc. Am.* **35**, 856-865 (1963).

365. Hellman, R. and Zwislocki, J., Loudness function of a 1000 cps tone in the presence of a masking noise. *J. Acoust. Soc. Am.* **36**, 1618-1627 (1964).

366. Hellman, R. and Zwislocki, J., Loudness determination at low sound frequencies. *J. Acoust. Soc. Am.* **43**, 60-63 (1968).

367. Helper, M.M., "The Effects of Noise on Work Output and Physiological Activation." Rept. 270. U.S. Army Medical Research Laboratories, Fort Knox, Kentucky, 1957.

368. Hermann, E.R., An audiometric approach to noise control. *Am. Ind. Hyg. Assoc. J.* **24**, 344-356 (1963).

369. Hilgard, E.R., Pain as a puzzle for psychology and physiology. *Am. Psychol.* **24**, 103-113 (1969).

370. Hillquist, R.K., Objective and subjective measurement of truck noise. *Sound and Vibration* **1**, 8-13 (1967).

371. Hinchcliffe, R., Threshold of hearing as a function of age. *Acustica* **9**, 303-308 (1959).

372. Hinchcliffe, R., "Report from Wales: Some Relations between Aging Noise Exposure and Permanent Hearing Level Changes." Committee S3-W-40 of the American National Standards Institute, Anderson, Indiana, August 1969.

373. Hinterkeuser, E.G., and Sternfeld, H., Jr., "Public Response to Noise of V/STOL Aircraft." Rept. D8-0907, Contract NAS1-7083. The Boeing Company, Morton, Pennsylvania, 1968.

374. Hirsh, I.J., The influence of interaural phase on interaural summation and inhibition. *J. Acoust. Soc. Am.* **20**, 536-544 (1948).

375. Hirsh, I.J., "The Measurement of Hearing." McGraw-Hill, New York, 1952.
376. Hirsh, I.J., Monaural temporary threshold shift following monaural and binaural exposures. *J. Acoust. Soc. Am.* **30**, 912-914 (1958).
377. Hirsh, I.J., Auditory perception of temporal order. *J. Acoust. Soc. Am.* **31**, 759-767 (1959).
378. Hirsh, I.J. and Bowman, W.D., Masking of speech by bands of noise. *J. Acoust. Soc. Am.* **25**, 1175-1180 (1953).
379. Hirsh, I.J. and Burgeat, M., Binaural effects in remote masking. *J. Acoust. Soc. Am.* **30**, 827-832 (1958).
380. Hirsh, I.J., Bilger, R.C., and Detherage, B.H., The effect of auditory and visual background on apparent duration. *Am. J. Psychol.* **69**, 561-574 (1956).
381. Hirsh, I.J., Reynolds, E.G., and Joseph, M., Intelligibility of different speech materials. *J. Acoust. Soc. Am.* **26**, 530-538 (1954).
382. Hodge, D.C. and McCommons, B.C., Acoustical hazards of children's toys. *J. Acoust. Soc. Am.* **40**, 911-918 (1966).
383. Hodge, D.C. and McCommons, B.M., Reliability of TTS from impulse-noise exposure. *J. Acoust. Soc. Am.* **40**, 911 (1966).
384. Hodge, D.C. and McCommons, R.B., "Some Studies of Temporary Hearing Losses Resulting from Repeated Exposure to Gunfire Noise." Rept. N67-27968, AD 634690. Human Engineering Laboratories, Aberdeen Proving Ground, Maryland, 1966.
385. Hodge, D.C. Gates, H.W., Soderholm, R.B., Helm, C.P., Jr., and Blackner, R.F., "Preliminary Studies of the Impulse Noise Effects on Human Hearing." Proj. HUMIN, TM 15-64, AD 618327. Human Engineering Laboratories, Aberdeen Proving Ground, Maryland, 1964.
386. Hoffman, H.S. and Fleshler, M., Startle reaction: modification by background acoustic stimulation. *Science* **141**, 928-930 (1963).
387. Hoffman, H.S. and Searle, J.L., Acoustic variables in the modification of startle reaction in the rat. *J. Comp. Physiol. Psychol.* **60**, 53-58 (1965).
388. Hood, J.D., Fatigue and adaptation of hearing. *Brit. Med. Bull.* **12**, 125-130 (1956).
389. Hoople, G.D., Noise induced hearing loss: diagnosis, presbycusis, susceptibility. *Laryngoscope* **68**, 477-486 (1958).
390. Hormann, H. and Todt, E., Noise and learning. *J. Exp. Appl. Psychol.* **7**, 422-426 (1960).
391. House, A.S., Williams, C.E., Hecker, M.H.L., and Kryter, K.D., Articulation-testing methods: consonantal differentation with a closed-response set. *J. Acoust. Soc. Am.* **37**, 158-166 (1965).
392. Hubbard, H.H., Nature of the sonic boom problem. *J. Acoust. Soc. Am.* **37**, 158-166 (1965).
393. Hubbard, H.H., Sonic booms. *Physics Today* **21**, 31-37 (1968).
394. Hubbard, H.H. and Maglieri, D.J., The nature, measurement and control of sonic booms. *In* "Acoustic Noise and Its Control." Publ. 26. Institute of Electrical Engineers, London, 1967.
395. Hubbard, H.H., Maglieri, D.J., and Coepland, W.L., Research approaches to alleviation of airport noise. *J. Sound Vib.* **5**, 377-390.
396. Hummel, W.F., Turner, J.W., and Cohen, A., "Vigilance Performance in Noise as Related to Personality and Noise Tolerance." Rept. TR-20. Occupational Health Research and Training Facility, Cincinnati, Ohio, 1965.
397. Ingerslev, F., Aircraft noise and the community. *In* "The Control of Noise: National Physical Laboratory Symposium Number 12," pp. 333-344. Her Majesty's Stationery Office, London, 1962.

398. Ingerslev, F., Measurement and description of aircraft noise in the vicinity of airports. *J. Sound Vib.* **3**, 95-99 (1966).
399. Jahn, G., Kurven gleicher Lautstärke für Oktavband-Passrauschen. (Curves of equal loudness for octave band noise.) *Hochfrequenztech. U. Elektroakust.* **67**, 187-189 (1959).
400. Jahn, G., Der Lautstärkeunterschied zwischen ebenem frontalen und diffuse Schalleinfall. "Proceedings of the Third International Congress on Acoustics," Vol. 1, pp. 21-25 (L. Cremer, ed.). Elsevier, New York, 1961.
401. Jansen, G., Effects of noise on the vegetative nervous system of man. *Ger. Med. Month.* **61**, 12-13 (1961).
402. Jansen, G., Adverse effects of noise on iron and steel workers. *Stahl. Eisen.* **81**, 217-220 (1961).
403. Jansen, G., Noise effect during physical work. *Intern. Z. Angew. Physiol.* **20**, 233-239 (1964).
404. Jansen, G., Effects of noise on physiological state. "National Conference on Noise as a Public Health Hazard." American Speech and Hearing Association, Washington, D.C., February, 1969.
405. Jansen, G. and Klensch, H., The influence of the sound stimulus and music on the ballistogram. *J. Appl. Physiol.* **20**, 258-270 (1964).
406. Jansen, G. and Schulze, J., Beispiele von Schlafstorungen durch Gerausche. *Klin. Wachr.* **3**, 132-134 (1964).
407. Janssen, J.H., A method for the calculation of the speech intelligibility under conditions of reverberation and noise. *Acustica* **7**, 305-310 (1957).
408. Jeffress, L.A., "Masking and Binaural Phenomena." Rept. DRL-A-245. Defense Research Laboratory, University of Texas, 1965.
409. Jeffress, L.A., Blodgett, H.C., and Deatherage, B.H., The masking of tones by white noise as a function of the interaural phases of both components. I. 500 cycles. *J. Acoust. Soc. Am.* **24**, 523-527 (1952).
410. Jepsen, O., Middle ear muscle reflexes in man. *In* "Modern Developments in Audiology," Chap. 6 (J. Jerger, ed.). Academic Press, New York, 1963.
411. Jerger, J.F., Auditory adaptation. *J. Acoust. Soc. Am.* **29**, 357-363 (1957).
412. Jerger, J.F. and Carhart, R., Temporary threshold shift as an index of noise susceptibility. *J. Acoust. Soc. Am.* **28**, 611-613 (1956).
413. Jerger, J.F., Tillman, T.W., and Peterson, J.L., Masking by octave bands of noise in normal and impaired ears. *J. Acoust. Soc. Am.* **32**, 385-390 (1960).
414. Jerison, H.J., Paced performances on a complex counting task under noise and fatigue conditions. *Am. Psychol.* **9**, 399-400 (1954).
415. Jerison, H.J., "Effect of a Combination of Noise and Fatigue on a Complex Counting Task." Rept. WADC-TR-55-360, AD 95232. Wright-Patterson Air Force Base, Ohio, 1955.
416. Jerison, H.J., "Differential Effects of Noise and Fatigue on a Complex Counting Task." Rept. WADC-TR-55-359, AD 110506. Wright Air Development Center, Wright-Patterson Air Force Base, Ohio, 1956.
417. Jerison, H.J., Performance on simple vigilance tasks in noise and quiet. *J. Acoust. Soc. Am.* **29**, 1163-1165 (1957).
418. Jerison, H.J., Effects of noise on human performance. *J. Appl. Psychol.* **43**, 96-101 (1959).
419. Jerison, H.J. and Arginteanu, J., "Time Judgments, Acoustic Noise, and Judgment Drift." Rept. WADC-TR-57-454, AD 130963. Wright Air Development Center, Wright-Patterson Air Force Base, Ohio, 1958.

420. Jerison, H.J. and Smith, A.K., "Effect of Acoustic Noise on Time Judgment." Rept. WADC-TR-55-358, AD 99641. Wright Air Development Center, Wright-Patterson Air Force Base, Ohio, 1955.

421. Jerison, H.J. and Wallis, R.A., "Experiments on Vigilance: Performance on a Single Vigilance Task in Noise and in Quiet." Rept. WADC-TR-57-318, AD 118337. Wright Air Development Center, Wright-Patterson Air Force Base, Ohio, 1957.

422. Jerison, H.J. and Wing, S., "Effects of Noise and Fatigue on a Complex Vigilance Task." Rept. WADC-TR-57-14, AD 110700. Wright Air Development Center, Wright-Patterson Air Force Base, Ohio, 1957.

423. Jerison, H.J., Crannel, C.W., and Pownall, D., "Acoustic Noise and Repeated Time Judgments in a Visual Movement Projection Task." Rept. WADC-TR-57-54, AD 118004. Wright Air Development Center, Wright-Patterson Air Force Base, Ohio, 1957.

424. Johnson, D.R. and Robinson, D.W., The subjective evaluation of sonic bangs. *Acustica* 18 (1967).

425. Jones, G.H.S. and Muirhead, J.C., "Model Studies of Blast Effects. IV. A Study of Blast Induced Resonance in Single Air Columns." Suffield Tech. Paper 174. Defence Research Board, Department of National Defense, Canada. Suffield Experimental Station, Ralston, Alberta, 1960.

426. Jonsson, E. and Sörensen, S., On the influence of attitudes to the source on annoyance reactions to noise. *Nord. Hyg. Tidskr.* 48, 35-45 (1967).

427. Jonsson, E., Kajland, A., Paccagnella, B., and Sörensen, S., "Annoyance Reactions to Traffic Noise in Italy and Sweden." Karolinska Institutet, University of Stockholm and University of Ferrara, 1968.

428. Jouvet, M., A study of the neurophysiological mechanisms of dreaming. *Electroencephalo. Clin. Neurophysiol.* Suppl. 24, 133-157 (1963).

429. Karplus, H.B. and Bonvallet, G.L., A noise survey of manufacturing industries. *Am. Ind. Hyg. Assoc. Quart.* 14, 235-263 (1953).

429a. Kenshalo, D. and Kryter, K.D., Middle ear infection and sound induced seizures in rats. *J. Comp. Physiol. Psychol.* 42, 328-331 (1949).

430. Kerrick, J.S., Nagel, D.C., and Bennett, R.L., Multiple ratings of sound stimuli. *J. Acoust. Soc. Am.* 45, 1014-1017 (1969).

431. Kirk, R.E. and Hecht, E., Maintenance of vigilance by programmed noise. *Percep. Motor Skills.* 16, 553-560 (1963).

432. Klumpp, R.G. and Webster, J.C., Physical measurements of equal speech-interfering navy noises. *J. Acoust. Soc. Am.* 35, 1328-1338 (1963).

433. Knight, J.J. and Coles, R.R.A., Determination of the hearing thresholds of naval recruits in terms of British and American standards. *J. Acoust. Soc. Am.* 32, 800-804 (1960).

434. Knudsen, V.O. and Harris, C.M., "Acoustical Designing in Architecture." Wiley, New York, 1950.

435. Korn, T.S., Effect of psychological feedback on conversational noise reduction in rooms. *J. Acoust. Soc. Am.* 26, 793-794 (1954).

436. Kosten, C.W. and Van Os, G.J., Community reaction criteria for external noises. *In* "The Control of Noise: National Physical Laboratory Symposium Number 12," pp. 373-387. Het Majesty's Stationery Office, London, 1962.

437. Kovrigin, S.D. and Mikheyev, A.P., "The Effect of Noise Level on Working Efficiency." Rept. N65-28297. Joint Publications Research Service, Washington, D.C., 1965.

438. Krauskopf, J. and Coleman, P.D., "The Effect of Noise on Eye Movements." U.S. Army Medical Research Laboratory, Fort Knox, Kentucky, 1956.

439. Kreul, E.J., Nixon, J.C., Kryter, K.D., Bell, D.W., Lang, J.S., and Schubert, E.D., A proposed clinical test of speech discrimination. *J. of Speech Hearing Res.* **11**, 536-552 (1968).

440. Kryter, K.D., Effects of ear protective devices on the intelligibility of speech in noise. *J. Acoust. Soc. Am.* **18**, 413-417 (1946).

441. Kryter, K.D., Loudness and annoyance-value of bands of noise. "Oralism and Auralism: Transactions of the 30th Annual Meeting of the National Forum on Deafness and Speech Pathology, St. Louis, Missouri, 1948," pp. 26-28.

442. Kryter, K.D., The effects of noise on man. Monograph Supplement 1. "Journal of Speech and Hearing Disorders." American Speech and Hearing Association, Washington, D.C., 1950.

443. Kryter, K.D., "Speech Communication in Noise." Rept. AFCRC TR 54-52. Operational Applications Laboratory, Air Force Cambridge Research Center, Bolling Air Force Base, Washington, D.C. 1955.

444. Kryter, K.D., Scaling human reactions to the sound from aircraft. *J. Acoust. Soc. Am.* **31**, 1415-1429 (1959).

445. Kryter, K.D., "The Measurement of Noises of U.S. Army Weapons." Rept. 744. Bolt Beranek and Newman Inc., Cambridge, Massachusetts, 1960.

446. Kryter, K.D., Damage-risk criteria for hearing. *In* "Noise Reduction," Chap. 19. McGraw-Hill, New York, 1960.

447. Kryter, K.D., The meaning and measurement of perceived noise level. *Noise Control* **6, 5**, 12-27 (1960); Addendum—*Noise Control* **7**, No. 2, 48 (1961).

448. Kryter, K.D., Methods for the calculation and use of the articulation index. *J. Acoust. Soc. Am.* **34**, 1689-1697 (1962).

449. Kryter, K.D., Validation of the articulation index. *J. Acoust. Soc. Am.* **34**, 1698-1702 (1962).

450. Kryter, K.D., Exposure to steady-state noise and impairment of hearing. *J. Acoust. Soc. Am.* **31**, 1515-1525 (1963).

451. Kryter, K.D., Hearing impairment for speech. *Arch. Otolaryngol.* **77**, 598-602 (1963).

452. Kryter, K.D., Damage risk criterion and contours based on permanent and temporary hearing loss data. *Am. Ind. Hyg. Ass. J.* **26**, 34-44 (1965).

453. Kryter, K.D., Temporary threshold shifts in hearing from acoustic impulses of high intensities. *Int. Audiol.* **5**, 323-330 (1966).

454. Kryter, K.D., "Review of Research and Methods for Measuring the Loudness and Noisiness of Complex Sounds." Rept. CR 442. National Aeronautics and Space Administration, Washington, D.C., 1966.

455. Kryter, K.D., Psychological reactions to aircraft noise. *Science* **151**, 1346-1355 (1966).

456. Kryter, K.D., Laboratory tests of physiological-psychological reactions to sonic booms. *J. Acoust. Soc. Am.* **39**, S65-S72 (1966).

457. Kryter, K.D., "Sonic Boom—Results of Laboratory and Field Studies." Paper presented at the National Conference on Noise as a Public Health Hazard (sponsored by the American Speech and Hearing Association), Washington, D.C., June 13-14, 1968.

458. Kryter, K.D., Concepts of perceived noisiness, their implementation and application. *J. Acoust. Soc. Am.* **43**, 344-361 (1968).

459. Kryter, K.D. (chairman), "Methods for the Calculation of the Articulation Index." S3-W-36, USASI Stand. S3,5-1969, American National Standards Institute, New York.

460. Kryter, K.D., Sonic booms from supersonic transport. *Science* **163**, 359-367 (1969).

461. Kryter, K.D., Evaluation of exposures to impulse noise. "Sixth American Medical Association Congress on Environmental Health, Chicago, 1969;" and *Arch. Environ. Health* **20**, 624-635 (1969).

462. Kryter, K.D., "Possible Modifications to Procedures for the Calculation of Perceived Noisiness." NASA Rept. CR-1635. Stanford Research Institute, Menlo Park, California, 1969.

462a. Kryter, K.D., General procedures for estimating and evaluating damage risk to hearing from noise. *J. Acoust. Soc. Am.* (submitted).

463. Kryter, K.D. and Garinther, G., Auditory effects of acoustic impulses from firearms. *Acta Oto-Laryngol.* Suppl. **211** (1966).

464. Kryter, K.D. and Licklider, J.C.R., Speech communication. *In* "Human Engineering Guide to Equipment Design," Chap. 4 (C. Morgan, J. Cook, A. Chapanis and M. Lund, eds.). McGraw-Hill, New York, 1963.

465. Kryter, K.D. and Pearsons, K.S., Judgment tests of the sound from piston, turbojet, and turbofan aircraft. *Sound* **1**, No. 2, 24-31 (1962).

466. Kryter, K.D. and Pearsons, K.S., Some effects of spectral content and duration on perceived noise level. *J. Acoust. Soc. Am.* **35**, 866-883 (1963).

467. Kryter, K.D. and Pearsons, K.S., Modification of noy tables. *J. Acoust. Soc. Am.* **36**, 394-397 (1964).

468. Kryter, K.D. and Pearsons, K.S., Judged noisiness of a band of random noise containing an audible pure tone. *J. Acoust. Soc. Am.* **38**, 106-112 (1965).

469. Kryter, K.D. and Whitman, E.C., Some comparisons between rhyme and PB-word intelligibility tests. *J. Acoust. Soc. Am.* **37**, 1146 (1965).

470. Kryter, K.D. and Williams, C.E., "Some Factors Influencing Human Response to Aircraft Noise: Masking of Speech and Variability of Subjective Judgments." Rept. FAA-ADA-42. Federal Aviation Administration, Washington, D.C., 1965.

471. Kryter, K.D. and Williams, C.E., Masking of speech by aircraft noise. *J. Acoust. Soc. Am.* **39**, 138-150 (1966).

472. Kryter, K.D., Ball, J.H., and Stuntz, S.E., SCIM—a meter for measuring the performance of speech communication systems. Paper 19 in "Transactions of the Third Canadian Symposium on Communications, IEEE, Montreal, Quebec, September 25-26, 1964."

473. Kryter, K.D., Johnson, P.J., and Young, J.R., "Psychological Experiments on Sonic Booms Conducted at Edwards Air Force Base." Final Rept., Contract AF 49(638)-1758, Stanford Research Institute. National Sonic Boom Evaluation Office, Arlington, Virginia, 1968.

474. Kryter, K.D., Johnson, P.J., and Young, J.R., "Judgment Tests of Flyover Noise from Various Aircraft." NASA Rept. CR-1635. Stanford Research Institute. National Aeronautics and Space Administration, Washington, D.C., 1969.

475. Kryter, K.D., Licklider, J.C.R., and Stevens, S.S., Premodulation clipping in AM voice communication. *J. Acoust. Soc. Am.* **19**, 125-134 (1947).

476. Kryter, K.D., Weisz, A.Z., and Wiener, F.M., Auditory fatigue from audio analgesia. *J. Acoust. Soc. Am.* **34**, 383-391 (1962).

477. Kryter, K.D., Williams, C.E., and Green, D.M., Auditory acuity and the perception of speech. *J. Acoust. Soc. Am.* **34**, 1217-1223 (1962).

478. Kryter, K.D., Ward, W.D., Miller, J.D., and Eldredge, D.H., Hazardous exposure to intermittent and steady-state noise. *J. Acoust. Soc. Am.* **39**, 451-464 (1966).

479. Kwiek, M., Badania przebiegu czulosci sluchu na natezenie tonu sinusowego metoda rozniczkowa. (Investigation of the relation between the hearing sensitivity and the intensity of sinusoidal tones by differential methods.) *The Friends of Sciences Society, Poznán,* Ser. A. **6**, 329 (1953).

480. Kwiek, M., Zagadnienia czulosci sluchu. (The problems of hearing sensitivity.) "Works of the Third Open Acoustical Seminar, University of Poznán, Poznán, Poland, 1959."

481. Kylin, B., Studies on the temporary hearing threshold shift at different frequencies after exposure to various octave bands of noise. *Acta Oto-Laryngol.* **50**, 531-539 (1959).

482. Kylin, B., Temporary threshold shift and auditory trauma following exposures to steady-state noise. An experimental and field study. *Acta Oto-Laryngol.* Suppl. **152**, 1-93 (1960).

483. Laird, D.A. and Coye, K., Psychological measurements of annoyance as related to pitch and loudness. *J. Acoust. Soc. Am.* **1**, 158-163 (1929).

484. Lane, H.L., Catania, A.C., and Stevens, S.S., Voice level: autophonic scale, perceived loudness, and effects of sidetone. *J. Acoust. Soc. Am.* **33**, 160-167 (1961).

485. Langdon, F.J. and Scholes, W.E., The traffic noise index: a method of controlling noise nuisance. *Architects' J.* **147**, (April 1968).

486. Lawrence, M. and Blanchard, C., Prediction of susceptibility to acoustic trauma by determination of threshold of distortion. *Ind. Med. Surg.* **23**, 5 (1954).

487. Lawrence, M. and Yantis, P.A., Overstimulation, fatigue and onset of overload in the normal human ear. *J. Acoust. Soc. Am.* **29**, 265-274 (1957).

488. Lebo, C.P. and Oliphant, K.P., Music as a source of acoustic trauma. *Laryngoscope* **78**, 1211-1218 (1968).

489. Lebo, C.P., Oliphant, K.P., and Garrett, J., Acoustic trauma from rock and roll music. *Calif. Med.* **107**, 378-380 (1967).

490. Legget, R.F. and Northwood, T.D., Noise surveys of cocktail parties. *J. Acoust. Soc. Am.* **32**, 16-18 (1960).

491. Lehmann, G., Autonomic reactions to hearing impressions. *Stud. Gen. (Berlin)* **18**, 700-703 (1965).

492. Lehmann, G., "Sick People and Noise." Max-Planck-Institut für Arbeitsphysiologie, Dortmund, Germany (undated).

493. Lehmann, G. and Meyer-Delius, J., "Gefässreaktionen der Körperperipherie bei Schalleinwirkung." Forschungsberichte des Wirtschafts-und-Verkehrsministeriums Nordrhein-Westfalen, Nr. 517, 1958.

494. Lehmann, G. and Tamm, J., Changes of circulatory dynamics of resting men under the effect of noise. *Intern. Z. Angew. Physiol.* **16**, 217-227 (1956).

495. Lehnhardt, D., Die Berufsschäden des Ohres. *Arch. Orhen-Nasen Kehlkopfheilk* **185**, 11 (1965).

496. Levi, R., Sao Paulo legislation on city noises and protection of public quietness and well-being. "Proceedings of the Third International Congress on Acoustics," Vol. 2, pp. 1088-1089 (L. Cremer, ed.). Elsevier, New York, 1961.

497. Levi, L., Physical and mental stress reactions during experimental conditions simulating combat. *Forsvarsmed.* **2**, 3 (1966).

498. Levi, L., Sympatho-adrenomedullary responses to emotional stimuli: methodologic, physiologic and pathologic considerations. *In* "An Introduction to Clinic Neuroendocrinology" (E. Bajusz ed.). S. Karger, Basel, 1967.

499. Levitt, H. and Rabiner, L.R., Predicting binaural gain in intelligibility and release from masking for speech. *J. Accoust. Soc. Am.* **42**, 820-829 (1967).

500. Licklider, J.C.R., Effects of amplitude distortion upon the intelligibility of speech. *J. Acoust. Soc. Am.* **18**, 429-434 (1946).
501. Licklider, J.C.R., The influence of interaural phase relations upon the masking of speech by white noise. *J. Acoust. Soc. Am.* **20**, 150-159 (1948).
502. Licklider, J.C.R., Basic correlates of the auditory stimulus. *In* "Handbook of Experimental Psychology," Chap. 25 (S.S. Stevens, ed.). Wiley, New York, 1951.
503. Licklider, J.C.R., Three auditory theories. *In* "Psychology: A Study of a Science" (S. Koch, ed.). McGraw-Hill, New York, 1959.
504. Licklider, J.C.R. and Guttman, N., Masking of speech by line-spectrum interference. *J. Acoust. Soc. Am.* **29**, 287-296 (1957).
505. Licklider, J.C.R. and Miller, G.A., The perception of speech. *In* "Handbook of Experimental Psychology," Chap. 26 (S.S. Stevens, ed.). Wiley, New York, 1951.
506. Licklider, J.C.R., Bindra, D., and Pollack, I., The intelligibility of rectangular speech-waves. *Am. J. Psych.* **61**, 1-20 (1948).
507. Licklider, J.C.R., Bisberg, A., and Schwartzlander, H., An electronic device to measure the intelligibility of speech. *Proc. Natl. Electron. Conf.* **15**, 1-6 (1959).
508. Lienard, P., Critères de détermination d'un niveau de bruit dangereux pour une exposition continué à un poste de travail. *Ann. Telecomm.* **17**, 268-270 (1962).
509. Lienert, G.A. and Jansen, G., Effect of noise on test performance. *Intern. Z. Angew. Physiol.* **20**, 207-212 (1964).
510. Lierle, D.M., Guide for the evaluation of hearing impairment: a report of the committee on conservation of hearing. *Trans. Am. Acad. Ophthalmol. Otolaryngol.* 236-238 (1959).
511. Lierle, D.M. and Reger, S., The effect of tractor noise on the auditory sensitivity of tractor operators. *Ann. Otol. Rhinol. Laryngol.* **67**, 372-388 (1958).
512. Lippert, S. and Miller, M.M., An acoustical comfort index for aircraft noise. *J. Acoust. Soc. Am.* **23**, 478 (1951).
513. Little, J.W., Human response to jet engine noise. *Noise Control* 7, No. 3, 11-14 (1961).
514. Littler, T.S., Techniques of industrial audiometry. "The Control of Noise: National Physical Laboratory Symposium Number 12," pp. 285-296. Her Majesty's Stationery Office, London, 1962.
515. Loeb, M., "A Further Investigation of the Influence of Whole-Body Vibration and Noise on Tremor and Visual Acuity." Rept. 165. U.S. Army Medical Research Laboratories, Fort Knox, Kentucky, 1954.
516. Loeb, M., "The Effects of Intense Stimulation on the Perception of Time." Rept. 269. U.S. Army Medical Research Laboratories, Fort Knox, Kentucky, 1956.
517. Loeb, M., "Comparison of Attenuation of Three Helmets and a Pair of Muffs by the Threshold Shift Method." Rept. 429, AD 240006. U.S. Army Medical Research Laboratories, Fort Knox, Kentucky, 1960.
518. Loeb, M. and Fletcher, J.L., Contralateral threshold shift and reduction in temporary threshold shift as indices of acoustic reflex action. *J. Acoust. Soc. Am.* **33**, 1558-1560 (1961).
519. Loeb, M. and Fletcher, J.L., Reliability and temporal cause of temporary and contralateral threshold shifts. *J. Speech and Hearing Res.* **5**, 284-291 (1962).
520. Loeb, M. and Fletcher, J.L., Temporary threshold shift in successive sessions for subjects exposed to continuous and periodic intermittent noise. *J. Auditory Res.* **3**, 213-220 (1963).
521. Loeb, M. and Fletcher, J.L., Temporary threshold shift for "normal" subjects as a function of age and sex. *J. Auditory Res.* **3**, 65-72 (1963).

522. Loeb, M., Fletcher, J.L., and Benson, R.W., Some preliminary studies of temporary threshold shift with an arc-discharge impulse-noise generator. *J. Acoust. Soc. Am.* **37**, 313-318 (1965).

523. Loeb, M. and Jeantheau, G., The influence of noxious environmental stimuli on vigilance. *J. Appl. Psychol.* **42**, 47-49 (1958).

524. Loeb, M. and Riopelle, J., Influence of loud contralateral stimulation in the threshold and perceived loudness of low frequency tones. *J. Acoust. Soc. Am.* **32**, 602-610 (1960).

525. Loeb, M., Richmond, G., and Jeantheau, G., "The Influence of Intense Noise on Performance of a Precise Fatiguing Task." Rept. 268. U.S. Army Medical Research Laboratories, Fort Knox, Kentucky, 1956.

526. Loye, D.P., Legal aspects of noise control. *Noise Control* **2**, No. 4, 56-60 (1956).

527. Lübeke, E., Mittag, G., and Port, E., Subjektive und objektive Bewertung von Maschinengeräuschen. *Acustica* **14**, 105-114 (1964).

528. Lukas, J.S. and Kryter, K.D., "Awakening Effects of Simulated Sonic Booms and Subsonic Aircraft Noise on 6 subjects, 7 to 72 Years of Age." NASA Contract NAS1-7592. Stanford Research Institute, Menlo Park, California, 1969.

529. Lundberg, B.K., "The Menace of the Sonic Boom to Society and Civil Aviation." FFA Memo. PE-19. The Aeronautics Research Institute of Sweden, 1966.

530. Maier, B., Bevan, W., and Behar, I., The effect of auditory stimulation upon the critical flicker frequency for different regions of the visible spectrum. *Am. J. Psychol.* **74**, 67-73 (1961).

531. Marill, T., "Detection Theory and Psychophysics." Rept. 319. Research Laboratory for Electronics, Massachusetts Institute of Technology, Cambridge, Massachusetts, 1956.

532. McBain, W.N., Noise, the "arousal hypothesis" and monotonous work. *J. Appl. Psychol.* **45**, 390-317 (1961).

533. McCann, P.H., The effects of ambient noise on vigilance performance. *Human Factors* **11**, 251-256 (1969).

534. McCroskey, R.L., Jr., "The Effect of Specified Levels of White Noise upon Flicker Fusion Frequency." Rept. 80, Contract N6-ONR-22525, AD 211759. U.S. Naval School Aviation Medicine, Bureau of Medical Surgery, Pensacola, Florida, 1958.

535. McKay, R., Plumbing noise control. "Proceedings of the Audio Engineering Society, 36th Convention, Hollywood, California, 1969."

536. McKennell, A.C., Aircraft annoyance around London Heathrow Airport. "U.K. Government Social Survey for the Wilson Committee on the Problem of Noise." Cmnd. 2056. Her Majesty's Stationery Office, London, 1963.

537. Mech, E.V., Factors influencing routine performance under noise. I. The influence of "set." *J. Psychol.* **35**, 283-298 (1953).

538. Melzack, R., Perception of pain. *Sci. Am.* **204**, 41-49 (1961).

539. Mercer, D.M.A., Noise-damage criterion using A weighting levels. *J. Acoust. Soc. Am.* **43**, 636-637 (1968).

540. Meyer-Delius, J., The effect of noise on men. *Automobile Tech. J.* **59**, 293 (1957).

541. Mikeska, E.E., Noise in the modern home. *Noise Control* **4**, No. 3, 38-41 (1958).

542. Miller, G.A., The masking of speech. *Psychol. Bull.* **44**, 105-129 (1947).

543. Miller, G.A., Sensitivity to changes in the intenstiy of white noise and its relation to masking and loudness. *J. Acoust. Soc. Am.* **19**, 609-619 (1947).

544. Miller, G.A., The perception of short bursts of noise. *J. Acoust. Soc. Am.* **20**, 160-170 (1948).

545. Miller, G.A., "Language and Communication." McGraw-Hill, New York, 1951.

545a. Miller, G.A. and Garner, W.R., Effect of random presentation on the psychometric function: Duplications for a quantal theory of discrimination. *Am. J. Psychol.* **57**, 451-467 (1944).

546. Miller, G.A. and Garner, W.R., The masking of tones by repeated bursts of noise. *J. Acoust. Soc. Am.* **20**, 691-696 (1948).
547. Miller, G.A. and Licklider, J.C.R., The intelligibility of interrupted speech. *J. Acoust. Soc. Am.* **22**, 167-173 (1950).
548. Miller, G.A., and Nicely, P.E., An analysis of perceptual confusions among some English consonants. *J. Acoust. Soc. Am.* **27**, 338-352 (1955).
549. Miller, G.A., Heise, G.A., and Lichten, W., The intelligibility of speech as a function of the context of the test materials. *J. Exp. Psychol.* **41**, 329-335 (1951).
550. Miller, H., Effects of high intensity noise on retention. *J. Appl. Psychol.* **41**, 370-372 (1957).
551. Miller, J.D., Temporary threshold shift and masking for noise of uniform spectrum level. *J. Acoust. Soc. Am.* **30**, 517-522 (1958).
552. Miller, J.D., Temporary hearing loss at 4000 cps as a function of a three-minute exposure to a noise of uniform spectrum level. *Laryngoscope* **68**, 660-671 (1958).
553. Miller, J.D., Damage to organ of corti from TTS. "Conference on Noise." CHABA, University of Syracuse, Syracuse, New York, 1968.
554. Miller, J.D., Watson, C.S., and Covell, W.P., Deafening effects of noise on the cat. *Acta Oto-Laryngol.* Suppl. **176** (1963).
555. Miller, L.N. and Beranek, L.L., Noise levels in the Caravelle during flight. *Noise Control* **4**, No. 5, 19-21 (1958).
556. Miller, L.N., Beranek, L.L., Pietrasanta, A.C., Clark, W.E., Doelling, N., and Kryter, K.D., "Studies of Noise Characteristics of the Boeing 707-120 Jet Airliner and of Large Conventional Propeller-Driven Airliners." Rept. 606. Port of New York Authority. Bolt Beranek and Newman Inc., Van Nuys, California, 1958.
556a. Mills, J.H., Geugel, R.W., Watson, C.S., and Miller, J.D., Temporary changes of the auditory system due to a prolonged exposure to noise. "78th Meeting of the Acoustical Society of America, San Diego, November 5, 1969."
557. Mills, C.H.G., The measurement of traffic noise. "The Control of Noise: National Physical Laboratory Symposium Number 12," pp. 345-357. Her Majesty's Stationery Office, London, 1962.
558. Mills, C.H.G. and Robinson, D.W., The subjective rating of motor vehicle noise. *Engineer* **211**, 1070-1074 (1961).
559. Minckley, B.B., A study of noise and its relationship to patient discomfort in the recovery room. *Nurs. Res.* **17**, 247-250 (1968).
560. Miskolczy-Fodor, F., Relation between loudness and duration of tonal pulses. II. Response of normal ears to sounds with noise sensation. *J. Acoust. Soc. Am.* **32**, 482-486 (1960).
561. Mohr, G.C., Cole, J.N., Guild, E., and von Gierke, H.E., Effects of low frequency and infrasonic noise on man. *Aerospace Med.* **36**, 817-824 (1965).
562. Møller, A.R., The sensitivity of contraction of the tympanic muscles in man. *Ann. Otol. Rhinol. Laryngol.* **71**, 86-95 (1962).
563. Moser, H.M., "Research Investigations on Voice-Communication in Noise." Rept. CSD-TDR-62-5, Contract AF 19(604)-6179, AD279870. Ohio State University Research Foundation, Columbus, Ohio, 1961.
564. Moser, H.M., Dreher, J.J., O'Neill, J.J., and Oyer, H.J., "Comparison of Mouth, Ear, and Contact Microphones." Tech. Rept. 37, AFCRC TN-56-68, AD 98820. U.S. Air Force Operational Applications Laboratory, Bolling Air Force Base, Washington, D.C. 1956. Rev. 1958.
565. Mrass, H. and Diestel, H.G., Bestimmung der Normalhörschwelle für reine töne bei einohrigem Hören mit Hilfe eines Kopfhörers. *Acustica* **9**, 61-64 (1959).

566. Mullins, C.J. and Bangs, J.L., Relations between speech discrimination and other audiometric data. *Acta Oto-Laryngol. (Stockholm)* **47**, 149-157 (1957).

567. Munson, W.A., The loudness of sounds. *In* "Handbook on Noise Control," Chap. 5 (Harris, C.M. ed.). McGraw-Hill, New York, 1957.

568. Munson, W.A. and Karlin, J.E., Isopreference method for evaluating speech-transmission circuits. *J. Acoust. Soc. Am.* **34**, 762-774 (1962).

569. Munson, W.A. and Wiener, F.M., In search of the missing 6 dB. *J. Acoust. Soc. Am.* **24**, 498-501 (1952).

570. Murray, N.E. and Piesse, R.A., "Aircraft Takeoff and Landing Noise Annoyance Criteria." Rept. 22. Commonwealth Acoustic Laboratories, Sydney, Australia, 1964.

571. Murray, N.E. and Reid, G., Temporary deafness due to gunfire. *J. Laryngol. Otol.* **61**, 95-130 (1946).

572. Murray, N.E. and Reid, G., Experimental observations on the aural effects of gunblast. *Med. J. Aust.*, **33**, 611 (1946).

573. Murray, N.E., Piesse, R.A., and Rose, J.A., "Noise and Hearing Conservation in the R.A.A.F." Rept. 15. Commonwealth Acoustic Laboratories, Sydney, Australia, 1959.

574. Nagel, D.C., Parnell, J.E., and Parry, H.J., "The Effects of Background Noise upon Perceived Noisiness." Rept. 1614, DS-67-22, AD 663902. Federal Aviation Administration. Bolt Beranek and Newman, Inc., Van Nuys, California, 1967.

574a. Nakamura, S. and Katano, Y., Relationship between TTS and duration of noise exposure. *J. Aud. Res.* **7**, 401-442 (1967).

575. Nestle *et al.* vs. City of Santa Monica, California (1968).

576. Nett, E., Doerfler, L.G., and Matthews, J., "The Relationship between Audiological Measures and Actual Social-Psychological-Vocational Disability." Proj. SP-167. Sponsored by the U.S. Department of Health, Education and Welfare, Office of Vocational Rehabilitation, at the University of Pittsburgh, Pennsylvania, (undated).

577. Niese, H. Proposal for a loudness meter capable of measuring noise with impulses in any type of sound field. *Hochfrequeuztech. Elektroakust.* **66**, 125-139 (1958).

578. Niese, H., Die lautstarke von Geraeuschen und ihre Annaeherung durch Mess-und Berechungmethoden. (Sound levels of noises and their approximation by methods for measurement and calculation.) *Hochfrequenztech. Elektroakust.* **73**, 3-14 (1963).

579. Niese, H., Loudness sensitivity, measurement, and trauma caused by impulse noise. *I.Z. Laryng. Rhinol. Otol.* **44**, 209-217 (1965).

580. Nelson, H.A., Legal liability for loss of hearing. *In* "Handbook of Noise Control," Chap. 38 (C.M. Harris, ed.). McGraw-Hill, New York, 1957.

581. Nixon, C.W. and Borsky, P.N., Effects of sonic boom on people: St. Louis, Missouri 1961-1962. *J. Acoust. Soc. Am.* **39**, S51-S58 (1966).

582. Nixon, C.W. and Hubbard, H.H., "Results of the USAF-NASA-FAA Flight Program to Study Community Responses to Sonic Booms in the Greater St. Louis Area." Tech. Note D-2705. National Aeronautics Space Administration, Langley Research Center, Hampton, Virginia, 1965.

583. Nixon, C.W., Harris, C., and von Gierke, H., "Rail Test to Evaluate Equilibrium in Low-level Wideband Noise." Rept. AMRL-TR-66-85. U.S. Air Force Aerospace Medical Research Laboratories, Wright-Patterson Air Force Base, Ohio, 1966.

584. Nixon, C.W., von Gierke, H.E., and Rosinger, G., Comparative annoyances of "approaching" versus "receding" sound sources. (A) *J. Acoust. Soc. Am.* **45**, 330 (1969).

585. Nixon, J.C. and Glorig, A., Noise-induced permanent threshold shift at 2000 cps and 4000 cps. *J. Acoust. Soc. Am.* **33**, 904-908 (1961).

586. Nixon, J.C., and Glorig, A., Noise-induced temporary threshold shift vs. hearing level in four industrial samples. *J. Auditory Res.* **2**, 125-138 (1962).

587. Nixon, J.C., Glorig, A., and Bell, D.W., Predicting hearing loss from noise induced TTS. *Arch. Otolaryngol.* **81**, 250-256 (1965).

588. Office of Naval Research. "An Annotated Bibliography and Critical Review of Voice Communications." Rept. ACR-26, PB 131584, AD 152263. U.S. Navy, Office of Naval Research, Washington, D.C., 1958.

589. Ohwaki, Y., Effects of intermittent noise on physical and mental works. *Tohoku Psychol. Folia.* **18**, 27-43 (1960).

590. Ogilive, J.C., Effect of auditory flutter on the visual critical flicker frequency. *Can. J. Psychol.* **54**, 167-170 (1956).

591. O'Hare, J., Intersensory effects of visual stimuli on the minimum audible threshold. *J. Gen. Psychol.* **54**, 167-170 (1956).

592. Ollerhead, J.B., "Subjective Evaluation of General Aircraft Noise." Tech. Rept. 68-35. Contract, FA67WA-1731. Wyle Laboratories, Federal Aviation Administration, Washington, D.C., 1968.

592a. Ollerhead, J.B., "The Noisiness of Diffuse Sound Fields at High Intensities." Final Rept. FAA-NO-70-3. Wyle Laboratories, Hunstville, Alabama, for the Federal Aviation Administration, Washington, D.C., 1970.

593. Olsen, J. and Nelson, E.N. Calming the irritable infant with a simple device. *Minn. Med.* **44**, 527-529 (1961).

594. Olson, H.F. and May, E.G., Electronic sound absorber. *J. Acoust. Soc. Am.* **25**, 1130-1136 (1953).

595. Olynyk, D. and Northwood, T.D., Subjective judgments of footstep-noise transmission through floors. *J. Acoust. Soc. Am.* **38**, 1035-1039 (1969).

596. Oppliger, G., and Grandjean, E., Vasomotor reactions of the hand to noise stimuli. *Helv. Physiol. Pharmacol. Acta* **17**, 275-287 (1957).

597. Oppliger, G. and Grandjean, E., "Vasomotorial Reactions of the Hand to Noise Stimuli." Rept. FTD-TT-63-591, AD 467646L. Foreign Technical Division, Wright-Patterson Air Force Base, Ohio, 1965.

598. Osipov, G., Moscow street noises and their hygenic evaluation. *Gig. Sanit.* **23**, 21-27 (1968).

599. Osipov, G.L. and Kovrigin, S.D., "Standardization of Automobile Traffic Noise." Joint Publ. Res. Serv. Transl. JPRS 4371. U.S. Department of Commerce, OTS, Washington, D.C., February 1961.

600. Oswald, I., Taylor, A.M., and Treisman, M., Discriminative responses to stimulation during human sleep. *Brain* **83**, 440-453 (1960).

601. Parnell, J.E., Nagel, D.C., and Parry, H.S., "Growth of Noisiness for Tones and Bands of Noise at Different Frequencies." Rept. 1521, DS-67-21, AD 663904. Federal Aviation Administration. Bolt Beranek and Newman Inc., Van Nuys, California, 1967.

602. Park, J.L., Jr. and Payne, M.C., Jr., Effects of noise level and difficulty of task in performing division. *J. Appl. Psychol.* **47**, 367-368 (1963).

603. Parrack, H.O., Community reaction to noise. *In* "Handbook of Noise Control," Chap. 36 (C.M. Harris, ed.). McGraw-Hill, New York, 1957.

604. Parrack, H.O., Effect of air-borne ultrasound on humans. *Intl. Aud.* **5**, 294-308 (1966).

605. Parry, H.J., The noisiness of steady-state equal-attribute contours. Rept. 1523, General Electric Company. Bolt Beranek and Newman Inc., 1967.

606. Passhier-Vermeer, W., "Hearing Loss Due to Exposure to Steady-State Broadband Noise." Rept. 35. Instituut voor Gezondheidstechniek, Netherlands, 1968.

607. Pascal, G.R., The effect of disturbing noise on the reaction time of mental defectives. *Am. J. Mental Def.* **54**, 691-699 (1953).

607a. Patton, R.A. Purulent otitis media in albino rats susceptible to sound induced seizures. *J. Psychol.* **24**, 313-317 (1947).

608. Pearsons, K.S., "Loudness of Sounds in the Presence of a Masking Noise." M.S. thesis, Massachusetts Institute of Technology, Cambridge, Massachusetts, 1959.

609. Pearsons, K.S., "The Effects of Duration and Background Noise Level on Perceived Noisiness." Rept. FAA-ADS-78. Bolt Beranek and Newman Inc., Van Nuys, California, 1966.

610. Pearsons, K.S., "Noisiness Judgments of Helicopter Flyovers." Rept. DS 67-1, Contract FA65WA-1260. Bolt Beranek and Newman Inc., Van Nuys, California, 1967.

611. Pearsons, K.S., Assessment of the validity of pure-tone corrections to perceived noise level. "Conference on NASA Research Relating to Noise Alleviation of Large Subsonic Jet Aircraft." NASA Langley Research Center, Virginia, 1968.

611a. Pearsons, K.S., Woods, B., and Kryter, K.D., "Preliminary Study on the Effect of Multiple and Modulated Tones on Perceived Noisiness." Rept. 1265, NASA CR-1117. National Aeronautics and Space Administration. Bolt Beranek and Newman Inc., Van Nuys, California, 1967.

612. Pearsons, K.S. and Horonjeff, R.D., "Category Scaling Judgment Tests on Motor Vehicle and Aircraft Noise." Rept. DS-67-8, Contract FA-65-WA-1260. Bolt Beranek and Newman Inc., Van Nuys, California, 1967.

613. Pearsons, K.S. and Kryter, K.D., "Laboratory Tests of Subjective Reactions to Sonic Boom." NASA Contract Rept. NASA CR-187, National Aeronautics and Space Administration, Washington, D.C., 1965.

613a. Pearsons, K.S. and Bennett, R.L., "The Effect of Temporal and Spectral Combinations on the Judged Noisiness of Aircraft Sounds." Rept. FAA-69-3. Bolt Beranek and Newman Inc., Van Nuys, California, 1969.

614. Pearsons, K.S., Horonjeff, R.D., and Bishop, D.E., "The Noisiness of Tones Plus Noise." Rept. 1520, Contract NAS1-6364. Bolt Beranek and Newman Inc., Van Nuys, California, 1967.

615. Perret, E., Grandjean, E., and Lauber, A., Subjective rating of aircraft noises by students during a lecture. *Ergonomics* **6**, 307 (1963).

616. Peters, R.W., "Research on Psychological Parameters of Sound." Rept. WADD TR 60-249, AD 240814. Wright Air Development Center Aerospace Medical Laboratory, Wright-Patterson Air Force Base, Ohio, 1960.

617. Peyser, A., Audiometric fatigue test for the purpose of occupational selection. *Acta Oto-Laryngol.* **41**, 158-168 (1952).

618. Pickett, J.M., Effects of vocal force on the intelligibility of speech sounds. *J. Acoust. Soc. Am.* **28**, 902-905 (1956).

619. Pickett, J.M., Limits of direct speech communication in noise. *J. Acoust. Soc. Am.* **30**, 278-281 (1958).

620. Pickett, J.M., Backward masking. *J. Acoust. Soc. Am.* **31**, 1613-1615 (1959).

621. Pickett, J.M., Low-frequency noise and methods for calculating speech intelligibility. *J. Acoust. Soc. Am.* **31**, 1259-1263 (1959).

622. Pickett, J.M. and Pollack, I., Prediction of speech intelligibility at high noise levels. *J. Acoust. Soc. Am.* **30**, 955-963 (1958).

623. Pickett, J.M. and Pollack, I., Intelligibility at high voice levels and the use of a megaphone. *J. Acoust. Soc. Am.* **30**, 1100-1104 (1958).

624. Piesse, R.A., Rose, J.A., and Murray, N.E., "Hearing Conservation in Industrial Noise." C.A.L. Rept. 19. Commonwealth Acoustic Laboratories, Sydney, Australia, 1962.

625. Pietrasanta, A.C. and Stevens, K.N., Noise exposure in communities near jet air bases. *Noise Control* **42**, 29-36 (1958).

626. Pietrasanta, A.C. and Stevens, K.N., "Noise Guide for Air Base Planners and Installations Officers." U.S. Air Force Rept. 511. Bolt Beranek and Newman Inc., Cambridge, Massachusetts, 1959.

627. Pinto, R.M.N., Sex and acoustic trauma: audiologic study of 199 Brazilian airline stewards and stewardesses (VARIG). *Rev. Brasil Med. (Rio)* 19, 326-327 (1962).

628. Plomp, R., Hearing losses induced by small arms. *Intl. Audiology* 6, 31-36 (1967).

629. Plomp, R. and Levelt, W.J.M., Tonal consonance with critical bandwidth. *J. Acoust. Soc. Am.* 38, 548-560 (1965).

630. Plutchik, R., "A Critical Analysis of the Literature Dealing with the Effect of Intermittent Sound Stimuli on Performance, Feeling, and Physiology, with Preliminary Work toward an Experimental Analysis of the Problem." Contract Nonr-225201, AD149589. Office of Naval Research, Washington, D.C., 1957.

631. Plutchik, R., Effect of high intensity intermittent sound on compensatory tracking and mirror tracing. *Percept. Motor Skills* 12, 187-194 (1961).

632. Plutchik, R., Physiological responses to high intensity intermittent sound. *Psychol. Rec.* 13, 141-148 (1963).

633. Pollack, I., Monaural and binaural threshold sensitivity for tones and for white noise. *J. Acoust. Soc. Am.* 20, 52-57 (1948).

634. Pollack, I., On the measurement of the loudness of white noise. *J. Acoust. Soc. Am.* 23, 654-657 (1951).

635. Pollack, I., The loudness of bands of noise. *J. Acoust. Soc. Am.* 24, 533-538 (1952).

636. Pollack, I., On the effect of frequency and amplitude distortion on the intelligibility of speech in noise. *J. Acoust. Soc. Am.* 24, 538-540 (1952).

637. Pollack, I., Masking of speech by repeated bursts of noise. *J. Acoust. Soc. Am.* 26, 1053-1055 (1954).

638. Pollack, I., Masking by a periodically interrupted noise. *J. Acoust. Soc. Am.* 27, 353-355 (1955).

639. Pollack, I., Speech communications at high noise level: The roles of a noise-operated automatic gain control system and hearing protection. *J. Acoust. Soc. Am.* 29, 1324-1327 (1957).

640. Pollack, I., Message procedures for unfavorable communication conditions. *J. Acoust. Soc. Am.* 30, 196-201 (1958).

641. Pollack, I., Speech intelligibility at high noise levels: Effect of short-term exposure. *J. Acoust. Soc. Am.* 30, 282-285 (1958).

642. Pollack, I., Loudness of periodically interrupted white noise. *J. Acoust. Soc. Am.* 30, 181-185 (1958).

643. Pollack, I., Binaural communication systems: Preliminary examination. *J. Acoust. Soc. Am.* 31, 81-82 (1959).

644. Pollack, I., Discrimination of sharp spectral changes within broad-band noises. *J. Auditory Res.* 3, 165-168 (1963).

645. Pollack, I., Onset discrimination for white noise. *J. Acoust. Soc. Am.* 35, 607-609 (1963).

646. Pollack, I., Interaction of forward and backward masking. *J. Auditory Res.* 4, 63-67 (1964).

647. Pollack, I. and Pickett, J.M., Effect of noise and filtering on speech intelligibility at high levels. *J. Acoust. Soc. Am.* 29, 1328-1329 (1957).

648. Pollack, I. and Pickett, J.M., Interaural effects upon speech intelligibility at high noise levels. *J. Acoust. Soc. Am.* 30, 293-296 (1958).

649. Pollack, I. and Pickett, J.M., Stereophonic listening and speech intelligibility against voice babble. *J. Acoust. Soc. Am.* 30, 131-133 (1958).

650. Pollack, I. and Pickett, J.M., Intelligibility of peak-clipped speech at high noise levels. *J. Acoust. Soc. Am.* 31, 14-16 (1959).

651. Pollack, I., Rubenstein, H., and Decker, L.R., Intelligibility of known and unknown message sets. *J. Acoust. Soc. Am.* **31**, 273-279 (1959).

652. Port, E., Über die lautstarke cinzelner kurzer Schallimpulse. (Concerning the intensity of separate, short, sound pulses.) *Acustica* **13**, 212-223 (1963).

653. Prather, W.C., Shifts in loudness of pure tones associated with contralateral noise simulation. *J. Speech Hear. Res.* **4**, 182-193 (1961).

654. Poulton, E.C. and Stevens, S.S., On the halving and doubling of the loudness of white noise. *J. Acoust. Soc. Am.* **27**, 329-331 (1955).

655. Quietzsch, G., Objektive and subjektive Lautstärkemessungen. (Objective and subjective loudness measurements.) *Acustica* **1**, 49-66 (1955).

656. Quiggle, R.R., Glorig, A., Delk, J.H., and Summerfield, A.B., Predicting hearing loss for speech from pure tone audiograms. *Laryngoscope* **67**, 1-15 (1957).

657. Quist-Hansson, S., Noise induced hearing loss amongst engineroom personnel on board Norwegian merchant ships. *Acta Oto-Laryngol. (Stockholm)* Suppl. **196** (1964).

658. Quist-Hanssen, S. and Steen, E., Observed and calculated hearing loss for speech in noise-induced deafness. *Acta Oto-Laryngol. (Stockholm)* Suppl. **158**, 277-281 (1960).

659. Raab, D.H., Forward and backward masking between acoustic clicks. *J. Acoust. Soc. Am.* **33**, 137-139 (1961).

660. Raab, D.H. and Soman, E., Effect of masking noise on lateralization and loudness of clicks. *J. Acoust. Soc. Am.* **34**, 1620-1624 (1962).

661. Rademacher, H.J., Die Lautstarke von Kraftfahrzeuggerauschen. (Loudness of vehicle noise.) *Acustica* **9**, 93-108 (1959).

662. Rechtschaffen, A., Hauri, P., and Zeitlin, M., Auditory awakening thresholds in REM and NREM sleep stages. *Percept. Motor Skills* **22**, 927-942 (1966).

663. Reese, T.W., Kryter, K.D., and Stevens, S.S., "The Relative Annoyance Produced for Various Bands of Noise." PB 27307. IC-65 Psycho-Acoustic Laboratory, Harvard University. U.S. Department of Commerce, Washington, D.C., 1944.

664. Reger, S.N., Effect of middle ear muscle action on certain psycho-physical measurements. *Ann. Otol. Rhinol. Laryngol.* **69**, 1179-1198 (1960).

665. Reger, S.N., Menzel, O.J., Ickes, W.K., and Steiner, S.J., Changes in air conduction and bone conduction sensitivity associated with voluntary contraction of middle ear musculature. *In* "Seminar on Middle Ear Function," pp. 171-180 (J.L. Fletcher, ed.). Rept. 576. U.S. Army Medical Research Laboratory, Fort Knox, Kentucky, 1963.

666. Reid, G., "'Permanent' Deafness Due to Gunfire." Rept. 12. Acoustic Research Laboratory, New Medical School, University of Sydney, Sydney, Australia, 1945.

667. Reid, G., Further observations on temporary deafness following exposure to gunfire. *J. Laryngol. Otol.* **61**, 609-633 (1946).

667a. Reiter, H.H., Effects of noise on discrimination reaction time. *Percep. Motor Skills* **17**, 418 (1963).

668. Regier, A.A., Mayes, W.H., and Edge, P.M., Jr., Noise problems associated with launching large space vehicles. *Sound* **1**, No. 6, 7-12 (1962).

669. Rettinger, M., Noise level reduction of "depressed" freeways. *Noise Control* **53**, 12-14 (1959).

670. Reynolds, G.S., and Stevens, S.S., Binaural summation of loudness. *J. Acoust. Soc. Am.* **32**, 1337-1344 (1960).

671. Rice, C.G. and Coles, R.A., "Impulsive Noise Studies and Temporary Threshold Shift." Fifth International Congress of Acoustics, Liége, Belgium, 1965.

672. Rice, C.G. and Zepler, E.E., Loudness and pitch sensations of an impulse sound of very short duration. *J. Sound Vib.* **5**, 285-289 (1967).

672a. Rice, C.G., Ayley, J.B., Bartlett, B., Bedford, W., Gregory, W., and Hallum, G., "A Pilot Study on the Effects of Pop Group Music on Hearing." I.S.V.R. Memo 266. University of Southampton Press, Southampton, England, 1968.

673. Richards, D.L., A development of the Collard principle of articulation calculation. *Proc. Inst. Elect. Engrs.* **103**, 679-691 (1956).

674. Richards, D.L. and Swaffield, J., Assessment of speech communication links. *Proc. Inst. Elect. Engrs.* **26**, 77-89 (1959).

675. Richards, E.J., Aircraft noise–mitigating the nuisance. *Astronaut.Aeronaut.* **5**, No. 1, 34-44 (1967).

676. Richter, R., "Sleep Disturbances Which We Are Not Aware of, Caused by Traffic Noise." EEG Station of the Neurological University Clinic, Basel (undated).

677. Riley, E.C., Critique on the concept of audiometer zero. *Arch. Otolaryngol. (Chicago)* **81**, 139-144 (1965).

678. Riley, E.C., Sterner, J.H., Fassett, D.W., and Sutton, W.L., Ten years' experience with industrial audiometry. *Am. Ind. Hyg. Assoc. J.* **22**, 151-159 (1961).

679. Robertson, D.W. and Stuckey, C.W., "Investigation and Evaluation of the Gel Speech System Test Set." Rept. RADC TR 61-88. Rome Air Development Center, Griffiss Air Force Base, New York, 1961.

680. Robinson, D.E. and Jeffress, L.A., Effect of varying the inter-aural noise correlation on the detectability of tonal signals. *J. Acoust. Soc. Am.* **35**, 1947-1952 (1963).

681. Robinson, D.W., The subjective loudness scale. *Acustica* **7**, 217-233 (1957).

682. Robinson, D.W., A new determination of the equal-loudness contours. *IRE Trans. Audio.* **AU-6**, 6-13 (1958).

683. Robinson, D.W., Statistical aspects of the relation between binaural and monaural thresholds. *Acustica* **11**, 185-190 (1961).

684. Robinson, D.W., The loudness of directional sound fields. "Proceedings of the Third International Congress on Acoustics," Vol. 1, pp. 89-92 (L. Cremer, ed.). Elsevier, New York, 1961.

685. Robinson, D.W., Subjective scales and meter readings. "The Control of Noise: National Physical Laboratory Symposium Number 12," pp. 243-261. Her Majesty's Stationary Office, London, 1962.

686. Robinson, D.W., "The Relationships between Hearing Loss and Noise Exposure." NPL AERO Rept. AC 32. National Physical Laboratory, Teddington, Middlesex, England, 1968.

686a. Robinson, D.W., "The Concept of Noise Pollution Level." NPL Aero Rept. AC 38. National Physical Laboratory, Teddington, Middlesex, England, 1969.

687. Robinson, D.W. and Bowsher, J.M., A subjective experiment with helicopter noises. *J. Royal Aeron. Soc.* **65**, 635-637 (1961).

688. Robinson, D.W. and Cook, J.P., "The Quantification of Noise Exposure." NPL AERO Rept. AC 31. National Physical Laboratory, Teddington, Middlesex, England, 1968.

689. Robinson, D.W. and Dadson, R.S., A re-determination of the equal-loudness relations for pure tones. *Brit. J. Appl. Phys.* **7**, 166-181 (1956).

690. Robinson, D.W. and Dadson, R.S., Threshold of hearing and equal-loudness relations for pure tones, and the loudness function. *J. Acoust. Soc. Am.* **29**, 1284-1288 (1957).

691. Robinson, D.W. and Whittle, L.S., The loudness of directional sound fields. *Acustica* **10**, 74-80 (1960).

692. Robinson, D.W. and Whittle, L.S., The loudness of octave-bands of noise. *Acustica* **14**, 24-35 (1964).

693. Robinson, D.W., Bowsher, J.M., and Copeland, W.C., On judging the noise from aircraft in flight. *Acustica* **13**, 324-336 (1963).

694. Robinson, D.W., Copeland, W.C., and Rennie, A.J., Motor vehicle noise measurement. *Engineer* **211**, 493-497 (1961).

695. Robinson, D.W., Whittle, L.S., and Bowsher, J.M., The loudness of diffuse sound fields. *Acustica* **11**, 397-404 (1961).

696. Robson, J.G. and Davenport, H.T., The effects of white sound and music upon the superficial pain threshold. *Can. Anesthesiol. Soc. J.* **9**, 105-108 (1962).

697. Roggevsen, L.S. and van Dishoeck, H.A., Vestibular reactions as a result of acoustic stimulation. *Pract. Otol. Rhinol. Laryngol.* **18**, 205-213 (1956).

698. Rosen, M., A method for calculating the average decibel loss of hearing in workman's compensation. *Trans. Am. Acad. Ophthalmol. Otolaryngol.* **61**, 507-509 (1957).

699. Rosen, S., Bergman, M., Plestor, D., El-Mofty, A., and Satti, M., Presbycusis study of a relatively noise-free population in the Sudan. *Ann. Otol. Rhinol. Laryngol.* **71**, 727-743 (1962).

700. Rosenblith, W.A., "The Relations of Hearing Loss to Noise Exposure." Exploratory Subcommittee Z24-X-2 of Sectional Committee on Acoustics, Vibration, and Mechanical Shock, American Standards Association, New York, 1954.

701. Rosenblith, W.A., Establishment of criteria based on the concept of noise exposure. *Laryngoscope* **68**, 497-507 (1958).

702. Rosenblith, W.A., Stevens, K.N., and Staff of Bolt Beranek and Newman Inc., "Handbook of Acoustic Noise Control. II. Noise and Man." Rept. WADC TR 52-204. U.S. Air Force Aerospace Medical Laboratory, Wright-Patterson Air Force Base, Ohio, 1953.

703. Rosenwinkel, N.E. and Stewart, K.C., The relationship of hearing loss to steady state noise exposure: A report of an industrial survey. *Am. Ind. Hyg. Assoc. Quart.* **18**, 227-230 (1957).

704. Rosenwinkel, N.E. and Stewart K.C., Hearing loss related to nonsteady noise exposure. *Am. Ind. Hyg. Assoc. J.* **20**, 290-293 (1959).

705. Rossi, L., Oppliger, G., and Grandjean, E., Neurovegetative effects of man of noises superimposed on a background noise. *Med. Lavoro.* **50**, 332-377 (1959).

706. Rowland, V., Differential electroencephalographic response to conditioned auditory stimuli in arousal from sleep. *Electroencephalog. Clin. Neurophysiol.* **9**, 585-594 (1957).

707. Rozhanskaya, E.V., On the question of the mathematical foundation of the theory of intelligibility. *Akad. Nauk. S.S.S.R. Komissiya Akustika. Trudy.* **7**, 53-60 (1953). U.S. Department of Commerce, OTS, Translation 62-11581.

708. Rubenstein, M.K., "Interaction between Vision and Audition." Rept. 151. U.S. Army Medical Research Laboratories, Fort Knox, Kentucky, 1954.

709. Rudmose, W., Hearing loss resulting from noise exposure. *In* "Handbook of Noise Control" (C.M. Harris, ed.). McGraw-Hill, New York, 1957.

710. Ruedi, L. and Furrer, W., Preliminary report on a new device protecting against noise. *Pract. Oto-Rhino. Laryngol. (Basel)* **6**, 255-265 (1944).

711. Saito, S. and Watanabe, S., Normalized representation of noiseband masking and its application to the prediction of speech intelligibility. *J. Acoust. Soc. Am.* **33**, 1013-1021 (1961).

712. Salmon, V., Surface transportation noise–review. *Noise Control* **2**, No. 4, 21-27 (1956).

713. Samojlova, I.K., The masking effect of short signals as a function of the time between the masked and masking sound. *Biofizika* **4**, 550-558 (1959).

714. Sampson, W.W. and Holder, L., A preliminary study in sonic pollution with especial reference to animal noise. *Sanitarian* **23**, 29-36 (1960).

715. Sanders, A., The influence of noise on two discrimination tasks. *Ergonomics* **4**, 253-258 (1961).

716. Sataloff, J., "Industrial Deafness. Hearing, Testing and Noise Measurement." McGraw-Hill, New York, 1957.

717. Sataloff, J., "Hearing Loss." Lippincott, Philadelphia, 1966.

718. Sataloff, J., Menduke, H., and Hughes, A., Temporary threshold shift in normal and abnormal ears. *Arch. Otolaryngol.* **76**, 52-54 (1962).

719. Sataloff, J., Vassallo, L., Vallotti, J., and Menduke, H., Long-term study relating temporary and permanent hearing loss. *Arch. Environ. Health (Chicago)* **13**, 637-640 (1966).

720. Sataloff, J., Vassallo, L., and Menduke, H., Hearing loss from exposure to interrupted noise. *AMA Arch. Environ. Health* **18** (1969).

721. Sato, T., Some results of the articulation tests for various combinations of bandpass filters. *J. Acoust. Soc. (Japan)* **14**, 159-164 (1958).

722. Saul, E.V. and Jaffe, J., "The Effects on Markmanship Performance of Reducing Gun Blast by Ear Defenders." Rept. 4, Contract DA-19-020-ORD-3461, AD 66894. Human Engineering Services in the Design of Small Arms, Tufts University, Medford, Massachusetts, 1955.

723. Saunders, W.H., The areal ratio and variations in normal hearing. *Laryngoscope* **71**, 1073-1078 (1961).

724. Schafer, T.H., Gales, R.S., Shewmaker, C.A., and Thompson, P.O., The frequency selectivity of the ear as determined by masking experiments. *J. Acoust. Soc. Am.* **22**, 490-496 (1950).

725. Scharf, B., Complex sounds and critical bands. *Psychol. Bull.* **58**, 705-717 (1961).

726. Schroeder, M.R. and Logan, B.F., The sound of rain. *Frequenz* **13**, 229-234 (1959).

727. Schubert, E.D., The effect of a thermal masking noise on the pitch of a pure tone. *J. Acoust. Soc. Am.* **22**, 497-499 (1950).

728. Schuknecht, H.F. and van den Ende, H., Acoustic trauma from a dog training whistle. *Henry Ford Hosp. Med. Bull.* **9**, 374-378 (1961).

729. Schultz, T.J., Noise-criterion curves for use with the USASI preferred frequencies. *J. Acoust. Soc. Am.* **43**, 637 (1968).

730. Selters, W., Prediction of temporary threshold shift after noise increase. *J. Acoust. Soc. Am.* **35**, 99-103 (1963).

731. Shambaugh, J.C., Jr., "An Investigation of Certain Individual Differences under the Stress of High Intensity Sound." Ph.D. thesis, Pennsylvania State University, University Park, 1950.

732. Sharpless, S. and Jasper, H.H., Habituation of the "arousal" reaction. *Brain* **79**, 655-680 (1956).

733. Shapely, J.L., Reduction in the loudness of a 250 cycle tone in one ear following the introduction of a thermal noise in the opposite ear. *Proc. Iowa Acad. Sci.* **61**, 417-422 (1954).

734. Shatalov, N.N., Saitanov, A.O., and Glotova, K.V., "On the State of the Cardiovascular System under Conditions of Exposure to Continuous Noise." Rept. T-411-R, N65-15577. Defense Research Board, Toronto, Canada, 1962.

735. Shepherd, L.J. and Sutherland, W.W., "Relative Annoyance and Loudness Judgments of Various Simulated Sonic Boom Waveforms." NASA Contract NAS1-6193. Lockheed California Company, Burbank, California, 1968.

736. Sherrick, C.E., Jr. and Mangabeira-Albernaz, P.L., Auditory threshold shifts produced by simultaneously pulsed contralateral stimuli. *J. Acoust. Soc. Am.* **33**, 1381-1385 (1961).

737. Shoenberger, R.W. and Harris, C.S., Human performance as a function of changes in acoustic noise levels. *J. Eng. Psychol.* **4**, 108-119 (1965).

738. Shoji, H., Yamamoto, T., and Takagi, K., Studies on the critical band for TTS. *J. Acoust. Soc. Japan,* **22**, 350-361 (1966).

739. Shoji, H., Yamamoto, T., and Takagi, K., Studies on TTS due to exposure to octave-band noise. *J. Acoust. Soc. Japan* **22**, 340-349 (1966).

740. Skillern, C.P., Human response to measured sound pressure levels from ultrasonic devices. *Am. Ind. Hyg. Assoc. J.* **26**, 132-136 (1965).

741. Simmons, F.B., Individual sound damage susceptibility: role of middle ear muscles. *Trans. Am. Otol. Soc.* **60**, 128-147, 147-149 (1963).

742. Simmons, F.B., Electrical stimulation of the auditory nerve in man. *Arch. Otolaryngol.,* **78**, 24-54 (1966).

743. Simmons, F.B., Middle ear muscle acoustic reflex as index of cochlear sensitivity in auditory experiments: Some technical notes. *J. Auditory Res.* **4**, 255-260 (1964).

744. Simon, C.W. and Emmons, W.H., "A Critical Review of the 'Learn-While-You-Sleep' Studies." Rept. P-534. The Rand Corporation, Santa Monica, California, 1954.

745. Sivian, L.J. and White, S.D., On minimum audible sound fields. *J. Acoust. Soc. Am.* **4**, 288-321, (1933).

746. Slavin, I.I., "Industrial noise and its reduction." Special Libraries Association, New York, 1959.

747. Slavin, I.I., Soviet tentative standards and regulations for restricting noise in industry (translated by C.R. Williams). *Noise Control* **5**, No. 5., 44-49, 64 (1959).

748. Slavin, I.I., "Standards and Regulations for Restricting Noise in Industry." Translation JPRS/NY-6-71 CSO-1476. Joint Publications Research Service, New York (undated).

749. Slavin, I.I., Objective noise gauges with a loudness scale. *Tr. Kom. Akustike Akad. Nauk S.S.S.R.* **7** (1963).

750. Small, A.M., Jr., Pure-tone masking. *J. Acoust. Soc. Am.* **31**, 1619-1625 (1959).

751. Small, A.M., Jr., Auditory adaptation. *In* "Modern Developments in Audiology," Chap. 8 (J. Jerger, ed.). Academic Press, New York, 1963.

752. Small, A.M., Jr., Bacon, W.E., and Fozard, J.L., Intensive differential thresholds for octave-band noise. *J. Acoust. Soc. Am.* **31**, 508-510 (1959).

753. Smith, K.R., Intermittent loud noise and mental performance. *Science* **114**, 132-133 (1951).

754. Smith, M.G. and Goldstone, G., "A Pilot Study of Temporary Threshold Shifts Resulting from Exposure to High-Intensity Impulse Noise." Rept. TM-19-61. U.S. Army Ordinance, Human Engineering Laboratories, Aberdeen Proving Ground, Maryland, 1961.

755. Society of Automotive Engineers, "Technique for Developing Noise Exposure Forecasts." Rept. DS67-14, Contract with Federal Aviation Administration, Washington, D.C., 1967.

756. Society of Automotive Engineers, 865, "Definitions and Procedure for Computing the Perceived Noise Level of Aircraft Noise." ARP 865. Society of Automative Engineers, New York, 1965.

757. Society of Automotive Engineers, Measurement of truck and bus noise. Standard J672. "1965 SAE Handbook," pp. 887-889. Society of Automotive Engineers, New York, 1965.

758. Sperry, W.C., "Aircraft Noise Evaluation." Rept. FAA-68-34. Federal Aviation Administration, Washington, D.C., 1968.

759. Spieth, W., Annoyance threshold judgments of bands of noise. *J. Acoust. Soc. Am.* **28**, 872-877 (1956).

760. Spieth, W., Downward spread of masking. *J. Acoust. Soc. Am.* **29**, 502-505 (1957).

761. Spieth, W. and Trittipoe, W.J., Temporary threshold elevation produced by continuous and "impulsive" noise. *J. Acoust. Soc. Am.* **30**, 523-527 (1958).

762. Spieth, W. and Trittipoe, W.J., Intensity and duration of noise exposure and temporary threshold shifts. *J. Acoust. Soc. Am.* **30**, 710-713 (1958).

763. Steinberg, J.C., Montgomery H.C., and Gardner, M.B., Results of the world's fair hearing tests. *J. Acoust. Soc. Am.* **12**, 291 (1940).

764. Stern, R.M., Effects of variation in visual and auditory stimulation on gastrointestinal motility. *Psychol. Rep.* **14**, 799-802 (1964).

765. Sterner, J.H., Noise and its effects on hearing. *Am. Ind. Hyg. Assoc. J.* **19**, 387-388 (1958).

766. Sternfeld, H., Jr., Spencer, R.H., and Schaeffer, E.G., "Study to Establish Realistic Acoustic Design Criteria for Future Army Aircraft." Rept. 192, Contract DA 44-177-TC-562, AD 267562. Vertol Division, The Boeing Company, Morton, Pennsylvania, 1961.

767. Steudel, U., Über Empfindung und Messung der Lautstärke. *Hochfrequenztechnik Elektroakustik* **41**, 116-128 (1933).

768. Stevens, K.N., Community noise and city planning. *In* "Handbook of Noise Control," Chap. 35 (C.M. Harris, ed.). McGraw-Hill, New York, 1957.

769. Stevens, K.N. and Pietrasanta, A.C., "Procedures for Estimating Noise Exposure and Resulting Community Reaction from Air Base Operations." Rept. WADC TN 57-10, AD 110705. Wright Air Development Center, Wright-Patterson Air Force Base, Ohio, 1957.

770. Stevens, K.N., Hecker, M.H.L., and Kryter, K.D., "An Evaluation of Speech Compression Systems." Rept. TDR 62-171, Rome Air Development Center, Griffiss Air Force Base, New York, 1962.

771. Stevens, K.N., Rosenblith, W.A., and Bolt, R.H., A community's reaction to noise: Can it be forecast? *Noise Control* **1**, No. 1, 63-71 (1955).

772. Stevens, S.S. (ed.), "Handbook of Experimental Psychology." Wiley, New York, 1951.

773. Stevens, S.S., The measurement of loudness. *J. Acoust. Soc. Am.* **27**, 815-829 (1955).

774. Stevens, S.S., The calculations of the loudness of complex noise. *J. Acoust. Soc. Am.* **28**, 807-832 (1956).

775. Stevens, S.S., Concerning the form of the loudness function. *J. Acoust. Soc. Am.* **29**, 603-606 (1957).

776. Stevens, S.S., Calculating loudness. *Noise Control* **3**, No. 5, 11-22 (1957).

777. Stevens, S.S., On the validity of the loudness scale. *J. Acoust. Soc. Am.* **31**, 995-1003 (1959).

778. Stevens, S.S., Ratio scales, partition scales, and confusion scales. "Psychological Scaling: Theory and Applications," pp. 49-66. Wiley, New York, 1960.

779. Stevens, S.S., Procedure for calculating loudness: Mark VI. *J. Acoust. Soc. Am.* **33**, 1577-1585 (1961).

780. Stevens, S.S., The basis of psychophysical judgments. *J. Acoust. Soc. Am.* **35**, 611-612 (1963).

780a. Stevens, S.S., Calculation of the perceived level in PLdB. (A) *J. Acoust. Soc. Am.* **47**, 88 (1970).

780b. Stevens, S.S., "Assessment of Noise: Calculation Procedure Mark VII." Paper 355-128. Laboratory of Psychophysics, Harvard University, Cambridge, Massachusetts, December 1969.

781. Stevens, S.S. and Davis, H., "Hearing." Wiley, New York, 1938.

782. Stevens, S.S. and Poulton, E.C., The estimation of loudness by unpracticed observers. *J. Exp. Psychol.* **51**, 71-78 (1956).

783. Strakhov, A.B., The effect of intensive noise on certain functions of the body. *Gig. Sanit.* **4**, 29-37 (1964).

784. Strakhov, A.B., "Some Questions of the Mechanism of the Action of Noise on an Organism." Rept. N67-11646. Joint Publications Research Service, Washington, D.C., 1966.

785. Strasberg, M., "Criteria for Setting Airborne Noise Level Limits in Shipboard Spaces." Rept. 371-N-12. U.S. Department of the Navy, BuShips, Washington, D.C., 1952.

786. Sumby, W.H. and Pollack, I., Visual contribution to speech intelligibility in noise. *J. Acoust. Soc. Am.* **26**, 212-215 (1954).

787. Summerfield, A., Glorig, A., and Wheeler, D., Is there a suitable industrial test of susceptibility to noise-induced hearing loss? *Noise Control* **4**, 40-46 54 (1958).

788. Symons, N.S., Workmen's compensation benefits for occupational hearing loss. *Noise Control* **4**, No. 5, 28-32 (1958).

789. Swets, J.A., Green, D.M., and Tanner, W.P., Jr., On the width of critical bands. *J. Acoust. Soc. Am.* **34**, 108-113 (1962).

790. Tamm, J., On the effect of noise on men. *J. Sci. Work* **10**, 97-99 (1956).

791. Tanner, W.P., Jr., What is masking? *J. Acoust. Soc. Am.* **30**, 919-921 (1958).

792. Taylor, W. and Pearson, J., A study of the hearing threshold level of dentists exposed to air-rotor drill noise. *J. Dent. Res.* **43**, 962-963 (1964).

793. Taylor, W., Pearson, J., Mair, A., and Burns W., Study of noise and hearing in jute weaving. *J. Acoust. Soc. Am.* **38**, 113-120 (1965).

794. Teichner, W.H. and Sadler, E., Loudness adaptation as a function of frequency, intensity and time. *J. Psychol.* **62**, 267-278 (1966).

795. Teichner, W.H., Arees, E., and Reilley, R., Noise and human performance, a psychophysiological approach. *Ergonomics* **6**, 83-97 (1963).

796. Terkildsen, K., The intra-aural muscle reflexes in normal persons and in workers exposed to intense industrial noise. *Acta Oto-Laryngol.* **52**, 384-396 (1960).

797. Thackray, R.I., Correlates of reaction time to startle. *Human Factors* **7**, 75-80 (1965).

798. Thiessen, G.J. and Subbarao, K., Effect of reverberation on assessment of repetitive impulse noise. *J. Acoust. Soc. Am.* **34**, 1761-1763 (1962).

799. Thiessen, G.J., Community noise—surface transportation. (A) *J. Acoust. Soc. Am.* **42**, 1176 (1967).

800. Thomas G., Volume and loudness of noise. *Am. J. Psychol.* **65**, 588-593 (1952).

801. Thwing, E.J., Effect of repetition on articulation scores for PB words. *J. Acoust. Soc. Am.* **28**, 302-303 (1956).

802. Tkachenko, A.D., Tonal method for determining the intelligibility of speech transmitted by communication channels. *Sov. Phys. Acoust.* **1**, 182-191 (1955).

803. Tondel, L.M., Jr., Noise litigation at public airports. "Alleviation of Jet Aircraft Noise Near Airports," pp. 117-131. A report of the Jet Aircraft Noise Panel, Office of Science and Technology, Executive Office of the President, 1966.

804. Tomatis, A., Somatic and psychic reactions to industrial noise. *Med. Aeron.* **14**, 163-178 (1959).

805. TRACOR, "Public Reactions to Sonic Booms." NASA Contract NASW1804. TRACOR Company, Austin, Texas, 1969.

805a. TRACOR, "Community Reaction to Airport Noise." Final Rept., NASA Contract NASW 1549, TRACOR Company, Austin, Texas, 1970.

806. Trittipoe, W.J., Residual effects at longer pre-exposure durations. *J. Acoust. Soc. Am.* **31**, 244-246 (1959).

807. U.S. Air Force Medical Service, "Hazardous noise exposure." Air Force Regulation 160-3. U.S. Department of the Air Force, Washington, D.C., 1956.

808. U.S. Public Health Service, "National Health Survey (1935-1936): Preliminary Reports, Hearing Study Series." Bulletins 1-7. U.S. Public Health Service, Washington, D.C., 1938.

809. U.S. Veterans Administration, Audiology: The determination of hearing loss. "Department of Medicine and Surgery Information Bulletin," pp. 10-115. Government Printing Office, Washington, D.C., 1960.

810. Vallee, J.P., Deafness in agriculture tractor drivers. *Gaz. Med. France* **72**, 3193-3194 (1965).

811. van Bergeijk, W.A., Pierce, J.R., and David, E.E., Jr., "Waves and the Ear." Anchor Books, Doubleday, New York, 1960.

812. van der Waal, J. and van Dishoeck, H., Individual susceptibility to occupational deafness and the audiometric tests. *Acta Otol. Rhinol. Laryngol.* (*Belgium*) **14**, 213-223 (1960).

813. van Dishoeck, H.A.E., Masking fatigue, adaptation and recruitment as stimulation phenomena of the inner ear. *Acta Oto-Laryngol.* **43**, 167-175 (1953).

814. van Leeuwen, H.A., "Noise Spectra and Audiograms in Industries of the Netherlands." Gezondheidsorganisatie TNO, Koningskade 12, 'S-Gravenhage, Netherlands, 1963.

815. van Os, G.J., Recent experiences with noise acceptability criteria for dwellings. "Fifth International Congress of Acoustics, Liege, Belgium, September 1965."

816. Veneklasen, P.S., City noise—Los Angeles. *Noise Control* **2**, No. 4, 14-19 (1956).

817. von Gierke, H.E., Personal protection. *Noise Control* **2**, No. 1, 37-44 (1956).

818. von Gierke, H.E., Vibration and noise problems expected in manned space craft. *Noise Control* **5**, No. 3, 8-16 (1959).

819. von Gierke, H.E., On noise and vibration exposure criteria. *Arch. Environ. Health* **11**, 327-339 (1965).

820. von Gierke, H.E. and Pietrasanta, A.C., "Acoustical Criteria for Work Spaces, Living Quarters, and Other Areas on Air Bases." Rept. WADC TN 57-248. Wright Air Development Center, Wright-Patterson Air Force Base, Ohio, 1957.

821. Walker, E.L. and Sawyer, T.M., Jr., The interaction between critical flicker frequency and acoustic stimulation. *Psychol. Rec.* **11**, 187-191 (1961).

822. Walker, W.H., Effect of certain noises upon detection of visual signals. *J. Exp. Psychol.* **67**, 72-75 (1964).

823. Wallach, H., Newman, E.B., and Rosenzweig, M.R., The procedence effect of sound localization. *Am. J. Psychol.* **62**, 315-336 (1949).

824. Ward, W.D., Latent and residual effects in temporary threshold shift. *J. Acoust. Soc. Am.* **32**, 135-137 (1960).

825. Ward. W.D., Noninteraction of temporary threshold shifts. *J. Acoust. Soc. Am.* **33**, 512-513 (1961).

826. Ward, W.D., Studies of the aural reflex. I. Contralateral remote masking as an indicator of reflex activity. *J. Acoust. Soc. Am.* 33, 1034-1045 (1961).

827. Ward, W.D., Effect of temporal spacing on temporary threshold shift from impulses. *J. Acoust. Soc. Am.* 34, 1230-1232 (1962).

828. Ward, W.D., Damage-risk criteria for line spectra. *J. Acoust. Soc. Am.* 34, 1610-1619 (1962).

829. Ward, W.D., Auditory fatigue and masking. *In* "Modern Developments in Audiology," Chap. 7 (J. Jerger, ed.). Academic Press, New York, 1963.

830. Ward, W.D., Studies on the aural reflex. II. Reduction of temporary threshold shift from intermittent noise by reflex activity; implications for damage-risk criteria. *J. Acoust. Soc. Am.* 34, 234-241 (1962).

831. Ward, W.D., Relation between noise and deafness. "Proceedings of the Eighth International Congress of Otorhinolaryngology, Tokyo." Reprint from "Excerpta Medica International Congress." Series No. 113. New York, October 1965.

832. Ward, W.D., The concept of susceptibility to hearing loss. *J. Occup. Med.* 7, 595-607 (1965).

833. Ward, W.D., Temporary threshold shifts following monaural and binaural exposure. *J. Acoust. Soc. Am.* 38, 121-125 (1965).

834. Ward, W.D., Temporary threshold shift in males and females. *J. Acoust. Soc. Am.* 40, 478-485 (1966).

835. Ward, W.D., Use of sensation level in measurements of loudness and of temporary threshold shifts. *J. Acoust. Soc. Am.* 39, 736-740 (1966).

836. Ward, W.D., Proposed damage-risk criteria for intermittent noise exposure. "Proceedings of the International Congress on Occupational Health, Vienna, September 19-24, 1966."

837. Ward, W.D., The use of TTS in the derivation of damage risk criteria for noise exposure. *Intl. Audiol.* 5, 309-313 (1966).

838. Ward, W.D., Relation between temporary and permanent noise-induced hearing losses. "Proceedings of the International Congress on Occupational Health, Vienna, September 19-24, 1966."

839. Ward, W.D., Adaptation and fatigue. *In* "Sensorineural Hearing Processes and Disorders," Chap. 9 (A.B. Graham, ed.). Little, Brown, Boston, 1967.

840. Ward, W.D., Susceptibility to auditory fatigue. *In* "Advances in Sensory Physiology," Vol. 3 (W.D. Neff, ed.). Academic Press, New York, 1967.

841. Ward, W.D. (ed.), "Proposed Damage-Risk Criterion for Impulse Noise (Gunfire)." Rept. of Working Group 57, Contract NONR 2300(05). National Academy of Sciences, National Research Council, Committee on Hearing Bioacoustics and Biomechanics, Washington, D.C., July 1968.

842. Ward, W.D. and Glorig, A., The relation between vitamin A and temporary threshold shift. *Acta Oto-Laryngol.* 52, 72-78 (1960).

843. Ward, W.D. and Glorig, A., A case of firecracker-induced hearing loss. *Laryngoscope* 71, 1590-1596 (1961).

844. Ward, W.D. and Nelson, D., "On the Equal-Energy Hypothesis Relative to Damage Risk Criteria in the Chichilla: Conference on Occupational Hearing Loss, British Acoustical Society, London, March 1970." (See also W.D. Ward, "The Effects of Noise on Hearing Thresholds: National Conference on Noise as a Public Health Hazard." American Speech and Hearing Association, Washington, D.C., 1969.)

845. Ward, W.D., Fleer, R.E., and Glorig, A., Characteristics of hearing losses produced by gunfire and by steady noise. *J. Auditory Res.* 1, 325-356 (1961).

846. Ward, W.D., Glorig, A., and Selters, W., Temporary threshold shift in a changing noise level. *J. Acoust. Soc. Am.* **32**, 235-237 (1960).

847. Ward, W.D., Glorig, A., and Sklar, D.L., Dependence of temporary threshold shift at 4 kc on intensity and time. *J. Acoust. Soc. Am.* **30**, 944-954 (1958).

848. Ward, W.D., Glorig, A., and Sklar, D.L., Temporary threshold shift from octave-band noise: Applications to damage-risk criteria. *J. Acoust. Soc. Am.* **31**, 522 (1959).

849. Ward, W.D., Glorig, A., and Sklar, D.L., Relation between recovery from temporary threshold shift and duration of exposure. *J. Acoust. Soc. Am.* **31**, 600-602 (1959).

850. Ward, W.D., Glorig, A., and Sklar, D.L., Temporary threshold shift produced by intermittent exposure to noise. *J. Acoust. Soc. Am.* **31**, 791-794 (1959).

851. Ward, W.D., Glorig, A., and Sklar, D.L., Susceptibility and sex. *J. Acoust. Soc. Am.* **31**, 1138-1140 (1959).

852. Ward, W.D., Selters, W., and Glorig, A., Exploratory studies on temporary threshold shift from impulses. *J. Acoust. Soc. Am.* **33**, 781-793 (1961).

853. Warren, C.H.E., "A Preliminary Analysis of the Results of Exercise Crackerjack and Their Relevance to Supersonic Transport Aircraft." Tech. Rept. Aero-2789. Ministry of Aviation, Royal Aircraft Establishment, Farnborough, England, 1961.

854. Waterhouse, R.V., Noise-control requirements in building codes. *In* "Handbook of Noise Control," Chap. 40 (C.M. Harris, ed.). McGraw-Hill, New York, 1957.

855. Wathen-Dunn, W. and Lipke, D.W., On the power gained by clipping speech in the audio band. *J. Acoust. Soc. Am.* **30**, 36-40 (1958).

856. Webb, D.R.B. and Warren, C.H.E., "An Investigation of the Effect of Bangs on the Subjective Reaction of a Community." Tech. Rept. 66072. Royal Aircraft Establishment, Jarnborough, England, 1966.

857. Webster, J.C., Important frequencies in noise-masked speech. *Arch. Otolaryngol.* **80**, 494-504 (1964).

858. Webster, J.C., Relations between speech-interference contours and idealized articulation-index contours. *J. Acoust. Soc. Am.* **36**, 1962-1969 (1964).

859. Webster, J.C., Effects of noise on speech intelligibility. "Proceedings of the National Conference on Noise as a Public Health Hazard." American Speech and Hearing Association, Washington, D.C., 1968.

860. Webster, J.C., Noise, you get used to it. *J. Acoust. Soc. Am.* **45**, 330 (1969). See also Webster, J.C. and Lepor, M., Noise, you get used to it. *J. Acoust. Soc. Am.* **45**, 751-757 (1969).

861. Webster, J.C. and Klumpp, R.G., Effects of ambient noise and nearby talkers on a face-to-face communication task. *J. Acoust. Soc. Am.* **34**, 926-941 (1962).

862. Webster, J.C., and Klumpp, R.G., Articulation index and average curve-fitting methods of predicting speech interference. *J. Acoust. Soc. Am.* **35**, 1339-1344 (1963). See also "Speech Interference Aspects of Navy Noises." Rept. 1314. U.S. Navy Electronics Laboratory, San Diego, California, 1965.

863. Webster, J.C., Davis, H., and Ward, W.D., Everyday-speech intelligibility. *J. Acoust. Soc. Am.* **38**, 668 (1965).

864. Webster, J.C., Muller, P.H., Thompson, P.O., and Davenport, E.W., The masking and pitch shifts of pure tones near abrupt changes in a thermal noise spectrum. *J. Acoust. Soc. Am.* **24**, 147-152 (1952).

865. Wegel, R.L., and Lane, C.E., The auditory masking of one pure tone by another and its probable relation to the dynamics of the inner ear. *Phys. Rev.* **23**, 266-285 (1924).

866. Weissler, P.G., International standard reference zero for audiometers. *J. Acoust. Soc. Am.* **44**, 264-275 (1968).

867. Welch, B.L., Psychophysiological response to the mean level of environmental stimulation, a theory of environmental integration. "Symposium on Medical Aspects of Stress in the Military Climate, Walter Reed Medical Center, Washington, D.C., 1964." U.S. Government Printing Office, Washington, D.C., 1965.

868. Wells, R.J., Recent research relative to perceived noise level. (A) *J. Acoust. Soc. Am.* **42**, 1151 (1967).

869. Wells, R.J. and Blazier, W.E., Jr., "A Procedure for Computing the Subjective Reaction to Complex Noise from Sound Power Data." Paper 124 presented at the Fourth International Congress on Acoustics, Copenhagen, 1962.

870. Wersall, R., The tympanic muscles and their reflexes. *Acta Oto-Laryngol.* Suppl. **139** (1958).

870a. West, L.J., Psychopathology produced by sleep deprivation. *In* "Proceedings of the Association for Research in Nervous and Mental Disease, Sleep and Altered States of Consciousness" (S.S. Kety, ed.). Williams and Wilkins, Baltimore, Maryland (1967).

870b. Weston, P.B., Miller, J.D. and Hirsh, I.J., Release from masking for speech. *J. Acous. Soc. Am.* **6**, 1053-1054 (1965).

871. Wheeler, D.E., Detection of noise susceptible ears. *Laryngoscope* **59**, 1328-1338 (1959).

871a. Wheeler, D.E., Measurement of industrial hearing loss. *Noise Control* **1**, No. 4, 9-15 (1955).

872. Wheeler, J. and Dickson, E.D.D., The determination of the threshold of hearing. *J. Laryngol. Otol.* **66**, 379-395 (1952).

873. Whittle, L.S. and Robinson, D.W., British normal threshold of hearing. *Nature* **189**, 617-618 (1961).

874. Wiener, F.M., On the diffraction of a progressive sound wave by the human head. *J. Acoust. Soc. Am.* **19**, 143-146 (1947).

875. Wiener, F.M., Experimental study of the airborne noise generated by passenger automobile tires. *Noise Control* **6**, 13-16 (1960).

876. Wiethaup, H., Offensive noise caused by domestic animals. *Med. Klin.* **60**, 1377-1378 (1965).

877. Wilkinson, R.T., Interaction of noise with knowledge of results and sleep deprivation. *J. Exp. Psychol.* **66**, 332-337 (1963).

878. Williams, C.E., Stevens, K.M., Hecker M.H.L., and Pearsons, K.S., "The Speech Interference Effects of Aircraft Noise." Rept. FAA DS-67-19, Contract FA66WA-1566, Federal Aviation Administration, Washington, D.C., 1967.

878a. Williams, C.E., Stevens, K.N., and Klatt, M., Judgment of the acceptability of aircraft noise in the presence of speech. *J. Sound Vib.* **9**, 263-275 (1969).

879. Williams, H.L., Hammack, J.T., Daly, R.L., Dement, W.C., and Lubin, A., Responses to auditory stimulation, sleep loss, and the EEG stages of sleep. *Electroencephalog. Clin. Neurophysiol.* **16**, 269-279 (1964).

880. Wilson, A.H., "Noise." Her Majesty's Stationery Office, London, 1963.

881. Wilson, W.H., Determination of susceptibility to abnormal auditory fatigue. *Ann. Oto-Laryngol.* **59**, 399-405 (1950).

882. Wolsk, D., Discrimination limen for loudness under varying rates of intensity change. *J. Acoust. Soc. Am.* **36**, 1277-1282 (1964).

883. Woodhead, M.M., The effects of bursts of loud noise on a continuous visual task. *Brit. J. Ind. Med.* **15**, 120-125 (1958).

884. Woodhead, M.M., Effect of brief loud noise on decision making. *J. Acoust. Soc. Am.* **31**, 1329-1331 (1959).

885. Woodhead, M.M., Value of ear defenders for mental work during intermittent noise. *J. Acoust. Soc. Am.* **32**, 682-684 (1960).

886. Woodhead, M.M., Visual searching in intermittent noise. *J. Sound Vib.* **1**, 157-161 (1964).

887. Woodhead, M.M., The effects of bursts of noise on an arithmetic task. *Am. J. Psychol.* 77, 627-633 (1964).
888. Woodhead, M.M., Performing a visual task in the vicinity of reproduced sonic bangs. *J. Sound Vib.* 9, 121-125 (1969).
889. World Health Organization, Noise: An occupational hazard and public nuisance. *World Health Org. Chron.* 20, 191-203 (1966).
890. Wright, H.N., Temporal summation and backward masking. *J. Acoust. Soc. Am.* 36, 927-932 (1964).
891. Wyatt, S. and Langdon, J.H., Industrial Health Research Board Rept. No. 77. Her Majesty's Stationery Office, London, 1935.
892. Yaffe, C.D. and Jones, H.H., Noise and hearing. "Relationship of Industrial Noise to Hearing Acuity in a Controlled Population." Public Health Service Publ. 850. U.S. Department of Health, Education, and Welfare, Washington, D.C., 1961.
893. Young, J.R., Attenuation of aircraft noise by woodsided and brick-veneered houses. NASA Rept. CR-1637 Stanford Research Institute, Menlo Park, California, 1969.
894. Young, J.R., Johnson, P.J., Kryter, K.D., and Aron, W.A., "Energy Spectral Density of Sonic Booms. Annex F of Sonic Boom Experiments at Edwards Air Force Base." Stanford Research Institute Interim Rept. NSBEO-1, Contract AF 49(638)-1758. Clearinghouse for Federal Scientific and Technical Information, U.S. Department of Commerce, Springfield, Virginia 1967.
895. Young, M.A. and Gibbons, E.W., Speech discrimination scores and threshold measurements in a non-normal hearing population. *J. Auditory Res.* 2, 21-33 (1962).
896. Young, R.W., Don't forget the simple sound-level meter. *Noise Control,* 4, No. 3, 42-43 (1958).
897. Young, R.W., Comments on "effect of environment on noise criteria." *Noise Control* 5, No. 6, 376 (1959).
898. Young, R.W. and Peterson, A., On estimating noisiness of aircraft sounds. *J. Acoust. Soc. Am.* 45, 834-838 (1969).
899. Zepler, E.E. and Harel, J.R.P., The loudness of sonic booms and other impulsive sounds. *J. Sound Vib.* 2, 249-256 (1965).
900. Zwicker, E., Über psychologische und methodische Grundlagen der Lautheit. (Psychological and methodical foundations of loudness.) *Acustica* 3, 237-258 (1958).
901. Zwicker, E., Ein Verfahren zur Berechnung der Lautstarke. (A means for calculating loudness.) *Acustica* 10, 304 (1960).
902. Zwicker, E., Subdivision of the audible frequency range into critical bands. (Frequenzgruppen). *J. Acoust. Soc. Am.* 33, 248 (1961).
903. Zwicker, E., Über die Lautheit von ungedrosselten und gedrosselten Schallen. (On the loudness of unmasked and masked sounds.) *Acustica* 13, 194-211 (1963).
904. Zwicker, E., Temporal effects in simultaneous masking and loudness. *J. Acoust. Soc. Am.* 38, 132-141 (1965).
905. Zwicker, E. and Feldtkeller, R., "Das Ohr als Nachrichtenempfanger," Auflage, 2, S. Hirzel, Stuttgart, 1967.
906. Zwicker, Von E. and Scharf, B., A model of loudness summation. *Psychol. Rev.* 72, 3-26 (1965).
907. Zwicker, Von E., Flottorp, G., and Stevens, S.S., Critical bandwidth in loudness summation. *J. Acoust. Soc. Am.* 29, 548-547 (1957).
908. Zwislocki, J.J., Acoustic filters as ear defenders. *J. Acoust. Am.* 23, 36-40 (1951).
909. Zwislocki, J.J., New types of ear protectors. *J. Acoust. Soc. Am.* 24, 762-764 (1952).
910. Zwislocki, J.J., Ear protectors. *In.* "Handbook of Noise Control," Chap. 8 (C.M. Harris, ed.). McGraw-Hill, New York, 1957.

911. Zwislocki, J.J., Theory of temporal auditory summation. *J. Acoust. Soc. Am.* **32**, 1046-1060 (1960).
912. Zwislocki, J.J., "Acoustics of the Middle Ear." Rept. 576. U.S. Army Medical Research Laboratories, Fort Knox, Kentucky, 1963.
913. Zwislocki, J., An acoustic method for clinical examination of the ear. *J. Speech and Hearing Res.* **6**, 303-314 (1963).
914. Zwislocki, J., Analysis of some auditory characteristics. *In* "Handbook of Mathematical Psychology," Vol. 3, Chap. 15, (R.D. Luce, R.R. Bush, and E. Galanter, eds.) Wiley, New York, 1965.